Volumes Already Published in this Series:

Volume 1: Cell Components
1985, ISBN 3-540-15822-7

Volume 2: Nuclear Magnetic Resonance
1986, ISBN 3-540-15910-X

Volume 3: Gas Chromatography/
Mass Spectrometry
1986, ISBN 3-540-15911-8

Volume 4: Immunology in Plant Sciences
1986, ISBN 3-540-16842-7

Volume 5: High Performance Liquid Chromatography
in Plant Sciences
1987, ISBN 3-540-17243-2

Volume 6: Wine Analysis
1988, ISBN 3-540-18819-3

Volume 7: Beer Analysis
1988, ISBN 3-540-18308-6

Volume 8: Analysis of Nonalcoholic Beverages
1988, ISBN 3-540-18820-7

Volume 9: Gases in Plant and Microbial Cells
1989, ISBN 3-540-18821-5

Volume 10: Plant Fibers
1989, ISBN 3-540-18822-3

Forthcoming:

Volume 11: Physical Methods in Plant Sciences
ISBN 3-540-50332-3

Plant Fibers

Edited by
H.F. Linskens and J.F. Jackson

Contributors

Jun-ichi Azuma N.K. Bansal R.M. Faulks D.J. Frost S.C. Fry F. Grolig T. Hayashi T. Higuchi U.A. Hurley P.P. Jablonski H. Kauss P. Komalavilas M. Kuwahara K. Kuwano D.T.A. Lamport S.G. Lawson T.L. Mason D.W. McCurdy T.A. Mitchell P.J. Moore A.J. Mort T. Nakamura J.S.G. Reid P.S. Rodis G.L. Rorrer R.D. Sabin R.R. Selvendran M.E. Sloan M.S. Sodha L.A. Staehelin L. da Silveira Lobo Sternberg K.M.M. Swords T. Umezawa A.V.F.V. Verne B.P. Wasserman G.O. Wasteneys R. Wells R.E. Williamson T. Yoshida

With 96 Figures and 41 Tables

Springer-Verlag
Berlin Heidelberg New York
London Paris Tokyo

Professor Dr. Hans-Ferdinand Linskens
Goldberglein 7
D-8520 Erlangen

Professor Dr. John F. Jackson
Department of Biochemistry
Waite Agricultural Research Institute
University of Adelaide
Glen Osmond, S.A. 5064
Australia

ISBN-13: 978-3-642-83351-9 e-ISBN-13: 978-3-642-83349-6
DOI: 10.1007/978-3-642-83349-6

Softcover reprint of the hardcover 1st edition 1989

2131/3145-543210 – Printed on acid-free paper

Introduction

Modern Methods of Plant Analysis

When the handbook *Modern Methods of Plant Analysis* was first introduced in 1954 the considerations were:

1. the dependence of scientific progress in biology on the improvement of existing and the introduction of new methods;
2. the difficulty in finding many new analytical methods in specialized journals which are normally not accessible to experimental plant biologists;
3. the fact that in the methods sections of papers the description of methods is frequently so compact, or even sometimes so incomplete that it is difficult to reproduce experiments.

These considerations still stand today.

The series was highly successful, seven volumes appearing between 1956 and 1964. Since there is still today a demand for the old series, the publisher has decided to resume publication of *Modern Methods of Plant Analysis*. It is hoped that the New Series will be just as acceptable to those working in plant sciences and related fields as the early volumes undoubtedly were. It is difficult to single out the major reasons for success of any publication, but we believe that the methods published in the first series were up-to-date at the time and presented in a way that made description, as applied to plant material, complete in itself with little need to consult other publications.

Contributing authors have attempted to follow these guidelines in this New Series of volumes.

Editorial

The earlier series *Modern Methods of Plant Analysis* was initiated by Michel V. Tracey, at that time in Rothamsted, later in Sydney, and by the late Karl Paech (1910–1955), at that time at Tübingen. The New Series will be edited by Paech's successor H.F. Linskens (Nijmegen, The Netherlands) and John F. Jackson (Adelaide, South Australia). As were the earlier editors, we are convinced "that there is a real need for a collection of reliable up-to-date methods for plant analysis in large areas of applied biology ranging from agriculture and horticultural experiment stations to pharmaceutical and technical institutes concerned with raw material of plant origin". The recent developments in the fields of plant biotechnology and genetic engineering make it even more important for workers in the plant sciences to become acquainted with the more sophisticated methods, which sometimes come from biochemistry and biophysics, but which also have been developed in commercial firms, space science laboratories, non-university research institutes, and medical establishments.

Concept of the New Series

Many methods described in the biochemical, biophysical, and medical literature cannot be applied directly to plant material because of the special cell structure, surrounded by a tough cell wall, and the general lack of knowledge of the specific behavior of plant raw material during extraction procedures. Therefore all authors of this New Series have been chosen because of their special experience with handling plant material, resulting in the adaptation of methods to problems of plant metabolism. Nevertheless, each particular material from a plant species may require some modification of described methods and usual techniques. The methods are described critically, with hints as to their limitations. In general it will be possible to adapt the methods described to the specific needs of the users of this series, but nevertheless references have been made to the original papers and authors. While the editors have worked to plan in this New Series and made efforts to ensure that the aims and general layout of the contributions are within the general guidelines indicated above, we have tried not to interfere too much with the personal style of each author.

There are several ways of classifying the methods used in modern plant analysis. The first is according to the technological and instrumental progress made over recent years. These aspects were used for the first five volumes in this series describing methods in a systematic way according to the basic principles of the methods.

A second classification is according to the plant material that has to undergo analysis. The specific application of the analytical method is determined by the special anatomical, physiological, and biochemical properties of the raw material and the technology used in processing. This classification will be used in Volumes 6 to 8, and for some later volumes in the series. A third way of arranging a description of methods is according to the classes of substances present in the plant material and the subject of analytic methods. The latter will be used for later volumes of the series, which will describe modern analytical methods for alkaloids, drugs, hormones, etc.

Naturally, these three approaches to developments in analytical techniques for plant materials cannot exclude some small overlap and repetition; but careful selection of the authors of individual chapters, according to their expertise and experience with the specific methodological technique, the group of substances to be analyzed, or the plant material which is the subject of chemical and physical analysis, guarantees that recent developments in analytical methodology are described in an optimal way.

Volume Ten – Plant Fibers

Cellulose fibers have not always been one of the major end products of carbon dioxide fixation through photosynthesis. The process of photosynthesis is thought to have begun some 2000 million years ago in primitive aquatic plant cells which did not have the need for large amounts of the structurally strong cellulose polymers. It was not until the late Devonian and Carboniferous times, approximately 300 million years ago, when plants had evolved sufficiently to have emerged from their aquatic environment, that there was a need for a strenghtening material, such as fibers of cellulose, to enable each plant to stand up and lift the growing protoplasmic mass above the horizon and compete for sunlight and simultaneously gain a larger area for absorption of gaseous raw materials. Cellulose has great crystallinity and tensile strength and together with lignin plays a major role as a structural component of living plant cells.

Mankind has gradually learnt to make use of these structural components. Thus around 3500 years B.C. the Egyptians were making papyrus parchments utilizing the strength of contained polymeric cellulose, cotton was used in clothing as long as 3000 years B.C., and by 2600 years B.C. the Chinese were using silk fibers, consisting in this case of another polymer, protein. The Chinese are also accredited with inventing paper around 100 years B.C., but it was not until the nineteenth century A.D. in Germany that a technology was invented to extract cellulose fibers from groundwood and these then formed into paper as we know it today. The importance of wood was thus enhanced, for nor only was it useful as a fuel and as a structural material for building, but now its use was increased again through paper manufacture.

Wood is essentially 50% cellulose, the fiber of importance for paper manufacture, and provides us with a relatively cheap and renewable source of raw material for industry. In paper-making, hydrogen bonding between the cellulose fibers holds the fibers together as the wet paper sheet is dried, a process that is easily reversible to permit easy reclamation of fibers for further use. The chain-like structure of cellulose consists of 1,4-β-D linked glucopyranose residues and that used in paper-making may bw 1000 to 15,000 glucose units in length. The hydroxyl groups of glucose provide plenty of scope for hydrogen bonding between fibers, while it is known that the tensile strength of the fiber is greater the longer the cellulose chain.

Forest trees are able to fix more carbon per hectare without fertilizer than almost any other group of plants, including those that man has cultivated, often in place of original forests. Annually, 4 to 9×10^{13} kg of carbon is fixed by photosynthesis and approximately half of this is converted to wood tissue. Production of dry matter by pine forests is 3180 g/sq. m/year, which compares with wheat at 344, corn 790 and sugar beet 1470. Only sugar cane at 6700 g/sq. m/year has a superior production. The importance of forest trees for efficient production of cellulose fibers can be drawn from these figures, essentially because a greater proportion of the available solar energy is captured and held by forests where the biomass is largely woody support tissue. The fibers produced in the woody tissue are long and hence capable of giving greater tensile strength, and are vastly superior to the cellulosic fibers extractable from annual plants, such as sugar cane, wheat, etc.

World consumption of paper and paperboard was 130 million tons in 1970 and growing fast, so that 420 million tons is estimated for the year 2000. More than 90% of this 1970 production was from wood (tree boles), and so the importance of the constituent fibers to human activity cannot be overstressed.

For the above reasons the fibers of wood, such as cellulose, find an important place in this volume, taking up almost half of the chapters presented, and if considered together with other materials making up and derived from plant cell walls, comprise three quarters of the volume. Investigational methods for the constituents of plant cell walls presented in this volume include biosynthesis of plant cell wall polysaccharides, immunogold techniques for cell wall components, extensin structure in cell walls, cross-linking in cell wall components, and treatments of lignin and callose and other 1,3-β-glucans.

An interesting aspect of plant cellulose analysis presented in this volume is the measurement of oxygen and hydrogen isotope ratios in plant cellulose, and the possibility that this parameter can be used as an indicator of plant productivity under drought stress. Cotton has been used by man for many centuries, it is another cellulosic fiber with many desirable features for use in the textile industry. It is believed that in cotton the degree of fibrillar aggregation is high, giving greater protection to the interlinking regions between cellulose crystallites. The way in which the cell wall lays down cellulose fibrils in a criss-crossed helical manner has a great effect on physical and chemical properties. The tensile strength increase in cotton on wetting and ammonia treatment can be traced to the swelling compression exerted on the fibrils in the inner secondary wall layers. Methods involved in cotton (lint) production, quality and yield estimation thus find a place in this volume.

No treatment of plant fibers would be complete without reference to insoluble dietary fibers, so important as a disease "protectant" in Western diets. Methods for analysis of these fibers in foods are therefore presented, the fibers treated including, again, cellulose, but also hemicellulose and lignin. Food involves carbohydrates not only as insoluble dietary fibers, but also as an energy source and sweetener, as well as a structural role. Thus in baking, we use flour and water, which gives a dough of desired consistency and which is influenced not only by protein, arabinoxylan and mixed-linkage β-glucan content, but also by the level of damaged starch granules. The latter absorb water and gelatinize to yield a rigid network which prevents loaf collapse on cooling. A chapter on the monitoring and controlling of the quality of bread made from wheaten flour is therefore pertinent. Finally, in modern times, mention of food without dealing with soybean protein in some way would be missing an important ingredient. Protein itself is of course a fiber of another type and in this case the soybean protein has a structural role in foods as well. Soybean protein products have found application as ingredients in many fabricated and processed foods due to both their nutritional value and functional properties. Functionally, the soybean proteinaceous hydrocolloid macromolecules act as viscosity enhancing and gelling agents in these foods. This volume is rounded off then by inclusion of a chapter presenting methods of analysis involved in soybean protein thermal gelation.

Acknowledgments. The editors express their thanks to all contributors for their efforts in keeping to production schedules, and to Dr. Dieter Czeschlik, Ms. K. Gödel, Ms. J. v. d. Bussche and Ms. E. Göhringer of Springer-Verlag for their cooperation with this and other volumes in Modern Methods of Plant Analysis. The constant help of José Broekmans is gratefully acknowledged.

Nijmegen, Siena and Adelaide, Spring 1989

H. F. LINSKENS
J. F. JACKSON

Contents

Biosynthesis of Cell-Wall Polysaccharides: Membrane Isolation, in Vitro Glycosyl Transferase Assay and Enzyme Solubilization
B. P. WASSERMAN, D. J. FROST, S. G. LAWSON, T. L. MASON, P. S. RODIS R. D. SABIN, and M. E. SLOAN (With 3 Figures)

Analysis of Cross-Links in the Growing Cell Walls of Higher Plants
S. C. FRY (With 5 Figures)

Fluorometric Measurement of Callose and Other 1,3-β-Glucans
H. Kauss

Measuring β-Glucan Deposition in Plant Cell Walls
T. Hayashi (With 7 Figures)

Methods Used in the Chemistry of Lignin Biodegradation
T. Umezawa and T. Higuchi (With 5 Figures)

Measuring Lignin Degradation
M. Kuwahara

Methods for Analysis of Dietary Fibre
R. R. SELVENDRAN, A. V. F. V. VERNE, and R. M. FAULKS
(With 3 Figures)

Methods Used in the Investigation of Insoluble Dietary Fiber
T. YOSHIDA and K. KUWANO (With 3 Figures)

Measurement of Lint Production in Cotton and Factors Affecting Yield
R. WELLS (With 1 Figure)

Analysis of Carbohydrates Conferring Hardness on Seeds
J. S. G. REID (With 3 Figures)

Methods Used in Monitoring and Controlling the Quality of Bread with Particular Reference to the Mechanical Dough Development Process
T. A. MITCHELL

Analytical Methods for Gelation of Soybean Proteins
T. NAKAMURA (With 3 Figures)

Techniques of Solar Crop Dryers
N. K. BANSAL and M. S. SODHA (With 15 Figures)

List of Contributors

AZUMA, JUN-ICHI, Department of Wood Science and Technology, Faculty of Agriculture, Kyoto University, Oiwake-cho, Sakyo-ku, Kyoto 606, Japan

BANSAL, NARENDRA K., Centre of Energy Studies, Indian Institute of Technology, Hauz Khas, New Delhi-110016, India

FAULKS, RICHARD M., AFRC Institute of Food Research, Norwich Laboratory, Colney Lane, Norwich NR4 7UA, Great Britain

FROST, DAVID J., Department of Food Science, New Jersey Agricultural Experiment Station, Rutgers University, Cook College, College Farm Road, New Brunswick, NJ 08903, USA

FRY, STEPHEN C., Department of Botany, University of Edinburgh, The King's Buildings, Mayfield Road, Edinburgh EH9 3JH, Great Britain

GROLIG, FRANZ, Botanisches Institut I, Justus-Liebig-Universität, Senckenbergstr. 17–21, D-6300 Giessen, FRG

HAYASHI, TAKAHISA, Basic Research Laboratory, Central Research Laboratories, Ajinomoto Co. Inc., 1-1 Suzuki-cho, Kawasaki 210, Japan

HIGUCHI, TAKAYOSHI, Research Section of Lignin Chemistry, Wood Research Institute, Kyoto University, Gokasho, Uji-shi Kyoto-fu 611, Japan

HURLEY, URSULA A., Plant Cell Biology Group, Research School of Biological Sciences, Australian National University, P.O. Box 475, Canberra City, A.C.T. Australia

JABLONSKI, PETER P., Plant Cell Biology Group, Research School of Biological Sciences, The Australian National University, P.O. Box 475, Canberra City, A.C.T. 2601, Australia

KAUSS, HEINRICH, Fachbereich Biologie der Universität Kaiserslautern, Postfach 3040, D-6750 Kaiserslautern, FRG

KOMALAVILAS, PADMINI, Botany and Plant Sciences, University of California, Riverside, CA 92521, USA

KUWAHARA, MASAAKI, Department of Bioresource Science, Kagawa University, Miki-cho, Kagawa 761-07, Japan

KUWANO, KAZUTAMI, Department of Home Economics, Tokyo Kasei Gakuin Junior College, Sanban-Cho 22, ChYoda-ku, Tokyo 102, Japan

LAMPORT, DEREK T. A., MSU-DOE Plant Research Laboratory, Michigan State University, East Lansing, MI 48824-1312, USA

Lawson, Stephen G., 7 Eton Place, Clark, NJ 07066, USA

Mason, Theresa L., Department of Food Science, Rutgers University, New Brunswick, NJ 08903, USA

McCurdy, David W., Plant Cell Biology Group, Research School of Biological Sciences, The Australian National University, P.O. Box 475, Canberra City, A.C.T. 2601, Australia

Mitchell, T. A., Wheat Research Institute, Department of Scientific and Industrial Research, P.O. Box 29-182, Christchurch, New Zealand

Moore, Patricia J., Department of Cell Biology and Anatomy, Northwestern University Medical School, 303 E. Chicago Avenue, Chicago, IL 60611, USA

Mort, Andrew J., Oklahoma Agricultural Experiment Station, Department of Biochemistry, Oklahoma State University, Stillwater, OK 74078-0454, USA

Nakamura, T., Hohnen Oil Co., Ltd., Research and Development, 1-2-3 Otemachi, Chiyoda-ku, Tokyo 100, Japan

Reid, J. S. Grant, School of Molecular and Biological Sciences, University of Stirling, Stirling FK9 4LA, Great Britain

Rodis, Panayotis S., Agricultural University of Athens, Dept. of Food Science and Technology, Votanikos, Athens 11855, Greece

Rorrer, Gregory L., Department of Chemical Engineering, Michigan State University, East Lansing,MI 48824, USA

Sabin, Robert D., Department of Food Science, Rutgers University, New Brunswick, NJ 08903, USA

Selvendran, Robert R., AFRC Institute of Food Research, Norwich Laboratory, Colney Lane, Norwich NR4 7UA, Great Britain

Sloan, Margaret E., Department of Food Science, 205 Alison Hall, University of Delaware, Newark, DE 19716, USA

Sodha, M. S., Vice Chancellor, Devi Ahilya University, Indore, India

Staehelin, L. Andrew, Department of Molecular, Cellular and Developmental Biology, Box 347, University of Colorado, Boulder, CO 80309, USA

Sternberg, L. da Silveira Lobo, Department of Biology, University of Miami, Coral Gables, FL 33124, USA

Swords, Kathleen M. M., Department of Molecular, Cellular and Developmental Biology, University of Colorado, Boulder, CO 80309-0347, USA

Umezawa, Toshiaki, Research Section of Lignin Chemistry, Wood Research Institute, Kyoto University, Gokasho, Uji-shi, Kyoto-fu 611, Japan

Verne, A. Verena F. V., AFRC Institute of Food Research, Norwich Laboratory, Colney Lane, Norwich NR4 7UA, Great Britain

Wasserman, Bruce P., Department of Food Science, Rutgers University, New Brunswick, NJ 08903, USA

Wasteneys, Geoffrey O., Plant Cell Biology Group, Research School of Biological Sciences, The Australian National University, P.O. Box 475, Canberra City, A.C.T. 2601, Australia

WELLS, RANDY, North Carolina State University, P.O. Box 7620, Raleigh, NC 27695-7620, USA

WILLIAMSON, RICHARD E., Plant Cell Biology Group, Research School of Biological Sciences, Australian National University, P.O. Box 475, Canberra City, A.C.T. 2601, Australia

YOSHIDA, TSUTOMU, Department of Food and Nutrition, Tachikawa College of Tokyo, Azumacho, Akishima-shi, Tokyo 196, Japan

Biosynthesis of Cell-Wall Polysaccharides: Membrane Isolation, in Vitro Glycosyl Transferase Assay and Enzyme Solubilization

B. P. WASSERMAN, D. J. FROST, S. G. LAWSON, T. L. MASON, P. S. RODIS, R. D. SABIN, and M.E. SLOAN

1 Introduction

The biosynthesis of plant cell-wall polysaccharides has been studied at the enzymological level for over 30 years (Delmer et al. 1985; Kauss 1985; Wasserman et al. 1986; Delmer 1987; Wasserman and Sloan 1989). The enzymes believed to be important for generating important polymers, such as cellulose, β-(1,3)-glucan (callose), xyloglucans, galacturonans, etc., are all thought to be membrane-bound glycosyl transferases. These enzymes catalyze the stepwise transfer of monosaccharides from nucleoside diphosphate sugars to an acceptor molecule such as the growing polysaccharide chain or a protein which may serve as a "primer". No direct evidence for the existence of protein primers has yet been found.

Understanding the enzymological and genetic regulation of cell-wall polysaccharide biogenesis would be greatly facilitated if appropriate genetic and immunological probes were available. Therefore, much effort has been focused on identifying the catalytic and regulatory subunits of various glycosyl transferases. The instability of many glycosyl transferases, particularly when detergents are introduced, has made this a challenging task. Yet, some systems, such as the β-(1,3)-glucan synthase from red beet storage tissue, have shown good stability and reasonable enrichments of activity have been obtained (Wasserman et al. 1985; Eiberger et al. 1985; Wasserman and McCarthy 1986; Eiberger and Wasserman 1987; Sloan et al. 1987). This chapter will describe methods used to prepare subcellular fractions, assay glucan synthase activity, and to effectively obtain solubilized enzyme preparations of high specific activity.

2 Membrane Isolation

Our studies have primarily utilized red beet (*Beta vulgaris* L.) and carrot (*Daucus carota* L.) storage tissue. These tissues are advantageous since they contain high levels of reasonably stable enzyme activity, large quantities of membranes may be harvested in a single day, and they are readily available. Plasma membrane activity from red beet is somewhat more stable than carrot. Both are easily propagated in suspension culture, making it possible to study *in vivo* regulation of activity.

Although a sucrose gradient centrifugation technique is described here, recently developed methods of plasma membrane isolation, such as free flow electrophoresis (Sandelius et al. 1986) and aqueous two-phase partition (Fink et al. 1987), may in the future become useful alternatives for obtaining glycosyl transferase preparations. When isolating membranes from a tissue source for the first time or if adopting a new method, it is advisable to conduct a complete set of marker enzyme assays to confirm the subcellular localization of activity.

The procedures described below have been adopted for the large-scale isolation and assay of membrane fractions from red beet storage tissue. Most steps are similar with carrot (Lawson et al. 1989). However, in those instances where reagents or parameters differ slightly, appropriate notations have been made in the text.

2.1 Isolation of Crude Membrane Fractions

Plant material is obtained from local farms or markets. The tops are removed and the roots may be stored in moist vermiculite for up to 3 weeks at 4° C until use. Microsomal membranes are isolated by differential centrifugation (Briskin and Poole 1983; Wasserman et al. 1985) as follows: Plant tissue is surface sterilized in a 0.5% sodium hypochlorite solution for 10 min. After rinsing with distilled water, the tissue is peeled, diced into 2–5 cm blocks and suspended in a homogenization buffer (all steps are conducted at 4° C) consisting of 250 mM sucrose, 3 mM EDTA, 0.5% polyvinylpyrolidone-40, 1 mM DTE and 70 mM HEPES-NaOH, pH 8.0 at a ratio of 50 ml buffer per 100 g fresh weight tissue. Rapid homogenization of tissue is achieved by passage through a Braun juice extractor. If large capacity GSA and Ti 45 (Beckman) rotors are available, 1500 g of tissue may be processed per day.

The extract is filtered through one layer of miracloth to remove residual particulate matter. Mitochondria and other large particles are then removed by centrifugation for 15 min at 13,000 g and the pellets are discarded. Microsomal membranes are recovered by centrifuging the supernatant for 30 min at 80,000 g. The pellets are then washed by diluting with Resuspension Buffer (RSB) containing 250 mM sucrose, 1 mM DTE and 10 mM HEPES-NaOH, pH 7.2 followed by recentrifugation in RSB. Microsomes are then resuspended to a protein concentration of 3–6 mg/ml in Storage Buffer (SB) consisting of 1 mM DTE, 10 mM HEPES-NaOH, pH 7.2, 15% (v/v) glycerol and may be stored at −80° C for at least several months. Protein content is determined by the method of Bradford (1976) using bovine serum albumin as the protein standard.

2.2 Plasma Membrane Enriched Preparations

Plasma membranes are prepared by discontinuous sucrose gradient centrifugation (Briskin and Poole 1983; Wasserman and McCarthy 1986). If using a Beckman SW 28.1 swinging bucket rotor, sucrose step gradients are made containing 4.0 ml of 40% sucrose (w/w), 5.0 ml of 30% sucrose, and 4.0 ml of 20% sucrose. All sucrose solutions contain 10 mM HEPES-NaOH, pH 7.2, and 1 mM DTE. Microsomal membranes (≦12 mg protein in 4 ml) are layered onto the gradient and centrifuged at 80,000 g for 2 h. (Caution: Carrot microsomes do not resolve well if >6 mg protein are layered onto the gradient.) With larger SW 28 buckets, gradients may be scaled up appropriately with ≦29 mg of protein applied to each gradient. Plasma membranes are recovered from the 30% to 40% interface, combined into centrifuge tubes, diluted with RSB, and centrifuged for 30 min at 80,000 g. Membranes are resuspended to a protein concentration of 2–4 mg/ml in SB and may be stored at −80° C.

3 Assay Methods for Glycosyl Transferases

3.1 General Aspects

The substrates for virtually all glycosyl transferases are nucleotide diphosphate (NDP) sugars. UDP-linked sugars appear to be the most common substrates for the biosynthesis of cell-wall polymers, however some GDP-linked substrates have been shown to participate, albeit at much lower rates than the UDP-sugars (Ray et al. 1969; Henry and Stone 1982; Henry et al. 1983). All rapid assays require the availability of NDP-sugars radiolabeled in the sugar moiety. Although some of the more common NDP-sugars are commerially available in the radiolabeled form such as UDP[^{14}C]-glc, GDP[^{14}C]-glc, UDP[^{14}C]-xyl, ADP[^{14}C]-Glc, UDP[1^{4}C]-glucuronic acid, many are not. The limited availability of some NDP-sugars radiolabeled in the sugar moiety has restricted thorough substrate screening experiments. In some instances, such as with UDP[^{14}C]-galacturonic acid, the putative precursor for galacturonans (Villemez et al. 1966; Lin et al. 1966; Bolwell et al. 1985; Cumming and Brett 1986), it has been possible to synthesize radiolabeled substrate.

Glycosyl transferase assays generally involve quantitating the rate of formation of radiolabeled product. In all variations, reactions are terminated at a specified time, preferably during the linear portion of the time course. Polysaccharide products are then separated from unincorporated substrate. The most common method for capturing product is to use glass fiber filter disks (Tsai and Hassid 1971; Smith and Stone 1973). Samples are spotted onto GF/A filter disks and unincorporated substrate removed by ethanol washing. Ethanol insoluble product remains on the filter. Product solubility characteristics may be ascertained by comparing the effects of different wash treatments, e.g., cold and hot water, alkali, chloroform/methanol, etc. With this adaptation, over a hundred assays may be performed in a working day. To avoid losses of glucan, some laboratories will apply biosynthesized glucan to filter disks with carrier cellulose (Hayashi et al. 1987) or to conduct precipitations with enzymatically inactive membranes (Ray 1980).

Variations of the above method are to separate NDP-sugars from various products by thin layer or paper chromatography (Ephritikhine et al. 1980) or by paper electrophoresis (Dalessandro and Northcote 1981). Polymers will generally remain at the origin, while NDP-sugars and derivatives thereof migrate in the mobile phase.

Assays that measure the biosynthesis of heteropolysaccharides such as xylo-glucan (Hayashi and Matsuda 1981) or glucuronoxylan (Waldron and Bret 1983) may require a more complex series of washes and separations to quantitate desired reaction products.

3.2 Glucan Synthase Assay

Enzyme assays are generally conducted in 100-μl assay mixtures in 1.5-ml capped polypropylene centrifuge tubes containing substrate, appropriate cofactors buffer and membranes. Concentrations of individual components will vary somewhat depending upon the tissue being analyzed. Specific levels used for red beet and carrot are given in Table 1. Protein levels may vary between 0.5 and 50 μg depending upon the relative purity of the sample to be assayed. Substrate levels may vary depending upon whether the so-called glucan synthase II (1 mM) or glucan synthase I (10 to 100 μM) activities are being assayed. Reaction mixtures are incubated at 30° C for 5 min, however the time may be extended in samples of dilute activity such as column fractions. Reactions are then terminated by heating at 90° to 100° C in a water-filled Dri-Bath for 5 min.

To isolate radiolabeled ethanol insoluble glucan, reaction mixtures are cooled and spotted in their entirety onto Whatman GF/A filters. Sample numbers are penciled onto the filters before spotting. The filters are dried under a heat lamp and placed onto a 12.5 cm Büchner Funnel. The filters are then washed successively with 66% (w/v) ethanol containing 0.85 mM EDTA, 66% ethanol and 70% ethanol. The filters are washed with acetone, dried, placed in scintillation vials with 5 ml of Liquiscint (National Diagnostics, Somerville, NJ) and counted. Quenching is corrected by counting known amounts of UDP[^{14}C]-glucose that have been spotted on filters and dried before counting. An additional set of filters is spotted with known amounts of UDP[^{14}C]-glucose. These are extracted with

Table 1. *β*-Glucan synthase assay system components. Reaction mixtures are typically 100 μl

Component	Red Beet	Carrot
Tris-HCl, pH 7.2	50 mM	–
Tris-HCl, pH 7.0	–	50 mM
EGTA	–	500 μM
$CaCl_2$	2 mM	750 μM[a]
$MgCl_2$	5 mM	–
Cellobiose	5 mM	5 mM
Digitonin	0.01%	0.01%
Membranes[b]	0.5 mg/ml	0.5 mg/ml
UDPG[c]	1 mM	1 mM

[a] Total concentration of Ca^{2+}. Free Ca^{2+} in the presence of 500 μM EGTA is approximately 250 μM.
[b] Protein levels may vary depending upon enzyme purity.
[c] Contains 0.1 mCi per mmol UDP-[^{14}C]-glucose (ICN Radiochemicals, Irvine, CA; 220 mCi/mmol). Approximately 30,000 cpm per 100 μl reaction mixture.

ethanol as above to determine wash efficiency and background binding of UDP[^{14}C]-glucose for reaction blanks. One enzyme unit is defined as the nmol glucose incorporated into ethanol insoluble glucan per min.

To prepare reaction products for size or linkage analysis, reaction mixtures are scaled-up approximately 20-fold. For methylation analysis, the use of plastic tubes is discouraged due to possible contamination of samples with plasticizers. Biosynthesized glucan is isolated by a series of extractions with centrifugations at 13,000 g. These consist of 2 extractions with water, 3 ethanol washes as described above and 2 extractions with chloroform/methanol (2/1). The glucan is finally dissolved in 100% DMSO for methylation analysis or gel filtration chromatography or d_6DMSO^{1}H-NMR. Methods of structural analysis are described elsewhere in this volume (Hayashi 1989).

4 Solubilization Methods

4.1 Overview

To gain additional insight into the regulation and molecular structure of membrane-bound glycosyl transferases, their separation from other membrane proteins is essential. Efficient solubilization techniques are a prerequisite toward purification of the enzyme complex and the generation of immunological and genetic probes. For many years, digitonin was the only detergent known to be able to solubilize glycosyl transferases such as β-(1,3)-glucan synthase. Various attempts to solubilize glycosyl transferases with digitonin and other detergents have been summarized (Wasserman et al. 1985; Wasserman and Sloan 1988).

There has been some progress in obtaining further purification from digitonin-solubilized enzyme (Eiberger and Wasserman 1987; Kauss and Jeblick 1987), however, difficulties have been encountered. One is stability, which varies greatly among tissue sources (Wasserman et al. 1985; Wasserman and Sloan 1988). Another has to do with the micellar nature of digitonin itself. For example, digitonin has an infinitesimally small critical micelle concentration (CMC) (0.02%) and a high micelle molecular weight (70,000) (Hjelmeland and Chrambach 1984). Since solubilization is conducted above the critical micelle concentration of digitonin, micelles are more likely to contain a heterogeneous population of proteins. This may confound purification efforts.

The detergent 3-[(3-cholamidopropyl)dimethylammonio]-1-propane sulfonate(CHAPS) has recently proven effective for solubilizing β-glucan synthase from red beet and carrot. CHAPS is a zwitterionic derivative of cholic acid (Hjelmeland 1980; Hjelmeland anc Chrambach 1984). The hydrophobic cholic acid portion of CHAPS bears structural similarity to the digitogenin portion of digitonin. The major advantages of this detergent over digitonin include a relatively high CMC (0.49% vs 0.02%), dialyzability, and a much lower micellar molecular weight (6,500 vs 70,000). It is possible to further enrich the CHAPS-solubilized enzyme by reconstitution into proteolipid vesicles by gel filtration chromatography (Sloan et al. 1987).

4.2 Solubilization Techniques

4.2.1 Red Beet Glucan Synthase

One-Step CHAPS Solubilization. To solutions containing membranes at protei concentrations of 2 to 3 mg per ml, an equal volume of buffer containing 1.2% (w/v) CHAPS (Sigma Chemical Co.), 2 mM EDTA, 2 mM EGTA and 100 mM Tris-HCl, pH 7.5, is added dropwise with stirring. The suspensions are incubate for 30 min on ice and then centrifuged at 80,000 g in a Beckman Type 40 roto for 30 min. The supernatants are recovered for further use. Pellets are resus pended in 0.6% (w/v) CHAPS, 1 mM EDTA, 1 mM EGTA and 50 mM Tris HCl, pH 7.5, and assayed. CHAPS-solubilized enzyme *requires* the presence o digitonin in assay mixtures for activity to be expressed (Fig. 1).

Percent solubilization is defined as the ratio of activity recovered in the super natant to the sum of activities recovered in the supernatant and resuspended pel let. Stability of activity to the solubilization process may be quantitated by takin the ratio of total activity recovered (sum of activities in the supernatant and re suspended pellet) to the total activity added to the solubilization mixture.

Two-Step CHAPS Solubilization. In this procedure, the specific activity of solu bilized enzyme may be approximately doubled by a preliminary membrane ex traction to remove contaminating proteins. This step utilizes a lower concentra tion of detergent and divalent cations, which suppresses glucan synthase solubi

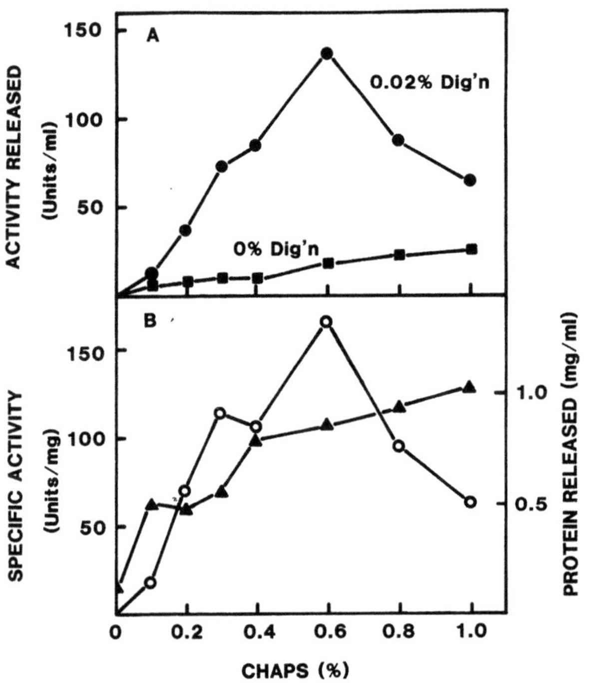

Fig. 1 A, B. Effect of CHAPS concentration on glucan syn thase solubilization. Red bee microsomes were solubilized by the one-step method in the presence of chelators as described. **A** ● CHAPS-solubilized enzyme activity as sayed with 0.02% digitonin. ■ CHAPS-solubilized enzym activity assayed without digi tonin. **B** ○ Specific activity o solubilized enzyme; ▲ Protei released

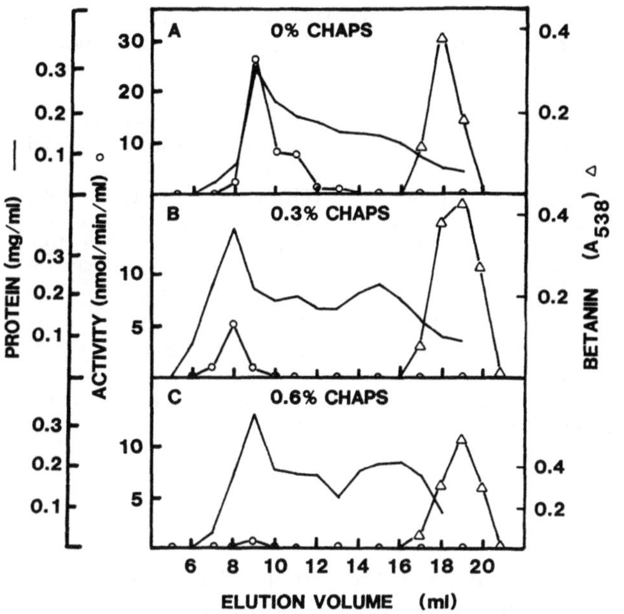

Fig. 2 A–C. Gel filtration profiles of CHAPS solubilized glucan synthase. Glucan synthase solubilized from microsomes by the one-step method was passed through gel filtration columns equilibrated with (**A**) 50 mM Tris-HCl, pH 7.5 (**B**) Buffer and 0.3% CHAPS and (**C**) Buffer and 0.6% CHAPS. o Activity; — Protein; △ Betanin. Samples were assayed at pH 7.5

lization. In the second step, solubilization is promoted by the inclusion of chelators (Sloan et al. 1987).

To solutions containing microsomal membranes at protein concentrations of 2 to 3 mg per ml or plasma membranes at 1–2 mg/ml, an equal volume of buffer containing 0.6% CHAPS, 10 mM $MgCl_2$ and 100 mM Tris-HCl, pH 7.5, is added. The suspensions are incubated on ice for 15–30 min and centrifuged at 80,000 g. Pellets are resuspended to a concentration of 2 to 3 mg per ml protein in 50 mM HEPES-NaOH, pH 7.5, and 15% glycerol and to this is added an equal volume of solubilization buffer consisting of 1.2% CHAPS, 2.0 mM EDTA, 2.0 mM EGTA and 100 mM Tris-HCl, pH 7.5. The suspensions are incubated on ice for 30 min and centrifuged as before.

Digitonin Solubilization. Microsomal or plasma membranes are solubilized as described above with a solubilization buffer containing 1.5% (w/v) digitonin (ICN Biochemicals, Lisle, IL), 5 mM EDTA and 100 mM Tris-HCl, pH 7.5.

Gel Filtration Chromatography and Functional Reconstitution. Solubilized enzyme is layered onto a gel filtration column packed with either Sepharose 4B or Sephacryl S-400. Sephacryl has the advantage of faster flow rates and less nonspecific interactions of enzyme with the column resins. Sample volume may range from 0.25–1.0 ml, depending upon whether a small (1.0 × 5.0 cm) or large (1.0 × 12 cm) column is used. Prior to column application, 500 μl of CHAPS-solubilized enzyme is diluted with 200 μl of 0.75 mg/ml asolectin, 0.6% CHAPS and 10% glycerol in 50 mM Tris-HCl, pH 7.5. For optimal activity recovery, the columns are eluted with 50 mM Tris-HCl, pH 7.5, containing 0.1% CHAPS, 0.1–0.15 mg/ml asolectin (Associated Concentrates, Woodside, NY), 1 mM DTE and 5% glyce-

Table 2. Sequential solubilization and functional reconstitution of red beet plasm: membrane glucan preparation and the sequential solubilization procedure were conducte as described in Materials and Methods. Elution buffer contained 0.1% CHAPS an 0.1 mg/ml asolectin

Fraction	Activity (Units)	Protein (mg)	Specific activity (U/mg)	Purification (fold)	Yielc (%)
Microsomes	295	5.45	54.1	–	100
Plasma membranes	155.1	1.32	117.9	2.2	53
CHAPS-Mg^{2+} pellet	134.8	0.71	189.9	3.5	46
CHAPS solubilized	117.4	0.15	782.4	14.5	39
Peak column fractions (2)[a]	62.0	0.055	1127.3	20.8	21

[a] Sum of activity and protein measurements in the two peak column fractions. The tot activity recovered in all fractions was 106 U representing 91% of the solubilized activit layered on the column.

rol (Fig. 2.). Fractions are assayed for activity and protein. Where possible, th elution profile of betanin (A_{538}), the soluble pigment entrapped within intac membrane vesicles, may be measured. Betanin elutes in the included volume an its measurement is useful for calibration purposes. Vesicles elute in the void vol ume (Sloan et al. 1987). They may be concentrated by centrifugation a 80,000 g.

Table 2 illustrates that over 20-fold purification relative to microsomes is ob tained when red beet plasma membranes are solubilized by the two-step metho and followed by reconstitution.

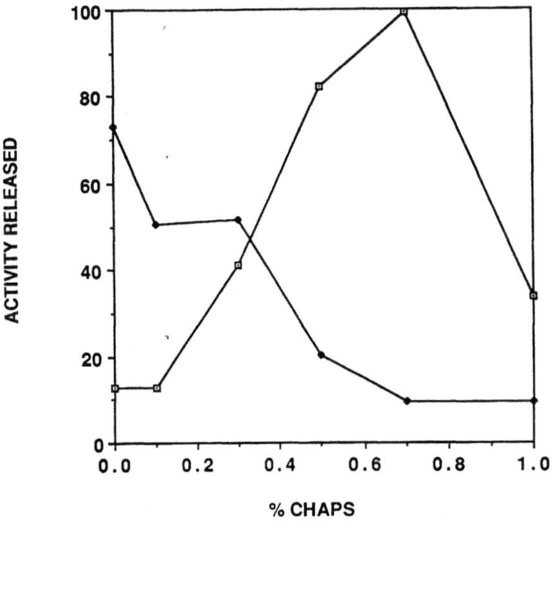

Fig. 3. Effect of CHAPS concen tration on solubilization of glucan synthase from carrot micro somes. ⊡ Activity released into supernatant; ◆ Activity remair ing in sediment

4.2.2 Carrot Glucan Synthase

The procedure is essentially similar to red beet, except that chelators should be omitted from the solubilization mixture (Lawson et al. 1989). To solutions containing membranes at protein concentrations of 2 to 3 mg per ml, an equal volume of buffer containing 1.2% (w/v) CHAPS, and 100 mM Tris-HCl, pH 7.5, is added dropwise with stirring. The suspensions are incubated for 30 min at 4° C and then centrifuged at 80,000 g. The effect of CHAPS concentration on glucan synthase solubilization is shown in Fig. 3.

Workers should be aware that we have found that there are some carrot cultivars which are easy to solubilize by the above procedure and others which are difficult. Packaged carrots were highly amenable to CHAPS solubilization, whereas commercially obtained "fresh" carrots with their greens attached did not solubilize well and were not as stable to detergent. The reasons for these differences are not clear and are the subject of further investigation.

5 Summary

The study of cell wall biosynthesis at the molecular level is an evolving field. Although several methods, such as membrane isolation, glycosyl transferase assay and, in some instances, solubilization and reconstitution, have been established, it is clear that additional tools are needed. These will include purification techniques (Wasserman and Sloan 1989) affinity probes (Read and Delmer 1987), the use of antibodies and the generation of appropriate genetic probes. A difficulty which may be possible to work around is the inherent instability of glycosyl transferases from some tissue types. Continued development and refinement of techniques used to isolate and analyze integral membrane proteins should greatly advance our knowledge of the mechanism by which cell wall polymers are assembled.

Acknowledgments. Support was provided in part by Grants 87-CRCR-1-2414 from the U.S.D.A., DMB 85-02523 from the National Science Foundation, a Busch Memorial Grant, the Rutgers Research Council and the New Jersey Agricultural Experiment Station with State and Hatch Act Funds. N.J.A.E.S. Publication No. D-10546-1-88. We thank Mrs. Marianne Bianco for providing excellent secretarial help.

References

Bolwell GP, Dalessandro G, Northcote DH (1985) Decrease of polygalacturonic acid synthase during xylem differentiation in sycamore. Phytochemistry 24:699–702

Bradford MM (1976) A rapid and sensitive method for the quantitation of microgram quantities of protein using the principles of protein dye-binding. Anal Biochem 72:143–147

Briskin DP, Poole RJ (1983) Characterization of a K^+-stimulated adenosine triphosphatase associated with the plasma membrane of red beet. Plant Physiol 71:350–355

Cumming CM, Brett CT (1986) A galacturonyltransferase involved in pectin biosynthesis In: Vian B, Reis D, Goldberg R (eds) Groupe Parois (France) Cell Walls '86 Proc 4t cell wall meeting Paris September 10th–12th, Université Pierre et Marie Curie, Ecol Normale Supérieure, p 360

Dalessandro G, Northcote DH (1981) Xylan synthetase activity in differentiated xyler cells of sycamore trees (*Acer pseudoplatanus*). Planta 151:53–60

Delmer DP (1987) Cellulose Biosynthesis. Annu Rev Plant Physiol 38:259–290

Delmer DP, Cooper G, Alexander D, Cooper J, Hayashi T, Nitsche C, Thelen M (198 New approaches to the study of cellulose biosynthesis. J Cell Sci Suppl 2:33–50

Eiberger LL, Wasserman BP (1987) Partial purification of digitonin-solubilized β-gluca synthase from red beet root. Plant Physiol 83:982–987

Eiberger LL, Ventola CL, Wasserman BP (1985) Solubilization of a digitonin-stable gluca synthase from red beet root. Plant Sci Lett 37:195–198

Ephritikhine G, Lamant A, Heller R (1980) Influence of sucrose on the characterizatio of higher plant membranes by β-glucan synthetase activity and its relation to osmoti pressure. Plant Sci Lett 19:55–64

Fink J, Jeblick W, Blaschek W, Kauss H (1987) Calcium ions and polyamines activate th plasma membrane-located 1,3-β-glucan synthase. Planta 171:130–135

Hayashi T (1989) Measuring β-glucan deposition in plant cell walls. In: Linskens HF, Jack son JF (eds) Modern methods of plant analysis, vol X. Springer, Berlin Heidelber New York Tokyo, p 138–160

Hayashi T, Matsuda K (1981) Biosynthesis of xyloglucan in suspension-cultured soybea cells. Occurrence and some properties of xyloglucan 4-β-D-glucosyltransferase and 6 a-D-xylosyltransferase. J Biol Chem 256:11117–11122

Hayashi T, Read SM, Bussell J, Thelen M, Lin FC, Brown RM (1987) UDP-glucose: (1,3 β-glucan synthases from mung bean and cotton. Differential effects of Ca^{2+} and Mg^{2} on enzyme properties and on macromolecular structure of the glucan product. Plar Physiol 83:1054–1062

Henry RJ, Stone BA (1982) Factors influencing β-glucan synthesis by particulate enzyme from suspension-cultured *Lolium multiflorum* endosperm cells. Plant Physiol 69:632 636

Henry RJ, Schibeci A, Stone BA (1983) Localization of β-glucan synthases on the mem branes of cultured *Lolium multiflorum* (ryegrass) endosperm cells. Biochem J 209:627 633

Hjelmeland LM (1980) A nondenaturing zwitterionic detergent for membrane biochemi try: design and synthesis. Proc Natl Acad Sci USA 77:6368–6370

Hjelmeland LM, Chrambach A (1984) Solubilization of functional membrane protein Meth Enzymol 104:305–308

Kauss H (1985) Callose biosynthesis as a Ca^{2+}-regulated process and possible relations t the induction of other metabolic changes. J Cell Sci Suppl 2:89–103

Kauss H, Jeblick W (1987) Solubilization, affinity chromatography and Ca^{2+}/polyamir activation of the plasma membrane-located 1,3-β-D-glucan synthase. Plant Sci 48:63 69

Lawson S, Mason TL, Sabin RD, Sloan ME, Drake RR, Haley BE, Wasserman BP (198 UDP-glucose (1,3)-β-glucan synthase from *Daucus carota* L.: Characterization, affinit labelling and solubilization. Plant Physiol. In press

Lin TY, Elbein AD, Su JC (1966) Substrate specificity in pectin synthesis. Biochem Bi phys Res Commun 22:650–657

Ray PM (1980) Cooperative action of β-glucan synthetase and UDP-xylose xylosyl tran ferase of golgi membranes in the synthesis of xyloglucan-like polysaccharide. Biochi Biophys Acta 629:431–444

Ray PM, Shininger TL, Ray MM (1969) Isolation of β-glucan synthetase particles fro plant cells and identification with Golgi membranes. Proc Natl Acad Sci USA 64:60 612

Read SM, Delmer DP (1987) Inhibition of mung bean UDP-glucose: (1,3)-β-glucan sy thase by UDP-pyridoxal. Evidence for an active-site amino group. Plant Physi 85:1008–1015

Sandelius AS, Penel C, Auderset G, Brightman A, Millard M, Morre DJ (1986) Isolation of highly purified fractions of plasma membrane and tonoplast from the same homogenate of soybean hypocotyls by free-flow electrophoresis. Plant Physiol 81:177–185

Sloan ME, Rodis P, Wasserman BP (1987) CHAPS solubilization and functional reconstitution of β-glucan synthase from red beet (*Beta vulgaris* L.) storage tissue. Plant Physiol 85:516–522

Smith MM, Stone BA (1973) β-glucan synthesis by cell-free extracts from *Lolium multiflorum* endosperm. Biochim Biophys Acta 313:72–94

Tsai CM, Hassid WZ (1971) Solubilization and separation of uridine diphospho-D-glucose:β-(1,4) glucan and uridine diphospho-D-glucose: β-(1,3)glucan glucosyltransferases from coleoptiles of *Avena sativa*. Plant Physiol 47:740–744

Villemez CL, Swanson AL, Hassid WZ (1966) Properties of a polygalacturonic acid-synthesizing enzyme system from *Phaseolus aureus* seedlings. Arch Biochem Biophys 116:446–452

Waldron KW, Brett CT (1983) A glucuronyltransferase involved in glucuronoxylan synthesis in pea (*Pisum sativum*) epicotyls. Biochem J 213:115–122

Wasserman BP, McCarthy KJ (1986) Regulation of plasma membrane β-glucan synthase from red beet root by phospholipids. Reactivation of Triton X-100 extracted glucan synthase by phospholipids. Plant Physiol 82:396–400

Wasserman BP, Sloan ME (1989) Molecular approaches for probing the structure and function of β-(1,3)-D-glucan and cellulose synthases. In: Weimer P, Haigler C (eds) Biosynthesis and biodegradation of cellulose and cellulosic materials. Marcel Dekker, New York

Wasserman BP, Ventola CL, Eiberger LL (1985) Glucan biosynthesis in red beet root microsomes and tissue slices. J Agric Food Chem 33:44–149

Wasserman BP, Eiberger LL, McCarthy KJ (1986) Biotechnological approaches for controlled cell wall glucan biosynthesis in fruits and vegetables. Food Tech 40,5:90–98

Analysis of Cross-Links in the Growing Cell Walls of Higher Plants

S. C. FRY

1 Background

1.1 Polymeric Components of the Growing Cell Wall

The growing cell walls of higher plants consist of a rather small number of major polymers (Table 1; for details, see Fry 1988). Most of these are polysaccharides, which fall into three major classes: cellulose [β–(1→4)–glucan, the skeletal frame-

Table 1. Major polymers of the growing cell wall

Polymer	Major sugar and amino acid residues	Solubility in water[a]	Net charge at pH 7	Approx % composition of growing wall of	
				Grasses	*Dicots*
Polysaccharides					
Cellulose	β–Glc*p*	Insoluble	0	30	30
Hemicelluloses					
Xyloglucan	β–Glc*p*, α–Xyl*p*, β–Gal*p*, α–Fuc*p*	Soluble[b]	0	4	25
Xylans	β–Xyl*p*, α–Ara*f*, α–Glc*p*A, [β–Gal*p*]	Soluble[b]	−ve	30	5
Mixed linked glucans	β–Glc*p*	Soluble	0	30	0?
Pectins					
Homogalacturonan	α–Gal*p*A, α–Rha*p*	Soluble[b]	−ve	5 (pectins total)	15
RG-I	α–Gal*p*A, α–Rha*p*, β–Gal*p*, α–Ara*f*, [Fuc*p*, Xyl*p*]	Soluble	−ve		15
RG-II	α–Gal*p*A, β–Rha*p*, α–Gal*p*, α–Fuc*p*, α–Ara*p*, Ara*f*, β–Gal*p*A, α–Rha*p*, Api*f*, β–Glc*p*A, KDO, Ace*f*A, Xyl*p*, Glc	Soluble	−ve		5
Glycoproteins					
Arabinogalactan-proteins	β–Gal*p*, α–Ara*f*, α–Ara*p*, Glc*p*A, Gal*p*A, [Rha, Man]; Hyp, Ser, Ala	Soluble	−ve	Variable	Variable
Extensin	β–Ara*f*, α–Ara*f*, α–Gal*p*; Hyp, Lys, Ser, Tyr, Val, [His]	Soluble	+ve	0.5	≃ 5

[a] The solubility indicated is that *after* extraction of the polymer from the wall.
[b] Solubility sometimes low unless ionic composition of solution optimized.

work of the wall] hemicelluloses (generally rigid, rod-shaped polysaccharides that hydrogen-bond to the cellulose and can be extracted from the wall with cold NaOH), and pectins (jelly–like, often acidic polysaccharides that are more loosely bound in the wall; a proportion can be extracted with chelating agents). In the growing cell walls of Dicotyledons, these three classes are about equally abundant, whereas grasses possess much less pectin (McNeil et al. 1984). All growing cell walls also contain some glycoproteins (Lamport 1977), and some contain large amounts of additional polymers – e.g., even *rapidly* growing epidermal walls contain cutin (Lendzian and Schönherr 1983), which is built up of a polyester of long aliphatic chains mixed, meshed or bonded with a network of phenolic rings (Holloway 1982).

1.2 The Value of Specific Degradative Techniques

We now have a reliable picture of the *primary* structures (sequence of monosaccharides and amino acis) of many wall polysaccharides (McNeil et al. 1984) and two glycoproteins [extensin (Chen and Varner 1985) and peroxidase (Wellinder 1976)]. Much interest therefore currently centers on higher orders of structure. However, it is relevant to consider here the major strategies and conclusions of the work on primary structure. In several wall carbohydrates, the primary structure includes sequences that must have been biosynthesized with great precision. For instance, rhamnogalacturonan-I has a repeating disaccharide as its backbone (McNeil et al. 1984); rhamnogalacturonan-II has two different, but very precise, heptasaccharide blocks (Melton et al. 1986); xyloglucan has a repeating nonasaccharide (Valent et al. 1980; Kato and Matsuda 1980); and extensin has hydroxyproline-linked tetra-arabinosides that are highly conserved (Akiyama et al. 1980). Detection of such repeating sequences was dependent on the use of suitably specific methods for cleaving the polymers' backbones at precisely defined sites (for a review, see Fry 1988). Thus, the repeating nonasaccharide of xyloglucan was discovered by treatment with cellulase, which specifically chops the xyloglucan backbone at *unbranched* (but not xylosylated) glucose residues; and the hydroxyproline–tetra-arabinoside of extensin was liberated by treatment with hot alkali, which cleaves peptide but not glycosidic bonds.

This chapter discusses attempts to develop similarly specific methods for cleaving the *cross-links* between wall polymers. The hope is that such methods will allow us to define the molecular architecture of the cell wall at the level of sophistication recently achieved for primary structures.

1.3 Cross-Links in the Assembly of a Growing Cell Wall

Except for cellulose, all the wall polymers listed in Table 1 are water-soluble: they are polyhydroxylated, hydrated molecules, and many of them are charged. Despite this, these polymers can organize themselves into a strong, water-insoluble structure – the cell wall. The cross-links that bind these otherwise water-soluble molecules together hold the key to our understanding of the structure and assembly of the plant cell wall. Such an understanding is essential for an explanation at the molecular level of how plants loosen and tighten their walls in order to promote or inhibit growth. It is also important to our attempts to manipulate the digestibility of plant cell walls.

2 Chemistry of Cross-Links

2.1 Chemistry of Noncovalent Cross-Links

2.1.1 Chemistry of Hydrogen-Bonds

An H-bond forms between the H atom of an –OH, –NH– or $-NH_2$ group and an accessible O or N atom, either in the same molecule (forming an intramolecular bond) or in a different molecule (forming a cross-link) (Schuster et al. 1976). Wall polymers can form H-bonds at many sites since they have large numbers of –OH and –O– groups. The polymers also form H-bonds with the H_2O that permeates the cell wall (Northcote 1972). Individual H-bonds are weak (bond energy 3–6 kcal/mol; cf. bond energies for covalent C–C, C–O, C–H and O–H bonds $\simeq$ 80–110 kcal/mol). They can therefore only contribute appreciably to the wall-binding of a polymer if many H-bonds can act in concert; this, however, is a common circumstance since wall polysaccharide molecules are large (molecular weights: cellulose $\simeq$400,000; xyloglucan $\simeq$200,000; rhamnogalacturonan-I $\simeq$300,000).

The orientation of the –OH and –O– groups in cellulose [β–(1$\rightarrow$4)–D–glucan] are such that large numbers of intra- and inter-molecular H-bonds can form: this accounts for the rigidity of the cellulose molecule and for its ability to form almost water-free microfibrils. Hemicelluloses of the growing wall resemble cellulose, being built with a backbone largely of β–(1$\rightarrow$4)–D–glucan or of the very similar β–(1$\rightarrow$4)–D–xylan. Hemicelluloses can therefore form H-bonds similar to those found in cellulose. However, most hemicelluloses differ from cellulose in possessing short side-chains (α–L–arabinose and α–D–glucuronic acid in arabinoxylan; α–D–xylose, β–D–galactose and α–L–fucose in xyloglucan) and in being acetylated. This means that they can form fewer H-bonds than cellulose and that they cannot form microfibrils. They can, nevertheless, H-bond on to the surface of a cellulose microfibril, where they probably form a molecular monolayer (Fry 1989).

The formation of H-bonds between newly secreted hemicelluloses and cellulose is presumed to be a nonenzymic process. In support of this conclusion, pure hemicelluloses will bind to a column of cellulose powder in vitro (Aspinall et al. 1969; Hayashi et al. 1987).

2.1.2 Monovalent Ionic Bonds

Ionic bonds are attributable to the electrostatic attraction between groups with strong positive and negative charges. The major negatively charged groups in wall polymers are the $-COO^-$ groups of galacturonic acid in pectins and glucuronic acid in xylans; these groups have pK_a values of $\simeq$3–5, and are thus only fully charged at pH values above this. No polysaccharides of higher plant cell walls have positive charges, but extensin has a very strong positive charge owing to the numerous lysine $-NH_2$ groups (charged at all biologically relevant pH values) and the histidine imidazole rings (only fully charged at pH values lower than 6). As with H-bonds, single ionic bonds are weak but if many of them can act in concert they may be responsible for holding a large polymer molecule firmly to its neighbors in the wall. This is the case with newly secreted extensin (Smith et al. 1984) and with many wall enzymes (Fry 1988).

The formation of ionic bonds is presumed to be nonenzymic: salt-extracted extensin will readily rebind to cell walls (and to artificial cation-exchange resins) in vitro.

2.1.3 Calcium Bridges

Calcium bridges are a special example of ionic bond, in which a single Ca^{2+} ion is attracted to two negatively charged groups (e.g., two $-COO^-$ groups) at the same time; if the two groups are on different polysaccharide molecules, a cross-link is formed. The bridging effect of Ca^{2+} may be greatly reinforced by the ability of the Ca^{2+} to form co–ordinate bonds with specific oxygen atoms on the surrounding sugar residues. In this way, homogalacturonan, a pectic polysaccharide, is cross-linked by Ca^{2+} ions in a precise manner known as the "egg-box" arrangement (Jarvis 1984).

The Ca^{2+}–bridging of newly secreted polysaccharides is thought to be a nonenzymic process since pure homogalacturonan is readily precipitated by added Ca^{2+}; however, a wall-bound enzyme (pectin methylesterase) may be required to convert pectins, which are initially deposited in the wall in a highly methyl-esterified form [R–CO–CH_3], into a free acid [R–COO^-] that can bind Ca^{2+}. In this sense, the formation of Ca^{2+}-bridges may be said to be enzyme-catalyzed. It is an important unanswered question whether the formation of Ca^{2+} bridges in the cell wall can also be controlled by the secretion of Ca^{2+} from the protoplast. Attempts in this connection to measure the Ca^{2+} content of cell walls are thwarted by the facts that, during the isolation of the wall, (a) Ca^{2+} may be liberated from the ruptured protoplasts and quickly bound by wall polymers, and (b) natural chelating agents, e.g. citrate, may be released from the protoplast and liberate some genuinely wall-bound Ca^{2+}.

2.2 Chemistry and Properties of Covalent Cross-Links

2.2.1 Phenolic Coupling Products

Phenolic coupling products are an important and interesting class of cross-link in the growing cell wall. Several wall polymers bear phenolic side-chains (Fry 1988), including (i) the abundant tyrosine residues of extensin (Lamport 1977), and (ii) hydroxycinnamates (=*p*–coumarate, ferulate etc.) and *p*–hydroxybenzoate, which are esterified to specific –OH groups on particular wall polysaccharides (Fry 1986 a). The main phenol-bearing polysaccharides in grass cell walls are arabinoxylans, which seem to be feruloylated exclusively on the O–5 position of their arabinofuranose side chains (Kato and Nevins 1985; Gubler et al. 1985; Ahluwalia and Fry 1986; Hartley 1987). In certain Dicotyledons, the ferulic acid groups are attached to the O–3 position of arabinopyranose residues and to the O–6 position of galactopyranose residues (Fry 1982a), both of which are nonreducing termini of neutral side-chains of acidic pectic polysaccharides. Arabinopyranose residues are rare in wall polysaccharides and it seems that over half of all these residues may be feruloylated (Fry 1983). The ferulate groups are transferred on to the polysaccharides intracellularly, probably while the polysaccharide is still nascent (Fry 1987b).

Phenols have a propensity to undergo oxidative coupling, a reaction that can be achieved either enzymically [e.g., with peroxidase $+H_2O_2$ (Gaspar et al. 1982)] or, in vitro, nonenzymically with oxidizing agents such as ferricyanide or tetranitromethane (Taylor and Battersby 1967). The oxidizing system removes a single H atom from each phenol, forming a free radical (Fig. 1, step *a*). The free radicals, after rapid mesomerization (steb *b*), pair off to form dimers linked by stable C–C (e.g., biphenyl) or C–O–C (e.g., diphenyl ether) bonds: one of the several possibilities is illustrated in step *c*. Any initially unstable coupling products then rapidly undergo rearrangements to yield more stable products (step *d*).

Any polysaccharides or glycoproteins that were attached to the coupling phenol molecules are thereby indirectly cross-linked. Cross-linking of wall polymers in this way could play a crucial rôle in the assembly of the growing cell wall. It also seems likely that an acceleration of the process of phenolic coupling could lead to a wall with altered properties, especially with a lower *extensibility* and thus decreased growth rate (Fry 1983, 1986a), and a lower *digestibility* and thus increased disease-resistance (Ride 1975) and decreased nutritional value (Fry 1984a). Conversely, the inhibition of otherwise constitutive phenolic coupling reactions in the cell wall could lead to a higher extensibility and digestibility (Fry 1979, 1980). Thus, it has been proposed that the rapid inhibitory effect of blue light on plant growth is due to increased phenolic coupling (Shinkle and Jones

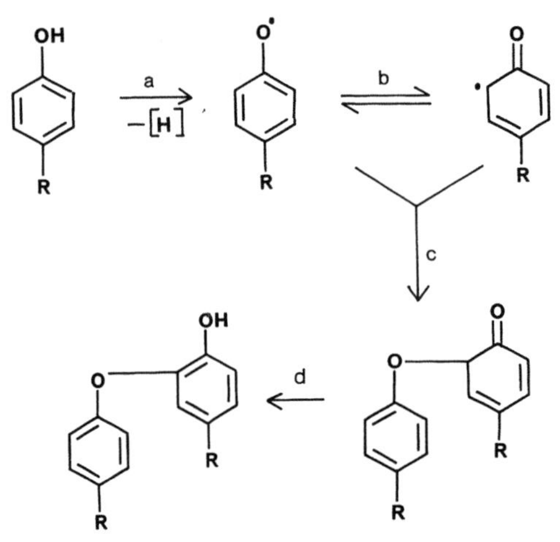

Fig. 1. Oxidative coupling of phenols, illustrated by the conversion of tyrosine to isodityrosine [–R = $-CH_2-CH(NH_2)-COOH$]. *a* Oxidation to form free radical [· = unpaired electron]; *b* Mesomerization; *c* Coupling of free radicals; *d* Rearrangement of initially-formed unstable structure

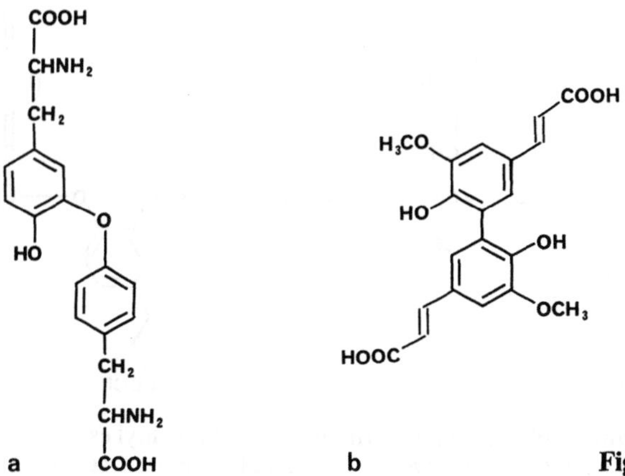

Fig. 2. a Isodityrosine; **b** Diferulic acid

1988), and that the long-term promotion of plant growth by gibberellic acid could be due to a suppression of phenolic coupling (Fry 1979, 1980). The possible biological rôles and regulation of phenolic coupling reactions in the growing plant cell wall are discussed in more detail by Biggs and Fry (1987).

Specific examples of phenolic coupling products that have been found in polymers of the plant cell wall (Fig. 2) include *isodityrosine,* which is a diphenylether derived from the tyrosine residues of extensin (Fry 1982 b) and *diferulic acid,* which is a biphenyl derived from ferulic acid (Markwalder and Neukom 1976; Hartley and Jones 1976; Carpita 1986). It is interesting that, although extensin contains the diphenyl ether (isodityrosine), it contains none of the biphenyl (dityrosine); in contrast, resilin (a protein in the insect cuticle) contains dityrosine but not isodityrosine (Andersen 1964). Evidently, the pairing of the mesomeric free radicals in the plant cell wall and insect cuticle is not random, but certain orientations of coupling are favored over others. The factor that forces the tyrosine residues of wall-bound extensin to couple as isodityrosine rather than dityrosine may be the presence of acidic polysaccharides ionically bound to the extensin (Fry 1987 a).

Some initially formed dimers, notably those produced from the hydroxycinnamic acids, are likely to be quinone-methides (Fig. 3). These are highly reactive molecules: they can react with H_2O to form benzyl alcohols, with carboxylic acids to form benzyl esters, and with alcohols to form benzyl ethers (Biggs and Fry 1987). In the cell wall there are so many polysaccharide-bound –OH groups that the formation of benzyl–polysaccharide ethers seems very likely to occur. Indeed, some of the hydroxycinnamate-related components of the wall (specifically labeled by pre-incubation of the living cells with [^{14}C]cinnamic acid) cannot be extracted and may therefore be linked by ether bonds such as those formed via quinone-methides (Whitmore 1976; Scalbert et al. 1985; Fry 1984 a, 1988; Carpita 1986).

The C–C and C–O–C bonds that directly join the coupled phenolics are highly stable under biological conditions, and also resist hot acids and alkalies [although prolonged heating with concentrated alkali will hydrolyze the diphenyl ether

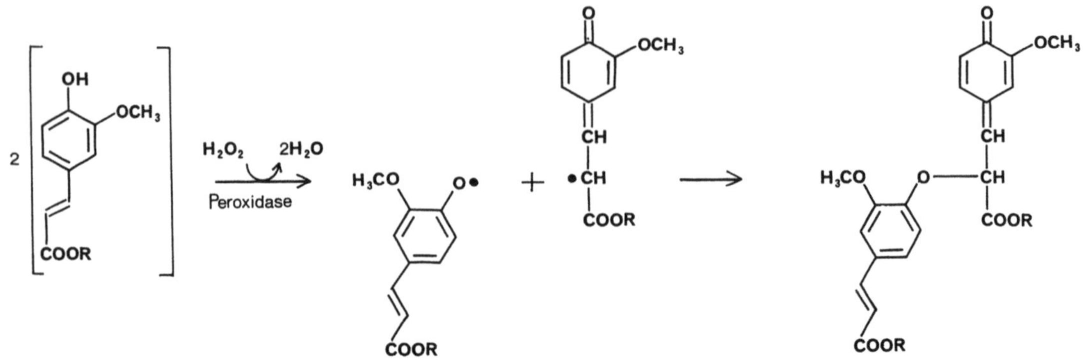

Fig. 3. The possible formation of a semi-stable quinone-methide dimer of feruloyl esters by peroxidase-catalyzed phenolic coupling (cf. Fig. 1). (*ROH* = an alcohol, e.g., a cell wall polysaccharide)

bond of isodityrosine (Fry 1982 b)]. The ester bonds that link hydroxycinnamates and *p*-hydroxybenzoate to polysaccharides are much more labile, and these bonds thus constitute the weakest link in an overall bridge of the type polysaccharide–diferulate–polysaccharide, but they are considerably less alkali-labile than the aliphatic uronoyl esters discussed below (Sect. 2.2.3). Benzyl–polysaccharide ether bonds tend to be more labile, both to acid and to alkali, than typical ether bonds (Enoki et al. 1983); they are nevertheless much more alkali-resistant than ester bonds and more acid-resistant than glycosidic bonds.

2.2.2 Glycosidic Cross-Links

Glycosidic bonds are the normal sugar–sugar bonds that build up the structure of all polysaccharides. The bond joins the potential reducing group (C=O) of one sugar unit to a different –OH group on the next sugar unit, forming an acetal. A *single* sugar unit per polysaccharide (the one at the right-hand end of the polysaccharide chain as conventionally drawn) has its potential C=O group free; this is the reducing terminus of the polysaccharide. It is possible that the reducing terminus of one polysaccharide, e.g., xyloglucan [1]

```
  X   X   X   X   X   X   X   X   X   X   X   X
  ↓   ↓   ↓   ↓   ↓   ↓   ↓   ↓   ↓   ↓   ↓   ↓
G→G→G→G→G→G→G→G→G→G→G→G→G→G→G→G→G→G→G→G→G→G→G→G→G*,        (Sch. 1)
    ↑       ↑       ↑       ↑       ↑       ↑
    X       X       X       X       X       X
```

[1] In this, as well as in the other schemes in this chapter, the structures do not represent precisely determined chemical sequences, but are generalized representations of the major characteristic features of the polymers concerned. Abbreviations used are: A = L-arabinose, G = D-glucose, R = L-rhamnose, U = D-galacturonic acid, X = D-xylose, →,↑,↓ = glycosidic bonds, * = reducing terminus, † = site of attack.

could form a glycosidic linkage to an –OH group on another polysaccharide molecule, e.g., pectin

(Sch. 2)

generating a "cross-link":

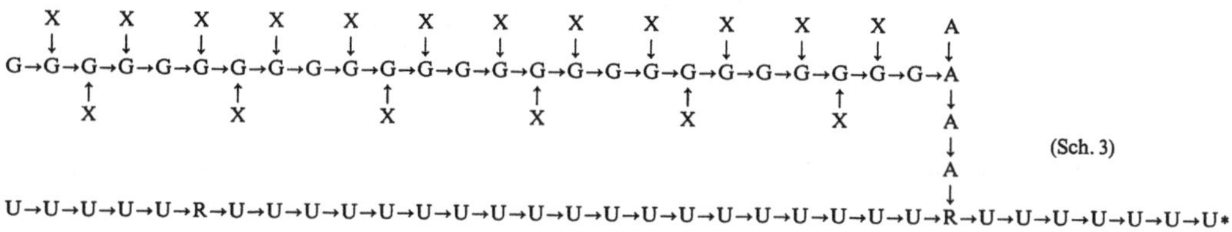

(Sch. 3)

Glycosidic cross-links of this type have been proposed (Bauer et al. 1973). Since each polysaccharide has only a single reducing terminus, this kind of cross-linking could not generate a network; at most it could form a highly complex tree structure. The product of glycosidic cross-linking would actually be a single polysaccharide molecule, since there would be no difference between the nature of the linkage between the two "halves"

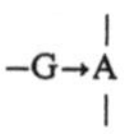

(Sch. 4)

and that of the linkages within each "half"

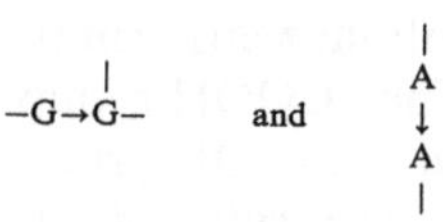

(Sch. 5)

However, if the two "halves" started out as qualitatively different polysaccharides (as shown in Scheme 3), the cross-link could be recognized as such since it would be a bond between two dissimilar sugars (in the example given above, these would be glucose, which is characteristic of hemicellulose, and arabinose, which is characteristic of pectins).

Normal glycosidic bonds of polysaccharides are formed at the Golgi or plasma membranes, using NDP sugars as high-energy sugar donors; any "inter-polysaccharide" glycosidic linkages might form in essentially the same way, in which case the wall polymers would be secreted already "cross-linked". Alternatively, they could form in the cell wall itself by transglycosylation. Since the cell wall is not known to contain any specialized high-energy sugar donors, cross-linking within the wall would require one intra-polysaccharide glycosidic bond († in

Scheme 6) to be cut, and the associated energy conserved in driving the formation of a new glycosidic linkage to an –OH group on another polymer molecule:

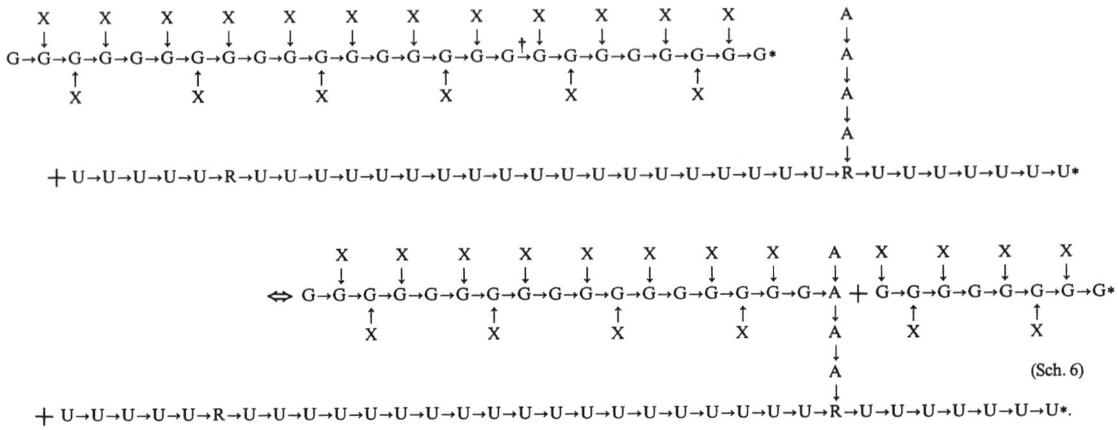

(Sch. 6)

The *average* molecular weight of the participating structures would thus be unchanged by transglycosylation, although in the extreme case (i.e., if the bond cut were not † but the one adjacent to the reducing terminus, G→G*) the products could be a polysaccharide dimer plus a single monosaccharide. Transglycosylation, if it occurs in vivo at all, is likely to be an enzyme-catalyzed reaction: some glycanases catalyze transglycosylation as well as hydrolysis if the conditions are right, at least in vitro (Fry 1988, 1989).

Glycosidic bonds are generally rather stable in aqueous alkali, but can be hydrolyzed with hot molar acid; also, many wall-bound enzymes are known that will catalyze the hydrolysis of glycosidic bonds (Fry 1985).

2.2.3 Possible Uronoyl–Sugar Ester Cross-Links

Ester bonds are another potential means of cross-linking two polysaccharide molecules if at least one of them possesses a uronic acid residue. The –COOH group of, say, a galacturonic acid residue of pectin could condense with an –OH group on a different polysaccharide (or protein) molecule, e.g., a glucose residue of cellulose, to form an ester linkage (R–CO–O–R′). The equilibrium of the reaction

$$R\text{–}COOH + R'\text{–}OH \Leftrightarrow R\text{–}CO\text{–}O\text{–}R' + H_2O$$

lies far to the left under biological conditions, and thus the synthesis of an ester requires the –COOH group to be activated. Intracellularly, this type of reaction often involves condensation of the –COOH group with CoA at the expense of ATP [or, in the special case of methyl-esterification, the alcohol (CH_3OH) is activated as *S*-adenosylmethionine]; in the cell wall, a possible mechanism is trans-esterification of the already methyl-esterified –COOH groups of galacturonic acid (Fig. 1):

$$R\text{–}CO\text{–}O\text{–}CH_3 + R'\text{–}OH \Leftrightarrow R\text{–}CO\text{–}O\text{–}R' + CH_3OH\,.$$

Although uronoyl-sugar ester bonds have been proposed, and would seem to be a plausible structure, no conclusive evidence has been presented to support their existence in plant cell walls (for a review, see Fry 1986 a).

Ester bonds are highly labile in cold dilute alkali, but somewhat more resistant to cold acid. The stability of feruloyl–polysaccharide esters to hot acid lies between the acid stability of furanose and that of pyranose glycosidic linkages (Fry 1983). Several wall-bound esterases are known which will catalyze the hydrolysis of various esters (Gahan and Bellani 1984); their precise functions in vivo are poorly understood, but if extracted and purified they might prove useful tools in the attempt to degrade specific ester cross-links in vitro.

3 Methods for Breaking Cross-Links

If a polymer can be extracted from the wall (or if an extracted polymer can be decreased in M_r) by a treatment known to split a specific type of cross-link without breaking the polymer's primary structure, then this can provide evidence for the presence of such a cross-link in vivo. Problems with this approach are that (i) few treatments are sufficiently specific that it can be stated with confidence that *no* other bonds are broken than the type targeted; and (ii) since polymers may be held in the wall by more than one type of cross-link, failure of a cleaving agent to solubilize a particular polymer does not prove that the targeted cross-link was not *partly* responsible for holding the polymer in the wall. Despite these shortcomings, the use of cross-link-cleaving agents is the most informative method currently available for investigation of wall cross-links, and some suitable agents are described below. A tabulation of cross-link-cleaving agents and their possible side-effects has been published (Fry 1986 a). One extension of this approach worthy of future study is the (simultaneous or sequential) application of various agents designed to cleave different types of cross-link.

3.1 Methods for Breaking Noncovalent Cross-Links

3.1.1 Methods for Breaking Hydrogen-Bonds

H-bonds are cleaved by low-M_r chaotropic agents – compounds that form stronger (and/or more numerous) H-bonds with the polymers than the polymers do with each other. Ideally, the chaotropic agent should be uncharged so that no cleavage of ionic bonds occurs; if this is impossible, a control should be run with a nonchaotropic solution of similar ionic strength.

One uncharged chaotropic agent that has been investigated in detail is 8 M urea. At neutral pH and 25° C (urea solutions turn alkaline on heating), urea is likely to be specific for the cleavage of H-bonds. It is highly effective at denaturing and solubilizing proteins, and will leach $\simeq 30\%$ of xyloglucan that has been experimentally H-bonded to pure cellulose (Aspinall et al. 1969). However, it is a poor extractant for cell wall polymers, solubilizing only about 2% of the wall even after the structure of the wall matrix has been opened up by enzymic removal of the pectin (Bauer et al. 1973). This may be because the hemicelluloses are held not only by H-bonds to the surface of the microfibrils but also by other bonds. Hayashi et al. (1987) have suggested that the xyloglucan weaves into the amorphous parts of the cellulose microfibrils as well as coating their surface. *A priori*, another possible explanation for the poor hemicellulose-extracting ability of chaotropic agents could be that the hemicellulose molecules are so intricately trapped in a pectin and/or extensin cage that they would not be able to escape from the wall even if *all* the H-bonds were broken. However, since removal of the pectins does not appreciably facilitate the extraction of hemicelluloses by 8 M urea, entrapment in a pectic cage seems improbable. A more detailed discussion of the possible arrangement of hemicellulose molecules relative to the cellulosic microfibrils and oxidatively coupled phenols is presented elsewhere (Fry 1989).

An ionic chaotropic agent that is somewhat more successful than urea at extracting wall-bound hemicelluloses, especially those containing mannose, is neutral 5.5 M guanidinium thiocyanate at 25° C (Selvendran et al. 1985). However, the small quantities of hemicellulose extracted compared with the large amounts in the wall again suggest that other bonds augment the hydrogen-bonds in cross-linking hemicellulose in the wall.

A third chaotropic agent that has attracted a great deal of attention recently is a 1:1 (mol/mol) mixture of water and 4-methylmorpholine-*N*-oxide (H_2O/MMNO). This is a solid at 25° C, and only melts above about 85° C; it is often used at about 120° C (Chanzy et al. 1982). Under these conditions, H_2O/MMNO solubilizes much of the primary cell wall, especially cellulose and hemicelluloses. However, the interpretation of results obtained with H_2O/MMNO requires caution: slight cleavage of glycosidic linkages has been detected by viscometric studies (Joseleau et al. 1980), and the rapid hydrolysis of esters has been detected (Fry 1986a). Thus, the solubilization of hemicelluloses by H_2O/MMNO does not prove that hydrogen-bonds are the sole means whereby hemicelluloses are held in the wall.

A fourth chaotropic agent, which is widely used on a day-to-day basis for the extraction of hemicelluloses, is 20% (w/v) NaOH (or KOH). Often $NaBH_4$ (e.g., 1%, w/v) is also added to minimize the occurrence of peeling (elimination) reactions (Fry 1988). Although highly effective for the practical goal of extracting hemicelluloses, NaOH cannot give evidence for H-bonding since it also destroys a number of other cross-links, both covalent (e.g., ester bridges) and noncovalent (e.g., ionic).

3.1.2 Methods for Breaking Ionic Bonds

Ionic bonds can be broken by addition of low-M_r salts that will themselves form stronger and/or more numerous ionic bonds with wall polymers than the polymers do with each other. Salts that have been used include 3 M LiCl, 1 M NaCl, 0.2 M $CaCl_2$ and 25–50 mM $LaCl_3$ or $AlCl_3$. A given ionic strength (and therefore ionic-bond-cleaving power) is reached with lower concentrations of di- and trivalent cations than with univalents. Salts of certain multivalent anions, e.g., Na_2HPO_4 and sodium citrate, have also sometimes been used, but these are not recommended as many of them are chelating agents or precipitants of Ca^{2+} and will thus break Ca^{2+}–bridges as well as ordinary ionic bonds.

Treatment of cell walls with low M_r salts solubilizes a substantial proportion of the wall-bound enzymes (Goldberg 1985; Fry 1985; Hatfield and Nevins 1987; Wallner and Nevins 1974; van der Wilden et al. 1983) and sometimes lectins (Haass et al. 1981); it also solubilizes a proportion of the wall-bound extensin (Smith et al. 1984). The solubilization often proceeds well at 0° C, so the extracted proteins retain enzyme activity. Solubilization may be rapid, e.g., all the extensin that can be eluted from tomato cell walls is eluted by dilute $AlCl_3$ within a few seconds (Smith et al. 1984); or it may be slow, e.g., the extraction of an endo–β–(1→3),(1→4)–glucanase from maize cell walls takes many hours' treatment with 3 M LiCl (Hatfield and Nevins 1987). Since the salts used have low chemical reactivity at 0–25° C, few if any side-effects are likely to occur. Indeed, many of the salts used are harmless enough to be applied to living cells (especially those in suspension culture), a technique that can provide strong evidence that a suspected wall protein is indeed extracellular. The proportion of extensin extracted with salt varies both between species and, within a single species, with physiological condition. This is most clearly seen in suspension-cultured cells: salt extracts almost 100% of the wall-bound extensin from spinach cultures, but virtually none from sycamore (S. C. Fry, unpublished observations). The proportion extracted from tomato cells varies during the culture cycle (lag phase – log phase – linear phase – stationary phase) (Smith et al. 1984). The extensin molecules extracted with salt are presumed to be those that have not acquired any inter-polypeptide isodityrosine cross-links (see Sect. 2.2.1).

3.1.3 Methods for Breaking Calcium Bridges

Calcium bridges are broken by any agent that removes Ca^{2+} from its attachment to the polymers. This can be achieved by treatment with 50 mM *trans*–1,2–diaminocyclohexane–*N,N,N′,N′*–tetra–acetic acid (CDTA) at 25° C for 6 h at pH 6.5 (it is preferable to use several changes of CDTA), and complete removal of Ca^{2+} can be verified by atomic absorbance spectrophotometry of the treated walls (Jarvis 1982). CDTA does solubilize a proportion of the wall-bound pectic polysaccharides, the actual percentage depending on the tissue: it is high for ripe fruit tissues, moderate for most vegetative plant tissues, and low for cell suspension cultures. This may indicate that in suspension cultures the pectin is held in the wall by other bonds in addition to Ca^{2+}-bridges; these additional bonds are not principally ionic as concentrated salts do not facilitate the extraction of pectin by CDTA (Jarvis 1982). It is possible that in ripe fruit other bonds also occur in addition to Ca^{2+}–bridges, but that partial breakdown of the pectin has taken place in the normal course of ripening, generating substantial pieces of pectin that are only Ca^{2+}–bridged or no longer wall-bound at all.

Other Ca^{2+}–chelating agents than CDTA have been used to solubilize pectic polysaccharides, but generally with heating. Examples are ethylenediaminetetraacetic acid (EDTA), oxalate, hexametaphosphate and citrate. Pectins are very susceptible to heat, breaking down by H^+-catalyzed hydrolysis at pH values below about 5 (Smidsrød et al. 1966), and breaking down by an OH^--catalyzed elimination reaction above about pH 4 (Aspinall 1973). Solubilization of pectins by *hot* chelating agents thus does not provide evidence for Ca^{2+}-bridges. Indeed, a similar degree of solubilization of pectins occurred with hot formate buffer (not a chelating agent) as did with hot oxalate at the same temperature and pH (Fry 1986a).

3.2 Methods for Breaking Covalent Cross-Links

3.2.1 Methods for Breaking Phenolic Coupling Products

Oxidatively coupled phenolics can in some cases be cleaved by mildly acidified sodium chlorite ($NaClO_2$). This is a strong oxidizing agent, and its mode of action is complex and incompletely understood. It evolves the poisonous and spontaneously explosive gas ClO_2, and should therefore always be used in open vessels in a fume cupboard. It can split open phenolic rings, chlorinate them, convert them to quinones, and de-alkylate them (Lindgren 1979; Dence et al. 1962). It can degrade lignin to soluble products. It will split open the diphenyl ether structure of isodityrosine (Fry 1982b); the products are varied, indicating that several reactions occur, but a major product is tyrosine, indicating that a genuine *splitting* of the dimer must take place. $NaClO_2$ is thus a cross-link-cleaving agent. Two plausible mechanisms for the splitting of isodityrosine by chlorite [by analogy with its action on model compounds such as vanillyl alcohol (Dence et al. 1962)] are shown in Fig. 4.

Treatment of cell walls with chlorite, under conditions that degrade isodityrosine, results in the solubilization of a substantial proportion of the salt-in-

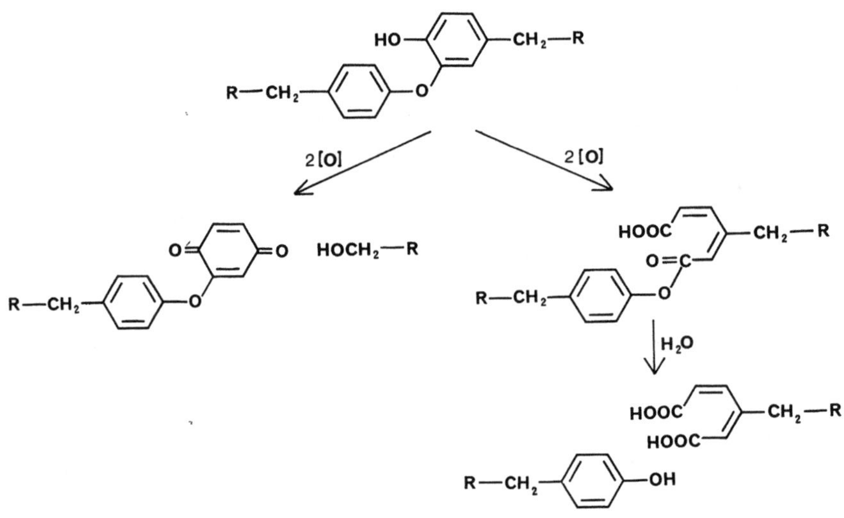

Fig. 4. Two plausible mechanisms for the oxidative cleavage of isodityrosine [–R = –CH(NH_2)–COOH] by acidified sodium chlorite. Readily identified products are serine (*left-hand scheme*) and tyrosine (*right-hand scheme*)

extractable extensin (O'Neill and Selvendran 1980). This is one of the main pieces of evidence suggesting that extensin is held in the wall by inter-polypeptide isodityrosine cross-links.

For chlorite to be a useful tool in the chemical dissection of plant cell walls, its action must be specific. What other bonds than phenolics are attacked by chlorite? Recent studies (S. C. Fry, unpublished results) have indicated a slight loss of viscosity of the soluble polysaccharide derivative, carboxymethylcellulose (CMC), indicating chlorite-cleavage of a very small proportion of the glycosidic bonds. Gel permeation chromatography studies of the model protein bovine serum albumin (BSA), as well as of salt-extracted extensin, have shown that chlorite can also cleave a small proportion of the peptide bonds in a protein. However, the degradation of peptide bonds in extensin can be prevented by addition of an excess of BSA, which is apparently attacked preferentially, "sparing" the extensin (K. J. Biggs, University of Edinburgh, pers. comm.).

Certain ether bonds between aromatics and carbohydrates can be cleaved by treatment with 2,3–dichloro–5,6–dicyanobenzo–1,4–quinone (DDQ) (Watanabe et al. 1986). The susceptible bonds are benzyl ethers and cinnamyl ethers in which the benzene ring is electron-donating. Although developed for studies of lignin–polysaccharide bonds in wood (see also Chap. Jun-Ichi Azuma of this book), DDQ would seem to have considerable potential for studies of the phenol-sugar bonds that occur in the growing cell wall.

3.2.2 Methods for Breaking Glycosidic "Cross-Links"

A glycosidic "cross-link" (i.e., an ordinary glycosidic bond that happens to link sugars in two portions of a polysaccharide that are so different that they are considered to count as two joined polysaccharides) can, like any other glycosidic bond, be cleaved by hot acid and by appropriate enzymes. The difficulty lies in restricting the cleavage to the key glycosidic bond that bridges the two polysaccharides. Since such glycosidic bonds are rare (one per polysaccharide-pair), it is unlikely that sufficiently specific microbial "restriction endoglycanases" will have evolved to hydrolyze them. This is therefore not a promising approach.

A better approach by which to study this type of cross-link is the converse strategy of degrading 99% of the chain-length of one of the polysaccharides, e.g., by means of a specific enzyme, and then finding the remaining 1% (oligosaccharide) of it firmly bound to the other polysaccharide. For example, cell walls are treated exhaustively with pure pectinase (endopolygalacturonase), and then the hemicelluloses are extracted with alkali and purified; if the structure shown in Scheme 3 were abundant, one would find traces of pectic monosaccharide residues (arabinose, rhamnose and galacturonic acid) firmly attached to the hemicellulose. There is some evidence for hybrid polysaccharides of this kind (Bauer et al. 1973), but the amounts do not appear to be large enough to make a significant contribution to the architecture of the cell wall (McNeil et al. 1979).

3.2.3 Methods for Breaking Ester Cross-Links

Ester cross-links can be cleaved by hot acid (not recommended since the acid stability of some esters exceeds that of some glycosidic bonds), cold base, or enzymes (esterases). Pure esterases are likely to be the most specific method; the difficulty lies in obtaining them in pure form. A broad-range carboxyesterase from mammalian liver is available commercially (e.g., Sigma Chemical Co.), but neither its effectiveness on esters likely to be found in the plant cell wall nor its ability to penetrate the wall matrix seems to have been explored. A narrower-range esterase from the fungus *Aspergillus niger* has been discovered that can hydrolyze low M_r hydroxycinnamoyl esters (e.g., *p*-coumaroyl quinic acid and caffeoyl quinic acid) (Varoquaux et al. 1982), but the pure enzyme is not available commercially and its activity on feruloyl-polysaccharides has not been investigated.

Cold base is the agent most frequently used to break ester bonds. Its disadvantages are that it tends to catalyze β-elimination reactions (both mid-chain cleavage of partially methyl-esterified pectic polysaccharides and end-wise "peeling" of many polysaccharides) and that aqueous base is a strong chaotropic agent so that base-extraction of a polysaccharide could indicate that it was held in the wall either by hydrogen-bonds or by ester-bonds. Specific conditions that go some way to circumventing these problems include:

(a) 0.05 M Na_2CO_3 at 2° C for 20 h, followed by 1 h at 21° C (Jarvis 1982). The preliminary treatment at 2° C ensures that ester hydrolysis greatly exceeds elimination-degradation of pectic polysaccharides, since the latter process has a high Q_{10} and is therefore retarded in the cold (Jarvis 1984).
(b) A series of progressively increasing concentrations of KOH (all in the presence of 0.3% $NaBH_4$ to suppress peeling reactions), e.g., 0.02, 0.03, 0.045, 0.06, 0.08, 0.10, 0.20, 0.30, 0.45, 0.60, 0.80, 1.0, 2.0, 3.0 and 4.0 M KOH, all at room temperature for 1 h (Carpita 1984). The lowest concentrations do not hydrolyze feruloyl esters, but the moderate concentrations do. Uronoyl esters are likely to be labile even in the mildest alkali.
(c) 0.25 M sodium methoxide in methanol at room temperature for 18 h. This has the advantage that H-bonds are not split, and thus the polysaccharides remain insoluble during the base treatment. The methoxide is then washed out with pure methanol, which still fails to extract any polymers. Finally, the cell walls are suspended in water at 65° C for 18 h: any polymers now found to be soluble [28% of the cell wall (Morrison 1977)] are concluded to have been solubilized (but not dissolved) by the base treatment. Since the base treatment was in methanol to prevent the breakage of H-bonds, the solubilizing effect of the base seems likely to have been the cleavage of ester-bonds (strictly by methyl-transesterification rather than hydrolysis).

Ester-bonds can also be broken by other nonhydrolytic treatments, for example by reductive cleavage using $LiBH_4$ or $LiAlH_4$:

$$R\text{–}CO\text{–}O\text{–}R' + 4[H] \Rightarrow R\text{–}CH_2\text{–}OH + R'\text{–}OH + H_2O\,.$$

$LiBH_4$ is likely to be more useful than $LiAlH_4$ because the former is more soluble in water. Little use appears to have been made of this approach to break ester cross-links in the plant cell wall, although $LiBH_4$ has been used to reduce polysaccharide-bound methylgalacturonate residues to galactose residues (Stevens and Selvendran 1984).

4 Authentic Low-Molecular-Weight Models of Possible Cross-Links

In the development of methods for the cleavage of cell wall cross-links, it is valuable to have low-molecular-weight soluble model compounds on which to test the effectiveness and specificity of new treatments. Compounds containing glycosidic bonds are readily available, e.g., disaccharides, methyl-glycosides, etc., as are certain sugar ethers [e.g., 3–*O*–methylglucose and 3–*O*–benzylglucose (from Koch-Light)] and sugar esters [e.g., glucose penta-acetate and tannic acid (=penta–*O*–galloylglucose)]. The following sections provide methods for the preparation of four authentic standards that are of particular significance in the plant cell wall and are not available commercially.

4.1 Synthesis of Artificial Hydroxycinnamoyl–Carbohydrate Esters

The following technique is based on that of Geissmann and Neukom (1971) and Geissmann (1970). It enables phenolic acids (including ferulic, *p*–coumaric and *p*–hydroxybenzoic acids) to be esterified essentially to any desired –OH group, e.g., on a carbohydrate. In the following example, ferulic acid is esterified to sor-

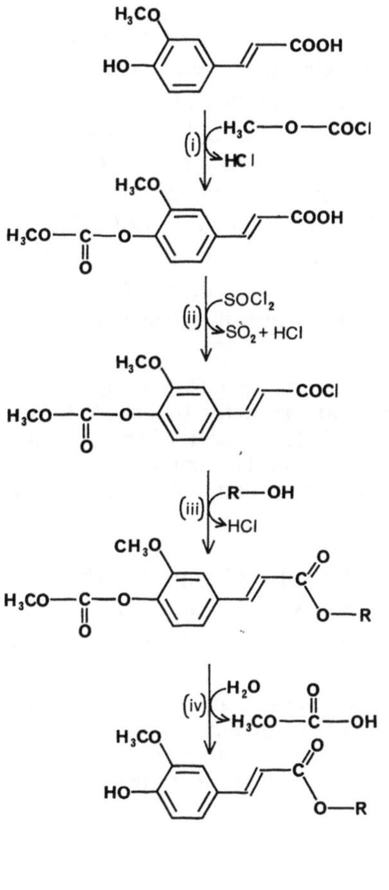

Fig. 5. Scheme for the synthesis of artificial feruloyl esters. *R–OH*=an alcohol, e.g., sorbitol or a polysaccharide (after Geissmann 1970). Steps (i) to (iv) are described in Sect. 4.1

bitol (=glucitol) or guaran (=galactomannan). The steps involved are summarized in Fig. 5. To maintain the ferulic acid as the *trans*-isomer, the synthesis should be conducted in subdued light. For an alternative technique, see Birkofer et al. (1966).

Step (i). Protection of phenolic –OH group. Suspend 10 g of *trans*-ferulic acid in 100 ml water and, while stirring, adjust the pH to 11.5 by dropwise addition of 10 M NaOH. Cool the clear solution to 0° C and, with continued stirring and cooling in an ice bath, add twelve 0.5-ml portions of methylchloroformate (=methoxyformyl chloride) at ≃2-min intervals. (A proportion of the carbomethoxyferulic acid produced will precipitate.) Continue stirring for a further 10 min. Adjust the pH to 3.5 with dilute HCl to complete the precipitation of the product. Collect the precipitate by filtration, rinse in water, and redissolve in a minimal volume of acetone. Recrystallize the product by addition of water. Dry the crystals in vacuo. Yield=82.4% theoretical.

If required, monitor progress of carbomethoxylation reaction: add a 10-μl sample of the suspension (before addition of HCl) to 30 μl acetone and load 1 μl of the resulting clear solution on to a silica-gel TLC plate (with fluorescent indicator); develop in toluene / acetic acid (9:1, v/v). The compounds are detected as dark absorbing spots against a fluorescent background upon exposure to a short-wavelength (254 nm) UV lamp. R_F values are: ferulic acid=0.39, carbomethoxyferulic acid=0.43.

Step (ii). Activation of –COOH group. Suspend 10 g of the dry carbomethoxyferulic acid in 100 ml of alcohol-free (=shaken six times with water), dry (=stored for a week over anhydrous $CaCl_2$ in the dark) chloroform. While stirring in a reflux apparatus at 50° C, add dropwise 4 ml of thionyl chloride. Continue stirring at 50° C to give a clear solution. Dry in vacuo to remove SO_2, HCl and any remaining $SOCl_2$, redry from 5 ml alcohol-free dry chloroform, and redissolve in 100 ml of the same. Add 600 ml of carbon tetrachloride, cool the flask on solid CO_2 until a slight turbidity appears, and then store at 0° C overnight to yield a heavy precipitate of carbomethoxyferuloyl chloride. Collect the precipitate, wash in carbon tetrachloride, dry in vacuo, and store in a desiccator over anhydrous $CaCl_2$. Yield=74% theoretical.

If required, monitor the progress of acyl chloride formation: dry a 50-μl sample of the chloroform solution, add 0.5 ml of 35% (w/v) ammonia, and store for 16 h at 25° C in a tightly capped tube. (The carbomethoxyferulic acid will yield ferulic acid, whereas the carbomethoxyferuloyl chloride will give ferulamide.) Apply 2.5 μl of the ammonia solution to a TLC plate as in step (i) and develop in butan–1–ol / 35% (w/v) ammonia (7:2, v/v). R_F values are: ferulic acid=0.23, ferulamide=0.58.

Step (iii). Carbomethoxyferuloylation of sorbitol. Dissolve 67 mg of sorbitol in 5 ml of dry dimethylformamide / pyridine (2:1, v/v). Add 33 mg of carbomethoxyferuloyl chloride and incubate at 25° C for ≃10 min. The reaction occurs very quickly.

The product, carbomethoxyferuloyl sorbitol, need not be isolated, although it can be isolated by paper chromatography as described in step (iv) if the ammonia treatment is omitted; R_F=0.71; blue-green fluorescence appearing only after 10–15 minutes' exposure of the test strip to ammonia vapor. The delay is due to the gradual occurrence in situ on the paper of the reaction described in step (iv).

Step (iv). Removal of carbomethoxy group by mild alkaline hydrolysis. To the 5 ml of dimethylformamide / pyridine solution from step (iii), add 5 ml water plus 0.3 ml 35% (w/v) ammonia. Incubate at 25° C for 10 min, then neutralize with acetic acid.

Step (v). Purification of the product. Load the whole ≃10 ml as a 40 × 2-cm streak on to Whatman 3MM chromatography paper, and develop overnight by the descending method in butan–1–ol / acetic acid / water (12:3:5, v/v/v).

The major product, feruloyl sorbitol, has $R_F = 0.47$ and shows blue fluorescence when viewed under a long-wavelength (366 nm) UV lamp; the color changes immediately to bright blue-green upon exposure of the paper to ammonia vapor. {Other R_F values are: sorbitol = 0.16 [not fluorescent; detect with $AgNO_3$ / NaOH– see §4.3 step (ii)], ferulic acid = 0.86, ferulamide = 0.77.} Elute the feruloyl sorbitol from the paper in 50% methanol by the method of Eshdat and Mirelman (1972) [pack the appropriate zone of paper into a 5-ml disposable plastic syringe barrel, hang this by its lugs into a centrifuge tube such that the nozzle has at least 1 cm clearance from the bottom of the tube, moisten the paper (with 50% methanol in this case), and bench–centrifuge the assembly for 5 min; remoisten the paper and re-centrifuge four more times; the solutes which are eluted from the paper are collected in the bottom of the tube in a minimal volume of solvent.] Dry in vacuo. Yield = 50%–70% theoretical.

For carbomethoxyferuloylation of polysaccharides, which are not readily soluble in dimethylformamide / pyridine, step (iii) may need to be conducted at 80° C (Geissmann 1970), and it is especially important to avoid even the slightest contamination with water: the polysaccharide must be thoroughly dried at high vacuum. After step (iii), the polysaccharide is precipitated by addition of 5 vol of acetone, collected by filtration, washed with more acetone to remove free phenolics, dried, redissolved in water, treated for 10 min with ammonia and neutralized with acetic acid [see step (iv)], dialyzed against water, and freeze–dried. To measure the ferulate content of the polysaccharide, a portion is dissolved or suspended in 1 M NaOH, incubated at 25° C for 30 min preferably under N_2 and then centrifuged; the A_{343} of the supernatant is a measure of ferulate concentration [$\log \varepsilon\ (M^{-1}\ cm^{-1}) = 4.37$].

4.2 Isolation of Naturally Occurring Feruloyl Disaccharides

The following method is for the enzymic dissection of cell wall polysaccharides to release feruloyl oligosaccharides. Further details are given elsewhere (Fry (1988).

Step (i). Purification of "Driselase". Stir 5 g Driselase (a mixture of enzymes from the fungus *Irpex lacteus*, available through Sigma Chemical Co.) for 15 min in 50 ml of 50 mM acetate (Na^+) buffer, pH 5.0. Bench-centrifuge for 5 min, and collect the clear supernatant. Make the volume up to 50 ml. With constant stirring, add 26 g of ammonium sulfate and, when this has dissolved, stand the solution at 0° C for 15 min to complete the precipitation of the enzymes. Centrifuge the suspension as before, but reject the supernatant. Resuspend the pellet in 50 ml of 52% (w/v) ammonium sulfate, and centrifuge again. Redissolve the pellet in 5 ml of water, and desalt the solution by passage through a 50-ml bed-volume column of Sephadex G-25 (medium grade), equilibrated and eluted with distilled water. Collect and freeze-dry the fastest-eluting brown material (void volume). Yield ≃0.5 g. Store below 0° C.

Step (ii). Isolation of cell walls. Collect 100 g (fresh weight) of cultured spinach cells by gentle squeezing through muslin, and stir vigorously overnight in 500 ml of 80% phenol [=phenol / water (4:1, w/w), available from BDH, Poole, U.K.] plus 200 ml acetic acid. Filter on sintered glass, and wash with 800 ml of 80% phenol / acetic acid / water (5:2:1, v/v/v), followed by 1 l of 70% (v/v) ethanol, and finally 1 l of acetone. Dry the cell walls.

Similar results can be obtained with spinach leaves or sugar beet roots that have been liquidized in a food mixer. If suspension cultures are used, it is convenient to label the feruloyl polysaccharides by pre-incubation of the living cells for a few days with *trans*-[^{14}C]cinnamic acid or L-[^{3}H]arabinose (Fry 1987b).

Step (iii). Enzymic digestion. Shake 1 g of spinach cell walls in 50 ml of pyridine/ acetic acid / water (1:1:23, v/v/v) containing 25 mg of chlorbutol and 0.25 g of purified Driselase at 25° C for 16 h. [If step (iv) is not to be carried out the same day, store the products frozen.]

Step (iv). Purification of products. The major products of Driselase digestion of plant cell walls are monosaccharides and disaccharides (Fry 1987b); the principal feruloylated products [feruloyl disaccharides (Fry 1982a)] are quantitatively minor components. Bench-centrifuge the sample to remove any undigested wall material. Pass the supernatant (≃50 ml) through a 500-ml bed-volume column of Bio-Gel P-2, equilibrated and eluted with pyridine / acetic acid / water (1:1:23, v/v/v). Collect about seventy-five 10-ml fractions. Dry 20 µl of each fraction on to Whatman no. 1 filter paper, and examine under a long-wavelength (366 nm) UV lamp ± fuming with ammonia vapor. Mark the fractions containing feruloyl esters, which give blue fluorescence, turning bright blue-green with ammonia vapor. Using the same paper, stain the sugars with aniline hydrogen phthalate [dip through a solution containing 1.6 g of phthalic acid in 49 ml of acetone, 49 ml of diethyl ether, 2 ml of water and 0.5 ml of aniline, dry for at least 5 min, and heat (ideally at 105° C for 5 min)]. Mono- and disaccharides stain various shades of red or brown and elute at ≃300–400 ml, whereas the feruloyl disaccharides (which are present in the 20-µl sample in such small quantities that they do not stain with aniline hydrogen phthalate) elute at ≃400–500 ml because they adsorb to the Bio-Gel. From the filter paper, deduce which fractions are rich in feruloyl disaccharides and relatively uncontaminated by free mono- and disaccharides; pool and freeze–dry these fractions. Redissolve the products in a minimal volume of 50% methanol, and apply as a 40 × 2-cm streak to Whatman 3MM paper. Develop by the descending method overnight in butan–1–ol / acetic acid / water (12:3:5, v/v/v), and locate the two major feruloyl disaccharides by UV fluorescence. R_F values are 3–*O*–(3–*O*–feruloyl–α–L–arabinopyranosyl)–L–arabinose = 0.76, and 4–*O*–(6–*O*–feruloyl–β–D–galactopyranosyl)–D–galactose = 0.47. Elute the appropriate zones with 50% methanol by the method of Eshdat and Mirelman (1972) [see Sect. 4.1, step (v)], and store at −40° C. Separation of the feruloyl disaccharides from any *p*-coumaroyl disaccharides is possible by re-chromatography in butan–1–ol / ethanol / water (20 : 5 : 11, v/v/v). The *p*-coumaroyl esters do not fluoresce without ammonia, but fluoresce blue in its presence (Fry 1982 a); they have slightly higher R_F values than the corresponding feruloyl esters.

4.3 Synthesis of Uronoyl–Sugar Esters

These compounds do not seem to have been reported in the literature, and the following is a preliminary small-scale preparation recently developed in the author's laboratory for the synthesis of one representative, 6?–*O*–galacturonoyl–*β*–methylglucoside. The yield is low, but adequate for initial characterization of this class of compound. In the preparation, the –OH groups of the galacturonic acid are not protected, but the large excess of *β*-methylglucoside over galacturonic acid will ensure that negligible depside (*O*–galacturonoyl-galacturonic acid) is formed.

Step (i). Condensation reaction. Dissolve 1 g of *β*–methylglucoside (available from Sigma Chemical Co.) in 1 ml of dimethylformamide with warming. Cool to 25° C. To 1 ml of the solution add 50 mg of galacturonic acid monohydrate plus 50 μl of pyridine (added as a catalyst) and stir to dissolve. Add 50 mg of dicyclohexylcarbodi-imide (DCC) and incubate at 25° C in a tightly capped tube for 2 h.

Step (ii). Purification of products. Add 9 ml of ice-cold pyridine / acetic acid / water (1 : 1 : 10, v/v/v) to the reaction mixture and incubate at 0° C for at least 15 min. Bench-centrifuge and apply 2 ml of the supernatant as a 40 × 2-cm streak to Whatman 3MM paper. Develop overnight by the descending method in butan–1–ol / pyridine / water (4 : 3 : 4, v/v/v), dry, and stain a guide-strip with aniline hydrogen phthalate [see Sect. 2.4.2, step (iv)]. The major stainable compounds are: galacturonic acid [orange-brown, R_F = 0.19–0.26 (slightly streaked)], unknown "A" [pale yellow-brown, R_F = 0.30, a product of galacturonic acid (also formed in the absençe of *β*–methylglucoside)], and putative esters "B" and "C" [pale yellow-brown, R_F values = 0.45 and 0.52 respectively (only formed when all three reagents – *β*–methylglucoside, galacturonic acid and DCC – are present together)]. The *β*–methylglucoside itself does not stain with aniline hydrogen phthalate, but it can be located at R_F = 0.61–0.70 on staining with $AgNO_3$ / NaOH [dip the paper through 0.75% (w/v) $AgNO_3$ in acetone (containing the minimum of water required to make the $AgNO_3$ dissolve), dry, dip through 0.125 M NaOH in ethanol and redry; brown spots appear within a few minutes at 25° C.] Esters "B" and "C" both yield *β*–methylglucoside plus galacturonic acid on alkaline hydrolysis, suggesting that they are ester-linked conjugates of these two components. Ester "B" may be 6–*O*–galacturonoyl *β*–methylglucoside and "C" may be a mixture of 2-, 3- and 4–*O*–galacturonoyl *β*–methylglucosides, since esters of primary alcohols tend to have lower R_F values than those of secondary alcohols

(Birkofer et al. 1966). The molar yield of esters "B" + "C" is approximately 20% of the initial galacturonic acid. The yield of ester "B" slightly exceeds that of ester "C".

4.4 Synthesis of Isodityrosine

The diphenyl-ether dimer, isodityrosine (Fry 1982b), is not available commercially but can be synthesized in small quantities by the following method. A larger-scale method is described elsewhere (Fry 1984b).

Step (i). Oxidation of tyrosine. To 0.4 g of L-tyrosine in 10 ml of 17% (w/v) ammonia add 0.67 ml of 0.75 M potassium ferricyanide. Stir at 25° C for 3 h in a tightly capped vessel. The resultant solution contains a low concentration of isodityrosine and dityrosine and some larger coupling products. This solution can be used in chromatography as a marker mixture without further purification. However, small quantities of isodityrosine can be purified as follows.

Step (ii). Purification of isodityrosine. Evaporate off some of the ammonia by standing the solution in a shallow dish in a draught of air. Filter off the crystals of unreacted tyrosine that appear. Load the whole filtrate on to Whatman 3MM paper as a 40 × 2-cm streak and develop overnight by the descending method in butan–1–ol / acetic acid / water (12:3:5, v/v/v). Isodityrosine approximately co-chromatographs with dityrosine (R_F=0.30–0.33); the latter can be seen under long-wavelength (366 nm) UV light as a bright-blue fluorescent band upon fuming with ammonia vapor. Elute this zone with 3.5% (w/v) ammonia by the method of Eshdat and Mirelman (1972) [see Sect. 4.1, step (v)], re-apply as a 10 × 2-cm streak to Whatman 3MM paper, and develop overnight by the descending method in butan–1–ol / pyridine / water (4:3:4, v/v/v). Locate the dityrosine (R_F=0.22) as before. Stain a narrow guide-strip for total phenolics [spray with Folin and Ciocalteu's phenol reagent (available from BDH, Poole, UK), de-colorize the yellow background by exposure of the paper to ammonia vapor, and record the blue-black zones]; isodityrosine has R_F=0.35. Elute this zone as before, and dry in vacuo. Recrystallize from hot water.

5 Alternative Methods

The focus of the present chapter has been the use of cross-link-cleaving reagents to dissect the chemical architecture of the cell wall. Other approaches (Fry 1988) that are available include:

a) A search for intact cross-links by treatment of cell walls with reagents that specifically degrade the *backbones* of the wall polymers without affecting their cross-links. In the case of extensin, this is possible by treatment with hot acid or alkali, either of which will cause hydrolysis of the peptide bonds, but not of the diphenyl-ether cross-link of isodityrosine (Fry 1982 b). Other possible approaches include the use of anhydrous HF to cleave glycosidic linkages without affecting ester cross-links (Mort and Bauer 1982), the use of lithium in ethylene diamine to cleave uronic acids without affecting other sugars (Mort and Bauer 1982), the use of base to cleave the aliphatic backbone of cutin without affecting the coupled phenolics (Holloway 1982), and the use of specific glycanases and lyases to split polysaccharide backbones without affecting nonglycosidic bonds (Fry 1984 a). This approach is unsuitable for the detection of *non*-covalent cross-links because although large numbers of these when acting in conert are adequate to cross-link polymer molecules, they are too weak to cross-link oligomers.

b) An attempt to solubilize wall polymers under relatively mild conditions that do not cleave cross-links, and then to show that two "individual" polymers are inseparable by all nondegradative techniques e.g. gel-permeation chromatography, electrophoresis, differential precipitation, lectin affinity chromatography, and comparable techniques. It is essential (and very difficult) in this approach to show that the two polymers are not simply co-chromatographing.

Acknowledgments. I am grateful to Mrs Janice Miller for excellent technical assistance and to the Agricultural and Food Research Council for a research grant in support of the previously unpublished work.

References

Ahluwalia B, Fry SC (1986) Barley endosperm cell walls contain a feruloylated arabinoxylan and a non-feruloylated β-glucan. J Cer Sci 4:287–295

Akiyama Y, Mori M, Katô K (1980) ^{13}C-NMR Analysis of hydroxyproline arabinosides from *Nicotiana tabacum*. Agric Biol Chem 44:2487–2489

Andersen SO (1964) The cross-links in resilin identified as dityrosine and trityrosine. Biochim Biophys Acta 93:213–215

Aspinall GO (1973) Degradation of polysaccharides. In: Bentley KW, Kirby GW (eds) Elucidation of organic structures by physical and chemical methods. Wiley, New York, pp 379–450

Aspinall GO, Molloy JA, Craig JWT (1969) Extracellular polysaccharides from suspension-cultured sycamore cells. Can J Biochem 47:1063–1070

Bauer WD, Talmadge KW, Keegstra K, Albersheim P (1973) The structure of plant cell walls. II. The hemicellulose of suspension-cultured sycamore cells. Plant Physiol 51:174–184

Biggs KJ, Fry SC (1987) Phenolic cross-linking in the plant cell wall. In: Cosgrove DJ, Knievel DP (eds) Physiology of cell expansion during plant growth. Am Soc Plant Physiol, Rockville, MD, pp 46–57

Birkofer L, Kaiser C, Kosmol H, Romussi G, Donike M, Michaelis G (1966) D-Glucose- und L-Rhamnoseester der *p*-Coumar- und Ferulasäure. Ann Chem 699:223–231

Blakeney AB, Harris PJ, Henry RJ, Stone BA (1983) A simple and rapid preparation of alditol acetates for monosaccharide analysis. Carbohydr Res 113:291–299

Carpita NC (1984) Fractionation of hemicelluloses from maize cell walls with increasing concentrations of alkali. Phytochemistry 23:1089–1093

Carpita NC (1986) Incorporation of proline and amino acids into the cell walls of maize coleoptiles. Plant Physiol 80:660–666

Chanzy H, Chumpitazi B, Peguy A (1982) Solutions of polysaccharides in *N*-methyl morpholine *N*-oxide (MMNO). Carbohydr Polym 2:35–42
Chen J, Varner JE (1985) An extracellular matrix protein in plants: characterization of a genomic clone for carrot extensin. EMBO J 4:2145–2151
Dence CW, Gupta MK, Sarkanen KV (1962) Studies on oxidative delignification mechanisms. Part II. Reactions of vanillyl alcohol with chlorine dioxide and sodium chlorite. TAPPI 45:29–38
Enoki A, Yaku F, Koshijima T (1983) Synthesis of LCC model compounds and their chemical and enzymatic stability. Holzforschung 37:135–141
Eshdat Y, Mirelman D (1972) An improved method for the recovery of compounds from paper chromatograms. J Chromatogr 65:458–459
Fry SC (1979) Phenolic components of the primary cell wall and their possible rôle in the hormonal regulation of growth. Planta 146:343–351
Fry SC (1980) Gibberellin-controlled pectinic acid and protein secretion in growing cells. Phytochemistry 19:735–740
Fry SC (1982a) Phenolic components of the primary cell wall: feruloylated disaccharides of D-galactose and L-arabinose from spinach polysaccharide. Biochem J 203:439–504
Fry SC (1982b) Isodityrosine, a new cross-linking amino acid from plant cell wall glycoprotein. Biochem J 204:449–455
Fry SC (1983) Feruloylated pectins from the primary cell wall: their structure and possible functions. Planta 157:111–123
Fry SC (1984a) Incorporation of [^{14}C]cinnamate into hydrolase-resistant components of the primary cell wall. Phytochemistry 23:59–64
Fry SC (1984b) Isodityrosine – its detection, estimation and chemical synthesis. Methods Enzymol 107:388–397
Fry SC (1985) Primary cell wall metabolism. Oxford Surv Plant Mol Cell Biol 2:1–42
Fry SC (1986a) Cross-linking of matrix polymers in the growing cell walls of angiosperms. Annu Rev Plant Physiol 37:165–186
Fry SC (1986b) In-vivo formation of xyloglucan nonasaccharide: a possible biologically active cell-wall fragment. Planta 169:443–453
Fry SC (1987a) Formation of isodityrosine by peroxidase isozymes. J Exp Bot 38:853–862
Fry SC (1987b) Intracellular feruloylation of pectic polysaccharides. Planta 171:215–211
Fry SC (1988) The growing plant cell wall: chemical and metabolic analysis. Longman, London
Fry SC (1989) Cellulases, hemicelluloses and auxin-stimulated growth: a possible relationship. Physiol Plant 75 (in press)
Gahan PB, Bellani LM (1984) Identification of shoot apical meristem cells committed to form vascular elements in *Pisum sativum L.* and *Vicia faba* L. Ann Bot 54:837–841
Gaspar T, Penel C, Thorpe T, Greppin H (1982) Peroxidases 1970–1980: a survey of the biochemical and physiological roles in higher plants. Université de Genève, Geneva
Geissmann T (1970) Untersuchungen über die oxidative Gelierung von Weizenmehlpentosanen und Phenolcarbonsäureestern von Polysacchariden. Thesis, Eidgenössische Technische Hochschule, Zürich
Geissmann T, Neukom H (1971) Vernetzung von Phenolcarbonsäureestern von Polysacchariden durch oxydative phenolische Kupplung. Helv Chim Acta 54:1108–1112
Goldberg R (1985) Cell-wall isolation, general growth aspects. In: Linskens HF, Jackson JF (eds) Modern methods in plant analysis, new series, vol 1. Springer, Berlin Heidelberg New York Tokyo
Gubler F, Ashford AE, Bacic A, Blakeney AB, Stone BA (1985) Release of ferulic acid esters from barley aleurone. II. Characterization of feruloyl compounds released in response to GA_3. Aust J Plant Physiol 12:307–317
Haass D, Frey R, Thiesen M, Kauss H (1981) Partial purification of a haemagglutinin associated with cell walls from hypocotyls of *Vigna radiata*. Planta 151:490–496
Hartley RD (1987) HPLC for the separation and determination of phenolic compounds in plant cell walls. In: Linskens HF, Jackson JF (eds) Modern methods in plant analysis, new series, vol 5. Springer, Berlin Heidelberg New York Tokyo

Hartley RD, Jones EC (1976) Diferulic acid as a component of cell walls of *Lolium multiflorum*. Phytochemistry 15:1157–1160

Hatfield RD, Nevins DJ (1987) Hydrolytic activity and substrate specificity of an endoglucanase from *Zea mays* seedling cell walls. Plant Physiol 83:203–207

Hayashi T, Marsden MPF, Delmer DP (1987) Pea xyloglucan and cellulose. V. Xyloglucan-cellulose interactions in vitro and in vivo. Plant Physiol 83:384–389

Holloway PJ (1982) The chemical constitution of plant cutins. In: Cutler DF, Alvin KL, Price CE (eds) The plant cuticle. Academic Press, London, pp 45–85

Jarvis MC (1982) The proportion of calcium-bound pectin in plant cell walls. Planta 154:344–346

Jarvis MC (1984) Structure and properties of pectin gels in plant cell walls. Plant Cell Env 7:153–164

Joseleau J-P, Chambat G, Chumpitazi-Hermoza B (1980) Solubilization of cellulose and other plant structural polysaccharides in 4-methylmorpholine-*N*-oxide: an improved method for the study of cell wall constituents. Carbohydr Res 90:339–344

Kato Y, Matsuda K (1980) Structure of oligosaccharides obtained by controlled degradation of mung bean xyloglucan with acid and *Aspergillus oryzae* enzyme preparation. Agr Biol Chem 44:1751–1758

Kato Y, Nevins DJ (1985) Isolation and identification of *O*-(5-*O*-feruloyl-α-L-arabinofuranosyl)-(1→3)-*O*-β-D-xylopyranosyl-(1→4)-D-xylopyranose as a component of *Zea* shoot cell walls. Carbohydr Res 137:139–150

Lamport DTA (1977) Structure, biosynthesis and significance of cell wall glycoproteins. In: Loewus FA, Runeckles VC (eds) The structure, biosynthesis and degradation of wood. Recent Adv Phytochem 11, Plenum, New York, pp 79–115

Lamport DTA, Epstein L (1983) A new model for the primary cell wall: a concatenated extensin-cellulose network. Proc Ann Plant Biochem Physiol Symp 2:73–83. University of Missouri Press, Columbia, Missouri

Lendzian KJ, Schönherr J (1983) In-vivo study of cutin synthesis in leaves of *Clivia miniata* Reg. Planta 158:70–75

Lindgren BO (1979) Reactions of lignin during bleaching with chlorine and chlorine dioxide (in Swedish). Sven Papperstidn 82:126–130

Markwalder H-U, Neukom H (1976) Diferulic acid as a possible crosslink in hemicelluloses from wheat germ. Phytochemistry 15:836–837

McNeil M, Darvill AG, Albersheim P (1979) The structural polymers of the primary cell walls of Dicots. Prog Chem Org Nat Prods 37:191–249

McNeil M, Darvill AG, Fry SC, Albersheim P (1984) Structure and function of the primary cell walls of plants. Annu Rev Biochem 53:625–663

Melton LD, McNeil M, Darvill AG, Albersheim P, Dell A (1986) Structural characterization of oligosaccharides isolated from the pectic polysaccharide Rhamnogalacturonan II. Carbohydr Res 146:279–305

Morrison IM (1977) Extraction of hemicelluloses from plant cell-walls with water after preliminary treatment with methanolic sodium methoxide. Carbohydr Res 57:C4–C6

Mort AJ, Bauer WD (1982) Application of two new methods for cleavage of polysaccharides into specific oligosaccharide fragments: structure of the capsular and extracellular polysaccharides of *Rhizobium japonicum* that bind soybean lectin. J Biol Chem 257:1870–1875

Northcote DH (1972) Chemistry of the plant cell wall. Annu Rev Plant Physiol 23:113–132

O'Neill MA, Selvendran RR (1980) Glycoproteins from the cell wall of *Phaseolus coccineus*. Biochem J 187:53–63

Ride JP (1975) Lignification in wounded wheat leaves in response to fungi and its possible rôle in resistance. Physiol Plant Pathol 5:125–134

Scalbert A, Monties B, Lallemand J-Y, Rolando C (1985) Ether linkage between phenolic acids and lignin fractions from wheat straw. Phytochemistry 24:1359–1362

Schuster P, Zundel G, Sandorfy C (1976) The hydrogen bond (3 vols), North Holland, Amsterdam

Selvendran RR, Stevens BJH, O'Neill MA (1985) Developments in the isolation and analysis of cell walls from edible plants. In: Brett CT, Hillman JR (eds) Biochemistry of plant cell walls. Cambridge Universi ty Press, pp 39–78
Shinkle JR, Jones RL (1988) Inhibition of stem elongation in *Cucumis* seedlings by blue light requires calcium. Plant Physiol 86:960–966
Smidsrød O, Haug A, Larson B (1966) The influence of pH on the rate of hydrolysis of acidic polysaccharides. Acta Chem Scand 20:1026–1034
Smith JJ, Muldoon EP, Lamport DTA (1984) Isolation of extensin precursors by direct elution of intact tomato cell suspension cultures. Phytochemistry 23:1233–1240
Stevens BJH, Selvendran RR (1984) Pectic polysaccharides of cabbage (*Brassica oleracea*). Phytochemistry 23:107–115
Taylor WI, Battersby AR (eds) (1967) Oxidative Coupling of Phenols, Dekker, New York
Valent BS, Darvill AG, McNeil M, Robertsen BK, Albersheim P (1980) A general and sensitive chemical method for sequencing the glycosyl residues of complex carbohydrates. Carbohydr Res 79:165–192
van der Wilden W, Segers JHL, Chrispeels MJ (1983) Cell walls of *Phaseolus vulgaris* leaves contain the Azocoll-digesting proteinase. Plant Physiol 73:576–578
Varoquaux P, Labavitch JM, Strand LL, Varoquaux-Tournemelle F (1982) Partial characterization of hydroxycinnamoylquinate esterase from *Aspergillus niger*. Lebensm-Wiss Technol 15:39–41
Wallner SJ, Nevins DJ (1974) Changes in cell walls associated with cell separation in suspension cultures of Paul's Scarlet rose. J Exp Bot 25:1020–1029
Watanabe T, Kaizu S, Koshijima T (1986) Binding sites of carbohydrate moieties toward lignin in "lignin-carbohydrate complex" from *Pinus densiflora* wood. Chem Lett 1986:1871–1874
Wellinder KG (1976) Covalent structure of the glycoprotein horseradish peroxidase (EC 1.11.1.7). FEBS Lett 72:19–23
Whitmore FW (1976) Binding of ferulic acid to cell walls by peroxidases of *Pinus elliottii*. Phytochemistry 15:375–378

Anhydrous Hydrogen Fluoride and Cell-Wall Analysis

A. J. Mort, P. Komalavilas, G. L. Rorrer, and D. T. A. Lamport

1 Introduction

A venerable tradition for hydrogen fluoride use in wood analysis dates back to Gore (1869) who determined the formula as HF and also observed that in anhydrous HF: "Paper, cotton-wool, calico, gelatin, and parchment were instantly converted into glutinous substances and dissolved." Later workers (e.g., Clark 1962) used strong aqueous HF to macerate woody tissues and estimate lignin content, while the recent revival of Fredenhagen and Cadenbach's (1933) early work using anhydrous HF to "crack" cellulose, is a promising chemical alternative to microbial degradation of cellulosics for fuels and chemicals (Smith et al. 1983); ultimately biomass will be too valuable to burn! However, only much more recently have we appreciated anhydrous HF as a very highly selective solvolysis reagent, triggered by its use in peptide chemistry as a "deprotecting" reagent. By analogy this led to the use of anhydrous HF as an effective and facile method for the chemical deglycosylation of glycoproteins and mucoproteins (Mort and Lamport 1977). Indeed the selectivity for solvolysis of different glycosidic bonds is quite remarkable at subzero temperatures, so that, as described here, it becomes possible to dissect difficult and complex saccharide structures with an almost surgical precision.

This review of current HF methodologies for cell-wall analysis therefore falls quite naturally into three sections: first a brief summary of its use for deglycosylating glycoproteins; second, selective glycosidic bond cleavage, and third, the practical and theoretical ramifications of vapor-phase HF solvolysis.

2 Glycoprotein Deglycosylation

Chemical deglycosylation of glycoproteins and mucoproteins is generally more than 90% complete after 1 h at 0° C for O-glycosidic linkages of neutral sugars, while O-glycosidic linkages of amino sugars require upward of 3 h at 23° C. One must emphasize that anisole does not effectively scavenge the released sugar fluorides; anhydrous methanol (10% v/v) is far better because the products, O-methyl glycosides, are water-soluble and dialyzable.

Generally, we recommend the deglycosylation procedures described earlier (Mort and Lamport 1977; Sanger and Lamport 1983) with the exception that we now quench the reaction mixture rapidly in a ten fold excess of ice-water followed by dialysis and freeze-drying. For example, after HF-deglycosylation, extensin monomers gave reproducible tryptic peptide maps and amino acid sequences showing that extensins largely consist of a few highly repetitive peptides (Smith et al. 1986). Using mild 1 h 0° C solvolysis, there is no evidence of HF-induced peptide cleavage (although we have not tested the very acid sensitive Asp-Pro linkage) but 3 h at 23° C does tend to give (unquantified) peptide cleavage. However, this degradation is not necessarily a disadvantage when raising antibodies against the core protein of heavily glycosylated mucoglycoproteins for subsequent screening of cDNA libraries and isolation of cDNA clones corresponding to the core of otherwise intractable proteins, such as heavily glycosylated cancer-related mucins (Gendler et al. 1987). The very high degree of glycosylation prevents cross-reactivity between antibodies raised against the native and deglycosylated mucins. However, we predict that for normal glycoproteins antibodies raised against both the native and the deglycosylated glycoprotein will cross-react; the first demonstration of such cross-reactivity involved extensin monomers (Kieliszewski and Lamport 1986) and the general method now makes it possible for the first time to quantify the contribution of both glycosylated and non-glycosylated epitopes to the antigenicity of a glycoprotein.

Interestingly, although deglycosylation of extensin monomers may lead to a decrease in polyproline II helix content (Van Holst and Varner 1984), the molecule can still be visualized electron microscopically by low-angle shadowing, as a flexuous rod (Heckman et al. 1988) which, however, then no longer behaves as a substrate for the cross-linking enzyme, extensin peroxidase (Everdeen et al. 1988).

Finally, we should mention anhydrous trifluoromethane sulfonic acid as a possible alternative (Edge et al. 1981; Woodward et al. 1987) to the use of anhydrous HF which reportedly "requires special handling, large quantities of material and gives poor yields," an observation neither documented (Sojar and Bahl 1987) nor confirmed by our experience at the microgram level (Sanger and Lamport 1983). However, the use of anhydrous TFMSA may be a useful addition to the glycoprotein chemist's toolbox.

3 Selective Cleavage of Glycosidic Linkages

During preliminary studies on the deglycosylation of glycoproteins by Mort and Lamport (1977) it was found that there was a very marked difference between the stability of neutral sugar linkages and amino sugar linkages to HF solvolysis in the oligosaccharide side-chains of fetuin and pig submaxillary mucin. Chitin (poly-N-acetyl glucosamine) from crab shells remained polymeric in HF at 0° C after 1 h, whereas cellulose, starch, and other neutral sugar polymers were completely depolymerized within 15 min. This led to the hope that there may be other distinct differences in labilities of sugars at other temperatures.

Various experiments with bacterial polysaccharides composed of structurally identified repeating units showed this to be the case (Mort and Bauer 1982; Mort 1983; Mort et al. 1983; Kuo and Mort 1986; Kuo et al. 1986). As an example, the extracellular polysaccharide of *Bradyrhizobium japonicum* 3 I 1b 138 (1) could be selectively cleaved at three glycosyl linkages at −23° C in HF in 15 min to yield a trisaccharide (2) (Mort and Bauer 1982).

α-D-Galp4Me
↓
6
[→3)-β-D-Glcp-(1→3)-α-D-GalpA4Ac-(1→3)-α-D-Manp-(1→3)-α-D-Glcp-(1→]$_n$
1 (n≫2)

β-D-Glcp-(1→3)-α-D-GalpA4Ac-(1→3)-α,β-D-Manp
2

α-D-Galp4Me
↓
6
β-D-Glcp-(1→3)-α-D-GalpA4Ac-(1→3)-α-D-Manp-(1→3)-α,β-D-Glcp
3 (Sch. 1)

This showed that under these conditions the α-galactosyl (4-O-Methyl) linkage, the α-mannosyl linkage, and the α-glucosyl linkage were labile, yet the β-glucosyl and the α-galacturonosyl (4-O-Acetyl) linkages were stable.

The generation of the trisaccharide could be almost quantitative. Thus, the stability and labilities mentioned above must be almost absolute at this temperature in HF. In addition to the stability of two of the glycosidic linkages, the acetate ester linkage was stable, since the trisaccharide also retained the acetate ester, which was shown to be on the galacturonic acid. When the temperature of the reaction was lowered to −40° C, more of the glycosidic linkages were stable and differential stabilities were less pronounced. At −40° C, partial solvolysis of α-linkages of neutral sugars led to the generation of a series of oligomers containing one or multiples of the repeating pentasaccharide (3). There was also partial cleavage of the galactosyl side-chains. However, from the product of the reaction it was easy to purify a good yield of the repeating unit and a useful amount of the repeating unit without its galactosyl side-chain (Mort et al. 1983).

Because of the great variety of sugars, linkages, and ring forms in plant cell walls, and in addition, various esters, there is great scope for the use of liquid HF in the generation of specific oligosaccharides for structure determination and for the investigation of their possible biological activities.

3.1 Apparatus Necessary

There is no commercially available apparatus for performing the experiments described in this section. Most of the components, however, are commercially available. For careful control of reaction temperatures and time it is necessary to be able to start the reaction with reagents pre-equilibrated to the desired temperature and to stop the reaction rapidly after the desired reaction time. To accomplish this, one must be able to transfer HF as a liquid from an equilibration vessel to

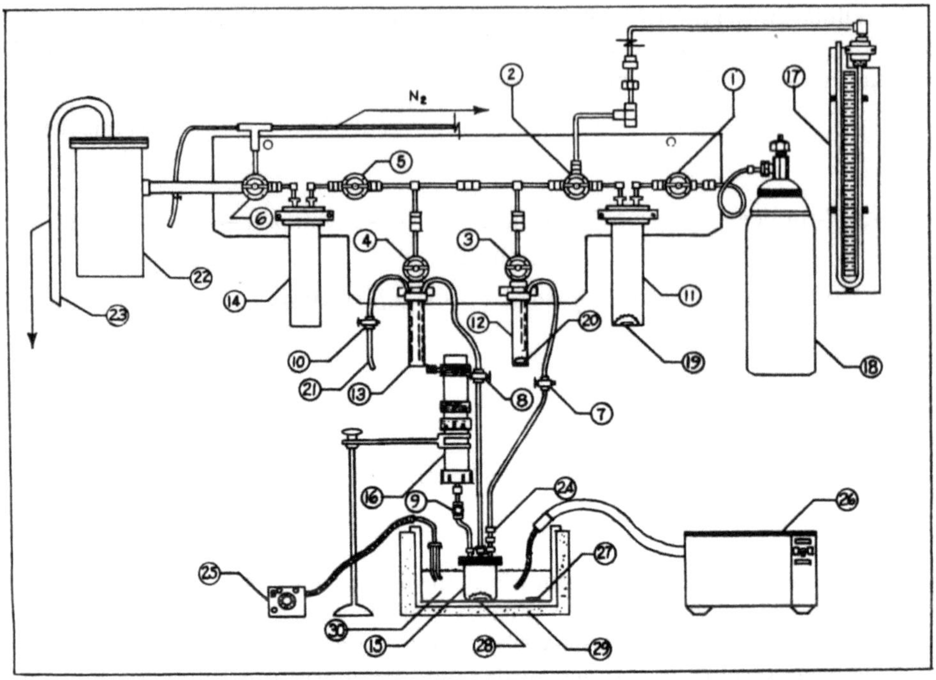

Fig. 1. Schematic representation of the hydrogen fluoride solvolysis apparatus. *1–8, 10* stopcocks; *9* teflon needle valve; *11–16* teflon and Kel-F vessels; *17* manometer; *18* hydrogen fluoride tank; *19, 20* stirrer bars; *21* exit to the sink for pressure release, if necessary; *22* calcium oxide trap; *23* connection to the vacuum pump; *24* 3-mm to 6-mm adaptor; *25* heater / regulator; *26* immersion cooler; *27, 28* stirrer bars; *29* insulated container; *30* 95% ethanol

the reaction vessel. A suitable apparatus was described recently (Mort 1983). In this apparatus the HF is transferred from vessel to vessel by means of 3-mm (o.d.) Teflon tubes reaching to the bottom of the vessel, using slight N_2 pressure and vacuum as the driving force.

The apparatus is a modified version (refer to Fig. 1) of the HF Reaction Apparatus Type IB obtained through Peninsula Laboratories, Inc., San Carlos, California. It contains six Teflon or Kel-F vessels and Teflon or Kel-F tubing which connects these vessels. The two fittings into which the vessels (12, 13) of the commercial unit attach are replaced by ones machined from 4-cm Teflon rod. These differ from the original in that the stem is made 2.5 cm long rather than 1.25 cm to allow for insertion of, in one case one and in the other two, lengths of 3-mm Teflon tubing through tightly fitting holes drilled through the fittings. The small stopcocks, and the fluid-flow needle valve for making connections between the vessels are obtained from Berghof/America Inc., Raymond, New Hampshire.

Vessel 15, used as a reaction vessel is a Teflon container with 500-ml capacity from Savillex Corporation, Minnetonka, Minnesota. It has a screw-cap with three inlets. Through the first 6-mm inlet 3-mm Teflon tubing from vessel 12 is connected to vessel 15 using a 3-mm-to-6-mm adaptor (24). The other 6-mm inlet is connected to vessel 16 [ether reservoir composed of two Teflon column segments (Savillex Corporation) joined together] by 6-mm Teflon tubing with needle valve 9. Vessel 13 is connected to the reaction vessel 15 via the 3-mm hole through 3-mm Teflon tubing.

3.2 Transfer of HF from HF Tank to the HF Reservoir

Hydrogen fluoride needed for 2 or 3 experiments was transferred from an HF tank to the HF reservoir 11 for ready use by the following procedure: Stopcocks 3 and 4 were closed. Stopcocks 2, 5, and 6 were opened and the vacuum pump was turned on. Reservoir 11 was evacuated and any leaks in the system were detected by closing stopcock 5 and observing the manometer. There were no leaks if the manometer reading remained constant. Slowly, stopcock 1 was opened and the line to the HF tank was also evacuated for 5 min. The presence of leaks in the line was checked as before. Reservoir 11 was cooled by immersing it in liquid nitrogen. After 20–30 min stopcock 2 was closed and the valve on the HF tank was opened slowly. The manometer reading did drop at this point. The amount of HF transferred into reservoir 11 was checked occasionally. When enough HF was collected in reservoir 11 the HF tank was closed as was stopcock 1. The HF in reservoir 11 was allowed to come to room temperature before use.

3.3 HF Solvolysis of Cell Walls

The apparatus for solvolysis of cell walls was as shown in Fig. 1. At least 1 h before the reaction, the cooling system, a liquid immersion cooler (FTS System, Inc., Stone Ridge, NY) was turned on. The cooling liquid used was 95% ethanol. Before attaching the different parts together, all vessels were checked to see whether they were clean and completely dry. Approximately 500 mg of dry cell walls from Acala 44 cotton suspension cells were placed in reaction vessel 15 along with a 2.5-cm-long stirrer bar. Stopcocks 1, 2, 9, and 10 were closed. Stopcock 6 was opened to the vacuum pump. Stopcocks 3, 4, 5, 7, and 8 were opened. The whole apparatus was evacuated for 15 min. The system was checked for leaks by closing stopcock 5 and observing the reading on the manometer. When a leak was detected its source was found by closing stopcocks to particular vessels. For example, if stopcocks 4, 7 and 8 were closed and the mercury still dropped in the manometer then the leak was in vessel 12. Once the system was leakproof the next step was to transfer HF from reservoir 11 to vessel 12. For 500 mg cell walls 20 ml HF was required. Stopcocks 1, 2, 4, 7, 8, 9, and 10 were closed. Stopcocks 3, 5 and 6 were opened. Vessels 11 and 12 were stirred using magnetic stirrers. Vessel 12 was cooled by immersing it in a cooling bath prepared by adding dry ice to acetone. After about 20–30 min when vessel 12 was well cooled stopcock 5 was closed and stopcock 2 was opened slowly. HF was then distilled into vessel 12 from reservoir 11. The manometer reading dropped when stopcock 2 was opened. It took about 15 to 20 min to transfer 20 ml of HF into vessel 12. When 20 ml of HF had transferred, stopcocks 2 and 3 were closed. The acetone cooling bath was removed. Both vessel 12 and reaction vessel 15 were immersed in the cooling bath of ethanol, the temperature of which was maintained at −23° C using the immersion cooler offset by a regulated immersion heater (FTS Systems, Inc., Stone Ridge, NY). The cooling bath was stirred to ensure uniform temperature throughout. Stirring was continued for 30 min to obtain temperature equilibrium. During the time that temperature equilibrium was being reached stopcock 5 was opened partially very slowly to remove HF in the line. Stopcock 3 was closed during this procedure or else HF from vessel 12 would be sucked into the calcium oxide trap (22). The bubbles which formed in the manometer, which escape from the fluorocarbon oil trapped on the sides of the manometer, were released by tapping the manometer gently so as to prevent excessive bursts of gas which would propel the mercury out of the manometer. Once all the bubbles were gone, the vacuum was released and vacuum pump was turned off. Stopcock 6 was turned to the nitrogen line. The next step was to transfer HF from vessel 12 to the reaction vessel 15. This was achieved using slight nitrogen pressure. The stopcocks 4, 8, 9, and 10 were closed. Stopcock 7 was opened slowly and then stopcock 3 was opened. Stopcock 5 was opened to let nitrogen blow HF into the reaction vessel. It took only a few seconds for the transfer. Once all the HF was transferred into the reaction vessel, stopcock 7 was closed. The nitrogen was turned off and the reaction was timed. The reaction vessel was stirred and the reaction was allowed to continue for 30 min. Ten minutes before the end of the reaction, about 300 ml of anhydrous diethyl ether was added to vessel 16. Small pieces of dry ice were added to the vessel to cool the ether. Dry ice was added until the vessel was frosty. The HF reaction was stopped by opening stopcock 9 and allowing cold ether to enter vessel 15. Slight vacuum was used to enhance transfer of ether by opening stopcocks 7, 3, and 5 with the stopcock 6 opened to vacuum. The cooling bath was removed. The reaction mixture was allowed to stir for 30 min and warm to room temperature. After 30 min the reaction mixture was filtered using a Teflon filter (50 mm diameter, fine grade, Savillex Corporation).

3.4 Filtration of the Reaction Mixture

To filter the reaction mixture, vessel 15 was removed and an inline Teflon filter unit (Savillex Corporation) was attached to it. Nitrogen was used to push the liquid through the filter. The insoluble material was rinsed with ether. The HF / ether / sugar mixture (filtrate) was collected in a Teflon container. The filter unit was sealed using a sealing ferrule at the outlet and the whole apparatus was evacuated for 30 to 45 min to dry the insoluble material and to remove the ether left behind in the whole HF line. The vacuum was released and the residue on the filter unit was extracted with water thrice, and the water extract was freeze-dried to obtain the water-soluble oligomers. The water-insoluble residue was also freeze-dried.

3.5 Recovery of Sugars from the HF / Ether-Soluble Fraction

The sugars, mainly monosaccharides and some disaccharides, present in the HF / ether-soluble fraction (filtrate), were recovered by evaporating the mixture. The filtrate was put in the reaction vessel 15, never to more than half of its capacity. A stirrer bar was placed in the vessel. One 6-mm inlet and the 3-mm inlet were sealed using sealing ferrules. Vessel 15 was attached through the other 6-mm inlet directly to the main line of the HF apparatus where vessel 13 is normally attached, using an 8-mm to 6-mm adaptor. Stopcocks 2 and 3 were closed, and 4 and 5 were opened. Stopcock 6 was opened to the vacuum. The vessel 14 was immersed in liquid nitrogen. When it was well cooled (about 15 min), the vacuum pump was turned on. Vessel 15 was immersed in a warm water bath and stirred on the magnetic stir / heater plate. The HF / ether complex was evaporated from vessel 15 and condensed in vessel 14. Sugars were retained in vessel 15. When vessel 15 was dry, the vacuum was released, and the sugars were dissolved in water and freeze-dried. The HF / ether complex in vessel 14 was allowed to come to room temperature and then neutralized by pouring it into excess ($\sim$60 g) solid calcium carbonate.

At this point the starting material had separated into three parts (refer to Fig. 3): the HF / ether-soluble fraction (filtrate) containing mainly monosaccharides and some disaccharides, the water extract containing mono-, di-, tri-, and oligosaccharides, and the water-insoluble residue containing polysaccharides and other water-insoluble materials.

4 Characterization of Cell-Wall Fractions

Cell walls contain three major classes of polysaccharides: cellulose, hemicellulose, and pectins, and a variable amount of protein (McNeil et al. 1984). The primary structure of cellulose is well known as are the major features of the structure of the hemicelluloses. However, the structure of pectin is very complex and major gaps remain in our knowledge of its structure. McNeil et al. suggest that there are three distinct regions within the pectins, the interconnections of which are completely unknown. The three regions are a homogalacturonan region, a section based on a repeating disaccharide backbone of galacturonic acid and rhamnose with variable side-chains of predominantly arabinose and galactose [Rhamnogalacturonan I (RGI)], and a small section of complex structure containing some unusual sugars [Rhamnogalacturonan II (RGII)]. Another aspect of cell walls which remains unclear is how the different classes of polymers interact to form the three-dimensional structure of the wall (Fry 1986). We need to answer such questions as: What are the cross-links between pectin and hemicellulose? Is there a link between cell-wall polysaccharides and cell-wall proteins? If so, what is its character?

A major problem to be overcome to allow characterization of cell walls is their insolubility. Almost all of the analytical methods that are commonly applied to structure determination are done in solution. Thus, fragments must be solubilized from the walls before fractionation and characterization can begin. Clearly the specificity of the fragmentation will have an enormous influence on the subsequent success in obtaining purified, characterizable pieces. Great specificity of fragmentation can be obtained using carefully purified enzymes (Talmadge et al. 1973), but enzymes of suitable specificity are often not available. Chemical methods of fragmentation will probably never attain the same degree of specificity, but in some cases they can come close. In fact, specificity for a certain linkage class, but not restricted to a specific polymer, makes chemical methods more generally applicable.

By using anhydrous HF we can obtain very specific fragments from cell walls which were solubilized by cleavage of particular groups of glycosidic linkages. We have found one of the most powerful methods of fragment characterization is NMR spectroscopy. However, the relative insensitivity of NMR spectroscopy

Table 1. Mole percent of sugars in different fractions of cell-wall material after HF-solvolysis at −23° C

Material	Percent of starting weight recovered	Ara	Rha	Fuc	Xyl	GalA	Man	Gal	Glc
Whole cell wall	100	17.5	10.6	2.5	14.2	37.7	0.7	8.3	8.7
Cell wall*	74% as sugar	8.8	6.7	1.3	13.3	23.7	0.5	6.5	39.3
Cell wall HF/ether −23° residue	65	4.8	11.6	–	7.3	40.19	1.5	11.5	22.5
Ether-soluble	27	22.2	11.0	3.4	31.9	8.1	0.5	11.8	11.1
Water extract	21	4.9	6.78	–	7.5	29.9	0.8	6.8	43.3
Residue after water extraction	44	1.8	1.1	–	2.4	65.2	–	1.5	28.0
Imidazole extract	15	1.1	1.1	–	3.5	90.4	–	–	3.9
Residue after imidazole extraction	26	2.2	1.3	–	3.2	53.7	–	2.3	37.3
EDTA extract (0.1 M disodium salt)	7	0.5	0.7	–	1.5	93.5	–	–	3.9
Residue after EDTA extraction	12	Trace	Trace	–	6.6	21.7	–	3.7	68.1
Residue after EDTA extraction*		–	0.6	–	1.8	5.4	–	1.5	90.7
HF O° insoluble residue (protein)	2	–	–	–	3.6	87.6	–	–	8.8

Sugar compositions of each fraction were determined after methanolysis and trimethylsilylation (Chaplin 1982) except in the cases marked with an * in which the sample was given a 30-min treatment in HF at 0° C to solubilize cellulose and any other sugar polymers resistant to methanolysis. Thus, most of the compositions shown do not reflect the cellulose content of the sample in the values for glucose.

dictates that we obtain multiple milligram quantities of relatively pure oligomers. The larger the oligomer obtained, the more complicated the spectrum becomes. Thus, small, uniform fragments are desired initially. Once these are characterized, larger fragments are desirable. The application of HF solvolysis allows this progressive approach because in many cases by reducing the temperature fewer of the glycosidic linkages are cleaved, and larger fragments are obtained which exhibit in their NMR spectra the features of the simpler oligomers to which are added additional signals. A striking example of this was observed in our work on the extracellular polysaccharide of *R. japonicum* 3I 1B138 (Mort et al. 1983).

A partial description of the information obtained from the products of HF solvolysis of cell walls from Acala 44 suspension cultures at −23° C is given here to illustrate the potential of this method for obtaining characterizable cell-wall fragments. Initial fractionation was performed as outlined in Fig. 3. The molar sugar compositions of the various fractions obtained are shown in Table 1.

4.1 HF / Ether-Soluble Fraction

Most of the arabinose and xylose and a significant proportion of the galactose and glucose were soluble in the HF / ether fraction. Only monosaccharides and anomalously nonpolar disaccharides are freely soluble in this mixed solvent. Thus arabinans, the xylosyl side-chains of xyloglucan, the β(1→4) galactans and any starch [α(1→4) glucans] had been reduced to predominantly monosaccharides. From this fraction, gel filtration on BioGel P-2 followed by chromatography on QAE-Sephadex, an acidic disaccharide could be obtained consisting of galacturonic acid and rhamnose (Komalavilas and Mort, submitted). The structure of the disacharide was investigated by [^{1}H]NMR spectroscopy, 2D[^{1}H]homocorrelation NMR spectroscopy, 2D[^{1}H-^{13}C]heteronuclear correlation spectroscopy, and by laser desorption mass spectroscopy. It was found to be a mixture of an acetylated and nonacetylated disaccharide. Both disaccharides were derived from the backbone of what has been called rhamnogalacturonan I (RGI) (McNeil et al. 1980) by cleavage at the rhamnosyl linkage and removal of arabinosyl and galactosyl sidechains (see Scheme 2).

RGI (Sidechain)$_x$
↓
----α-[D-Gal*p*A-(1→2)-α-L-Rha*p*-]$_n$----
↑3
$(OAc)_{0.5}$

↓HF −23 °C

α-D-Gal*p*A-(1→2)-β-L-Rha*p*-(1→2)┐ (cyclic, returning to ↑ at GalpA)
↑3
$(OAc)_{0.5}$

+Ara-OH+Gal-OH (Sch. 2)

The location of the acetate group could be unambiguously determined by the NMR results. The 1D[^{1}H]spectrum (see right hand side of Fig. 2) shows the pres-

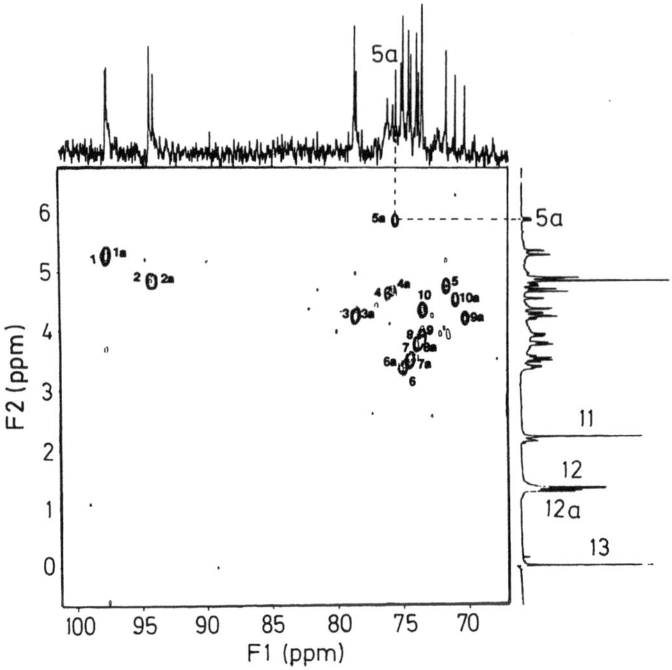

Fig. 2. [^{13}C-^{1}H]NMR 2D-heteronuclear correlation spectrum of the repeating disaccharide (galacturonosyl-rhamnose) isolated from cotton cell walls, together with the proton-decoupled 1D[^{13}C]spectrum (*above*) and the [^{1}H]spectrum (*right*). The *dashed lines* illustrate how the signals correspond to both ^{13}C signals. *Plain numbers* refer to signals from the nonacetylated disaccharide and the *numbers followed by an "a"* refer to signals in the acetylated disaccharide. Signals are identified as follows. *1* C-1, H-1, Gal A; *2* C-1, H-1 Rha; *3* C-2, H-2 Rha; *4* C-5, H-5 Gal A; *5* C-3, H-3 Gal A; *6* C-5, H-5 Rha; *7* C-4, H-4 Rha; *8* C-3, H-3 Rha; *9* C-2, H-2 Gal A; *10* C-4, H-4 Gal A; *11* CH_3 of Acetate; *12* CH_3 of Rha; *13* CH_3 of TSP. 2D nmr spectroscopy was performed as described by Gray (1983) The spectrum was recorded with a Varian XL-300 NMR spectrometer (300 MH$_z$) at 25° C, in D_2O using Sodium 2,2,3,3-tetradeuterio-4,4-dimethyl-4-sila-pentanoate (TSP) as internal standard at 0.00 ppm

ence of acetate by the signal at 2.18 ppm, and that it is ester-linked to a unique position by thé doublet of doublets at 5.83 ppm. Under the influence of an adjacent ester the chemical shift of a ring hydrogen atom is shifted approximately 1 ppm downfield from where it would be in the nonesterified case. From the splittings of the peak (10.63, 3.21 Hz) it is clear that the hydrogen at the acetylated position is axial and is adjacent to one axial and one equatorial hydrogen atom. This could only be the situation for the 3 or 2 position of the galacturonic acid residue or the 3 position of the rhamnose residue.

The identity of each ^{1}H signal was determined by examining the [^{1}H-^{1}H]homocorrelation spectrum. In this two-dimensional method each pair of hydrogen atoms which are spin-coupled gives rise to an off-diagonal peak and thus adjacent hydrogens can be identified. Once each hydrogen signal was identified, the exact location of the acetate ester in the disaccharide could be deduced. The position of the acetate group, anomeric configuration and linkages between the sugars could be confirmed by [^{13}C]NMR spectroscopy. Especially useful for identifying the carbon resonances was the [^{1}H-^{13}C]heterocorrelation spectrum (see Fig. 2). In this type of spectroscopy a signal at a coordinate given by a particular ^{1}H and ^{13}C chemical shift identifies directly hydrogen-carbon bonded pairs. Thus knowledge of the identity of either one unambiguously identifies the other.

No previous report has mentioned the possibility of acetylation of the backbone of RGI. As far as we know, HF solvolysis is the only method by which the acetylated disaccharide could be obtained.

From the chemical shifts of the carbon atoms in the disaccharides we could deduce that during the generation of the disaccharides the rhamnosyl fluoride which was the initial product of cleavage of the α-rhamnosyl linkage in the RGI glycosylated the 2 position of the galacturonic acid immediately adjacent to it via a β-glycosidic linkage.

This serves to illustrate that some care needs to be taken to prevent glycosidic linkages from being formed during the HF solvolysis reaction. In all of our experience we have seen the *formation* of glycosidic linkages *only* in three situations: (1) Reactions in which the concentration of sugar in the HF is 10% or more. (Glucose or any other neutral or acidic sugar can be polymerized in these concentrations to form mixed linked oligomers; Mort and Parker 1982; Kraska and Micheel 1976). (2) Reactions in which the glycosyl fluorides produced are held close to other sugars by solubility effects. (HF solvolysis of Gellan gum at −40° C even at low concentrations gave rise to a combination of the expected tetrasaccharide repeat unit and a polymer which appeared to be composed of the repeating units polymerized together via mixed linkages; Kuo et al. 1986.) (3) Reactions in which glycosyl fluorides are held sterically adjacent to a hydroxyl group, e.g., the generation of the cyclic disaccharide described above from RGI.

In the vast majority of cases this normally undesirabe side reaction can be completely avoided by using low concentrations of sugars in the HF. In the case of Gellan gum even using lower concentrations of sugar made no difference to the formation of polymers. However, performing the solvolysis in the presence of 10% methanol in the HF approximately doubled the yield of the expected tetrasaccharide, which in this case became the methyl glycoside (Kuo et al. 1986). We believe that due to insolubility of the gum in HF, the initial rhamnosyl fluoride which would be formed on the reducing end of the polymer where it had been cleaved was held close to other polymer chains and thus reacted rapidly before it could diffuse away. Methanol could compete in this reaction but not very effectively. In the case of RGI, 3D models predict that the fluoride on the rhamnose is placed perfectly to react with the 2 position of the galacturonic acid, and it did so essentially quantitatively. We have not seen competition for that reaction by methanol.

4.2 Water-Soluble Fraction

Fractionation of the water-soluble portion of the HF-treated cotton cell walls allowed us to isolate more of the disaccharide of galacturonic acid and rhamnose from RGI but this fraction contained the majority of the portion of the disaccharide that was not acetylated. (Acetylation seems to enhance solubility in the HF / ether mixture.) Another fraction which can be isolated by gel filtration (BioGel P-2 void volume) and ion exchange chromatography (retained on QAE Sephadex) is a homogalacturonan. Galacturonosyl linkages seem to be completely stable to solvolysis in HF at −23° C. By [^{1}H]NMR spectroscopy this was shown to contain a quite high proportion of methyl esterified (~40%) galacturonic acid. By HPLC gel filtration on tandem TSK SW4000 and SW2000 columns in 0.3M sodium acetate buffer this homogalacturonan fraction was found to have an average degree of polymerization greater than 20, as determined by the elution volume of a purified 20-residue-long oligogalacturonide. As has been shown previously (Fishmann et al. 1986) oligogalacturonides behave as though they are much larger than randomly coiled polysaccharides of equivalent molecular weight even at high ionic strength, but that at low ionic strength fractionation by gel filtration becomes futile because of aggregation of even short oligomers of DP2 and above.

Further analysis of this fraction will allow us to learn about the clustering of methyl esterification and strict homopolymer lengths in the homogalacturonan region of pectin.

4.3 Water-Insoluble Residue

Methanolysis of the residue from the cell walls after extraction with water yielded predominantly galacturonic acid and glucose (see Table 1).

Almost all of the galacturonic acid contained in the residue could be extracted using 0.5M imidazole buffer at pH 7.0 followed by extraction with 100 mM EDTA. Imidazole is the preferred extractant because it can be easily removed from the extract by dialysis against water, whereas even after prolonged dialysis some residual EDTA remains associated with the polygalacturonic acid. A similar situation occurs if CDTA is used in place of EDTA. Both of these chelating agents show peaks in the region of the [^{1}H]NMR spectrum in which the CH_3 of the methyl ester of galacturonic acid appears, and thus severely restrict qualitative estimation of the presence or absence of methyl esterification.

The relative size distribution of the polygalacturonic acid segments can be determined using the tandem gel filtration columns mentioned in the previous section. The apparent size of the polygalacturonic acid segments extracted by the imidazole and chelating agents is generally larger than that extracted by water from the HF-treated walls. However, there is some overlap. We suspect that differential extractability between the fractions is mostly because of the much lower affinity of the methyl-esterified regions for each other. We have found in experiments with wheat cell walls that even oligomers of nonmethyl-esterified galacturonic acid with a degree of polymerization of around 5 were not extracted from a residue with water but could be readily extracted using 0.5M imidazole buffer (Ryan and Mort unpublished). Once extracted they were perfectly water-soluble. This observation fits well with that of (Baydoun and Fry 1985) that even short oligogalacturonides are almost immobile in plants, presuming they would bind to polygalacturonic acid in the cell walls.

4.4 Final Residue

After the extractions, described above, the only sugars remaining are 5% of the original galacturonic acid and glucose. If the residue is subjected to a 0° C HF solvolysis prior to sugar analysis, it becomes apparent that the residue is predominantly cellulose. One might ask what proportion of the original cellulose remains in the residue and has any of it become less highly structured.

When purified cellulose fibers were treated with HF at −23° C for 30 min in HF as described for the cell walls, only two fractions were obtained. About 2% of the sample was recovered from the HF / ether fraction. The predominant sugars were xylose and mannose with a lesser amount of glucose. This composition indicates that only residual xylan and mannan (minor contaminants) were extracted. No sugar could be solubilized from the residue with water. Thus, cellulose is highly resistant to HF solvolysis at −23° C. However at 0° C, cellulose becomes 90% water-soluble within 5 min, and completely monomeric glucose and glucosyl fluoride within 15 min (Mort and Parker 1982). But there is a recent report (Franz et al. 1987) that between −15° C and −30° C in HF cellulose breaks down to water-soluble cellodextrins. A comparison of the methods that we use shows that Franz et al. did not strictly maintain the reaction temperature at the low temperatures. It appears that room temperature cellulose was added to liquid HF at somewhere between −10° C and 0° C and then the mixture was cooled to the temperatures noted. We suspect that the depolymerization took place very rapidly before the nonreactive temperatures were achieved.

We have noticed an increase in the glucose content that we measure by methanolysis followed by GLC analysis in cell walls after treatment with HF at −23° C. We hypothesize that the crystalline nature of the cellulose is disrupted somewhat by the strongly hydrogen-bonding properties of the liquid HF. Since the linkages between the glucose residues do not break, the glucan chains remain insoluble in water but the chains do become more accessible (in these disrupted regions) to the methanolic HCl.

4.5 Summary

From the results we have presented we hope to convince the reader that by treating cell walls at −23° C, fragments can be obtained from which information (perhaps not obtainable by other methods) can be derived. The major specificities obtained at this temperature are complete stability of uronic acid linkages, complete stability of β-glucosidic linkages, stability of methyl and acetyl esters and instability of all of the other glycosidic linkages. These specificities yield oligomers and polymers of uronic acids and cellulose, while destroying all of the other cell wall polysaccharides.

If the HF solvolysis is carried out at lower temperatures, i.e., −40° C or −73° C, fewer glycosidic linkages are broken. A compilation of our results with bacterial polysaccharides at various temperatures, plant cell walls at −23° C, and preliminary results from cell walls at lower temperatures is shown in Table 2.

Our scheme for the HF solvolysis of cell walls is summarized in Fig. 3.

Table 2. Expected lability of the various cell-wall sugar linkages to HF solvolysis at selected temperatures

Temperature °C	α-Ara*f*	Ara*p*	α-Fuc*p*	β-Gal*p*	α-Gal*p*A	α-Gal*p*A Me	α-Glc*p*	α-Glc*p*A	β-Glc*p*	β-Glc*p*A	β-Man*p*	α-Rha*p*	α-Xyl*p*	b-Xyl*p*
RT	+	?	+	+	+	+	+	?	+	?	+	+	+	+
O	+		+	+	±	+	+	?	+	?	+	+	+	+
− 6	+		+	+	−	+	+	?	+	?	+	+	+	+
−23	+		+	+	−	−	+	−	±	−	?	+	+	+
−40	+		+	±	−	−	±	−	−			±	+	±
−73	+		−	−	−	−	−	−	−			−	−	−

\+ Labile.
− Not labile.
RT Room temperature.
? Not tested.
± Depends on additional factors.

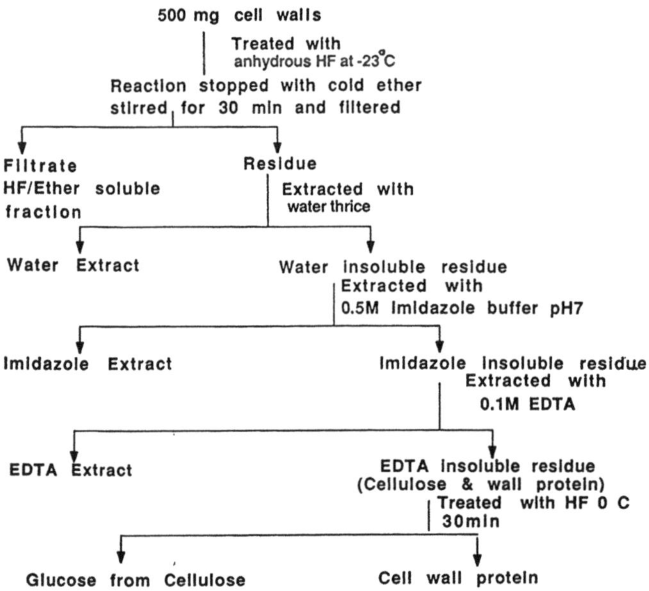

Fig. 3. Scheme for liquid-phase HF solvolysis of cell walls and isolation of cell wall polysaccharides

5 Vapor-Phase HF Solvolysis of Lignocellulose

This section describes the "vapor-phase HF solvolysis of lignocellulose", a novel and useful technique for (1) composition analysis of "hard-to-hydrolyze" secondary cell-wall polysaccharides, and (2) analysis of the secondary cell-wall lignin distribution.

The secondary cell wall of lignified plant tissue is an intricately constructed, biopolymer composite of cellulose, hemicellulose, and lignin. In general, it is difficult to accurately and simultaneously determine the carbohydrate composition and lignin distribution in lignified plant tissue because nature designed the lignocellulosic matrix to resist both enzymatic and chemical attack. The micropores in the secondary cell wall severely restrict the mobility of bulky cellulose- and lignin-degrading enzymes through the lignocellulosic matrix and thus hinder selective enzymatic degradation (Grethlein 1985). Although hemicellulose can be selectively removed from the lignocellulosic matrix by mild-acid hydrolysis (Fanta et al. 1984) or alkali extraction, cellulose or lignin cannot be completely and selectively solvent-extracted without first disintegrating the native architecture of the lignocellulosic matrix (Bjorkman 1956).

The cellulose microfibril also resists hydronium-ion catalyzed hydrolysis. The rigid hydrogen-bonding network within the paracrystalline cellulose microfibril severely dampens pyranosyl ring flexure to the cyclic oxonium / carbonium ion intermediate formed upon hydronium-ion catalyzed cleavage of the β-1,4-O-glycosidic bond (Harris 1975). Therefore, hydrolysis of cellulose from the lignocellulosic matrix by aqueous mineral acids (e.g., H_2SO_4 and HCl) occurs only at high temperatures or acid concentrations – conditions which promote dehydration of sugars (Harris 1975) and condensation of lignin (Lai and Sarkanen 1971).

Anhydrous HF liquid is a potent and selective reagent for the solvolysis of cell-wall polysaccharides. In fact, anhydrous HF liquid at 0° C readily dissolves even "hard-to-hydrolyze" crystalline cellulose from the secondary cell wall. The HF then solvolytically cleaves the β-1,4-O-glycosidic bonds joining anhydroglucose (glucan) units constituting the cellulose chain to release α-D-glucosyl fluoride in HF solution. This solvolysis proceeds rapidly to completion with little apparent glucoside dehydration (Hardt and Lamport 1982a; Defaye et al. 1983). The reaction chemistry of the liquid-phase HF-solvolysis of cellulose is well-documented (Hardt and Lamport 1982a, b; Defaye et al. 1982, 1983). A microtechnique for liquid-phase HF solvolysis of cell-wall polysaccharides is described by Sanger and Lamport (1983). Although the liquid-phase HF solvolysis is useful for determining the carbohydrate composition (Selke et al. 1982, 1983) and lignin composition (Clark 1962) in lignocellulose, the solvolysis tends to distintegrate the architecture of the lignocellulosic matrix (Selke et al. 1983).

Anhydrous HF vapor is also a potent reagent for the solvolysis of "hard-to-hydrolyze" secondary cell-wall polysaccharides (Rorrer et al. 1987, 1988a). The solvolysis of lignocellulose by anhydrous HF vapor (see Fig. 4) is briefly described here. Anhydrous HF vapor at ambient conditions (typically 30° C and 1.0 atm) readily penetrates the lignocellulose particle, physically adsorbs onto the lignocellulosic matrix, and then solvolytically cleaves cellulose and hemicellulose to sugar fluorides. The reaction is usually complete within 2 to 15 min, depending on reaction conditions. Since this reaction occurs statically (i.e., no mixing), the sugar fluorides reside on a anatomically intact residual lignin framework. These sugar fluorides are readily dissolved away from the residual lignin matrix by washing the HF-reacted solid in water. Thus, the technique for vapor-phase HF solvolysis of lignocellulose offers (1) quantitative determination of "hard-to-hydrolyze"

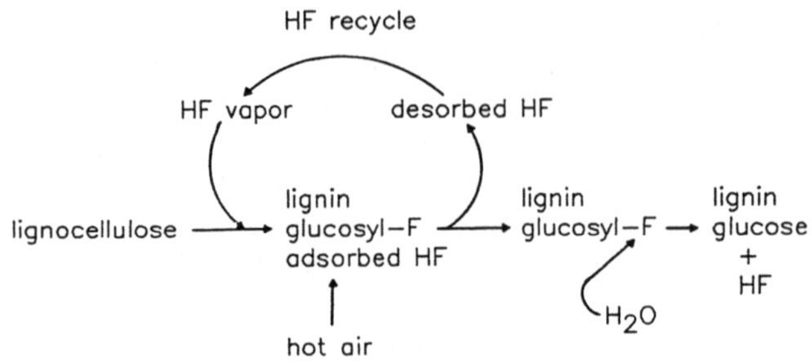

Fig. 4. Basic concept of vapor-phase HF solvolysis of lignocellulose

secondary cell-wall polysaccharides, (2) rapid and simple fractionation of polysaccharides from lignin, and (3) preparation of a anatomically intact residual lignin matrix free of polysaccharides. Furthermore, since HF vapor readily penetrates the lignocellulose particle, "whole" lignocellulose samples (not just finely ground material) can be analyzed. Therefore, the vapor-phase HF solvolysis technique has obvious advantages over enzymatic degradation, solvent extraction, and aqueous acid hydrolysis for selective removal and analysis of the secondary cell-wall components of lignified plant tissue.

The vapor-phase HF solvolysis technique evolved from recent laboratory-scale, reaction kinetics studies of Rorrer et al. (1986, 1987, 1988 a, b). In these studies, HF vapor was used to process lignocellulosic biomass (e.g., wood) to sugars (glucose and xylose) fermentable to the liquid fuel ethanol.

The present technique for the "vapor-phase HF solvolysis of lignocellulose" is presented in five parts: (1) lignocellulose sample preparation and anhydrous HF properties, (2) vapor-phase HF solvolysis apparatus and protocol, (3) sugar analysis, (4) sample results for the hardwood *Populus grandidentata*, and (5) microscopy of the residual lignin framework. Modifications of the technique for specific applications and limitations of the technique are also discussed. Although the hardwood *Populus grandidentata* is the substrate described in the protocol, the general technique is applicable to any lignified plant tissue.

5.1 Lignocellulose Sample Preparation and Anhydrous HF Properties

Lignocellulose Sample Preparation. Roughly-chipped bigtooth aspen wood (*Populus grandidentata*, Michx, 2.5 cm mesh, 6% moisture) is microtomed with the grain of the wood into wafers 0.5 mm thick. The wafers are then cut into squares of ca. 1.0 cm/side and dried *in vacuo* over phosphorous pentoxide. Typical weight of the lignocellulose chips (after drying) is 20 mg. The composition of *Populus grandidentata* is given in Table 3. If desired, the "extractives" in wood (resins, flavanoids, tannins, etc.) can be removed by soxlet-extraction in 2:1 v/v benzene / ethanol (ASTM 1984).

Anhydrous Hydrogen Fluoride. Anhydrous Hydrogen Fluoride (HF) is obtained in liquid form from Matheson Gas Products. The HF liquid is contained in a steel

Table 3. Composition of aspen wood (dry basis)

Component	Description	Wt %
Cellulose	Linear polymer of anhydroglucose (glucan) units joined by β-1,4-0-glycosidic linkages; cellulose chains hydrogen-bonded in parallel array to form the paracrystalline cellulose microfibril	50.0
Hemicellulose	Amorphous, short-chain polymer of anhydroxylose (xylan) units joined by β-1,4-0-glycosidic linkages; glucan, mannan, side-chains and uronic acid residues; hydrogen-bonded to outer surface of cellulose microfibril	29.0
Lignin	Amorphous, three-dimensional polymer based on the phenylpropanoid unit; encrusts cellulose and hemicellulose	16.6
Extractives	Resins, flavanoids, tannins, oils	4.1
Ash	Alkali metal oxides and salts	0.3

cylinder of 1600 mL capacity equipped with a Monel manual control valve (Matheson model 55-A). The cylinder is stored sealed under its own vapor pressure at room temperature within a fume hood.

Anhydrous HF vapor is potentially fatal if inhaled or ingested. HF also burns skin tissue, with the extent of damage dependent on the HF concentration, amount, and time of contact. HF toxicity levels, first-aid procedures, and safe HF handling procedures are well documented (Braker and Mossman 1971; Gall 1980; Allied Chemical Corporation 1978, 1984).

The physical properties of HF liquid and vapor are also well documented. The normal boiling point of HF is 19.54° C. HF vapor at 1.0 atm and 20° C to 80° C is nonideal, because HF molecules in the vapor state associate by hydrogen bonding to form oligomers ranging from 2 to 6 HF units in size (Smith 1958). HF liquid properties (including vapor pressure) are summarized by Simons (1950) and Gall (1980). Detailed thermodynamic properties of HF vapor are provided by Vanderzee and Rodenburg (1970).

5.2 Vapor-Phase HF Solvolysis Apparatus and Protocol

The vapor-phase HF solvolysis apparatus is illustrated in Fig. 5. A flowsheet summarizing the vapor-phase HF solvolysis protocol is presented in Fig. 6.

Materials of Construction. Anhydrous HF vapor is toxic and corrosive. Therefore, the entire apparatus is contained within a fume hood. The apparatus is designed to completely contain the HF vapor within the apparatus even when repeated access to the HF-reacting lignocellulose sample is desired. All HF-wetted tubing (0.64 cm o.d.), fittings (Swagelok, Inc.) and control valves (Whitney, Inc.) are fabricated from HF-resistant Monel (alloy 400) metal. The reactor is constructed of plain-carbon steel, which is suitable for HF vapor service provided water vapor is not present (Allied Chemical Corporation 1978). The tubular por-

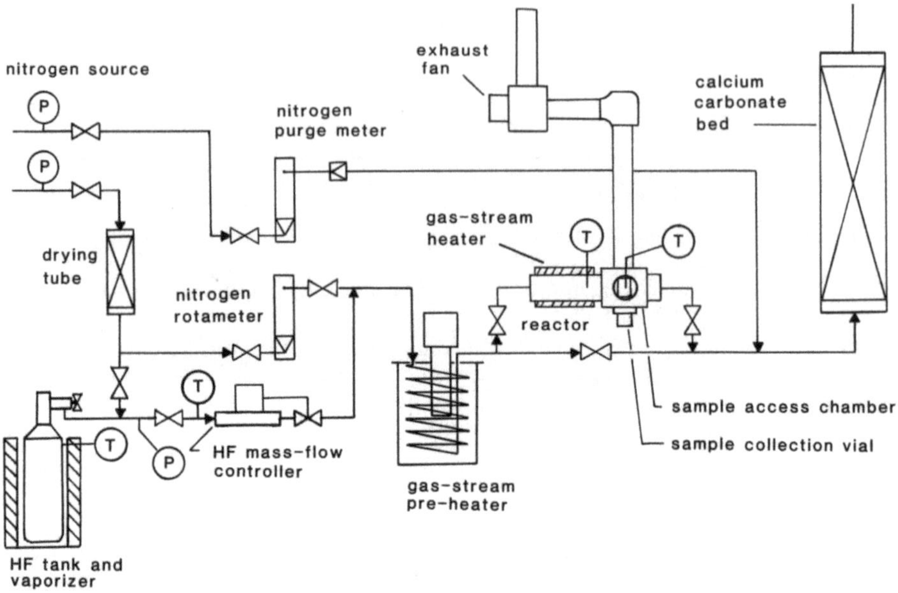

a

Fig. 5 a–c. Vapor-phase HF solvolysis apparatus. **a** Flow diagram; **b** reactor close-up; **c** reactor shell without Teflon sampling mechanism: *1* sample access chamber; *2* tubular portion of reactor (extends through sample access chamber); *3* sample collection vial (15-ml Teflon "tuft-tainer" vial screwed into base of sample access chamber)

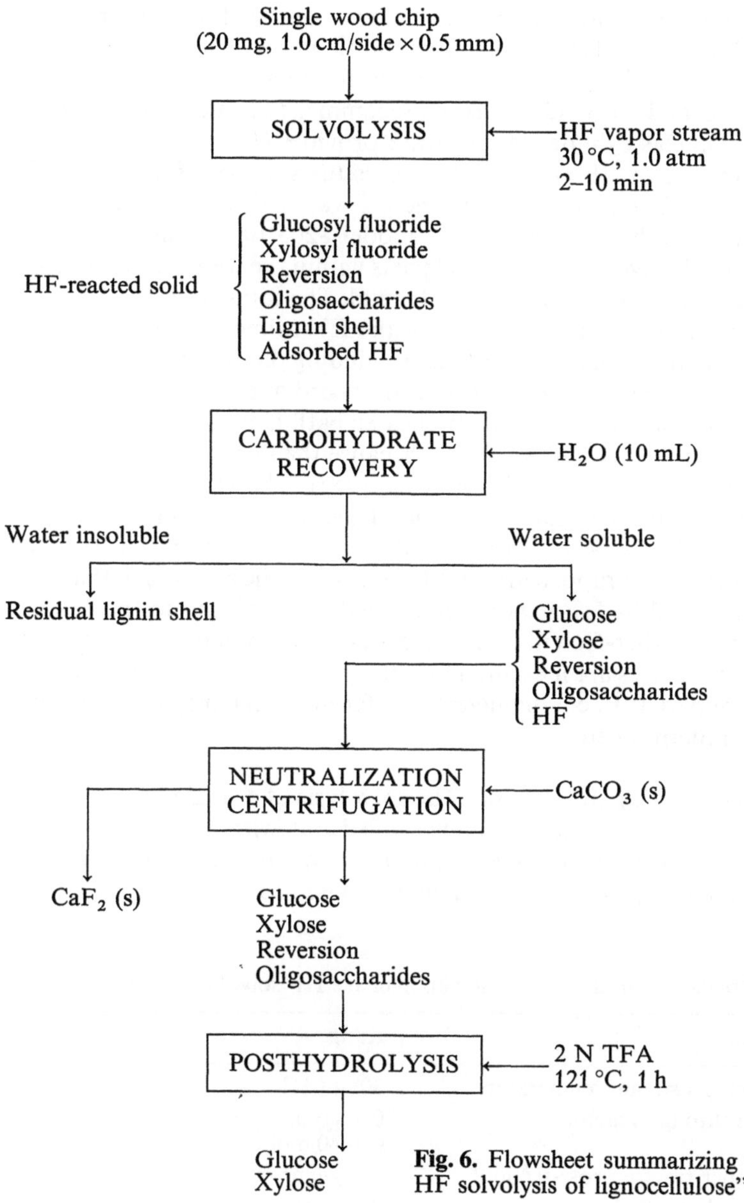

Fig. 6. Flowsheet summarizing the "vapor-phase HF solvolysis of lignocellulose" protocol

tion of the reactor is fabricated from ASTM schedule-40 (3.8 cm o.d., 3.2 cm i.d.) cold-rolled steel pipe. The calcium carbonate bed shell is fabricated from polyvinylchloride (PVC) pipe and bushings, and sealed with PVC-cement.

HF Vapor Delivery and Control. Although the flow design of the vapor-phase HF solvolysis apparatus is apparent from Fig. 5, several features merit special description. Anhydrous HF liquid (n.b.p. 19.54° C) is vaporized directly from the HF cylinder (1600 mL capacity) by a heating mantle under gentle heating. The cylinder skin temperature is monitored by a surface thermocouple and maintained between 27° C and 30° C (50° C is the recommended safety limit; Braker and Mossman 1971). This cylinder temperature provides (1) slow and even HF vapor evolution, and (2) an HF vapor delivery pressure sufficient for driving HF vapor across the mass-flow control valve. The HF vapor flowrate is monitored and controlled by a Matheson Instruments model 8205 (0–5 slpm) mass-flow transducer and control valve. All HF-wetted parts within the transducer and control valve are plated with HF-resistant Monel metal. The transducer is calibrated for mass flow by measuring the weight loss of the HF cylinder as a function of time at a given flow setting. All HF vapor lines (including the HF mass flow transducer and control valve) are wrapped with heating cord and heated to ca. 40° C to prevent HF vapor condensation. The preheater bath is used to reference the mixed HF vapor / nitrogen stream at 30° C, because the heat of dilution of HF vapor is endothermic (Briegleb and Strohmeier 1953). The HF / nitrogen gas-stream temperature within the tubular portion of the reactor is maintained by a 300-W heating coil (stainless-stell sheath of 0.16 cm diameter by 2 m straight length). The heating coil is wrapped around the outside of the reactor, so that corrosion of the heating coil by HF vapor is prevented. Power input to the heating coil is provided by a Barber-Coleman model 580A PID temperature controller. All gas-stream temperatures are measured by Omega Engineering type-J Inconel-sheathed thermocouples (0.16 cm diameter) interfaced to an Omega Engineering (model 660) digital thermometer.

Reactor Detail and Sampling Mechanism. The solvolysis of a single lignocellulose sample to solid sugars and residual lignin carried out within the reactor (see Fig. 5) of the vapor-phase HF solvolysis apparatus. Within the tubular portion of the reactor, a single lignocellulose sample is exposed to an HF vapor stream

Table 4. Range of reactor conditions for the vapor-phase HF solvolysis apparatus

Reaction Condition	Range
HF or HF/nitrogen gas-stream temperature	30°–80° C
HF vapor flowrate through reactor	0–8 g/min 0–0.40 mol/min
HF mass velocity (per reactor cross-section[a])	0–1 g/cm^2 min
N_2 diluent flowrate through reactor	0–5 slpm 0–6.25 g/min 0–0.22 mol/min
HF partial pressure[b]	0–1.0 atm
Total pressure	1.0 atm
Maximum lignocellulose sample loading	ca. 200 mg 1.0 cm/side × 0.5 cm thick

[a] Reactor cross-sectional area = 7.9 cm^2.
[b] HF partial pressure calculated by equation [1].

(in nitrogen diluent) of precisely controlled HF vapor flowrate, nitrogen diluent flowrate, and HF / nitrogen gas-stream temperature. Total pressure (P_T) in the reactor is always 1.0 atm. The complete range of operating conditions are summarized in Table 4. The HF partial pressure (P_{HF}) in the reactor is determined by

$$P_{HF} = P_T F_{HF} / (F_{HF} + F_{N_2}) \quad (1)$$

where F_{HF} and F_{N_2} are the molar flowrates of HF vapor and nitrogen respectively.

A Teflon sampling mechanism (see Fig. 7 a) loads a single lignocellulose wafer into the reactor, and then retrieves the HF-reacted solid for later sample processing and analysis. The sample access chamber contains the Teflon sampling mechanism and sample access port. The lignocellulose wafer (typically 20 mg) rests on the Teflon sample cartridge. The Teflon sample cartridge rests on the tip of the sampling probe. When the Teflon sampling mechanism is inserted into the tubular portion of the reactor (see Fig. 7 b, c), a Teflon plug (with polyethylene gasket not shown) mounted at the base of the sampling probe seals the reactor at the sample access port. The sample cartridge and lignocellulose wafer are now inserted within the tubular portion of the reactor. After a given reaction time, the sampling probe is unplugged and rotated 90° (see Fig. 7 d, e). The HF-reacted solid and Teflon sample cartridge are deposited into a 15-ml Teflon "tuft-tainer" vial (Pierce Chemical) screwed into the base of the sample access chamber. This vial contains 5.00 ml deionized / distilled water (to quench the reaction) and a known inclusion of myo-inositol (typically 6.00 mg), an internal standard used for later chromatographic quantification of the sugar products. Other HF-reacted sample retrieval options are possible, depending on the specific application desired.

HF Disposal. During sampling, the seal between the tubular portion of the reactor and the sampling mechanism is momentarily broken. An exhaust fan plumbed into the sample access chamber sucks up the fugative HF-vapors and vents them to the fume-hood exhaust. In normal operation, the reactor-effluent HF vapor is passed through a 5 kg bed of dry calcium carbonate chips (ca. 1.0 cm mesh). The calcium carbonate neutralizes the HF vapor to calcium fluoride (s) and carbon dioxide:

$$CaCO_3(s) + 2\,HF(g) \rightarrow CaF_2(s) + CO_2 + H_2O \quad (2)$$

By reaction stoichiometry, 5.0 kg calcium carbonate chips are sufficient to neutralize 2000 ml liquid HF. Usually, the calcium carbonate bed is recharged with fresh calcium carbonate chips whenever the HF cylinder (1600 ml capacity) is changed.

Standard Reaction Procedure. Lignocellulose wafers (1.0 cm/side, 0.5 mm thick, ca. 20 mg) are weighed to precision of 0.1 mg. The HF cylinder is heated to constant skin temperature of ca. 30° C. The HF vapor flowrate, nitrogen, flowrate, and HF / nitrogen gas-stream temperature are set and allowed to stabilize. A sin-

3
2
1
a
2
1
b
c

Fig. 7 a–e. Teflon sampling mechanism operation. **a** Teflon sampling mechanism (top view): *1* Teflon sample cartridge and wood chip; *2* sample access probe with sample access port plug; *3* outer plug for sealing sample access chamber; **b** Teflon sampling mechanism inserted in sample access chamber (top view): *1* Teflon sample cartridge and wood chip; *2* sample access port; **c** insertion of Teflon sample cartridge and wood chip into the tubular portion of the reactor and sealing of the sample access port (HF / nitrogen flow in tubular portion of reactor only); **d** retrieval of Teflon sample cartridge and wood chip from reactor and rotation of sampling probe; **e** Teflon sample cartridge and wood chip deposited into sample collection vial (vial normally contains 5.00 ml deionized / distilled water and a known amount of myo-inositol internal standard)

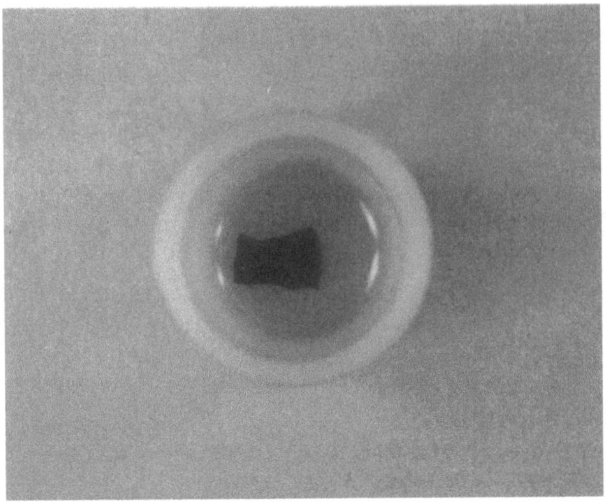

Fig. 8. HF-reacted residual lignin shell (after quenching). HF-reacted solid is dark brown; after quenching in water, the residual lignin shell (water-insoluble) is light brown

gle lignocellulose sample is then loaded into the reactor via the Teflon sampling mechanism just described. After a specified reaction time (typically within 15 min), the HF-reacted solid is removed from the reactor by the Teflon sampling mechanism and quenched in water in order to stop the reaction. The Teflon sample cartridge is washed with an additional 5.00 ml deionized / distilled water (washings added to reaction products) and then carefully dried with a cotton swab before reloading. The procedure is repeated until the desired number of HF-reacted lignocellulose samples are obtained. At the end of a series of reactions, the HF vapor flow is turned off and the apparatus flow system is flushed thoroughly with nitrogen at 2.00 slpm for at least 30 min.

When the HF-reacted sample is quenched in water, water-soluble sugar fluorides and oligomers are dissolved from the water-insoluble lignin residue. Any HF physically adsorbed on the HF-reacted sample is also dissolved. The water-insoluble residual lignin shell is shown in Fig. 8. The anatomically intact residual lignin shell (brown in color) is removed from the water-soluble products, rinsed with deionized / distilled water, and stored in 1.0 ml deionized / distilled water or 2:1 (v/v) acetone/water. The residual lignin shell is flaccid and easily fragmented, and therefore must be handled carefully. Calcium carbonate powder is added incrementally (10 to 20 mg) to the water-soluble product solution to precipitate dissolved HF to calcium fluoride. The pH-neutral sugar product solution is then centrifuged (400 rpm × 5 min), decanted from the insoluble calcium carbonate and calcium fluoride, and frozen for storage.

Variations on the Standard Reaction Procedure. Some applications may require that the HF-reacted sample be retrieved as a solid without water quenching. The reaction procedure is easily modified to obtain a solid HF-reacted sample. After a desired reaction time, the HF vapor flow is turned off and an ambient 100%

nitrogen stream at 2.00 slpm is allowed to pass over the HF-reacted sample for at least 30 min. The nitrogen stream desorbs the physically adsorbed HF from the sample. The entire sampling mechanism is then removed from the reactor and the HF-reacted solid sample is removed from the Teflon sample cartridge. This HF-reacted sample (solid sugars plus residual lignin matrix) is brittle and easily crumbled, and therefore must be handled carefully.

Delicate lignocellulose samples can also be accomodated. The lignocellulose sample is carefully mounted onto a polyethylene wafer of 1.0 cm/side × 1.0 mm thick. The polyethylene backing helps to maintain the overall structural integrity of the sample, particularly if the sample is very thin (say less than 50 μm). The polyethylene wafer (plus lignocellulose sample) is then inserted into the reactor via the Teflon sampling mechanism described earlier. The HF-reacted solid is HF-desorbed under flowing nitrogen and then retrieved from the reactor as described in the preceding paragraph. The water-soluble reaction products are then gently washed away from the residual lignin matrix by carefully pipetting deionized / distilled water onto the HF-reacted solid.

Safety Considerations. As previously mentioned, anhydrous HF is both toxic and corrosive. The vapor-phase HF solvolysis apparatus must be contained within a fume hood. Protective garments, including full face-shield, elbow-length impermeable gloves (Viton), and labcoat must be worn during all experimental work with HF. All personnel working with HF must be acquainted with safe HF-handling procedures, HF toxicity levels, and first-aid in the event of direct exposure to HF via inhalation or skin contact.

5.3 Sugar Analysis

Posthydrolysis. During both liquid and vapor-phase HF solvolysis of lignocellulose, glucosyl fluoride and other sugar fluorides repolymerize to short-chain, water-soluble reversion oligosaccharides (Defaye et al. 1982, 1983; Hardt and Lamport 1982b; Rorrer et al. 1987). In order to recover the parent glucose (and other monomer sugars) from these reversion oligosaccharides, the water-soluble products are hydrolyzed in dilute acid under mild conditions. Typically, a 50-μl aliquot of the water-soluble product solution (containing 0 to 50 μg each of sugar) is blown down to dryness over flowing nitrogen at 50° C, and then hydrolyzed in 200 μl of 2 N trifluoroacetic acid (TFA) for 1 h at 121° C within a septum-sealed 1.0-ml conical glass "reacta vial" (Pierce Chemical). These hydrolysis conditions are sufficient to hydrolyze sugar fluorides and reversion oligomers to momomer sugars (Rorrer et al. 1987). The hydrolyzed sample is then blown down to dryness over a nitrogen stream at 50° C in order to evaporate the volitile TFA (n.b.p. 72.4° C). The dry sample is then redissolved in 20 μl of degassed HPLC-grade water and vortexed prior to sugar analysis by HPLC or GLC / alditiol acetates.

Some sugar degradation occurs during 2 N TFA posthydrolysis at the conditions just described. The posthydrolysis correction factor $K_{f,s}$ is defined as

$$K_{f,s} = \text{g sugar before posthydrolysis/g sugar after posthydrolysis} \qquad (3)$$

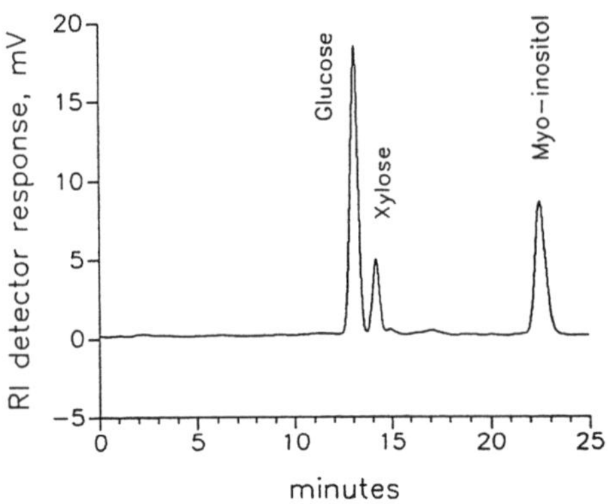

Fig. 9. Typical HPLC chromatogram. HPLC hardware, column, and operating conditions specified in text. RI detector range: 16×10^{-6} RIU/10 mV

Typical values of $K_{f,s}$ for glucose and xylose are 1.03 and 1.14 respectively (Rorrer et al. 1987; de Ruiter and Burns 1986).

HPLC. The sugar composition in the HF-reacted lignocellulose sample is conveniently assayed by High-Performance Liquid Chromatography (HPLC). A Spectra-Physics isocratic HPLC system, consisting of a SP8810 precision isocratic pump, Rheodyne 7125 injector (20-μl sample loop), SP8790 column heater, and SP6040 differential refractive index detector is suitable for carbohydrate analysis by HPLC. Glucose, xylose, mannose, etc. and myo inositol (the internal standard) are separated on a BIO-RAD Laboratories HPX-87P monosaccharide analysis HPLC column. The 20-μl assay sample (see posthydrolysis) is injected directly onto the column without derivatization. A BIO-RAD Laboratories "Microguard" deashing column is plumbed in front of the analysis column to remove particulates or residual acid (TFA) present in the 20-μl injection sample. The column is eluted with degassed HPLC-grade water at 0.6 ml/min, and the column temperature is isothermally maintained at 85° C. Column effluent sugars are detected by differential refractive index with respect to water. A typical HPLC chromatogram is presented in Fig. 9.

The yield (Y_s, g sugar/g dry wood) of each sugar released by vapor-phase HF solvolysis of lignocellulose is calculated by the "internal standard" method:

$$Y_s = \frac{A_s R_{F,I} w_I}{A_I R_{F,s} w_o} C \; K_{f,s} \tag{4}$$

where A_s is the peak area of a given sugar, A_I is the peak area of the internal standard, C is a conversion factor, $R_{F,s}$ is the detector response factor for a given sugar (μV-sec/μg), $R_{F,I}$ is the detector response factor for the myo-inositol internal standard (μV-sec/μg), w_I is the weight of the myo-inositol internal standard (mg) in the quenching solution, and w_o is the *dry* weight of lignocellulose chip before reaction (mg).

GLC/alditol-acetates. Gas-Liquid Chromatography (GLC) is a cumbersome but sensitive method for assaying the sugars released from the vapor-phase HF solvo-

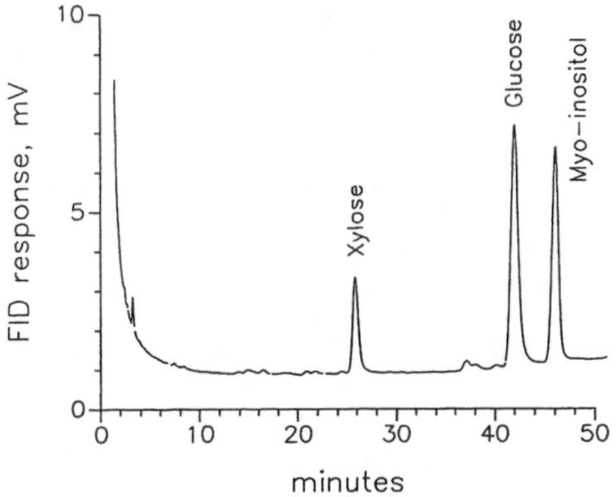

Fig. 10. Typical GLC / alditol acetates chromatogram. Column and operating conditions specified in text. Data acquired on Perkin Elmer 910 gas chromatograph: attenuation ×8, amplification ×10

lysis of lignocellulose. Before analysis, the sugars in the 20-μl assay sample (see posthydrolysis) are derivatized to their corresponding alditol acetates (Albersheim et al. 1967). The alditol acetates are separated on a 2 m×2 mm i.d. glass column packed with PEGS 224 (0.2% polyethylene glycol succinate, 0.2% polyethylene glycol adipate, 0.4% G.E. silicone XF 1150) on Gas Chrom Q. Typically, 1 μl sample (ca. 1 μg each sugar derivative) and 3 μl ethyl acetate solvent are injected directly onto the column. The column carrier gas is helium at 40 ml/min, and the column oven temperature is programmed as follows: 130° C initial, 1.0° C/min ramp to 180° C, 16 min hold at 180° C. The injector temperature is set at 210° C, and the detector temperature is set at 240° C. Column effluent components are detected by a Flame Ionization Detector (FID). A typical GLC / alditol-acetates chromatogram is presented in Fig. 10. Sugar yields are calculated by the "internal standard" method [Eq. (4)].

5.4 Sample Results for the Hardwood *Populus grandidentata*

The solvolysis rate of lignocellulose by anhydrous HF vapor is dependent on several conditions, namely (1) reaction time, (2) HF / nitrogen gas stream flowrate, (3) HF / nitrogen gas-stream temperature, (4) HF partial pressure, (5) lignocellulose wafer thickness (Rorrer et al. 1988a, b). Typical reaction conditions and sugar yields for the vapor-phase HF solvolysis of the hardwood *Populus grandidentata* (aspen) are presented in Table 5.

The carbohydrate composition of aspen wood as determined by vapor-phase HF solvolysis is compared to literature values (Timell 1957) for the carbohydrate composition of aspen wood in Table 6. In general, the glucan and xylan compositions determined by vapor-phase HF solvolysis are consistently low. HF-catalyzed, Friedel-Crafts condensation of sugar fluorides onto lignin may account for the discrepancy (Rorrer et al. 1987, 1988a).

Methanolysis of glucosyl fluoride to 1-O-methyl glucose is known to inhibit glucosyl fluoride reversion during liquid-phase HF solvolysis of cellulose (Franz et al. 1987). In situ methanolysis of glucosyl fluoride may likewise inhibit HF-catalyzed glucosyl fluoride / lignin condensation. In fact, when methanol-ab-

Table 5. Selected reaction conditions and water-soluble carbohydrate recoveries from the vapor-phase HF solvolysis of the hardwood *Populus grandidentata* (aspen wood). Detailed data given by Rorrer et al. (1987, 1988a, b)

Vapor-phase HF solvolysis conditions[a, b]				Yield[d]			
Reaction time min	Reaction temp. °C	HF partial pressure atm	Equilibrium HF loading g HF/g[c]	Glucose[e] mmol/g	Glucose[e] g/g	Xylose[f] mmol/g	Xylose[f] g/g
> 2	30	1.00	1.5	2.928 ±0.156	0.527 ±0.028	1.007 ±0.173	0.150 ±0.026
6[g]	30	1.00	–	3.125 ±0.062	0.562 ±0.011	1.047 ±0.050	0.157 ±0.008
>10	30	0.60	0.7	2.892 ±0.030	0.521 ±0.005	1.067 ±0.143	0.160 ±0.020
> 6	50	1.00	0.6	2.900 ±0.100	0.522 ±0.018	0.920 ±0.167	0.138 ±0.025

[a] Total HF/nitrogen flowrate fixed at 0.4 g/cm^2 min (3.2 g/min). Lower flowrates can be used, but the reaction time to maximum conversion must be increased accordingly because the reaction is subject to convective mass transfer resistances.
[b] Lignocellulose chip thickness fixed at 0.5 mm. Different lignocellulose particle thicknesses or diameters can be used, but the reaction time to maximum conversion must be adjusted accordingly because the reaction is subject to intraparticle mass-transfer resistances.
[c] Equilibrium HF loading determined by gravimetric HF adsorption in previous work (Rorrer et al. 1988a).
[d] All yield determinations based on g of original, *dry* lignocellulose (e.g., g glucose/g dry lignocellulose). ±1 std. dev., n = 7 determinations. "Extractives" were not removed prior to HF-solvolysis.
[e] Glucose = 180 g/mol, Glucan = 162 g/mol.
[f] Xylose = 150 g/mol, Xylan = 132 g/mol
[g] 20 µl methanol pipetted onto substrate (per 10 mg dry lignocellulose), then immediately reacted with HF vapor at given conditions; n = 3 determinations.

Table 6. Composition of aspen wood: vapor-phase HF solvolysis and literature values compared

Degradation method	Carbohydrate composition[a] (% of dry, extractive-free aspen wood)				
	Glucan	Xylan	Mannan	Galactan	Arabinan
72% H_2SO_4, 30° C, 1 h[b]	56.5	16.0	2.1	1.2	0.5
HF vapor	50.0	13.8	1.7	1.0	–
HF vapor, methanol added to substrate[c, d]	52.9	14.4	–	–	–

[a] Percent glucan and xylan calculated from data given in Table 5; values adjusted for extractive-free basis.
[b] Values given by Timell (1957) for hydrolysis of aspen wood in 72% H_2SO_4 at 30° C for 1 h, liquid:solid ratio 9:1.
[c] Minor wood sugars (mannose, galactose, etc.) obtained from the vapor-phase HF solvolysis of aspen wood were not quantified.
[d] See Table 5 for reaction conditions.

sorbed lignocellulose (ca. 20 μl/10 mg substrate) is reacted with HF vapor, the sugar yields are increased relative to dry lignocellulose, as shown in Tables 5 and 6. Therefore, the accuracy of carbohydrate composition determination via vapor-phase HF solvolysis is improved if methanol (20 μl/10 mg) is pipetted onto the lignocellulosic sample just before reaction with HF vapor.

Rorrer et al. (1988 a, b) previously showed that HF vapor must first physically adsorb onto the lignocellulosic matrix before HF-solvolysis of cellulose and hemicellulose can occur. The equilibrium HF loading on the lignocellulosic matrix decreases with decreasing HF partial pressure and increasing temperature. The minimum HF loading for complete reaction with cellulose and hemicellulose in *Populus grandidentata* is ca. 0.4 g HF/g dry wood. This value represents the HF required to saturate the hydroxyl groups and O-glycosidic linkages constituting cellulose and hemicellulose.

In general, the framework of the residual lignin matrix is best preserved if the equilibrium HF loading on the sample is between 0.4 and 1.0 g HF/g dry wood. Higher equilibrium HF loadings tend to disintegrate the lignocellulosic matrix. Therefore, for preparation of an anatomically intact residual lignin framework, HF vapor conditions of $P_{HF}=0.6$ atm, 30° C, or $P_{HF}=1.0$ atm, 50° C are recommended.

5.5 Microscopy of the Residual Lignin Framework

The technique for vapor-phase HF solvolysis of lignocellulose generates an anotomically intact residual lignin framework free of polysaccharides. A technique for preparation of the residual lignin framework for visualization by light microscopy is briefly described here.

A 2 mm × 2 mm-specimen is cut from the residual lignin shell. The specimen is dried in 1.0 ml acetone/1% 2,2-DMP for 24 h. The acetone is decanted, and the specimen is impregnated with 1.0 ml 1:1 v/v dry acetone / "spurrs" resin (22.5% wt% or vol% vinyl cyclohexane dioxide, 18.0% diglycidyl ether of propylene glycol, 56.6% nonenyl succinic anhydride, and 0.9% dimethylaminoethanol) for 24 h at room temperature under slow wrist-action shaking. The impregnation procedure is repeated with 1.0 ml 2:1 v/v dry acetone / "spurrs" resin, 1.0 ml 3:1 v/v dry acetone / "spurrs" resin, and finally, 1.0 ml "spurrs" resin. The specimen and resin are then poured into a casting mold, and the "spurrs" resin is cured in the casting mold within a vacuum oven at 60° C for 24 h. Several 2-μm-thick cross-sections are sliced from the imbedded specimen using a glass-blade mechanical ultramicrotome. The sections are stained successively with methylene blue and basic fuchsin (lignin – blue/purple, cellulose – red/pink) and visualized directly via light microscopy.

Sample light micrographs (200 ×) of the native lignocellulosic matrix and HF-reacted residual lignin shell are presented in Fig. 11 a and 11 b respectively.

In earlier work, Sachs et al. (1963) used aqueous Hydrogen Fluoride (5% to 80% wt) to dissolve polysaccharides away from the lignocellulosic matrix of pine wood. Sachs then observed the lignin distribution in the residual lignin framework by electron microscopy.

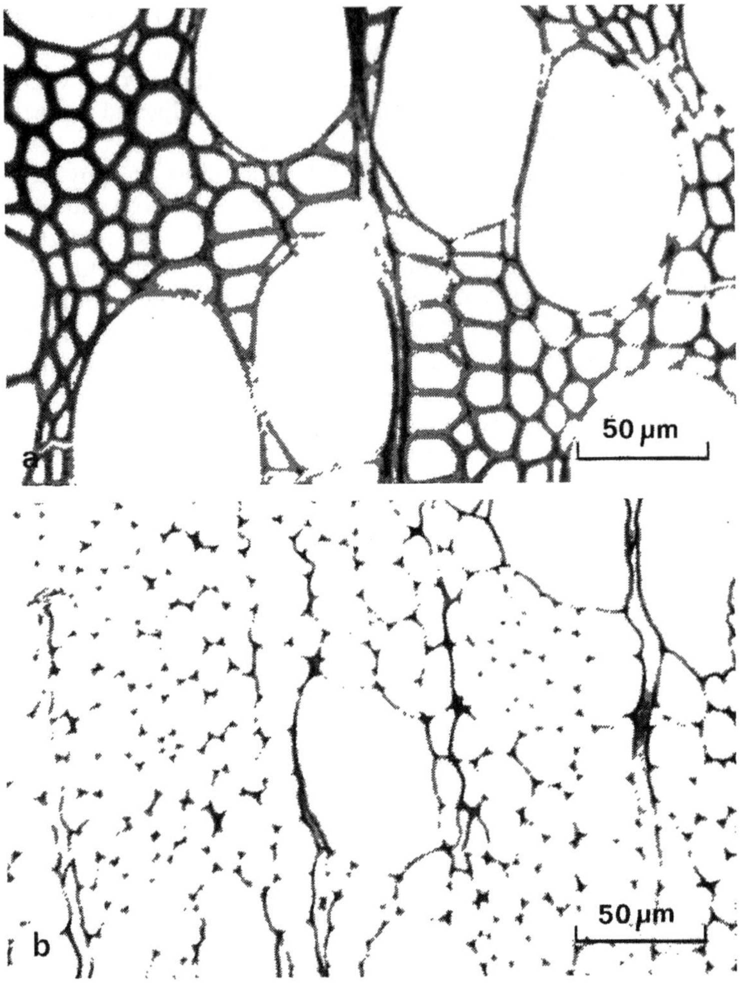

Fig. 11 a, b. Light micrographs (× 200) of the lignocellulosic matrix of the hardwood *Populus grandidentata*. **a** native lignocellulosic matrix; **b** HF-reacted residual lignin shell (water-soluble reaction products dissolved away from lignocellulosic matrix). Reaction conditions: 100% HF; 50° C, 3 min

The vapor-phase HF solvolysis technique is not useful for isolating chemically intact lignin from lignocellulose. Although residual lignin from vapor-phase HF solvolysis of lignocellulose has not yet been characterized, Smith et al. (1983) and Defaye et al. (1983) showed that residual lignin from liquid-phase HF solvolysis of wood is significantly condensed but not fluorinated. Clark (1962) showed that the Klason content of HF-lignin is at least 95%. Therefore, both liquid and vapor-phase HF solvolysis can be used to determine the overall lignin composition of lignified plant tissue, but not its native chemical structure.

Acknowledgments. Andrew J. Mort gratefully acknowledges partial support from the National Science Foundation via grant DMB-8603864 to obtain the Advanced Data System for the XL-300 NMR spectrometer. Andrew J. Mort also gratefully acknowledges support from the U.S. Department of Energy under contract DE FGD5 ER13496. Gregory L. Rorrer and Derek T. A. Lamport gratefully acknowledge support from the National Science Foundation via grant CPE-8315703 for construction of the vapor-phase HF solvolysis apparatus. The authors thank Dale Porter for drawing the diagram of the liquid-phase HF solvolysis apparatus.

References

Albersheim P, Nevins DJ, English PD, Karr A (1967) A method for the analysis of sugars in plant cell-wall polysaccharides by gas-liquid chromatography. Carbohydr Res 5:340–345

Allied Chemical Corporation (1978) Hydrofluoric acid. Industrial Chemicals Division, Morristown NJ, USA

Allied Chemical Corporation (1984) First aid treatment for hydrofluoric acid burns. Industrial Chemicals Division, Morristown, NJ, USA

American Society for Testing and Materials (ASTM) (1984) Standard method for preparation of extractive-free wood. Standard method #D1105-84, pp. 279–280

Bjorkman A (1956) Studies on finely divided wood. Part 1. Extraction of lignin with neutral solvents. Sven Papperstidn 59:477–485

Boydoun EAH, Fry SC (1985) The immobility of pectic substances in injured tomato leaves and its bearing on the identity of the wound hormone. Planta 165:269–276

Braker W, Mossman AL (1971) Hydrogen fluoride. In: Matheson Gas Data Book, Matheson Gas Products, East Rutherford, NJ, USA, pp 305–309

Briegleb G, Strohmeier W (1953) Über den Assoziationszustand des Fluorwasserstoffs im Gaszustand II. Z Elektrochem 57:668–674

Chaplin MF (1982) A rapid and sensitive method for the analysis of carbohydrate components in glycoproteins using gas-liquid chromatography. Anal Biochem 123:336–341

Clark IT (1962) Determination of lignin by hydrofluoric acid. TAPPI 45:310–314

Defaye J, Gadelle A, Pedersen C (1982) The behavior of cellulose, amylose, and xylan towards anhydrous hydrogen fluoride. Carbohydr Res 110:217–227

Defaye J, Gadelle A, Papadopoulos J, Pedersen C (1983) Hydrogen fluoride saccharification of cellulose and lignocellulosic materials. J Appl Poly Sci: Appl Poly Symp 37:653–670

de Ruiter JM, Burns JC (1986) Rapid determination of cell wall monosaccharides in flaccidgrass. J Agric Food Chem 34:780–785

Edge ASB, Faltynek CR, Hof L, Reichert LE, Weber P (1981) Deglycosylation of glycoproteins by trifluorómethanesulfonic acid. Anal Biochem 118:131–137

Everdeen DS, Kiefer S, Willard JJ, Muldoon EP, Dey PM, Li XB, Lamport DTA (1988) Enzymic cross-linkage of monomeric extensin precursors in vitro. Plant Physiol 87:616–621

Fanta GF, Abbott TP, Herman AI, Burr RC, Doane WM (1984) Hydrolysis of wheat straw hemicellulose with trifluoroacetic acid. Fermentation of xylose with Pachysolen tannophilus. Biotechnol Bioeng 26:1122–1125

Fishman ML, Pepper L, Damert WC, Philips JG, Barford RA (1986) A critical reexamination of molecular weight and dimensions for citrus pectins. In: Fishman ML, Jen JJ (eds) Chemistry and function of pectins. ACS symposium series 310

Franz R, Fritsche-Lang W, Deger HM, Erckel R, Schlingmann M (1987) The behavior and derivatizations of carbohydrates in hydrogen fluoride. J Appl Poly Sci 33:1291–1306

Fredenhagen K, Cadenbach G (1933) Der Abbau der Cellulose durch Fluorwasserstoff und ein neues Verfahren der Holzverzuckerung durch hochkonzentrierten Fluorwasserstoff. Angew Chem 46:113–117

Fry SC (1986) Cross-linking of matrix polymers in growing cell walls of angiosperms. Annu Rev Plant Physiol 37:165–186
Gall JF (1980) Hydrogen fluoride. In: Grayson M, Eckroth D (eds) Kirk-Othmer Encyclopedia of Chemical Technology, vol 10, 3rd edn. Wiley, New York, pp 733–753
Gendler SJ, Burchell JM, Duhig T, Lamport D, White R, Parker M, Taylor-Papadimitriou J (1987) Cloning of partial cDNA encoding differentiation and tumor-associated mucin glycoproteins expressed by human mammary epithelium. Proc Natl Acad Sci USA 84:6060–6064
Gore J (1869) On hydrofluoric acid. J Am Chem Soc 22:396–406
Gray GA (1983) Multinuclear two-dimensional NMR: Assignments of natural abundance polypeptide 13-C, 1-H and 15-N chemical shifts and demonstration of isomer interconversion. Org Mag Res 21:111–118
Grethlein HE (1985) The effect of pore size distribution on the rate of enzymatic hydrolysis of lignocellulosic substrates. Bio/Tech 3:155–160
Hardt H, Lamport DTA (1982a) Hydrogen fluoride saccharification of cellulose and xylan: isolation of α-D-glucopyranosyl fluoride and α-D-xylopyranosyl fluoride intermediates, and 1,6-anhydro-β-D-glucopyranose. Phytochemistry 21:2301–2303
Hardt H, Lamport DTA (1982b) Hydrogen fluoride saccharification of wood: lignin fluoride content, isolation of α-D-glucopyranosyl fluoride and posthydrolysis of reversion products. Biotechnol Bioeng 24:903–918
Harris JF (1975) Acid hydrolysis reactions for utilizing plant carbohydrates. J Appl Poly Sci: Appl Poly Symp 28:131–144
Heckman JW, Terhune BT, Lamport DTA (1988) Characterization of native and modified extensin monomers and oligomers by electron microscopy and gel filtration. Plant Physiol 86:848–856
Kieliszewski M, Lamport DTA (1986) Cross-reactivities of polyclonal antibodies against extensin precursors determined via ELISA techniques. Phytochemistry 25:673–677
Kraska U, Micheel F (1976) Synthetische Polysaccharide mit D-glucose und 1,6-anhydro-D-glucoseresten. Carbohydr Res 49:195–199
Kuo MS, Mort AJ (1986) Location and identity of the acyl substituents on the extracellular polysaccharides of *Rhizobium trifolii* and *Rhizobium leguminosarum*. Carbohydr Res 145:247–265
Kuo MS, Mort AJ, Dell A (1986) Identification and location of L-glycerate, an unusual acyl substituent on gellan gum. Carbohydr Res 156:173–187
Lai YZ, Sarkanen KV (1971) In: Sarkanen KV, Ludwig CH (eds) Lignins: occurrence, formation, structure, and reactions. Wiley Interscience, New York
McNeil M, Darvill AG, Albersheim P (1980) Structure of plant cell walls X. Rhamnogalacturonan I. A structurally complex pectic polysaccharide in the walls of suspension-cultured sycamore cells. Plant Physiol 66:1128–1134
McNeil M, Darvill AG, Fry SC, Albersheim P (1984) Structure and function of the primary cell walls of plants. Annu Rev Biochem 53:625–663
Mort AJ (1983) An apparatus for safe and convenient handling of anhydrous liquid HF at controlled temperatures and reaction times. Application to the generation of oligosaccharides from polysaccharides. Carbohydr Res 122:315–321
Mort AJ, Bauer WD (1982) Application of two new methods for cleavage of polysaccharides into specific oligosaccharide fragments. Structure of the capsular and extracellular polysaccharides of *Rhizobium japonicum* that bind soybean lectin. J Biol Chem 257:1870–1875
Mort AJ, Lamport DTA (1977) Anhydrous hydrogen fluoride deglycosylates glycoproteins. Anal Biochem 82:289–309
Mort A, Parker S (1982) The use of anhydrous HF solvolysis in biomass conversion to glucose. SERI report CP 232–1520
Mort AJ, Utille JP, Torri G, Perlin AS (1983) High selectivity in the partial degradation of an extracellular polysaccharide of *Rhizobium japonicum* with liquid hydrogen fluoride: a NMR spectroscopic study. Carbohydr Res 121:221–232
Rorrer GL, Hawley MC, Lamport DTA (1986) Reaction rates for gas-phase hydrogen fluoride saccharification of wood. I&EC Prod R&D 25:589–595

Rorrer GL, Ashour SA, Hawley MC, Lamport DTA (1987) Solvolysis of wood and pure cellulose by anhydrous hydrogen fluoride vapor. Biomass 12:227–246

Rorrer GL, Mohring WR, Hawley MC, Lamport DTA (1988 a) Adsorption and reaction processes of the solvolysis of wood and pure cellulose by anhydrous hydrogen fluoride vapor. Energy & Fuels 2:556–566

Rorrer GL, Mohring WR, Hawley MC, Lamport DTA (1988 b) A detailed kinetic and heat transport model for the hydrolysis of lignocellulose by anhydrous hydrogen fluoride vapor. Chem Eng Sci 43:1831–1836

Sachs IB, Clark IT, Pew JC (1963) Investigation of lignin distribution in the cell wall of certain woods. J Poly Sci: Part C, 2:203–212

Sanger MP, Lamport DTA (1983) A microapparatus for liquid hydrogen fluoride solvolysis: sugar and amino sugar composition of *Erysiphe graminis* and *Triticum aestivum* cell walls. Anal Biochem 128:66–70

Selke SM, Hawley MC, Hardt H, Lamport DTA, Smith G, Smith JJ (1982) Chemicals from wood via HF, I&EC Prod R&D 21:11–16

Selke SM, Hawley MC, Lamport DTA (1983) Reaction rates for liquid-phase hydrogen fluoride saccharification of wood. In: Soltes J (ed) Wood and agricultural residues: research on use for feed, fuels, and chemicals. Academic Press, New York, pp 329–349

Simons JH (1950) Hydrogen fluoride. In: Simons JH (ed) Fluorine chemistry. Academic Press, New York, pp 225–256

Smith DF (1958) Hydrogen fluoride polymer spectrum, hexamer and tetramer. J Chem Phys 28:1040–1056

Smith JJ, Lamport DTA, Hawley MC, Selke SM (1983) Feasibility of using anhydrous hydrogen fluoride to "crack" cellulose. J Appl Poly Sci: Appl Poly Symp 37:641–651

Smith JJ, Muldoon EP, Willard JJ, Lamport DTA (1986) Tomato extensin precursors P1 and P2 are highly periodic structures. Phytochemistry 25:1021–1030

Sojar HT, Bahl OP (1987) A chemical method for the deglycosylation of proteins. Arch Biochem Biophys 259:52–57

Talmadge K, Keegstra KK, Bauer WD, Albersheim P (1973) The structure of plant cell walls. I. The macromolecular components of the walls of suspension-cultured sycamore cells with detailed analysis of the pectic polysaccharides. Plant Physiol 57:158–173

Timell TE (1957) Carbohydrate composition of ten North American species of wood. TAPPI 40:568–572

Van Holst G-J, Varner JE (1984) Reinforced polyproline II conformation in a hydroxyproline-rich cell wall glycoprotein from carrot root. Plant Physiol 74:247–251

Vanderzee CE, Rodenburg WW (1970) Gas imperfections and excess thermodynamic properties of gaseous hydrogen fluoride. J Chem Therm 2:461–478

Woodward HD, Ringler NJ, Selvakumar R, Simet IM, Bhavanandan, Davidson EA (1987) Deglycosylation studies on tracheal mucin glycoproteins. Biochemistry 26:5315–5322

Immunogold Localization of Specific Components of Plant Cell Walls

P. J. MOORE

1 Introduction

Of central importance to the functional properties of cell walls is their chemical composition and the three-dimensional arrangement of these components in the wall. Thus, knowledge of cell-wall structure is essential to understanding cell-wall function. While considerable progress has been made in identifying cell-wall constituents and elucidating their primary structure (McNeil et al. 1984), much less is known of how these polymers are distributed throughout the wall. To begin to address this question, we have generated polyclonal antibodies against two cell-wall polymers, rhamnogalacturonan I (RG-I), a pectic polysaccharide, and xyloglucan (XG), a hemicellulose, and have used these antibodies to determine when and where RG-I and XG are synthesized and how they are distributed throughout the cell wall.

Previous efforts to localize polysaccharides within the cell wall have consisted of selectively staining for pectic polysaccharides (Albersheim et al. 1960), selectively extracting components of the cell walls using chemicals or enzymes and examining the residue (Deshpande 1976; Roland and Vian 1981; Roland et al. 1977), and by localizing the substrates of wall-degrading enzymes using enzyme-colloidal gold markers (Ruel and Joseleau 1984). These methods have been successful in demonstrating the location of classes of cell-wall polymers in the wall. In particular, they have demonstrated that pectic polysaccharides are concentrated in the middle lamella (Albersheim et al. 1960; Desphande 1976) and hemicelluloses are present mainly in the cellulose microfibril-containing region of the wall surrounding the cell and are excluded from the middle lamella (Roland et al. 1977; Ruel and Joseleau 1984). The limitation of such techniques is that they can localize classes of molecules, but not specific polymers within that class. Thus, studies on the localization of pectic polysaccharides have demonstrated pectins not only in the middle lamella, but also in the cellulose microfibril-containing region of the cell wall (Albersheim 1960; Desphande 1976; Roland and Vian 1981). Whether this is the distribution of all pectins, or if some pectic polysaccharides are present only in the middle lamella and others in the cell walls, cannot be ascertained in such studies.

Antibodies raised against polysaccharides were first used to study cell-wall structure in brown alga. Vreeland and co-workers have produced both polyclonal (Vreeland 1970, 1972) and monoclonal (Vreeland and Laetsch 1984; Vreeland et al. 1984) antibodies to carbohydrates extracted from the brown alga *Fucus distichus*, and have used these antibodies to localize specific carbohydrates in the cell wall and extracellular matrix regions of different cell types in this seaweed. Antibodies have also been used to localize the arabinogalactan proteins in the cell walls and intercellular matrix of the transmitting tissue of *Nicotiana alata* styles (Sedgley and Clarke 1986) and the cell-wall glycoprotein extensin in carrot tap root tissues (Stafstrom and Staehelin 1988; Swords and Staehelin, this volume).

We have used our anti-matrix polysaccharide antibodies to localize RG-I and XG in the cell wall of expanding and dividing cells of clover (*Trifolium pratense*). Our results show that RG-I is restricted to the middle lamella and the expanded region of middle lamella at the corner junctions of cells. RG-I is not present in the cellulose microfibril-containing region of the cell wall. XG, on the other hand, is restricted to the cellulose containing region of the wall surrounding the cell and is excluded from the middle lamella (Moore and Staehelin 1988). In the same study, we also used these antibodies to determine where in the cytoplasm XG and RG-I are assembled and have confirmed that the Golgi apparatus is the site of synthesis of these matrix polysaccharides. The timing of synthesis of RG-I and XG differ, however, RG-I being secreted only during cell expansion, while XG is secreted during both cell expansion and cytokinesis.

To be successful, an immunocytochemical study must balance the preservation of cellular structure and the antigenicity of the molecules to which the antibodies bind. One approach that overcomes directly the problem of retention of antigenicity is to immunostain ultrathin frozen sections (Tokuyasu 1980). However, the preparation of ultrathin frozen sections requires specialized equipment (Raikhel et al. 1984) and, in the case of plant cells, does not usually yield cells and tissues that are structurally well preserved at the electron microscope level. Antibodies can also be applied to tissues prior to embedding. However, pre-embedding labeling necessitates the diffusion of antibodies into the tissue (Craig et al. 1987). Diffusion of large molecules such as antibodies or antibody-colloidal gold probes into plant cells is constrained by the small pore size of the cell wall (Hoggart and Clarke 1984). The third approach, post-embedding immunolabeling of thin sections allows antibodies access to antigens exposed at the surface of the section. The drawback to this technique is the potential loss of antigenicity of the molecule of interest due to fixation, dehydration and embedding. Carbohydrate antigens may be less sensitive than protein antigens to commonly used fixation protocols, but may still be affected by removal of water and the heat of resin polymerization. Thus the choice of resin may also affect the levels of immunostaining. Often material embedded in resins which show good structural preservation is no longer able to bind antibodies. In resins for which processing is less harsh, thus preserving antibody binding activity, retention of structural details can be impaired. Thus investigators must choose between different methods of preparation, each of which has advantages and disadvantages. In this chapter we will describe in detail how we raised the anti-matrix polysaccharide antibodies. We will also describe the types of tests we performed in order to establish the specificity of the antibodies. Results of post-embedding immunolabeling of clover root tip cortical cells and spongy parenchyma cells will be presented, along with a discussion of how to optimize tissue fixation and embedding in order to obtain the best results.

2 Production of Polysaccharide-Specific Antibodies and Tests for Specificity of Antibodies

2.1 Preparation of Antibodies

To generate polysaccharide-specific polyclonal antibodies (Moore et al. 1986), polysaccharides are coupled to a protein carrier prior to injection into rabbits.

Since RG-I (prepared according to McNeil et al. 1980) is negatively charged, it is possible to couple it to methylated bovine serum albumin (BSA) by the method of Vreeland (1970). The negatively charged polysaccharide presumably forms an ionic complex with the positively charged protein. Equal weights of RG-I and methylated BSA are mixed. The mixture is then emulsified in Freund's complete adjuvant and injected subcutaneously in six sites along the back of the rabbit. Rabbits are boosted with the same material emulsified in incomplete Freund's adjuvant 1 week and 2 weeks after the first injection and again at 9 weeks. Antibodies can be detected 12 weeks following the initial injection.

Because XG is not negatively charged, a different reaction must be used to couple XG to a protein carrier. Purified XG (Bauer et al. 1973) is coupled to ovalbumin by the periodate-lysine fixation method of McLean and Nakane (1974) in which the periodate apparently oxidizes carbohydrate moities on the XG and on the oligosaccharide side-chains of ovalbumin to form aldehyde groups. These groups can then be cross-linked via lysine, a divalent amine. XG and ovalbumin are incubated together in the fixative (McLean and Nakane 1974) for 2 h at room temperature. To make the fixative solution 0.1M sodium phosphate (dibasic) is added to 0.2M lysine-HCl until the pH reaches 7.4. This solution is then diluted to 0.1M lysine with 0.1M sodium phosphate buffer, pH 7.4. Three parts of the lysine-phosphate buffer is mixed with one part distilled H_2O and solid sodium-m-metaperiodate is added to a concentration on 0.01M. Excess fixative is then removed by dialysis overnight. After coupling to ovalbumin, the antigen was emulsified in Freund's adjuvant and injected as above.

Vreeland has produced monoclonal antibodies against carbohydrates extracted from brown and red marine algae by both in vivo immunization of mice (Vreeland and Laetsch 1984) and in vitro immunization of mouse spleen cells (Vreeland et al. 1984), a technique that may also be applicable to making such antibodies against complex carbohydrates of higher plant cell walls. Monoclonal antibodies recognize a specific epitope on a complex polysaccharide and may be useful for identifying functionally important regions of plant cell-wall polysaccharides. Anderson et al. (1984) found that most of the monoclonal antibodies raised against style extracts recognize arabinose or galactose residues. Monoclonal antibodies which recognize a terminal α-L-arabinofuranosyl residue can be considered a specific probe for the arabinogalactan proteins in the style since greater than 90% of the total arabinosyl residues of style extracts are associated with soluble arabinogalactan proteins (Sedgley and Clarke 1986).

2.2 Specificity of Anti-Cell-Wall Matrix Polysaccharide Antibodies for Cell-Wall Polymers

Immunoglobulins are isolated from the sera by $(NH_4)_2SO_4$ precipitation (Mayer and Walker 1980). Antisera is screened for reactivity with the matrix polysaccharides using a Bio-Rad Immunoblot Assay kit with 100-ng samples of purified RG-I or XG dotted onto nitrocellulose paper. Our antibodies always show low levels of staining; this could be due to loss of the antigen from the nitrocellulose, or low affinity or low concentration of the antibodies.

RG-I RG-II XG Ext

A. Anti-RG-I

B. Anti-RG-I preimmune

C. Anti-XG

D. Anti-XG preimmune

Fig. 1 A–D. *Dot blot* showing specificity of the anti-cell wall matrix polysaccharide antibodies. Droplets of 1 μl containing 100 ng of RG-I, RG-II, XG or extensin are placed on nitrocellulose paper. Anti-RG-I antiserum reacts strongly with RG-I and to a much lesser extent with extensin (**A**). Anti-XG serum reacts only with XG (**C**). Preimmune serum from both rabbits did not recognize their antigens but does show slight reactivity with extensin (**B** and **D**). All sera were diluted 1 : 10

The antiserum raised against the carbohydrate coupled to the protein carrier not only contains antibodies which recognize the purified polysaccharide, but also contains antibodies which recognize the protein carrier. Anti-methylated BSA or anti-ovalbumin antibodies are removed from the serum by preabsorption with an excess of the protein carrier. Thus, antiserum is mixed with an equal volume of 1 mg/ml methylated BSA or ovalbumin and allowed to incubate overnight. If a precipitate is formed, it is removed from the serum by centrifugation. Otherwise the serum is used with the protein still in the serum.

To determine if the anti-RG-I and anti-XG antibodies cross-react with other cell-wall polymers, we test the antisera against other purified cell-wall molecules on dot blots. Thus, we test both antisera against XG, RG-I, and rhamnogalacturonan II (RG-II), another pectic polysaccharide, (all provided by Dr. P. Albersheim) and extensin, a cell-wall glycoprotein (provided by Dr. J. Stafstrom). The anti-RG-I antiserum specifically recognizes only the purified RG-I (Fig. 1 A). It cross-reacts slightly with the extensin, however the anti-RG-I preimmune and anti-XG preimmune serum (Fig. 1 C, D) also cross-react with extensin. Likewise, the anti-XG antiserum only recognizes its specific antigen, XG, and no other cell-wall matrix components (Fig. 1 B). One concern we had was that the anti-XG antibodies might cross-react with cellulose microfibrils since both have backbones of β-1,4 linked glucose residues. To determine if the anti-XG antibodies do cross-react with cellulose, we dotted purified sonicated ramie cellulose fibrils onto nitrocellulose paper and tested the serum with the immunoblot assay kit. The anti-XG antibodies do not cross-react with the cellulose microfibrils.

2.3 Specificity of Anti-Cell-Wall Matrix Polysaccharide Antibodies to Saccharides

To determine if our antibodies are simply recognizing single sugar residues that might be present in wall polymers other than RG-I and XG, we preabsorb the antisera with purified monosaccharides and polysacccharides that are found within either RG-I or XG. Equal volumes of the antiserum and 1 mg/ml sugar solution are mixed and the mixtures are allowed to incubate overnight at 4° C. The preabsorbed serum is then tested for reactivity on dot blots against 200 ng

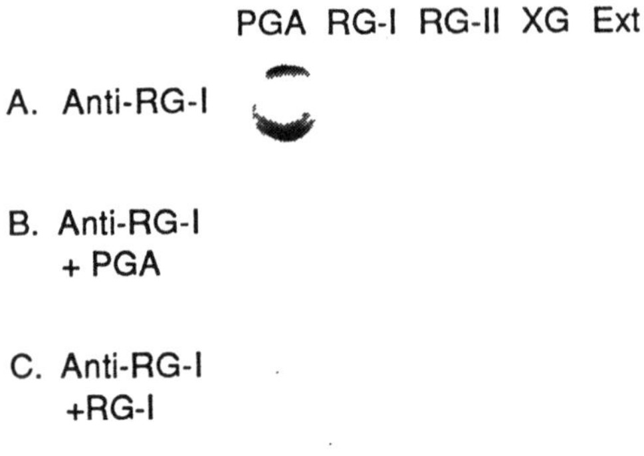

Fig. 2 A–C. *Dot blot* showing cross-reactivity of anti-RG-I antiserum to commercially available polygalacturonic acid (*PGA*). Droplets of 2 µl containing 200 ng of PGA, RG-I, RG-II, XG or extensin are placed on nitrocellulose paper. Anti-RG-I serum reacts strongly with PGA and RG-I (**A**). Anti-RG-I serum preabsorbed with PGA (**B**) or RG-I (**C**) does not react with either PGA or RG-I

of RG-I, XG, RG-II, extensin and polygalacturonic acid. Anti-RG-I is preabsorbed with L-arabinose, D-galactose, L-rhamnose and polygalacturonic acid. Anit-XG is preabsorbed with D-xylose and L-fucose.

Preabsorption of anti-RG-I with arabinose, galactose or rhamnose does not affect antibody binding to dot blots. Polygalacturonic acid, however, does cross-react with the anti-RG-I serum and preabsorption with commercial polygalacturonic acid abolishes antibody reactivity against RG-I on dot blots. Likewise, preabsorption of the anti-RG-I serum with purified RG-I eliminates staining of the polygalacturonic acid on dot blots (Fig. 2). Based on these findings we conclude that our anti-RG-I antibodies recognize the short stretches of polygalacturonic acid that may exist on the ends of the RG-I molecules. This result suggests that our anti-RG-I antibody would also recognize other pectic polysaccharides with long regions of repeating galacturonic acid residues, such as homogalacturonan. Preabsorption of anti-XG with xylose or fucose does not affect antibody binding to dot blots. Thus neither the anti-RG-I nor the anti-XG antibodies are recognizing single sugar residues in the XG side-chains.

X-ray crystallographic studies of specific antibody-antigen complexes have shown that the size of the contact region between the antibody and its antigen is sufficiently large (2.0 × 2.5 nm) to cover a domain of up to several amino acids or sugar residues (Amit et al. 1985). Thus antibodies against large polysaccharides like RG-I and XG are probably not recognizing single sugar residues, but a set of residues in a specific spatial arrangement. Based on current knowledge of the specific linkage groups of sugar oligomers derived from enzymatically degraded complex cell-wall polysaccharides such as RG-I and XG (McNeil et al. 1984), the spatial arrangement of sets of sugar residues should be unique for each cell-wall polymer. The lack of cross-reactivity between our anti-RG-I and anti-XG antibodies and other cell-wall polymers, and the inability of monosaccharides to inhibit binding of these antibodies to the cell wall, are thus consistent with the notion that our antibodies are indeed highly specific probes for the two cell-wall-matrix polysaccharides.

3 Immunolabeling with the Anti-Cell-Wall Matrix Polysaccharide Antibodies

3.1 Preparation of Plant Tissues for Immunolabeling

We use two tissues, root tips and expanding leaves, for our studies of cell wall structure and biosynthesis in red clover (*Trifolium pratense*). We found that the fixation conditions required to preserve the structure of the small, nonvacuolated root tip cells are different from those required to fix the larger, vacuolated spongy parenchyma cells of the leaf. This is probably due to the osmotic activity of the fixative agents and the greater sensitivity to changes in osmotic conditions of vacuolated cells. To optimize the fixation of other types of leaves or tissues the osmolarity of the buffer may need to be adjusted (Bullock 1984). Root tips are obtained by germinating clover seeds on moist filter paper for 4 days. The terminal 3 mm of the root tips are excised under fixative. The root tips are fixed in 2.5% glutaraldehyde in 10 mM sodium phosphate buffer, pH 7.2, for 2 h at room temperature. Next, the root tips are rinsed with buffer and postfixed in 1% osmium tetroxide in 10 mM sodium phosphate buffer, pH 7.2, for 1 h at room temperature. Expanding leaves of clover are harvested and fixed in 2% paraformaldehyde plus 0.1% glutaraldehyde in 100 mM sodium phosphate buffer, pH 7.2. The leaves are vacuum infiltrated with fixative for 10 min and then incubated in the fixative for 2 h more at room temperature. This mixture of formaldehyde and aldehyde gives the best preservation of the leaf tissue, with the least observable plasmolysis of the tissue. This is important, since, as discussed in greater detail below, plasmolysis can lead to labeling artifacts (Chap. 3.3). The leaves are then post-fixed in 1% osmium tetroxide in 10 mM sodium phosphate buffer, pH 7.2, for 1 h at room temperature.

One problem in immunocytochemistry is the preservation of structure without the loss of antigenicity of the molecule of interest. Most protein antigens are susceptibel to aldehyde and osmium fixation (Craig and Goodchild 1982), but in some cases the antigenic sites blocked during osmium fixation can be unmasked by treatment of the fixed tissue with sodium meta-periodate (Bendayan and Zollinger 1983). This treatment, however, destroys carbohydrate antigens. Carbohydrates, on the other hand, do not react to any great extent with either glutaraldehyde or osmium (Bullock 1984), and thus both of these fixatives can usually be used when localizing complex carbohydrate antigens. It should be considered, however, that carbohydrates could be lost during processing due to not being fixed. This is probably not a great problem in the study of cell-wall structure in that the wall is highly cross-linked and most components probably retain their normal distribution during processing for electron microscopy. This may not be the case, however for small, soluble components of the cell wall or for contents of cytoplasmic vesicles.

After fixation, both tissues are dehydrated in an ethanol series (30%, 50%, 70%, 90%, 100%; 10 min each step). The tissue is then infiltrated in the resin LR White (Craig and Miller 1984; Polysciences, Warrington, PA, USA) by treatment in 2:1 (v/v) ethanol: LR White for 1 h, 1:2 ethanol: LR White for 2 h and 100%

LR White overnight at 4° C (all other steps are carried out at room temperature). The infiltrated tissue is then embedded in gelatin capsules and polymerized overnight at 50° C.

If root tips are to be embedded in Lowicryl K4M (Carlemalm et al. 1982; Polysciences, Warrington, PA, USA), rather than LR White resin, they are processed as follows: the root tips are fixed in 2.5% glutaraldehyde in 10 mM sodium phosphate buffer, pH 7.2, for 2 h at room temperature. The root tips are not post-fixed in osmium tetroxide since osmium inhibits the polymerization of Lowicryl K4M (Carlemalm et al. 1982). Following glutaraldehyde fixation the root tips are washed in buffer and dehydrated through an acetone series (30%, 50%, 70%, 90%, and 100%; 10 min each step). The root tips are then stained en bloc with hafnium chloride (0.5% in 100% acetone; Aldrich Chemical Co., Milwaukee, WI, USA) for 20 min. After being rinsed twice with 100% acetone, the root tips are transferred to ethanol in a dilution series (2:1 acetone:ethanol, 1:2 acetone:ethanol, 100% ethanol; 10 min each step). The tissue is then infiltrated in the resin Lowicryl K4M. The infiltration is done at 4° C. The tissue is placed in 2:1 ethanol:resin for 2 h, then transferred to 1:2 ethanol:resin and left overnight. The tissue is then infiltrated in 100% Lowicryl K4M for a period of 1 to 2 weeks, with a fresh change of resin every day. Finally, the tissue is embedded in Beem (Polysciences, Warrington, PA, USA) capsules and polymerized overnight at −20° C with UV light.

3.2 On-Grid Immunolocalization with Anti-Cell-Wall Matrix Polysaccharide Antibodies

For immunolabeling of cell walls and cytoplasm, 70–100-nm sections (silver to gold) are cut. The sections are picked up on formvar-carbon-coated nickel grids (300 mesh). The grids are first treated with 0.1N HCl to remove glutaraldehyde from the sections (Craig and Goodchild 1984). All treatments are done in 5-μl droplets on parafilm in a moist chamber. The grids are then incubated in a freshly prepared blocking solution of 5% low fat dried milk powder (Carnation Co., Los Angeles, CA, USA) in PBST (10 mM sodium phophate buffer, pH 7.2, 500 mM NaCl, 0.1% Tween-20) for 20 min. Since protein A-gold will bind to formvar, we routinely sink our grids in the blocking solution so that all formvar surfaces are blocked if the grid sinks in subsequent solutions. Excess blocking solution is removed from the grids by blotting and the grids are then incubated for 1 h on the primary antiserum diluted 1:4 to 1:10 in PBST. The grids are then washed for 30 s in a continuous stream of PBST containing 0.5% Tween-20. An increased Tween-20 concentration in the wash buffer will reduce the amount of background staining. After the wash, the grids are placed on a solution of protein A-gold prepared as described in Slot and Geuze (1985) that is diluted to a very pale pink color with PBST. This method reliably produces monodisperse gold particles which are homogeneous in size. After 30 min in protein A-gold, the grids are washed with PBST containing 0.5% Tween-20, followed by a distilled water rinse. Up to the water wash, we never allow the grids to dry during processing.

Protein A is used as our secondary probe instead of goat-anti-rabbit IgG antibodies due to the stability of protein A and the ease of preparation of the protein A-colloidal gold complex (Roth et al. 1978; Roth 1986). Protein A sometimes reacts weakly with certain rat and mouse IgGs (Roth 1986). Thus, since monoclonal antibodies are raised in mice, a goat anti-mouse IgG probe may need to be substituted if monoclonal antibodies are employed. Colloidal gold is used as a marker for electron microscopy based on its high density, making it easily visible in the microscope even at relatively low magnifications (Horisberger and Rosset 1977). In addition, it is possible to prepare colloidal gold in many different sizes (Slot and Geuze 1985; Roth 1986). This permits two antigens to be labeled on a single section by means of two different antibodies marked by gold probes of different sizes. The double labeling procedure we use is a modification of the procedure outlined in Titus and Becker (1985). The grids are first treated with 0.1N HCl for 5 min. The grids are then blocked in 5% low-fat dried milk powder solution in PBST for 20 min. Following the milk block, the grids are incubated in the first antiserum (usually the serum which recognizes the least abundant antigen) for 1 h. The first antiserum is washed from the grids with PBST (0.5% Tween-20) and they are then treated with the smaller protein A-gold probe (7.5 nm). The grids are again rinsed thoroughly with PBST (0.5% Tween-20) and then incubated in an excess of protein A (0.2 mg/ml in PBST). The grids are then incubated in the second antiserum for 1 h, washed with PBST (0.5% Tween-20) and incubated in the larger protein A-gold probe (17 nm). Finally, the grids are washed in PBST (0.5% Tween-20), then water, and lastly are counterstained with uranyl acetate and lead citrate.

3.3 Results of Immunolabeling with Anti-Cell-Wall Matrix Polysaccharide Antibodies

Using on-grid immunolocalization we have localized RG-I and XG both in the plant cell wall and intracellularly during synthesis (Moore and Staehelin 1988). In the cell wall of root tip cortical cells and of spongy parenchyma cells of clover, anti-RG-I antibodies label the middle lamella (Figs. 3, 5). The label is concentrated over the expanded region of middle lamella at the corner junctions between cells, but label can also be seen between closely appressed cell walls. In leaf cells, a high density of label is also present over the edges of cell walls surrounding the air spaces. The labeling at the inner surface of the cell wall seen in a small region of Fig. 5 (star) is an artifact that occurs when the plasma membrane is pulled back from the cell wall because of plasmolysis. Because most electron microscope fixation, dehydration and plastic embedding schedules for plant tissues produce some plasmolysis, this kind of artifact is often difficult to circumvent. When the plasma membrane remains appressed against the cell wall, no such labeling of the inner cell-wall surface was observed.

Anti-XG antibodies also label clover root tip and spongy parenchyma cell walls. Anti-XG antibodies label the cellulose microfibril containing region of the cell wall surrounding the cells (Figs. 4, 6). Gold label is found only in very low levels in the expanded regions of middle lamella, and the label that is present is

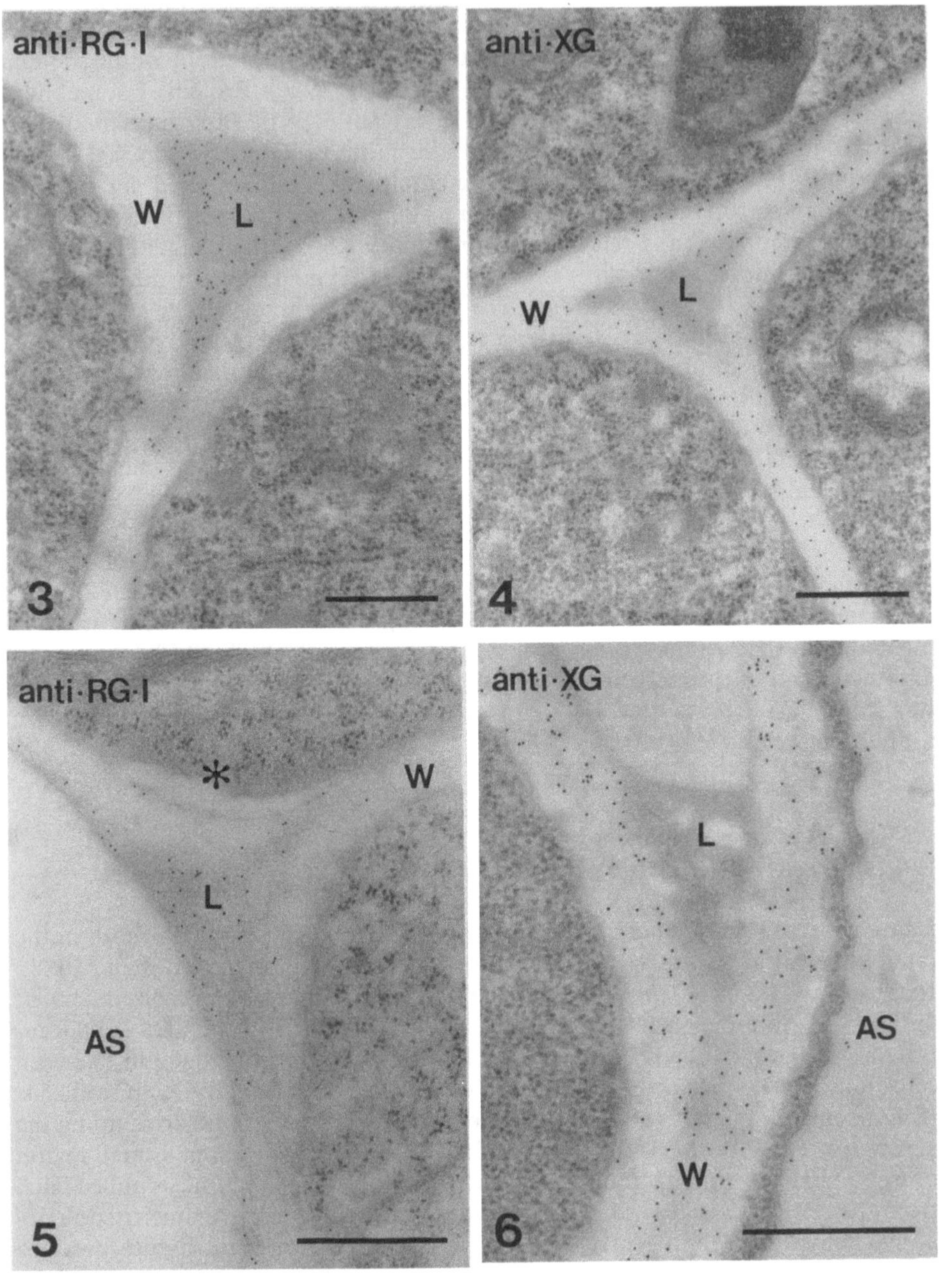

Figs. 3–6. Immunolabeling of root tip cortical cells and leaf parenchyma cells of red clover with anti-cell-wall matrix polysaccharides. **Fig. 3.** In root tip cortical cells the *anti-RG-I* antibodies label the expanded region of the middle lamella (*L*) as well as the closely appressed region of the middle lamella between cell walls. **Fig. 4.** In root tip cortical cells *anti-XG* antibodies label the cellulose microfibril-containing region of the cell wall (*W*) but not the expanded region of middle lamella (*L*). **Fig. 5.** Leaf parenchyma cells labeled with the *anti-*

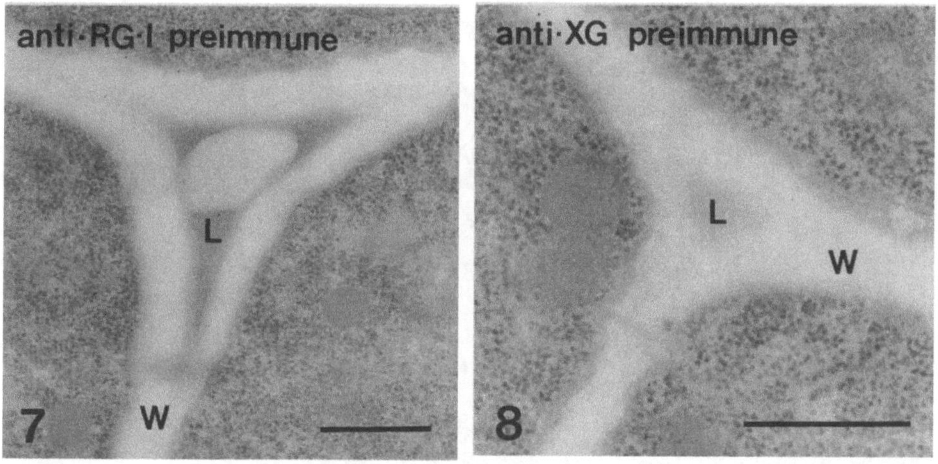

Figs. 7, 8. Immunolabeling of clover root tip cells with preimmune serum from rabbits producing anti-cell-wall matrix polysaccharide antibodies. Neither *anti-RG-I preimmune* serum (**Fig. 7**) nor *anti-XG preimmune* serum (**Fig. 8**) label either the cell wall (*W*) or the middle lamella (*L*); bar = 0.5 μm

restricted to the outer edges adjacent to the cell wall. Preimmune antiserum from the rabbits producing either the anti-RG-I or anti-XG antiserum did not label cell walls to any appreciable extent (Fig. 7, 8). These results, along with results of immunolabeling of cell walls with anti-extensin antibodies (Stafstrom and Staehelin 1988), are summarized in the diagram in Fig. 9.

In addition to labeling the cell wall, anti-RG-I and anti-XG antibodies label a restricted set of intracellular organelles in expanding cortical cells of clover root tips. Both antibodies label the Golgi stacks and associated vesicles but do not label the rough endoplasmic reticulum, mitochondria, nuclei or other organelles (Figs. 10–13). This confirms many biochemical studies which had identified the Golgi apparatus as the site of synthesis of the cell-wall matrix polysaccharides. However, immunocytochemistry adds details which are not evident in biochemical studies, such as the concentration of gold label over the margins of the cisternae, rather than the middle of the cisternae (Figs. 10, 11; arrows). This may indicate that the synthetic enzymes are preferentially localized in the margins of the cisternal stacks or that RG-I and XG become concentrated in the marginal regions immediately after synthesis.

RG-I antibodies show label over the middle lamella (*L*) and the material lining the air spaces (*AS*), but not the rest of the cell wall (*W*). Label is occasionally seen at the inner surface of the cell wall in regions where the plasma membrane has pulled away from the cell wall (*star*) because of plasmolysis of the tissue during fixation. **Fig. 6.** In leaf parenchyma cells, as in root tip cells, the *anti-XG* antibodies label the cell wall (*W*) but not the middle lamella. *AS* air space; *V* vacuole; bars = 0.5 μm

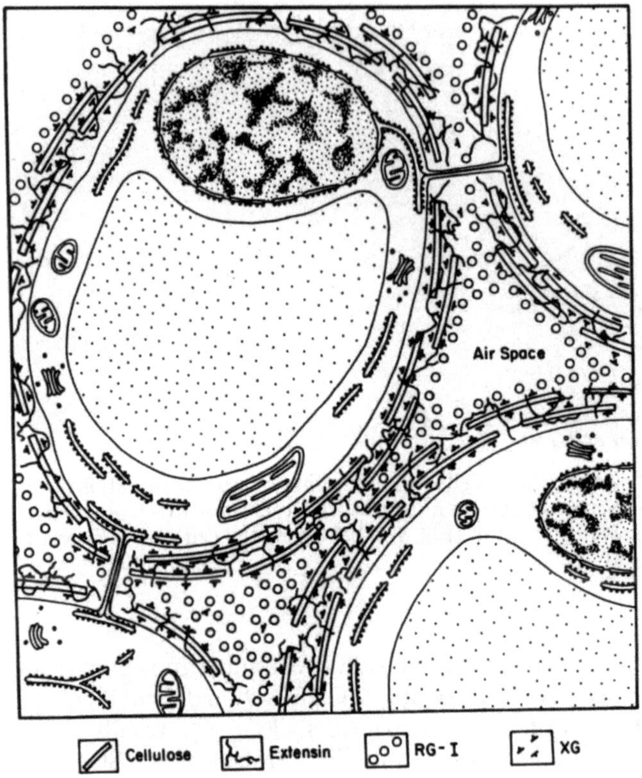

Fig. 9. Diagram summarizing the immunolocalization of the cell-wall matrix components RG-I, XG, and extensin. Both XG and extensin are present in the cellulose microfibril-containing region of the cell wall (Moore and Staehelin 1988; Stafstrom and Staehelin 1988) and are excluded from the middle lamella. RG-I, on the other hand, is present only in the middle lamella (Moore and Staehelin 1988)

One area in which immunolabeling results raise more questions than they answer is on the timing during the cell cycle of the secretion of the matrix polysaccharides. In cells undergoing cytokinesis, the cell plate is not labeled with the anti-RG-I antibodies (Fig. 14). This is surprising in that the cell plate historically has been thought to be rich in acidic pectic polysaccharides and similar in composition to the middle lamella. Anti-XG antibodies, on the other hand, do label the cell plate (Fig. 15). These results indicate that we may have to revise our ideas of how the newly synthesized cell wall is assembled. Another question raised by the absence of RG-I from the cell wall is how RG-I is transported through the cellulosic cell wall to the the middle lamella and corner regions during cell expansion, and why RG-I accumulates there.

To ensure that the antisera recognize only their specific antigens in the cell wall, the antisera were preabsorbed against their respective antigens. With both the anti-RG-I and anti-XG antisera, allowing the serum to stand overnight with an equal volume of 1 mg/ml RG-I and XG, respectively, completely prevents la-

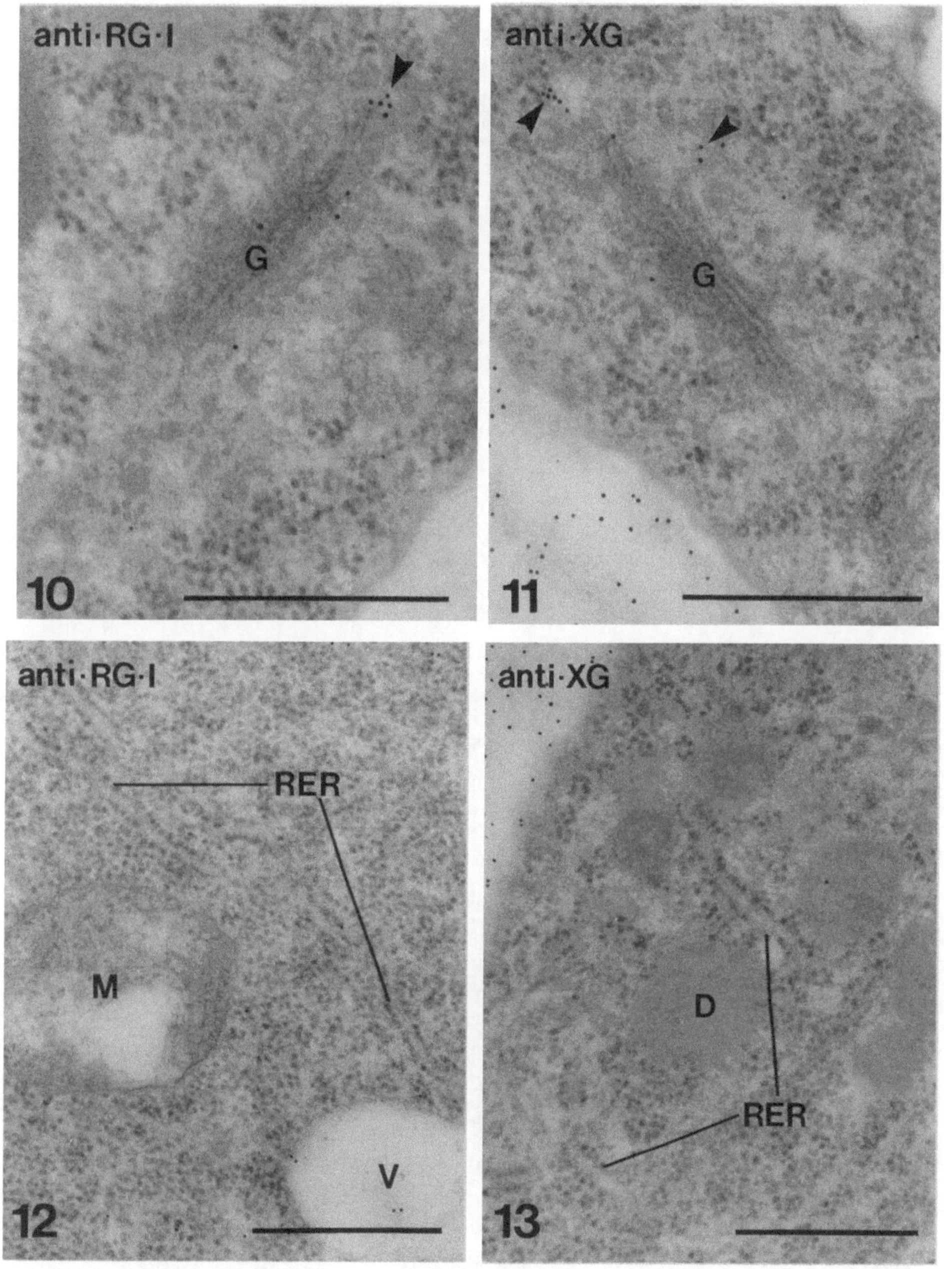

Figs. 10–13. Immunolabeling of the Golgi apparatus in expanding root tip cortical cells. Both *anti-RG-I* (**Fig. 10**) and *anti-XG* (**Fig. 11**) label the Golgi complex (*G*) and associated vesicles (**Figs. 10, 11**; *arrows*) in these cells. Other cellular organelles such as the rough endoplasmic reticulum (*RER*), mitochondria (*M*), vacuoles (*V*) and lipid droplets (*D*) are not labeled by either *anti-RG-I* (**Fig. 12**) or *anti-XG* (**Fig. 13**) antibodies; bar = 0.5 μm

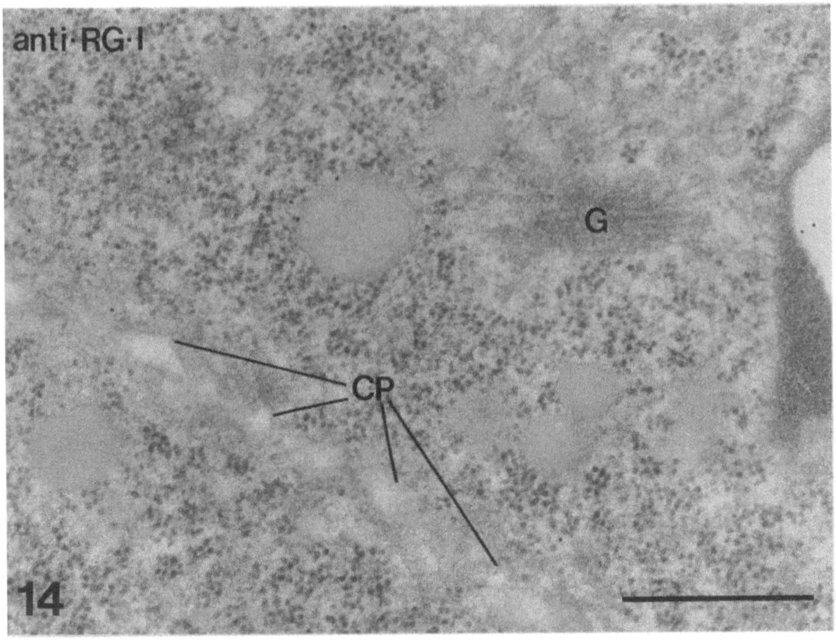

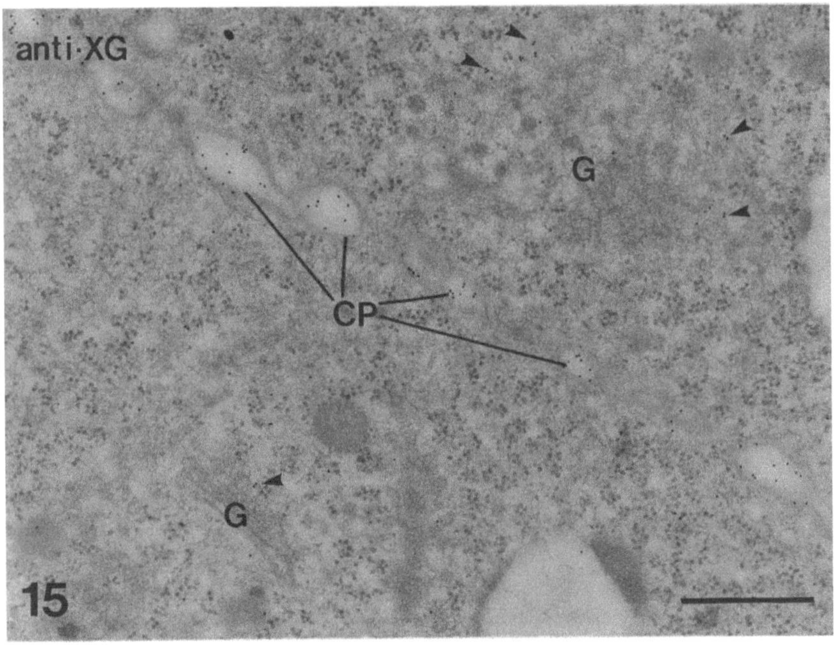

Figs. 14, 15. Immunolabeling of the forming cell plate in dividing root tip cortical cells. *Anti-RG-I* antibodies (**Fig. 14**) do not label either the cell plate (*CP*) or the Golgi complex (*G*) during cytokinesis. In contrast, *anti-XG* antibodies (**Fig. 15**) label both the cell plate (*CP*) and the Golgi stacks (*G*) and associated vesicles (*arrows*); bar = 0.5 μm

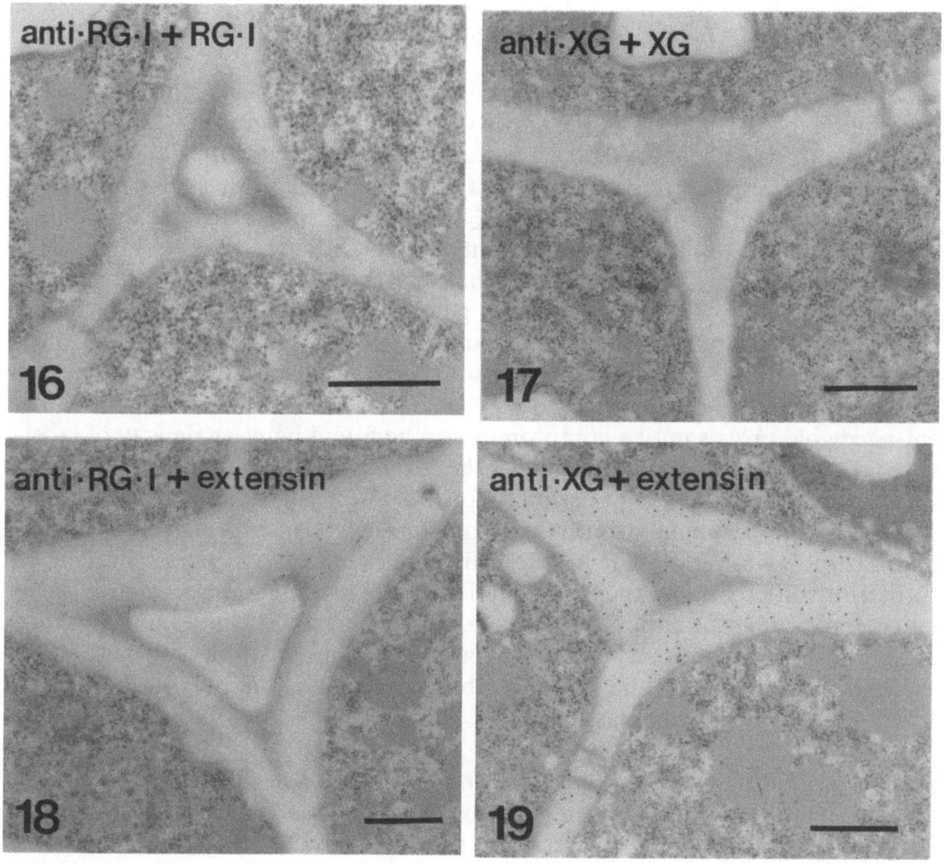

Figs. 16–19. Immunolabeling of root tip cortical cells with preabsorbed antiserum. *Anti-RG-I* antibodies preabsorbed with purified *RG-I* (**Fig. 16**) and *anti-XG* antibodies preabsorbed with purified *XG* (**Fig. 17**) no longer label the cell wall or the middle lamella. On the other hand, preabsorption of the *anti-RG-I* antiserum (**Fig. 18**) or the *anti-XG* antiserum (**Fig. 19**) with the cell wall glycoprotein extensin does not affect the labeling patterns of either antibody; bar = 0.5 μm

beling of the cell wall (Figs. 16, 17). Preabsorption with extensin, however, does not effect the labeling patterns of either the anti-RG-I or anti-XG antiserum (Figs. 18, 19). Anti-XG was also preabsorbed with cellulose microfibrils prior to section labeling. Again, cellulose has no effect on anti-XG antibody labeling. Preabsorption of anti-RG-I with commercial polygalacturonic acid, however, does eliminate section labeling. Thus the epitope this antiserum appears to recognize in the cell walls probably includes stretches of polygalacturonic acid, which Selvendran (1985) has previously hypothesized to be in the middle lamella.

3.4 Problems Encountered During Immunolabeling

Density of Immunolabeling. One problem we encountered while studying the intracellular labeling patterns of the anti-RG-I and anti-XG antiserum was the low levels of antibody labeling over such intracellular structures as the Golgi complex and secretory vesicles. In fact, the average number of gold particles per Golgi stack for anti-RG-I antibodies is 3.47 ± 1.66 (mean $\pm$ one standard deviation), while the anti-XG antiserum is only slightly better with 6.47 ± 1.81 gold particles per Golgi stack. The size of the gold particle attached to the protein A affects the density of labeling. The use of a small gold probe (such as 7.5 nm) results in heavier labeling of cellular structures than a larger gold particle (such as 17 nm), probably due to steric hindrances. Very small gold particles, however, can be difficult to see in the electron microscope. Thus one needs to experiment with different-sized gold probes until a compromise can be found between density of labeling and the ability to easily distinguish the marker at the optimal magnification for a given sample. Another possibility for the low density of immunolabeling is that due to the extreme flexibility of carbohydrates, the conformation of the dehydrated polysaccharide that has been fixed and embedded in plastic might be sufficiently different than the conformation of the hydrated polysaccharide-protein complex used for antibody production such that the effective concentration of the antiserum is reduced. That is, only a certain subpopulation of the anti-polysaccharide antibodies in the serum is able to recognize the polysaccharides in fixed tissues embedded in plastic.

Improvement of Cytoplasmic Labeling with the Resin Lowicryl K4M. To improve the density of intracellular labeling, we have tested Lowicryl K4M, a resin developed for embedding at low temperatures (Carlemalm et al. 1982) and thereby enhancing the preservation of antigenic sites for immunocytochemical studies (Roth et al. 1981).

Cortical root tip cells embedded in Lowicryl K4M show good cytoplasmic preservation and, when immunolabeled, show increased levels of antibody label over the Golgi apparatus (Figs. 20, 21). The increase in density of gold label varied from 2.5 times greater for the anti RG-I antibodies to 4.5 times greater for the anti-XG antibodies. The structural features of the Golgi apparatus, however, are often less distinct in Lowicryl K4M than in LR White, probably due to the absence of osmium tetroxide fixation. A major drawback to using Lowicryl for plant tissue is evident in Fig. 22. Thus, despite the long infiltration period, the walls in Lowicryl K4M embedded root tips tend to become wavy or even partly folded and often the sections tear at the interface between the cells and the cell walls. We are unsure why this is so, but it may be due to the inability of the resin to completely permeate the cell walls. Alternatively, the deformation of the wall may also reflect a general shrinkage of the tissue. Since the cell wall cannot shrink in length, newly formed walls become wrinkled as the cytoplasm contracts. Thicker walls, which have more strength, do not change shape as dramatically, but rather the cytoplasm rips away from them.

Although the use of Lowicryl K4M can be of great use, especially for studying an antigen of low abundance, or one that is particularly sensitive to heat, it is not

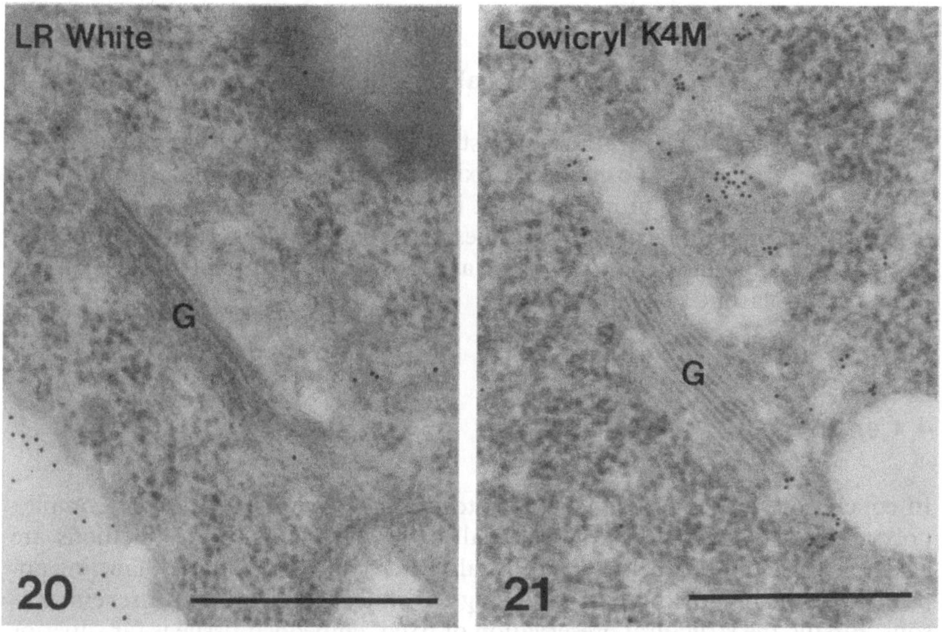

Figs. 20, 21. Effect of embedding medium on immunolabeling densities. Golgi stacks (*G*) in tissue embedded in the resin *LR White* (**Fig. 20**) show good structural preservation and are labeled with the anti-XG antibodies. The level of gold label, however, is much greater over Golgi stacks (*G*) in tissue embedded in the resin *Lowicryl K4M* (**Fig. 21**) although the structural features of the Golgi complex are less distinct; bar = 0.5 μm

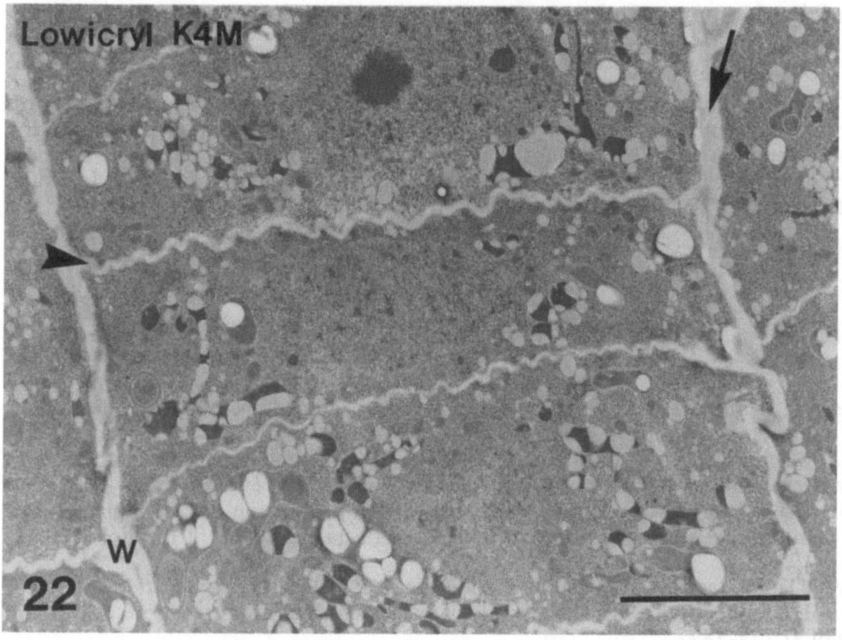

Fig. 22. Clover root tip embedded in the resin *Lowicryl K4M*. While the cytoplasm appears to be relatively well preserved in this resin, the cell walls are often distorted. This effect is more pronounced in the young, thin cross walls (*arrowhead*) than in the older, thicker longitudinal walls (*arrow*); bar = 0.5 μm

an ideal resin to use for analyzing cell-wall structure due to the distortions it produces. In contrast, samples embedded in the resin LR White exhibited walls that appeared rigid and straight, thus suggesting that they are well preserved in this resin. However, the use of osmium tetroxide and the heat of polymerization may destroy the antigenicity of molecules of interest. Thus, we always employ both resins for our immunolocalization studies, LR White to look at the distribution of different compounds in the cell walls, and Lowicryl K4M to probe the composition of cytoplasmic structures.

4 Conclusion

In conclusion, antibodies can be used to examine the distribution of complex polysaccharides throughout the cell wall of higher plants. Many methods are available for immunolabeling the cell wall. Post-embedding immunolabeling appears to be the best method for localizing cell-wall carbohydrates for several reasons. Firstly, the structural preservation of fixed, embedded tissue is far superior to that of ultrathin frozen sections. Secondly, complex polysaccharide antigens appear to be less sensitive than protein antigens to the fixatives commonly used in electron microscopy. Lastly, section labeling may allow antibodies greater access to antigens than diffusion through tissue prior to embedding. There are now several resins and probes available for immunocytochemistry, each with advantages and disadvantages. Only experimentation with different parameters will determine the best conditions for a given system.

Acknowledgments. Thanks are due to Dr. L. Andrew Staehelin, Kathleen M.M. Swords and Margaret A. Lynch for discussion, comments and critical reading of the manuscript, and to Maureen Gaines for preparing the artwork. The author's work cited here has been supported by National Science Foundation grant DCB 8615763 and by National Institutes of Health grant GM 18639 to L.A. Staehelin.

References

Albersheim P, Mühlethaler K, Frey-Wyssling A (1960) Stained pectin as seen in the electron microscope. J Biophys Biochem Cytol 8:501–506

Amit AG, Mariuzza RA, Phillips SEV, Poljak RJ (1985) Three-dimensional structure of an antigen-antibody complex at 6Å resolution. Nature 313:156–158

Anderson MA, Sandrin MS, Clarke AE (1984) A high proportion of hybridomas raised to a plant extract secrete antibody to arabinose or galactose. Plant Physiol 75:1013–1016

Bauer WD, Talmadge KW, Keegstra K, Albersheim P (1973) The structure of plant cell walls. II. The hemicellulose of the walls of suspension-cultured sycamore cells. Plant Physiol 51:174–187

Bendayan M, Zollinger M (1983) Ultrastructural localization of antigenic sites on osmium-fixed tissues applying the protein A-gold technique. J Histochem Cytochem 31:101–109

Bullock GR (1984) The current status of fixation for electron microscopy: a review. J Microsc 133:1–15
Carlemalm E, Garavito RM, Villiger W (1982) Resin development for electron microscopy and an analysis of embedding at low temperature. J Microsc 126:123–143
Craig S, Goodchild DJ (1982) Post-embedding immunolabeling. Some effects of tissue preparation on the antigenicity of plant proteins. Eur J Cell Biol 28:251–256
Craig S, Goodchild DJ (1984) Periodate-acid treatment of sections permits ongrid immunolocalization of pea seed vicilin in ER and Golgi. Protoplasma 122:35–44
Craig S, Miller C (1984) LR White resin and improved on-grid immunogold detection of vicilin, a pea seed storage protein. Cell Biol Int Rep 8:879–886
Craig S, Moore PJ, Dunahay TG (1987) Immunogold localization of intra- and extra-cellular proteins and polysaccharides in plant cells. Scan Electron Microsc 1:1431–1437
Deshpande BP (1976) Observations of the fine structure of plant cell walls. I. Use of permanganate staining. Ann Bot 40:433–437
Hoggart RM, Clarke AE (1984) Porosity of *Gladiolus* stigmatic papillae and pollen tube walls. Ann Bot 53:271–277
Horisberger M, Rosset J (1977) Colloidal gold, a useful marker for transmission and scanning electron microscopy. J Histochem Cytochem 25:295–305
Mayer RJ, Walker JH (1980) Immunocytochemical methods in the biological sciences. Academic Press, New York
McLean IW, Nakane PK (1974) Periodate-lysine-paraformaldehyde fixative. A new fixative for immunoelectron microscopy. J Histochem Cytochem 22:1077–1083
McNeil M, Darvill AG, Albersheim P (1980) The structure of plant cell walls. X. Rhamnogalacturonan I, a structurally complex pectic polysaccharide in the walls of suspension-cultured sycamore cells. Plant Physiol 66:1128–1134
McNeil M, Darvill AG, Fry SC, Albersheim P (1984) Structure and function of the primary cell walls of plants. Annu Rev Biochem 53:625–663
Moore PJ, Staehelin LA (1988) Immunogold localization of the cell wall matrix polysaccharides rhamnogalacturonan I and xyloglucan during cell expansion and cytokinesis in *Trifolium pratense* L.; implications for secretory pathways. Planta 174:433–445
Moore PJ, Darvill AG, Albersheim P, Staehelin LA (1986) Immunogold localization of xyloglucan and rhamnogalacturonan I in the cell walls of suspension-cultured sycamore cells. Plant Physiol 82:787–794
Raikhel NV, Mishkind M, Palevitz BA (1984) Immunocytochemistry in plants with colloidal gold conjugates. Protoplasma 121:25–33
Roland JC, Vian B (1981) Use of endopolygalacturonase for a topochemical study of elongating cell walls at the ultrastructural level. J Cell Sci 48:333–343
Roland JC, Vian B, Reis D (1977) Further observations on cell wall morphogenesis and polysaccharide arrangement during plant growth. Protoplasma 91:125–141
Roth J (1986) Post-embedding cytochemistry with gold-labelled reagents: a review. J Microsc 143:125–137
Roth J, Bendayan M, Orci L (1978) Ultrastructural localization of intracellular antigens by the use of protein A-gold complex. J Histochem Cytochem 26:1074–1081
Roth J, Bendayan M, Carlemalm E, Villiger W, Garavito M (1981) Enhancement of structural preservation and immunocytochemical staining in low temperature embedded pancreatic tissue. J Histochem Cytochem 29:663–671
Ruel JC, Joseleau JP (1984) Use of enzyme-gold complexes for the ultrastructural localization of hemicelluloses in the plant cell wall. Histochemistry 81:573–580
Sedgley M, Clarke AE (1986) Immunogold localization of arabinogalactan protein in the developing style of *Nicotiana alata*. Nord J Bot 6:591–598
Selvendran RR (1985) Developments in the chemistry and biochemistry of pectic and hemicellulosic polymers. J Cell Sci Suppl 2:51–88
Slot JW, Geuze HJ (1985) A new method of preparing gold probes for multiple labeling cytochemistry. Eur J Cell Biol 38:87–93
Stafstrom JP, Staehelin LA (1988) Antibody localization of extensin in carrot cell walls. Planta 174:321–332

Titus DE, Becker WM (1985) Investigation of the glyoxysome-peroxisome transition in germinating cucumber cotyledons using double-label immunoelectron microscopy. J Cell Biol 101:1288–1299
Tokuyasu KT (1980) Immunochemistry on ultra-thin frozen sections. Histochemistry 12:381–403
Vreeland V (1970) Localization of a cell wall polysaccharide in a brown alga with labelled antibody. J Histochem Cytochem 18:371–373
Vreeland V (1972) Immunocytochemical localization of the extracellular polysaccharide alginic acid in the brown seaweed *Fucus distichus*. J Histochem Cytochem 20:358–367
Vreeland V, Laetsch WM (1984) Monoclonal antibodies to seaweed carbohydrates. In: Colwell RR, Pariser ER, Sinsky AJ (eds) Biotechnology of marine polysaccharides. Hemisphere Press, New York
Vreeland V, Slomich M, Laetsch WM (1984) Monoclonal antibodies as molecular probes for cell wall antigens of brown alga *Fucus*. Planta 162:506–517

Oxygen and Hydrogen Isotope Measurements in Plant Cellulose Analysis

L. DA SILVEIRA LOBO STERNBERG

1 Stable Isotopes

Isotopes are atoms containing the same number of protons but a different number of neutrons in their nucleus. Most of us are familiar with unstable isotopes (e.g., Carbon-14) which decay to another element (e.g., Carbon-14 decays to Nitrogen-14). This chapter, however, is concerned with stable isotopes which do not decay and thus produce no radioactivity. The most abundant elements found in plant biomass have two or more stable isotopes. Carbon exists either as carbon-12 (6 protons and 6 neutrons) or carbon-13 (6 protons and 7 neutrons), hydrogen can be found either as deuterium (1 proton and 1 neutron) or protium (1 proton), and oxygen can be found either as oxygen-16 (8 protons and 8 neutrons), oxygen-17 (8 protons and 9 neutrons), or oxygen-18 (8 protons and 10 neutrons). Isotopes having the least number of neutrons are usually called the lighter isotopes, and those having more neutrons are called the heavier isotopes. The natural abundance of the isotope of those elements with the least mass (lower number of neutrons) is greater than those with the larger mass. Thus carbon in the biosphere is comprised of 98.89% carbon 12 and only 1.11% of carbon-13 (Nier and Gulbransen 1939). Hydrogen is comprised of 99.9844% protium and 0.0156% deuterium, and oxygen is comprised of 99.763% oxygen-16, 0.0375% oxygen-17 and 0.1995% oxygen-18 (Hoefs 1973). It is extremely difficult to measure absolute abundance of stable isotopes, thus they are usually measured and expressed relative to a standard. There is an extensive literature in methods of measuring carbon isotope ratios of plant matter. Thus, this chapter will only consider the methodology of measuring oxygen and hydrogen isotope ratios of plant matter. The most common unit of measurement for stable isotopes is the δ notation given as:

$$\delta‰ \equiv [(R_{sample}/R_{standard}) - 1] \times 10^3 \quad (1)$$

where R is either the D/H or $^{18}O/^{16}O$ ratios of the sample and standard respectively. Because of the greater abundance and ease of measurement, isotope ratios $^{18}O/^{16}O$ are usually measured in preference to $^{17}O/^{16}O$. The standard used for δD values is SMOW (Standard Mean Ocean Water). Oxygen isotope ratios can be expressed relative to SMOW or may be expressed relative to the PDB carbonate standard which is the $^{18}O/^{16}O$ ratio of carbon dioxide released from the reaction at 25° C of phosphoric acid and carbonate from the Pee Dee belemnite formation in South Carolina. Isotope ratios of hydrogen and oxygen are determined in a mass spectrometer. In order to measure hydrogen isotope ratios of cellulose, all carbon-bound hydrogens must be converted to hydrogen gas (H_2), whereby they are measured in a mass spectrometer as the abundance of molecules of mass 3 (HD) and of mass 2 (H_2) relative to standard hydrogen gas generated from SMOW. For oxygen isotope ratios determination, all oxygens in cellulose are combined with carbon to make carbon dioxide; the mass spectrometer measures abundances of the carbon dioxide molecule with a mass of 46 [$^{12}C^{18}O^{16}O$] and mass 44 [$^{12}C^{16}O^{16}O$] relative to a standard carbon dioxide gas. The mass spectrometer alternately ionizes the standard and sample gas, and accelerates the ions such that the ions containing heavier isotopes (HD^+ for hydrogen, $^{12}C^{18}O^{16}O^+$ and $^{13}C^{16}O^{16}O^+$, for carbon and oxygen) will have a lower velocity. This stream of ions is passed through a magnetic field. Ions having the heavier isotopes and thus moving at a lower velocity are deflected less than those moving at a higher velocity with the lighter isotopes. Thus the single ion beam is split by the magnetic field into two beams each having ions with different masses. The two beams are collected and their relative abundances converted to current by receptors known as Faraday's cup and this signal is amplified and converted to δ values via a microprocessor. Further information on the functioning of a mass spectrometer can be found in Hoefs (1973) and Gat and Gonfiantini (1981).

2 Isotope Ratios in Plants

Much work has concentrated on the interpretation of carbon isotope ratios in plants (see O'Leary 1981 for a review). Considerably less, however, has been reported in the interpretation of deuterium and oxygen-18 abundance of plant matter. This in part may be due to the complexity of hydrogen and oxygen isotope analysis of plant material which this chapter hopefully will clarify.

Hydrogen and oxygen isotope ratios in plants were initially used to interpret paleoclimates (Libby et al. 1976; Epstein et al. 1977; Schiegl 1972). The reasoning was that the δD and $\delta^{18}O$ values of water available to the plant varies with climate. Locations at higher latitudes with colder climates have water with less deuterium (lower δD values) and oxygen-18 contents (lower $\delta^{18}O$ values), and regions at lower latitudes with warm climates have water with a higher deuterium and oxygen-18 content (Craig 1961; Friedman 1953). It was proposed that if plants are recording the isotope ratios of water available for their growth during synthesis of cellulose, then it should be possible to determine δD and $\delta^{18}O$ values of cellulose from ancient wood and infer ancient values of fresh water (rainwater) and climate (Epstein et al. 1977). Inference of the isotope ratios of environmental waters however is complicated by several factors. For example, it has been observed that the incorporation of deuterium into cellulose is extremely susceptible to metabolic effects. DeNiro and Epstein (1981) have shown that temperature can affect the relationship between δD of cellulose and the δD of water. Estep and Hoering (1980) observed that light quality can affect the relationship between δD values of microalgae and the δD value of the water in which they grew. More recently, stable hydrogen and oxygen isotopes in cellulose have been used to determine physiological states of plants during cellulose synthesis. Ting and Rayder (1982) observed that there are metabolic states of plants which cannot be detected with carbon isotope ratios. Sternberg et al. (1984) and Sternberg and DeNiro (1983) have shown that δD values of cellulose can be of great help in deciphering these subtle metabolic differences between plants. Oxygen isotope ratios in plant matter have been little utilized in deciphering physiological states. However, measurement of oxygen isotope ratios in plant cellulose holds great promise, since it may record the productivity of plants under drought stress. Further information on applications of stable isotope analysis of plant cellulose can be found in Sternberg (1988).

3 Preparation of Samples for Combustion

3.1 Cellulose Extraction

Cellulose is prepared by the method described in Wise (1944). Two to 4 g of plant material is dried at 50° to 70° C and ground in a Wiley mill, after which it is delipified by using organic solvents such as benzene, methanol, or acetone. Delipification of plant material is specially important for tree trunks since they may contain a considerable amount of resins which are depleted in deuterium. After delipification ground material is boiled in 200 ml of distilled water for 2 h, and cooled to 70° C. All procedures in cellulose extraction from here on should be conducted under a well-vented hood. One g of sodium chlorite and 1 ml of acetic acid are added and the mixture kept at 70° C for 1 h. Sodium chlorite and acetic acid are added every hour for a period of about 5 h, until cellulose looks thoroughly bleached. The fuming solution is then cooled and cellulose is allowed to settle at the bottom of the beaker. After a few minutes the top solution is decanted and the residual cellulose is washed at least three times with distilled water. Cellulose is then soaked in a 17% sodium hydroxide solution for 45 min, which solubilizes and removes hemicelluloses. The sodium hydroxide solution is decanted and cellulose is washed several times with distilled water, after which it is incubated for 15 min in a 10% acetic acid solution. After 15 min cellulose is thoroughly washed and either freeze-dried or dried at 50° to 60° C in an oven. Cellulose may be further purified by dissolving it in concentrated phosphoric acid and reprecipitated with distilled water.

3.2 Cellulose Nitration

Analysis of hydrogen isotope ratios in plant cellulose is hampered by the presence of hydroxyl hydrogens. Hydroxyl hydrogens are easily exchangeable with water, and measurements of δD values on total cellulose will not only give δD values of carbon-bound hydrogens but will also give δD values of the hydrogens of the water to which the cellulose was most recently exposed. For this reason hydrogen isotope analysis of cellulose is done on cellulose that has its hydroxyl hydrogens removed by nitration.

3.2.1 Preparing Cellulose for Nitration

100 mg to 200 mg of dried cellulose is loaded in 125 ml ground glass stoppered 125-ml Erlenmeyer flasks. Thirteen to 15 samples can conveniently be done for every nitration. Before nitration, Erlenmeyer flasks with cellulose are kept for about 1 h in a large evacuated vessel to remove any moisture associated with the cellulose.

3.2.2 Nitration Solution

There are two common ways of nitrating plant cellulose. That most commonly used in my laboratory is the acetic anhydride method described by Timmel (1955). The other method is the phosphorus pentoxide method described by Alexander and Mitchell (1949).

A. Acetic Anhydride Method. The glassware for the acetic anhydride method is shown in Fig. 1. All procedures are carried out in a vented hood enclosure. The nitration solution is prepared in a 1000-ml round-bottom flask (Pyrex). This vessel has a stopper with a 125-ml separatory funnel and a glass tube about 7 cm long with a glass wool plug and silica gel which allows venting and yet minimizes moisture entering the nitration solution vessel. 269 ml of fuming nitric acid is added to the solution vessel. The vessel is tightly shut with the rubber stopper-

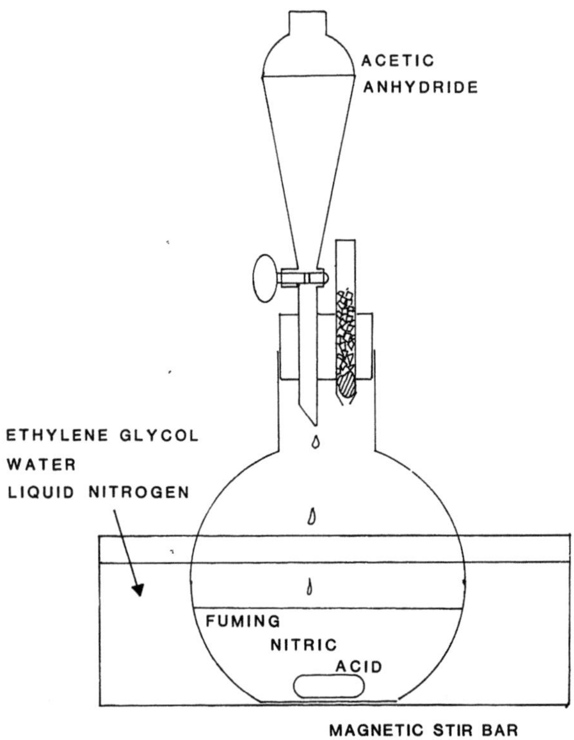

Fig. 1. Preparation of nitration solution. 450 ml of acetic anhydride is slowly dropped into 269 ml of cooled fuming nitric acid at a rate of about 2 drops per second. Nitric acid is stirred by means of a magnetic stir bar. The solution is kept cool with a slush of 50/50 ethylene glycol/water solution and liquid nitrogen

separatory funnel-glass tube assembly. The solution vessel is than immersed in an ethylene glycol / water bath (50% ethylene glycol, 50% water) with liquid nitrogen to the consistency of syrup and a temperature of about −20° C. This whole assembly is placed over a magnetic stirrer and the fuming nitric acid stirred and cooled. Acetic anhydride in the separatory funnel is slowly dropped into the cooled fuming nitric acid at a rate of about 1 to 2 drops per second. The solution is constantly kept cool by adding liquid nitrogen to the ethylene glycol / water bath. After adding 450 ml of acetic anhydride to the fuming nitric acid, the nitration solution is ready to be used and the stopper assembly with the separatory funnel and glass tube is replaced with a regular stopper. About 50 ml of this nitration solution is added to each Erlenmeyer flask containing 100 to 200 mg of cellulose, which is promptly shut and sealed on the outside with "Parafilm". The reaction is kept at 5° C for 4 h by immersing the reaction vessels in ice, and stopped by quickly adding an ice/water slush to the reaction vessels until full (it is important to do this under a vented hood). The washing and stabilization of the nitrated cellulose is accomplished by a vacuum filtration assembly of a 1000-ml vacuum filtration vessel and a fritted glass funnel. Contents of the reaction vessels are filtered through the fritted glass funnel and washed with 1 l of water, after which they are soaked in distilled water for 20 min. Cellulose nitrate is than immersed in methanol for another 20 min, washed thoroughly with distilled water, freeze-dried over night, and purified the next day.

Dry cellulose nitrate powder containing impurities and incompletely nitrated cellulose is dissolved in acetone in a 50-ml glass centrifuge tube and centrifuged at a sufficient speed to separate a pellet of undisolved matter. After centrifugation the acetone solution is decanted into a 100-ml beaker. The pellet is redisolved in acetone and the above procedure is repeated several times so as to thoroughly extract all cellulose nitrate. The beaker containing the cellulose nitrate-acetone solution is allowed to sit in a vented hood until the solution thickens to a consistency of syrup. At this stage the solution is vigorously squirted with a stream of water from a wash bottle. Cellulose nitrate which precipitates as a fibrous mass, is thoroughly washed and freeze-dried. The fibrous mass of cellulose nitrate may be made into small packets by compressing the fibrous mass with a pair of tweezers and a spatula.

B. Phosphorus Pentoxide Method. The major difference between this method and the previous one is the nitration solution. The nitration acid mixture is prepared by slowly adding 404 g of phosphorus pentoxide to 1000 g of cold 90% fuming nitric acid. The acid is kept cold by immersion in an ice-water bath, and swirled continuously during addition of phosphorus pentoxide. After adding phosphorus pentoxide, the acid mixture is filtered through glass wool into a glass-stoppered bottle and stored in a cool dark place (Alexander and Mitchell 1949).

Approximately 1 g of cellulose is introduced into 40 g of nitrating mixture and kept in a refrigerator for several hours. Neutralization and precipitation of cellulose nitrate after this reaction is similar to that of the acetic anhydride method.

4 Preparation of Gases for Isotopic Analysis

4.1 Hydrogen

D / H ratios of hydrogen from nitrated cellulose are measured with hydrogen gas. Cellulose nitrate is combusted with cupric oxide and hydrogen from the water of combustion is extracted with a hot uranium furnace and analyzed in the mass spectrometer. The exact method of preparation of this gas is described in the following paragraphs.

Several 9-mm OD pyrex breakseal tubes approximately 30 cm long are prepared ahead of time and annealed at 500° C for 3 h to remove any traces of impurities which could also produce water upon combustion. These tubes are numbered with a diamond pen in order to identify the samples. About 20 mg of cellulose nitrate and 1 g of cupric oxide is loaded into each tube. The tube is partially collapsed about 15 cm from the breakseal end after loading the cellulose nitrate and the cupric oxide. Tubes are connected to the vacuum system via a Cajon 6-UT-6 fitting and evacuated. Care must be taken so as not to have cupric oxide dust entering the vacuum system, and it is prudent to have glass wool between the sample tube and the vacuum system. The tubes are evacuated for a period of about 30 min. The breakseal ends are immersed about 5 cm deep in a boiling water bath for about 1 h with continuous vacuum pumping. A convenient way to accomplish this is to use immersion heaters, such as those used to make one cup of hot beverage, in beakers full of water. After vacuum pumping for about 2 h, sample tubes are sealed and inserted into a muffle furnace and heated to 520° C and maintained at that temperature for 3 h. Samples are ready to be processed in the vacuum system shown in Fig. 2 upon cooling. After combustion and cooling, sample tube is placed in a 14-mm OD pyrex tubing about 30 cm long with a magnetic stir bar on top of it and fused into the line (position A). All compartments of the line are evacuated including the uranium furnace (B), which should be at a temperature of 750° C. Never expose the uranium furnace to atmosphere when it is at 750° C; the uranium, being pyrophoric, will ignite and become completely oxidized. After all compartments are evacuated the beaded U tube trap is immersed in liquid nitrogen and the breakseal cracked with a magnetic stir bar. Carbon dioxide and water will freeze in this trap. The nitrogen can be either pumped out directly via the rotary vacuum pump or pumped via a Toepler pump (C) through a bypass of the uranium furnace (E) and its volume and yield measured. It is important to determine the yield of nitrogen because incomplete nitration (less than 93%) can interfere with the hydrogen isotope ratios (DeNiro 1981). The liquid nitrogen bath in the beaded U tube trap is replaced with a dry ice–organic slush bath after all of the nitrogen gas has been pumped out. Carbon dioxide will be released from the trap at this temperature. Here again it is strongly recommended that the carbon dioxide be saved and its yield measured. In addition, the carbon isotope ratios of cellulose nitrate (a highly purified fraction) can give important results. All gas distillation up to this point is done with the uranium furnace isolated from the whole system via stopcocks 3, 4, and 5. After nitrogen and carbon dioxide have been removed from the combustion products, it is time

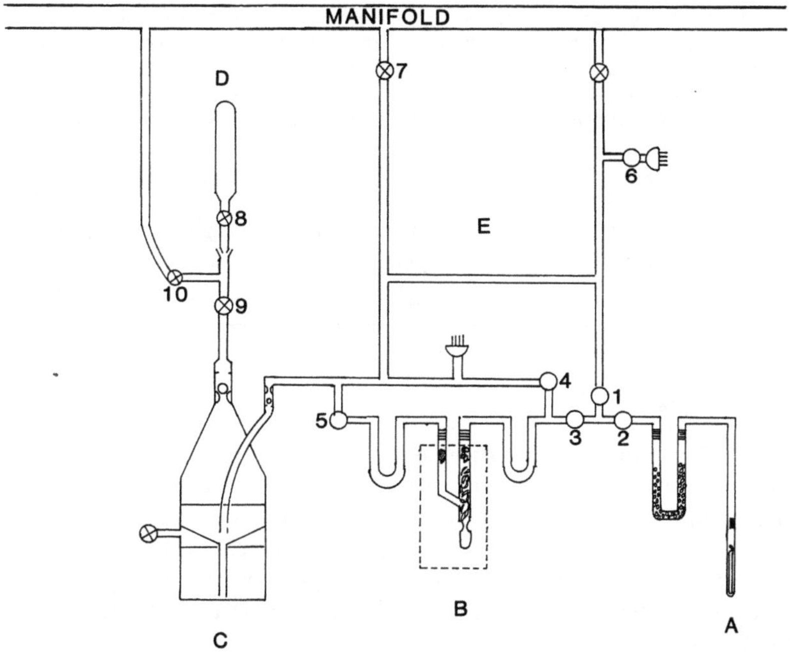

Fig. 2A–E. Vacuum system to separate hydrogen from water of cellulose nitrate combustion. Sample is introduced into vacuum by breakseal method (*A*). Water and carbon dioxide are frozen in beaded U tube trap with liquid nitrogen. Nitrogen is either pumped off or collected with Toepler pump via bypass (*E*) and its volume measured. CO_2 is released from U tube trap by immersing it on a dry ice–organic solvent slush. CO_2 is either pumped off or saved for carbon isotope analysis. Water is then transferred to the U tube trap to the right of the uranium furnace by heating beaded U tube trap. Water is passed through uranium furnace (*B*) by moving liquid nitrogen to the leftmost trap. Any water that is not converted is trapped in the left U tube trap. This procedure is repeated two or three times to insure complete conversion of water to hydrogen by the uranium. Hydrogen is pumped to the compartment (*D*) with the Toepler pump (*C*). *Circles* represent Teflon stopcocks, *circles with an x* represent ground glass stopcocks. *Half circles with lines* represent vacuum gauges

to use the uranium furnace and generate hydrogen gas by the following method. Stopcock 1 is closed and stopcock 3 is opened. The beaded U tube trap is heated with a flame and water released from this trap is frozen in the U tube trap to the right of the uranium furnace in about 3 min. Valve number 3 is shut and the liquid nitrogen bath is taken from the right U tube trap and placed on the left U tube trap, and valve 5 is open, whereupon the hydrogen gas is liberated from the water by the formation of uranium oxide. This gas is pumped with the Toepler pump into compartment D. Any water that has not decomposed will freeze in the leftmost trap and can be passed again through the uranium furnace by shifting the liquid nitrogen from the left to the right trap and opening and closing the appropriate valves. My experience has been that two passes usually quantitatively convert all the water to hydrogen.

Table 1. Percent by weight of nitrogen, carbon, hydrogen and oxygen for cellulose and cellulose nitrate respectively. Formulas to calculate expected percentage of carbon and hydrogen for incompletely nitrated cellulose are also shown

	Cellulose	100% nitration	Variable nitration
Nitrogen	0	14.14	$\frac{\text{moles of } N_2 \times 28}{\text{weight of sample (g)}}$
Hydrogen	6.17	2.36	-0.269 (% N)$+6.17$
Oxygen	49.38	59.26	–
Carbon	44.44	24.24	-1.428 (% N)$+44.44$

4.1.1 Calculation of Yields

After separation and manometric measurement of all gases from the cellulose nitrate combustion it is possible to check the effectiveness of the nitration. Table 1 shows how to calculate the % nitrogen by weight in cellulose nitrate. Complete nitration should give a value of 14.14%, however in most cases the percent nitrogen content is about 13.5%. In some cases the sample can be misweighed, and if that is the case another way of calculating the % nitrogen in the sample is given by the equation below

$$\%N = 47.13 / [(H/N) + 1] \tag{2}$$

where H / N is the volume ratio of hydrogen to nitrogen measured in the sample. This method has the advantage that it does not depend on weighing accuracy, nor manometric mis-calibrations since the H_2 and N_2 volumes are measured in the same manometer. The disadvantage of this system is that the assumption must be made that the nitration product is pure cellulose with various degrees of nitration. Purity of product can be checked by elemental analysis. Shown in Table 1 are various proportions of carbon and hydrogen expected, given that the product is pure cellulose with various degrees of nitration.

4.2 Oxygen

Oxygen isotope ratios are measured on cellulose purified as described in Sect. 3.1. Approximately 6 to 10 mg of cellulose and 0.375 g of mercuric chloride are loaded into Vycor breakseal tubes which have been previously annealed. Tubes containing samples are connected to the vacuum system and evacuated for ½ h at room temperature, and 1 h at 100° C as in the cellulose nitrate samples. Samples are then sealed and ready for combustion. Combustion of samples at 550° C must be done under a vented hood, since if per chance a sample tube breaks, toxic mercuric chloride fumes are given off. Oxygen isotope ratios of cellulose are measured on carbon dioxide generated from the combustion of cellulose and mercuric chloride at 550° C for 6 h. Three gases will be formed from this combustion: HCl, CO_2 and CO. HCl is eliminated by a reaction with isoquinoline, and carbon monoxide is converted to carbon dioxide in a spark chamber. The exact method of carbon dioxide preparation from this combustion follows.

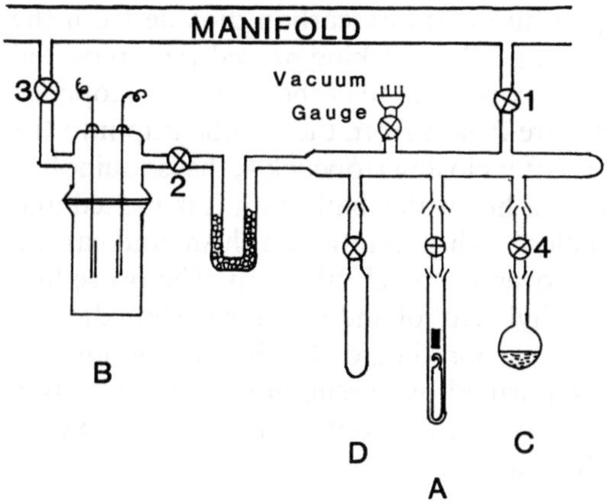

Fig. 3A–D. Vacuum system to process gaseous products of cellulose and mercuric chloride combustion. Sample is introduced to vacuum system by the breakseal method in vessel *A*. Vacuum gauge is shut during this procedure to protect it from HCL. HCL and CO_2 from combustion is frozen with liquid nitrogen in U tube trap. Carbon monoxide is introduced into spark chamber (*B*) with a 5000 V potential. The bottom of the spark chamber is immersed in liquid nitrogen. After complete conversion of carbon monoxide to carbon dioxide all gases are frozen with liquid nitrogen in the isoquinoline vessel (*C*). Isoquinoline reacts with HCL leaving only CO_2. This CO_2 is sealed in vessel *D* and further purified as described in text

Before samples can be processed in an oxygen line, isoquinoline must be degassed. A usual amount of 15 to 20 ml of isoquinoline is placed in a round-bottom flask having a volume of 125 ml as shown in Fig. 3. The flask is connected to the vacuum system through a stopcock (stopcock 4) as shown in Fig. 3. Isoquinoline is cooled to liquid nitrogen temperature and all noncondensible gases are pumped out to vacuum. Stopcock 4 is then shut and isoquinoline is then heated to 100° C by immersing vessel in boiling water. The vessel is then cooled to dry ice temperature with a dry ice–organic solvent slush and gases are pumped out again. The above procedure is repeated several times until noncondensible gases do not come out of the isoquinoline when the vessel is immersed in a liquid nitrogen–ethyl alcohol slush. Never let the isoquinoline vessel, with stopcock 4 open, to the vacuum system at room temperatures, since isoquinoline has a high vapor pressure. Once isoquinoline has been purified samples can be processed in the oxygen line.

The sample in a vycor breakseal tube is connected to the line as shown in Fig. 3 (position A). The line is evacuated and the U tube trap is immersed in liquid nitrogen. Stopcocks 1 and 2 are shut as well as the vacuum gauge. The sample tube is cracked with a magnetic stir bar. HCl and CO_2 will freeze in the U tube trap after about 1 or 2 min. Carbon monoxide, which does not condense at liquid nitrogen temperature, is introduced into the spark chamber by closing stopcock 3 and opening stopcock 2. The bottom of spark chamber is immersed in liquid nitrogen and voltage is first applied at low potential and gradually increased. The spark chamber should glow blue. After most of the carbon monoxide has been converted to carbon dioxide, the glow will flicker and disappear. The spark chamber is brought up to room temperature and any carbon dioxide from the spark chamber is refrozen in the U tube trap. The sparking procedure is repeated again. After all carbon monoxide has been converted to carbon dioxide, stopcock 2 is closed and carbon dioxide and HCl are frozen from the U tube trap into the isoquinoline vessel with liquid nitrogen. After closing stopcock 4, the isoquinoline vessel is brought up to about 90° C with a hot-water bath until it is melted and subsequently it is cooled with an ethyl alcohol-liquid nitrogen slush, and carbon dioxide from the isoquinoline vessel is frozen in the U tube trap. The procedure of melting isoquinoline, cooling it with ethyl alcohol and freezing carbon dioxide is repeated several times until there is no carbon dioxide left in the isoquinoline vessel. Carbon dioxide gas is once more purified by passing it through two ethyl alcohol / liquid nitrogen traps and its volume measured manometrically. Oxygen yields are calculated by the following formula

$$[(\text{Mol } CO_2 / \text{Grams of sample}) \times 64.8] \times 10^2 \quad (3)$$

The usual oxygen yields for cellulose are about 90% or greater, oxygen yields for sucrose are about 95%. After this purification the carbon dioxide is ready for mass spectrometry analysis.

References

Alexander WJ, Mitchell RL (1949) Rapid measurement of cellulose viscosity by the nitration method. Anal Chem 21:1497–1500

Craig H (1961) Isotope variations in meteoric water. Science 133:1702–1703

DeNiro MJ (1981) The effects of different methods of preparing cellulose nitrate on the determination of the D / H ratios of non-exchangeable hydrogen of cellulose. Earth Planet Sci Lett 54:177–185

DeNiro MJ, Epstein S (1981) Isotopic composition of cellulose from aquatic organisms. Geochim Cosmochim Acta 42:495–506

Epstein S, Thompson P, Yapp CJ (1977) Oxygen and hydrogen isotopic ratios in plant cellulose. Science 198:209–1215

Estep MP, Hoering TC (1980) Biogeochemistry of the stable hydrogen isotopes. Geochim Cosmochim Acta 44:1197–1206

Friedman I (1953) Deuterium content of natural waters. Geochim Cosmochim Acta 4:89–103

Gat R, Gonfiantini R (1981) Stable isotope hydrology: deuterium and oxygen-18 in the water cycle. International Atomic Energy Agency, Vienna, Austria Technical Report No 206

Hoefs J (1973) Stable isotope geochemistry. Springer, Berlin Heidelberg New York

Libby LM, Pandolfi LF, Payton PH, Marshall J, Bercker B, Sienhenlist VG (1976) Isotopic thermometers. Nature 261:284–288

Nier AO, Gulbransen EA (1939) Variations in the relative abundance of the carbon isotopes. J Am Chem Soc 61:697–698

O'Leary MH (1981) Carbon isotope fractionation in plants. Phytochemistry 20:553–567

Schiegl WE (1972) Deuterium content of peat as a paleoclimatic recorder. Science 175:512–513

Sternberg L da SL (1988) Oxygen and hydrogen isotope ratios in plant cellulose: mechanisms and application. In: Ehleringer J, Rundel P, Nagy KA (eds) Stable Isotopes in Ecological Research. Springer, Berlin Heidelberg New York Tokyo (in press)

Sternberg L, DeNiro MJ (1983) Isotopic composition of cellulose from C_3, C_4 and CAM plants growing in the vicinity of one another. Science 220:9947–948
Sternberg LO, DeNiro MJ, Ting IP (1984) Carbon, hydrogen and oxygen isotope ratios of cellulose from plants having intermediate photosynthetic modes. Plant Physiol 74:104–107
Timmel TE (1955) Chain-length distributions of native white spruce cellulose. Pulp Paper Mag Can 104–117
Ting IP, Rayder L (1982) Regulation of C_3 to CAM shifts. In: Ting IP, Gibbs M (eds) Crassulacean acid metabolism. Am Soc Plant Physiol, pp 677–679
Wise LE (1944) Wood chemistry. Reinhold, Washington D.C., USA

Analysis of Lignin-Carbohydrate Complexes of Plant Cell Walls

JUN-ICHI AZUMA

1 Introduction

The association between lignin and carbohydrates in the lignified plant cell walls is of fundamental importance for characterizing the structure and function of cell walls, improving the pulping and bleaching processes and utilizing lignocellulosic plants as biomass.

Numerous studies have been made in the attempt of finding a solution to this problem. Björkman (1957a, b) presented a method for the preparation of hemicelluloses accompanied by lignin, and designated this material as "lignin-carbohydrate complex" (LCC). The results of investigations on LCC have been reviewed by Merewether (1957), Gruschnikov and Shorygina (1970), Lai and Sarkanen (1971), Adler (1977), and Fengel and Wegener (1984). Recent evidence together with much of these information strongly supports the existence of covalent bonds between lignin and carbohydrates. Extensive model experiments on the synthesis of lignin-carbohydrate bondings yielded valuable information about the nature of the linkages (Freudenberg and Grion 1959; Freudenberg and Harkin 1960; Tanaka et al. 1976, 1979; Hemmingson et al. 1978; Hemmingson 1979; Košiková et al. 1979; Ohara et al. 1980; Katayama et al. 1980; Leary et al. 1983a, b; Enoki et al. 1983).

The proposed linkages between lignin and carbohydrates are classified into six types: (1) Glycosidic linkages between sugars and hydroxyl groups at the side-chain of lignin (Kawamura and Higuchi 1954; Smelstorius 1974; Enoki et al. 1983); (2) Phenolic glycosidic linkages (Hayashi and Tachi 1958; Hayashi 1961; Košiková et al. 1969, 1977); (3) Ester linkages between the carboxyl groups of uronic acids and the hydroxyl groups at β- or γ-position of lignin; (4) Benzyl ether linkages (Lundquist et al. 1979; Košiková et al. 1979; Eriksson et al. 1980; Minor 1982); (5) Benzyl ester linkages (Eriksson et al. 1980; Minor 1982; Meshitsuka et al. 1982); (6) Hemiacetal or acetal linkages between carbonyl groups at β-position of lignin and hydroxyl groups of sugar (Bolker 1963). Among these types, the possibilities of glycosidic linkages, benzyl ether and benzyl ester linkages, have been demonstrated with model compounds. However, there were considerable difficulties in obtaining unambiguous results on the nature and frequency of such linkages in the native plant cell walls. The usual strategy is to isolate LCC in a homogeneous state. Previously, it has been attempted to isolate LCC from various origins; such as (1) Finely-divided plant meal (Pew and Weyna 1962; Brownell 1971; Morrison 1973; Eriksson and Lindgren 1977; Joseleau and Gancet 1981; Minor 1982; Meshitsuka et al. 1982; Obst 1982; Ford 1986); (2) Milled wood lignin (MWL) (Wegener and Fengel 1979; Lundquist et al. 1979, 1980); (3) Residues left over after enzyme and chemical treatment of the plant cell walls (Kreicberga et al. 1974; Mouttet and Robert 1978; Sergeeva et al. 1979); (4) Chlorite and peroxyacetic acid delignification liquors (Wegener 1976; Glinski and Nicholls 1977); (5) Alkaline extracts from holocellulose and wood (Merewether and Samsuzzaman 1972; Smelstorius 1974; Fenckl and Fengel 1982; Linnell et al. 1966; Kringstad and Cheng 1969); (6) Pulps (Minor 1986; Iversen and Wännström 1986); (7) Pulping liquors (Simonson 1971a, b; Košiková et al. 1984; Iversen and Westermark 1985); (8) Derivatized wood (Košiková et al. 1969; Košikova et al. 1973); (9) In vitro synthesized dehydrogenative polymers (DHP) (Joseleau and Kesraoui 1986), (10) Cambium tissue (Fukuda and Tomimura 1976), and (11) Tissue culture (Fukuda and Kanda 1976; Whitmore 1978). Among these starting materials, the finely divided plant meal and MWL were superior to the others because of lack of any chemical and biochemical treatment which will cause splitting of the possible linkages between lignin and carbohydrates, although the breaking of some carbon-carbon bonds in the hexoses during milling could not be excluded (Iversen 1985). To studies in this direction a new field has been opened by the development of new convenient methods for the isolation and fractionation of LCC (Azuma and Koshijima 1985, 1988; Azuma et al. 1981, 1985 a, b; Mukoyoshi et al. 1981; Kato et al. 1984; Watanabe et al. 1987).

In the case of monocotyledons, including graminaceous plants (Harris and Hartley 1980), and rarely of dicotyledons (Hartley and Harris 1981), the linkage analysis is more complicated than in woody plants because of the presence of phenolic acids, such as ferulic and *p*-coumaric acids, in their cell walls (Hartley 1987, McNeil et al. 1984), leading to their limited rate of biodegradation (Hartley 1981; Neilson and Richards 1982; Brice and Morrison 1982). Most of the *p*-coumaric acid in bamboo and grass lignins was esterified to the γ-positions of the side-chains of lignin (Shimada et al. 1971; Nakamura and Higuchi 1976). It has also been suggested that *p*-coumaric acid is esterified at C-1 in xylan (Ford 1986). Ferulic acid has been shown to be ether-linked to lignin (Scalbert et al. 1985, 1986). It has also been suggested that these phenolic acids are esterified to polysaccharides (Markwalder and Neukom 1976; Hartley and Jones 1977). Fry (1982, 1983 a) has isolated two feruloylated disaccharides of D-galactose and L-arabinose from the primary cell wall of spinach. A feruloylated trisaccharide, *O*-[5-*O*-(*trans*-feruloyl)-α-L-arabinofuranosyl]-(1→3)-*O*-β-D-xylopyranosyl-(1→4)-D-xylopyranose, has been isolated from enzymatic hydrolysate of bagasse LCC (Kato et al. 1983), *Zea mays* shoots (Kato and Nevins 1985), barley aleurone layers (Smith and Hatley 1983) and straw (Mueller-Harvey et al. 1986), and wheat bran (Gubler et al. 1985). A *p*-coumaroylated trisaccharide, having the same sequence and linkages, was also isolated from barley straw (Mueller-Harvey et al. 1986). A new feruloylated tetrasaccharide, *O*-β-D-xylopyranosyl-(1→4)-*O*-[5-*O*-(*trans*-feruloyl)-α-L-arabinofuranosyl-(1→3)]-*O*-β-D-xylopyranosyl-(1→4)-D-xylopyranose, was recently isolated from the enzymatic hydrolysate of bagasse LCC (Kato et al. 1987 a). Since these phenolic acid-containing oligosaccharides contain the labile ester linkages, special care is needed for their isolation in the native state.

This chapter deals with new methods for isolation, fractionation and structural determination of LCC, including linkage analysis among lignin, phenolic acids and carbohydrates.

2 Isolation and Fractionation of Lignin-Carbohydrate Complexes from the Milled Wood Lignin Fraction

Of the following methods the procedure for LCC (Fig. 1) was also applicable to LCC containing phenolic acids.

2.1 Isolation of Lignin-Carbohydrate Complexes from Milled Wood Lignin Fraction

2.1.1 Preparation of Extractive-Free and Depectinated Plant Meal (Method of Koshijima et al. 1972; Azuma and Koshijima 1988)

Air-dried plant is milled to 24–80 mesh in a Willy Mill and extracted with ethanol-benzene (1 : 2, v/v) for 48 h by using soxhlet extraction apparatus. The extractive-free plant meal (1 kg) is then depectinated by extraction with 10 l of 0.25% (w/v)

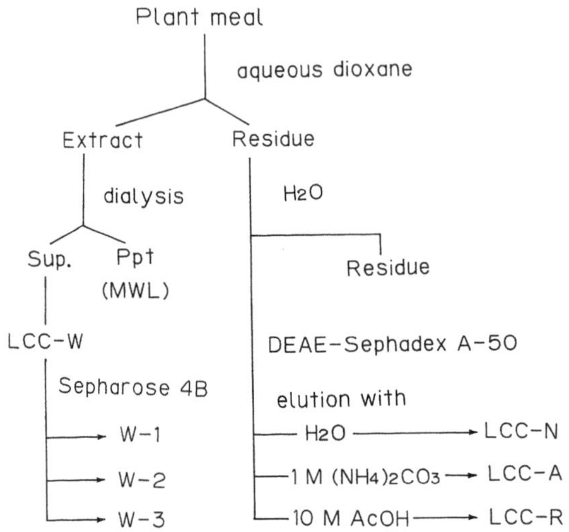

Fig. 1. Procedures for isolation and fractionation of lignin-carbohydrate complexes

aqueous potàssium acetate at 60° C for 24 h. The plant meal is recovered by filtration through cloth filter, washed with distilled water and acetone, and dried thoroughly at 40° C. This depectination procedure should be omitted if pectic materials are required for analysis. The extractive-free and depectinated plant meal was then treated in a vibratory ball-mill for 48 h under nitrogen with external cooling by tap water.

2.1.2 Extraction of Lignin-Carbohydrate Complexes from Milled Wood Lignin Fraction (Method of Azuma et al. 1981; Azuma and Koshijima 1988)

The finely divided plant meal (1 kg) was suspended in 10 l of 80% (v/v) aqueous 1,4-dioxane and stirred vigorously for 48 h at room temperature. The extracted material was recovered by filtration through filter paper (Whatman No. 2). The residual plant meal was washed and reextracted with the same solvent. The filtrates and washings were dialyzed against distilled water after concentration to a small volume by evaporation below 40° C. MWL precipitated during dialysis was removed by centrifugation at 8000 g for 15 min at 10° C and washed three times with distilled water. The clear brownish solution is concentrated to a small volume and lyophilized. This LCC was partially purified by solubilization in 50% (v/v) aqueous 1,4-dioxane and extraction with chloroform. The aqueous layer was dialyzed against distilled water and lyophilized. Purification was further carried out by solubilization of LCC (10 g) in 56 ml of pyridine-acetic acid-water (9:1:4, v/v) and extraction with 6 vol of chloroform (Lundquist and Simonson 1975). The aqueous solution was recovered by centrifugation at 8000 g for 15 min at 10° C after addition of 400 ml of distilled water. The chloroform and intermediate gelatinous layers were washed by repeated addition of distilled water and centrifugation. The combined aqueous solution was dialyzed against distilled water and lyophilized to give brownish LCC-W which was completely soluble in water.

2.1.3 Fractionation of Lignin-Carbohydrate Complexes

A. Fractionation of Lignin-Carbohydrate Complexes by Size Exclusion Chromatography (Method of Azuma et al. 1981; Azuma and Koshijima 1988). LCC-W (1 g) was solubilized in 15 ml of 0.025 M sodium phosphate buffer, pH 6.8, and applied on a column (5.2 × 95 cm) of Sepharose 4B preequilibrated with the same buffer. Elution was carried out at 60 ml/h at room temperature and fractions (20 ml) were collected. The lignin content was monitored by measuring absorbance at 280 nm with UV detector. A portion, usually 0.05–0.2 ml of each fraction, was analyzed for carbohydrate content by the phenol-sulfuric acid assay (Dubois et al. 1956). The separated LCC subfractions were pooled, dialyzed against distilled water and lyophilized (Fig. 2, Table 1).

B. Fractionation of Lignin-Carbohydrate Complexes by Hydrophobic Chromatography. Hydrophobic chromatography is a technique which allows the separation of biopolymers based on the differences in size and distribution of available hydrophobic regions on the biopolymers (Shaltiel 1974; Hjertén 1976; Hofstee 1976). Since LCC contains hydrophilic carbohydrate and hydrophobic lignin in the same molecule, it is expected to have amphipathic property. Yaku and Koshijima (1971) and Yaku et al. (1979, 1981) showed the presence of LCC and LCC fragment which form micells in an aqueous solution, and the occurrence of solubilization of lignin by LCC. Azuma et al. (1981) also showed that pine LCC-W

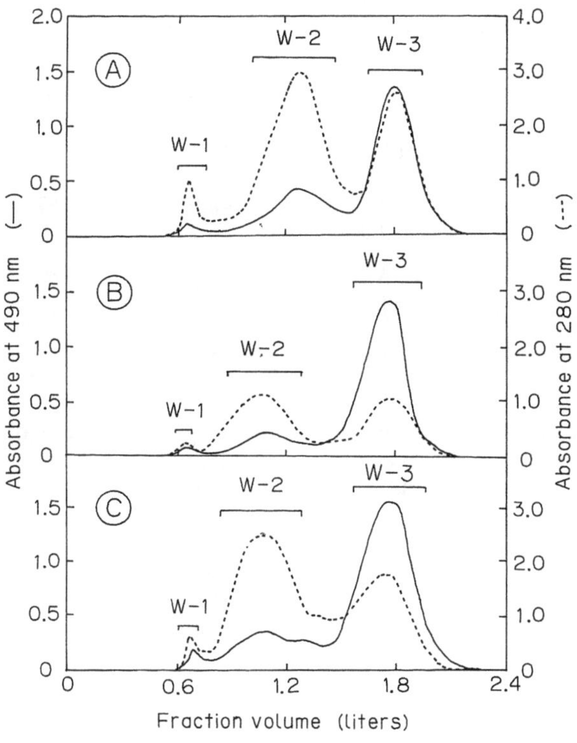

Fig. 2 A–C. Fractionation of lignin-carbohydrate complexes (LCC) by size exclusion chromatography on Sepharose 4B. LCC from Japanese cypress (*Chamaecyparis obtusa*): Cpd. **A**. LCC from Acacia (*Acacia auriculaeformis*): Cpd. **B**. LCC from bamboo (*Phyllostachys pubescens*): Cpd. **C**. Details see text (Sect. 2.1.3 A)

Table 1. Properties of the lignin-carbohydrate complexes isolated from the milled wood lignin fraction

Components	Sapwood of Japanese cypress (*Chamaecyparis obtusa*)		
	LCC-W	W-2	W-3
Distribution ratio	100	40.5	38.1
Carbohydrate content[b]			
Neutral sugar	45.1	24.5	48.2
Uronic acid	6.5	5.8	8.0
Lignin content[b]			
Klason	57.7	49.8	36.3
Acid soluble	4.7	4.5	5.5
Esterified acid content[b]			
Acetic acid	2.1	2.0	2.8
p-Coumaric acid	–	–	–
Ferulic acid	–	–	–
$[\alpha]_D^{25}$ (*c* 0.5, 50% 1,4-dioxane, degree)	−3.4	−4.4	−5.5
S°_{20} (S)	n.d.[d]	15.5	0.3
$\bar{M}w$ ($\times 10^{-4}$)	n.d.	43.0	0.4
Neutral sugar composition[c]			
L-Rhamnose	3.8	3.5	7.8
L-Arabinose	22.6	15.1	24.6
D-Xylose	19.8	17.1	13.4
D-Mannose	26.1	27.0	28.7
D-Galactose	17.3	24.1	15.5
D-Glucose	10.4	13.3	10.0

Components	Acacia (*Acacia auriculaeformis*)		
	LCC-W	W-2	W-3
Distribution ratio	100	23.6	45.1
Carbohydrate content[a]			
Neutral sugar	69.1	48.9	87.1
Uronic acid	9.5	6.5	9.6
Lignin content[b]			
Klason	40.1	49.2	27.6
Acid soluble	11.2	11.9	12.8
Esterified acid content[a]			
Acetic acid	5.9	1.7	7.9
p-Coumaric acid	–	–	–
Ferulic acid	–	–	–
$[\alpha]_D^{25}$ (*c* 0.5, 50% 1,4-dioxane, degree)	−30.1	−3.3	−35.0
S°_{20} (S)	n.d.[d]	20.0	0.4
$\bar{M}w$ ($\times 10^{-4}$)	n.d.	77.0	0.4
Neutral sugar composition[c]			
L-Rhamnose	1.9	1.6	1.8
L-Arabinose	1.3	2.0	0.9
D-Xylose	93.4	88.9	95.6
D-Mannose	0.5	1.0	0.2
D-Galactose	1.2	2.7	0.6
D-Glucose	1.6	3.7	0.8

Table 1 (continued)

Components	Bamboo *(Phyllostachys pubescens)*[f]		
	LCC-W	W-2	W-3
Distribution ratio	100	23.0	51.7
Carbohydrate content[a]			
Neutral sugar	48.0	34.7	67.9
Uronic acid	2.0	1.6	3.6
Lignin content[b]			
Klason	22.7	40.1	6.3
Acid soluble	3.1	3.3	3.4
Esterified acid content[a]			
Acetic acid	6.4	6.0	6.6
p-Coumaric acid	4.8	5.8	0.7
Ferulic acid	0.3	0.2	0.4
$[\alpha]_D^{25}$ (*c* 0.5% 1,4-dioxane, degree)	−37.7	−19.1	−51.1
S_{20}° (S)	n.d.[d]	17.5	1.0
$\bar{M}w$ ($\times 10^{-4}$)	n.d.	70.0	0.4
Neutral sugar composition[d]			
L-Rhamnose	t[e]	t	t
L-Arabinose	4.7	5.5	4.5
D-Xylose	94.2	94.3	95.0
D-Mannose	t	t	t
D-Galactose	0.7	0.1	0.3
D-Glucose	0.4	0.1	0.2

[a] Values are given on dry-weight basis.
[b] Values are determined by the Tappi standard methods of T. 222os-74 and UM250 and given on dry-weight basis.
[c] Values are given on relative percent-weight basis.
[d] n.d. = not determined.
[e] t = trace.
[f] (Azuma et al. 1985b).

and its subfractions are adsorbed on phenyl- and octyl-Sepharose CL-4B gels. Here, the fractionation of LCC based on the criteria of the adsorbing ability on these hydrophobic gel columns (15 × 1.8 cm) is presented (Takahashi et al. 1982; Azuma and Koshijima 1985). The columns were equilibrated with 0.025 M sodium phosphate buffer, pH 6.8, containing 0.8 M ammonium sulfate. One-hundred milligrams of LCC was solubilized in 5 ml of the equilibrating buffer and applied on the column. The column was thoroughly washed with the same buffer to elute the unadsorbed fraction. Ammonium sulfate, which promoted the adsorption of LCC, was omitted in the case of graminaceous and dicotyledonous plants because this salt prevented solubilization of LCC. The adsorbed materials were then stepwise eluted with 0.025 M sodium phosphate buffer containing (1) 15% of 2-ethoxyethanol and 0.6 M ammonium sulfate, (2) 30% of 2-ethoxyethanol and 0.4 M ammonium sulfate, (3) 45% of 2-ethoxyethanol and 0.2 M ammonium sulfate, and (4) 50% of 2-ethoxyethanol to obtain four fractions. Fractions (6 ml) were collected at a flow rate of 40 ml/h, and both lignin and carbohydrate content in each fraction were respectively determined by measuring absorbance

Table 2. Lignin content of the lignin-carbohydrate complex fractions separated by hydrophobic chromatography of the W-2 fraction of Japanese cypress *(Chamaecyparis obtusa)* (Fig. 1, Table 1)

Component	Original	Phenyl-Sepharose				
(%)	W-2	P-I	P-II	P-III	P-IV	P-V
Lignin content	49.8	11.2	18.3	34.3	56.4	67.1
		Octyl-Sepharose				
		O-I	O-II	O-III	O-IV	O-V
		15.3	28.5	41.6	57.0	60.8

at 280 nm, and by the phenol sulfuric acid method (Dubois et al. 1956) after 2-ethoxyethanol was removed prior to analysis. The eluted fractions were exhaustively dialyzed against distilled water and lyophilized (Table 2).

2.1.4 Comments on Hydrophobic Chromatography

The existence of lignin-carbohydrate bondings in LCC was supported by overlapping of the two elution profiles due to lignin and carbohydrates. When various denaturating reagents were used to desorb the adsorbed LCC, the desorbing ability increased in the following order: urea < guanidine hydrochloride < 1,2-ethanediol < 1,4-dioxane < 2-ethoxyethanol (Takahashi et al. 1982). The extent of adsorption by phenyl- and octyl-Sepharose CL-4B gels was exclusively dependent on their lignin content (Azuma et al. 1981, 1985 a, b; Azuma and Koshijima 1985; Kato et al. 1984, 1987 b; Takahashi et al. 1982; Mukoyoshi et al. 1981) (Table 6.2). The aromatic-aromatic interactions, together with hydrophobic interactions, contribute significantly to the bonds with phenyl-Sepharose in contrast to octyl-Sepharose in which hydrophobic interactions operated exclusively if the lignin content was high enough to interact with 1-octyl ligands. Gradient, instead of stepwise, elution has also been shown to be effective for fractionation of LCC (Tanabe and Kobayashi 1987).

2.2 Isolation of Lignin-Carbohydrate Complexes from the Residual Plant Meal Previously Extracted with Aqueous 1,4-Dioxane

Björkman (1957 a, b) presented a method for isolating LCC (Björkman LCC) from finely divided wood meals. Dimethyl formamide (DMF) or dimethylsulfoxide (DMSO) has been used to extract LCC. The extraction conditions of Björkman LCC have been fully analyzed and improved by Koshijima et al. (1972). The yield of Björkman LCC increased with increase of the milling time, but milling for more than 48 h resulted in an appreciable chain scission of the polymers in LCC. DMF has been shown to dissolve only lower-molecular-weight

Table 3. Properties of the lignin-carbohydrate complexes isolated from the residual pine wood *(Pinus densiflora)* meal previously extracted with aqueous 1,4-dioxane

Component	Lignin-carbohydrate complexes (LCC)[d]			
	Original	LCC-N	LCC-A	LCC-R
Distribution ratio	100	43.3	48.7	2.1
Carbohydrate content (%)[a]				
Neutral sugar	80.0	95.5	76.0	41.5
Uronic acid	4.2	–	6.4	1.9
Lignin content (%)[a]	17.9	3.7	26.6	43.6
Acetyl content (%)[a]	3.3	7.6	–	–
$[\alpha]_D^{25}$ (*c* 0.5, water, degree)	−15.5	−28.2	−11.4	− 8.0
S_{20}^{o}	n.d.[c]	0.9	0.8	n.d.
$\bar{M}w$ ($\times 10^{-4}$)	1.2	1.2	1.1	n.d.
Carbohydrate composition[b]				
L-Arabinose	3.8	–	6.7	5.2
D-Xylose	21.0	–	45.6	24.9
D-Mannose	52.3	74.0	29.8	33.0
D-Galactose	7.0	4.6	7.9	16.8
D-Glucose	15.9	21.5	10.0	20.0

[a] Values are given on dry-weight basis.
[b] Values are given on relative percent-weight basis.
[c] n.d. = not determined.
[d] (Watanabe et al. 1987).

components (Mn; 400–5,000) in comparison with DMSO (Mn; 10,000–15,000). Björkman's procedure for extraction of LCC was excellent and reliable but required time-consuming and rather tedious operations. To eliminate this drawback, Watanabe et al. developed a new convenient technique to isolate water-soluble LCC from the residual wood meal previously extracted with 80% aqueous 1,4-dioxane (Koshijima et al. 1984; Watanabe et al. 1987). The chemical properties of this LCC has been shown to be similar to those of the Björkman LCC (Table 3). The following recommended procedure (Fig. 1) was slightly modified by alteration in extraction temperature to minimize migration of esterified acetyl groups (Reicher et al. 1984).

2.2.1 Recommended Procedure for Isolation of Lignin-Carbohydrate complexes from the Finely Divided Plant Meal Previously Extracted with Aqueous 1,4-Dioxane

The plant meal (1 kg), previously extracted twice with 80% aqueous 1,4-dioxane (Sect. 2.1.2), was stirred in 10 l of distilled water for 24 h at 25° C. After filtration through filter paper (Whatman No. 2), the residue was washed three times with 1 l of distilled water and then reextracted with distilled water for 24 h at 25° C. Both extracts were combined and evaporated to a small volume and poured into 5 vol of ethanol. The precipitate formed was recovered by centrifugation, washed with ethanol, followed by petroleum ether to obtain water-soluble LCC as a white powder. LCC that remained in the residual plant meal could also be extracted further with DMSO or DMF following the procedure of Koshijima et al. (1972).

2.2.2 Fractionation of Lignin-Carbohydrate Complexes by Ion Exchange Chromatography

(Method of Tanaka and Koshijima 1972; Yaku et al. 1976; Koshijima et al. 1976)

Ion exchange chromatography of LCC was carried out on a column (10 × 80 cm) of DEAE-Sephadex A-50 (carbonate form). LCC (20 g) was solubilized in distilled water (70 ml), applied on a column, and eluted with distilled water at 500 ml/h until minus test by the phenol-sulfuric assay (Dubois et al. 1956). The eluate was concentrated to a small volume and poured into 5 vol of ethanol to precipitate the neutral fraction which was recovered by centrifugation to obtain a neutral fraction containing 1%–4% of lignin (LCC-N). The column was next eluted with 1 M ammonium carbonate solution. The eluate was concentrated to a small volume and subjected to ethanol precipitation to obtain acidic LCC (LCC-A). The column was then eluted with 10 M acetic acid. The eluate was dialyzed against distilled water, concentrated to a small volume, and subjected to ethanol precipitation to obtain residual acidic LCC containing 40%–51% of lignin (LCC-R).

2.3 Purity and Molecular Weight Determination of Lignin-Carbohydrate Complexes

Purity of the isolated LCC was checked by gel permeation chromatography (GPC), analytical ultracentrifugation and electrophoresis. GPC was conducted on TSK gel columns (30 × 7.8 cm) of G5000HXL, G4000HXL and G2000HXL (TOSOH) using DMF as an eluting solvent. The columns were calibrated against polystyrenes (TSK Standard Polystyrenes) and pullulans having known molecular weights (Shodex Standard P-82). Be careful not to use dextrans because of their low solubilities in DMF. Velocity sedimentation was performed in 0.025 M sodium phosphate buffer, pH 6.8, at 42,040 rpm and 20° C using a Beckman Model E ultracentrifuge equipped with Schlieren optics. Synthetic boundary cell (12 mm, SK-SB-1 capillary type) was used for lower-molecular-weight LCC (MW: $<2.0 \times 10^4$), while single-sector cell (12 mm, SK4) was used for higher-molecular-weight LCC (MW: $>2.0 \times 10^5$). The molecular weights of LCC were also obtainable by GPC and sedimentation equilibrium (Yphantis 1964). Electrophoresis of LCC was carried out according to the procedure of Dudman and Bishop (1968). LCC was pre-colored with Procion Blue MX-R and purified by Sephadex G-25 using distilled water as an eluent. The freshly stained LCC was then subjected to electrophoresis on glass-fiber filter paper (Whatman GF/C) in 0.05 M borate buffer at 50 mA for 20 min, and 0.1 M borax-sodium chloride buffer at 100 mA for 20 min.

3 Methods for Analyzing Constituent Units of Lignin-Carbohydrate Complexes

3.1 Determination of Component Sugars

Total carbohydrate and uronic acid contents were determined by the phenol sulfuric acid (Dubois et al. 1956) and harmine (Wardi et al. 1974) methods, respectively. The modified carbazole method (Galambos 1967) is also effective, but interference with neutral sugars is higher than with the harmine method.

For analysis of neutral sugar composition, 10 mg of LCC was hydrolyzed with 0.5 M sulfuric acid at 100° C for 6 h in a sealed tube. After cooling, a known amount of methyl *β*-D-glucopyranoside (Kusakabe et al. 1977) was added to the hydrolysate which was then neutralized with barium carbonate. The precipitated barium sulfate was removed by filtration and washed with distilled water. The filtrate and washings were combined and deionized with small columns of Dowex 50 × 8 (H^+ form) and Dowex 1 × 8 (acetate form) jointed in tandem and evaporated to dryness. The mixture of neutral sugars was reduced with sodium borohydride overnight at room temperature. After excess reagent was decomposed by acetic acid, the boric acid formed was converted into volatile methyl borate by addition of methanol, which was removed by repeated evaporation. The alditols were then acetylated with a mixture (1:1, v/v) of pyridine: acetic anhydride for 1 h at 100° C. After acetic anhydride was removed by co-distillation with toluene, alditol acetates were separated by gas-liquid chromatography (GLC) using a glass column (0.3 × 200 cm) of 3% ECNSS-M on Gas Chrom Q (80–100 mesh) at 180°–190° C (Sloneker 1972).

Identification of uronic acid was carried out by the method of Sone et al. (1978). LCC was hydrolyzed with 90% formic acid for 16 h at 100° C followed by 4 M trifluoroacetic acid for additional 6 h. The neutral monosaccharide and uronic acid were reduced with sodium borohydride and separated into two parts. One part was acetylated and the resulting alditol acetates were analyzed by GLC. The other part was heated with methanol in the presence of dried Dowex 50 × 8 (H^+ form) (ca. 100 mg/ml) for 1 h at 100° C. After the resin was removed, aldonic acid methyl esters were reduced with sodium borohydride. The resulting alditols were then acetylated and analyzed by GLC. The uronic acid was identified and quantitated by comparison of the two results obtained with and without methanolysis.

3.2 Determination of Configuration of Sugars

The mixtures of neutral sugars obtained after hydrolysis as described above were separated by preparative TLC with ethyl acetate-pyridine-water (3:2:1, v/v) as an irrigant on Silica Gel 60 (Merck No. 5745). After two developments, a small slice of silica gel plate having 1 cm in width was cut from both edges, and sugars were located by spraying with *p*-anisidine hydrochloride (Hough et al. 1950) followed by heating for 10 min at 120° C. The individual portions of silica gels corresponding to the separated monosaccharides were scraped off with a spatula and eluted with the developing solvent. Each eluant was evaporated to dryness to give individual monosaccharides, which were further purified by passing through a small charcoal column. The monosaccharides were recovered by washing with distilled water.

There are several reports on determination of the configuration of monosaccharides (Bebault et al. 1973; Gerwig et al. 1978, 1979; Leontein et al. 1978; Oshima and Kumanotani 1981; König et al. 1981; Little 1982; Schweer 1982; Oshima et al. 1983). The procedure described by Leontein et al. (1978) is recommended because of the stability of the derivatized products. One milligram of each monosaccharide was treated with 0.5-ml of (+)-2-octanol and one drop of trifluoroacetic acid overnight at 130° C in a sealed tube. After evaporation, (+)-2-octyl D- and L-glycosides were acetylated and separated on a column (25 m × 0.28 mm) of SP-1000 at 200° C.

3.3 Structural Determination of Monomeric Units of Lignin

Monomeric units of lignin were determined by alkaline nitrobenzene oxidation. The following method (Sudo et al. 1985) has an advantage over the conventional method (Leopold 1952) in its requirement of no – derivatization. LCC containing 40 mg of lignin was heated with nitrobenzene (0.24 ml) in 2 M sodium hydroxide (4.0 ml) at 170° C for 2 h using a stainless steel microbomb (capacity 10 ml). After cooling, the reaction mixture was taken out with distilled water and extracted with ethyl ether to remove excess nitrobenzene. The oxidation products were extracted with ethyl ether after acidification of the solution with 2 M hydrochloric acid and separated by high performance liquid chromatography (HPLC) on an octadecyl silica (ODS) gel column (25 × 0.46 cm) of Finepack SIL C-18 (JASCO) using water-acetonitrile-acetic acid (80:15:5, v/v) as an eluting solvent at 40° C.

3.4 Methylation Analysis

LCC was methylated by the modified method (Lindberg 1972) of Hakomori (1964) followed by the method of Kuhn (Wallenfels et al. 1963). The fully methylated LCC was hydrolyzed and the partially methylated alditol acetates were prepared according to Lindberg (1972). The partially methylated alditol acetates was separated by GLC on a column of SP-2330 (30 m × 0.28 mm) (Fox et al. 1983) and analyzed by gas-liquid chromatography–mass spectrometry (GC-MS) (Jansson et al. 1976) using a column (25 m × 0.22 mm) of CP-Sil88 (Lomax et al. 1985).

3.5 Periodate Degradation Analysis

LCC and deesterified LCC prepared by treatment of LCC with 0.25 N sodium hydroxide for 4 h at 56° C were stirred with 50 ml of 0.05 M sodium metaperiodate in the dark for 20 days at 5° C. Ethylene glycol was added to decompose excess periodate. Oxidation product was reduced with sodium borohydride, exhaustively dialyzed against distilled water and lyophilized. The product (5 mg) was hydrolyzed with 2 ml of 2 M trifluoroacetic acid for 1 h at 120° C. The hydrolysate was analyzed by GLC and GLC-MS after reduction and conversion into the corresponding alditol acetates.

3.6 Determination of Esterified Components

The isolated LCC contained esterified acetic acid due to partially O-acetylated hemicellulose and esterified phenolic acids, such as *p*-coumaric acid and ferulic acid in the case of graminaceous plants (Azuma et al. 1985a; Azuma and Koshijima 1988; Kato et al. 1984, 1987b). For analysis of O-acetyl groups, the sample (10 mg) was hydrolyzed with 1 M hydrochloric acid (0.3 ml) in a sealed tube for 2 h at 100° C and a portion of this hydrolyzed solution was directly applied on a GLC column (20% tetramethylcyclobutanediol adipate-4% phosphoric acid on Chromosorb W; 2 m × 0.3 cm) at 120° C (Tomoda et al. 1975) or HPLC column (Aminex HPX-87H; 30 cm × 7.8 mm; Bio Rad) using 10% acetonitrile in 0.02 M sulfuric acid as an eluent at 65° C.

Esterified phenolic acids were saponified with 1 M sodium hydroxide solution and separated by GLC (Hartley and Jones 1975) and HPLC (Hartley 1987).

3.7 Determination of Etherified Phenolic Acids

Ether-linked phenolic acid was released by acidolysis (Lundquist 1970), which was known to split arylglycerol-β-aryl ethers as well as benzyl aryl ethers (Adler

Table 4. Phenolic acid content of the lignin-carbohydrate complexes isolated from sugarcane bagasse

Phenolic acid	Lignin-carbohydrate complexes[a]						
	LCC-W	W-2	W-3	Björkman LCC	LCC-N[b]	LCC-A[b]	LCC-R[b]
Released by saponification							
p-Coumaric acid	4.82	6.15	1.43	3.02	2.54	0.72	4.26
Ferulic acid	1.34	0.74	1.68	0.64	0.50	1.13	0.59
Released by acidolysis after saponification							
p-Coumaric acid	0.10	0.09	0.04	0.09	0.09	0.05	0.10
Ferulic acid	0.15	0.10	0.12	0.16	0.07	0.26	0.22

[a] Values are given on percent-weight basis.
[b] Lignin-carbohydrate complex fractions separated from Björkman LCC.

et al. 1957; Lundquist and Lundgren 1972). The residue (20 mg) obtained after saponification (Sect. 3.6) was dissolved in 10 ml of a mixture of 1,4-dioxane-2 M hydrochloric acid (9:1, v/v) and refluxed for 2 h. The solution was poured into 100 ml of distilled water and the pH of the solution adjusted to 4 with saturated aqueous sodium hydrogen carbonate. After dioxane was removed by evaporation, phenolic acids were extracted with ethyl ether and analyzed by GLC and HPLC (Sect. 3.6). The result obtained in the case of sugar-cane bagasse is shown in Table 4 (Kato 1986).

3.8 Spectroscopic Analysis

Ultraviolet (UV) spectra were measured for solutions in 50%–80% aqueous 1,4-dioxane. Infrared (IR) spectra were measured for potassium bromide discs. The isolated LCCs were dissolved in deuterium oxide and their nuclear magnetic resonance (^{1}H- and ^{13}C-NMR) spectra were measured. The proton chemical shifts (δ) in ppm were given as downfield values from that of sodium 2,2,3,3-tetradeuterio-3-(trimethylsilyl)propionate (TSP) as internal standard. The carbon chemical shifts (δ) in ppm were given as relative values to that of internal 1,4-dioxane, which was determined as 67.4 ppm downfield from tetramethylsilane (TMS). The spectra due to lignin were recommended to be assigned on the basis of the accumulated results for UV (Goldschmid 1971), for IR (Hergert 1971), for ^{1}H-NMR (Ludwig 1971; Morohoshi 1973) and for ^{13}C-NMR (Lüdemann and Nimz 1974; Nimz and Lüdemann 1976; Himmelsbach and Barton II 1980; Nimz et al. 1981; Lapierre et al. 1982). The NMR and IR spectra due to carbohydrates were heavily overlapped with lignin and were difficult to assign except to anomeric regions, the signals of which could be assigned by comparing the spectra of cabohydrates (Barker et al. 1956; Gorin 1981; Azuma and Koshijima 1983; Bock et al. 1984; Bradbury and Jenkins 1984).

4 Linkage Analysis Between Lignin and Carbohydrates

4.1 Separation of LCC Fragments by Adsorption Chromatography (Method of Koshijima et al. 1984; Watanabe et al. 1985)

In the analysis of the nature of the chemical linkages between lignin and carbohydrates, considerable efforts has been undertaken to analyze the behavior against alkali,, acid, reducing reagent and enzymatic treatments (Kringstad and Cheng 1969; Brownell 1971; Merewether and Samsuzzaman 1972; Smelstorius 1974; Yaku et al. 1976; Eriksson and Lindgren 1977; Košiková et al. 1979; Eriksson et al. 1980; Meshitsuka et al. 1982; Lundquist et al. 1980; Joseleau and Gancet 1981; Minor 1982; Obst 1982; Iversen and Westermark 1985; Iversen and Wännström 1986; Iversen et al. 1987; Ford 1986; Minor 1986). Enzymolysis was effective in increasing the attaching points between lignin and carbohydrates. Enzymatic digest usually contained two types of LCC fragments, water-soluble and water-insoluble. Water-insoluble LCC fragments have been widely used to characterize lignin-carbohydrate bondings, because of the easiness of separation (Eriksson and Lindgren 1977; Eriksson et al. 1980; Joseleau and Gancet 1981; Minor 1986). Water-soluble LCC fragments, however, were difficult to separate and stop from contaminating large quantities of mono- and oligosaccharides. The procedures described below are recommended to isolate lignin-carbohydrate fragments from enzymatic digest by adsorption chromatography on Toyopearl HW40S (TOSOH).

The LCC was treated with cellulase "Meicelase" (Meiji Seika) for 48 h at 40° C at substrate and enzyme concentrations of 2.5% and 0.25%, respectively, in distilled water. The enzymatic hydrolysate was heated for 5 min in a boiling water bath to inactivate the enzymes and to facilitate coagulation of the water-insoluble materials. The enzymatic hydrolysate was then centrifuged to recover the supernatant, which was lyophilized and extracted with methanol to remove enzymes and higher-molecular-weight materials. The precipitate formed during enzymatic treatment was washed with distilled water, solubilized in 50% aqueous 1,4-dioxane and poured into water. The methanol-soluble materials were concentrated to dryness and dissolved in distilled water. The water-soluble materials were applied on a column (4.0 × 45 cm) of Toyopearl HW40S. After the unadsorbed materials were washed out thoroughly with distilled water, the adsorbed LCC fragments were eluted with 50% aqueous 1,4-dioxane.

4.2 Linkage Analysis Between Lignin and Carbohydrates by Oxidative Cleavage

Recently, Oikawa et al. (1982) reported that 2,3-dichloro-5,6-dicyanobenzoquinone (DDQ) splits the benzyl ether linkages of 4-methoxybenzyl and 3,4-dimethoxybenzyl ethers selectively at room temperature to give alcohols. DDQ also released 91%–93% of glucose from isoconiferin and its 4-O-methoxy derivative after 1-h incubation at 50° C in 50% aqueous 1,4-dioxane. However, DDQ was inert to acetyl groups and glycosidic linkages and could not split benzyl ether linkages if acetyl groups were introduced at phenolic hydroxyls of 4-hydroxy-3-methoxybenzyl ethers, due to electron-withdrawing effects of acetyl groups. Based on these results, Koshijima et al. (1984) and Watanabe et al. (1986) applied DDQ treatment to analyze the linkages between lignin and carbohydrates. The procedures described below are recommended to detect benzyl ether and conjugated-ether linkages in LCC.

LCC fragments were firstly methylated with diazomethane to block phenolic hydroxyl groups, and acetylated with pyridine-acetic anhydride. The methylated and acetylated LCC fragment (200 mg) was solubilized in 19 ml of dichloromethane : water (18 : 1, v/v) and treated with 50 mg of DDQ for 2 h at 50° C. The solution was applied to a column (1.5 × 15 cm) of Silica Gel 60 (Merck) and washed thoroughly with distilled water to remove unreacted DDQ and the corresponding hydroquinone. The Silica Gel 60 adsorbed 50 mg of 1,2,3,4,6-penta-*O*-acetyl glucose per gram of silica under this conditions. SEP-PAK Silica cartridge

Table 5. Partially methylated alditol acetates from DDQ treatment of the acetylated lignin-carbohydrate fragments

Partially methylated alditol acetate	Lignin-carbohydrate fragments[a]	
	Pine[c]	Beech[d]
5-Ara[b]	1.5	–
2-Xyl	11.6	71.6
3-Xyl	15.7	28.4
2-Man	1.0	–
3-Man	1.7	–
6-Man	30.6	–
2-, 3-Gal	t[e]	–
6-Gal	27.5	–
2-, 3-Glc	t	–
6-Glc	10.5	–

[a] Values are given on molar percent basis.
[b] 5-Ara = peracetate of 5-*O*-methyl arabinitol etc.
[c] Lignin-carbohydrate fragment from enzymatic treatment of pine *(Pinus densiflora)* LCC-W.
[d] Lignin-carbohydrate fragment from enzymatic treatment of beech *(Fagus crenata)* LCC-W.
[e] t = trace.

(Waters, Millipore) could be used alternatively but its capacity is 30 mg per 1 g of silica. The column was then eluted with methanol. The material eluted with methanol was evaporated to dryness, methylated by diazomethane-boron trifluoride etherate (Deferrari et al. 1972) or methyl trifluoromethane-sulfonate (Prehm 1980) and hydrolyzed with 90% aqueous formic acid followed by 0.25 M sulfuric acid. The partially methylated monosaccharide was converted into alditol acetate and analyzed by GLC and GC-MS (Sect. 3.4) (Table 5). The positions of methylation are indicative of attachment sites of lignin.

5 Linkage Analysis Between Phenolic Acids and Carbohydrates

Enzymatic treatment has been usually applied prior to clarifying linkages between phenolic acids and carbohydrates. The main point in the structural studies on the phenolic acid-carbohydrate linkages is the removal of simple sugars from phenolic acid-containing oligosaccharides. This could be achieved by adsorption chromatography on Sephadex LH-20 gels (Fry 1983b; Kato et al. 1987a). The esterified oligosaccharides were selectively adsorbed on these gels and eluted out after monosaccharides. Although Sephadex LH-60 gels adsorbed phenolic acid containing oligosaccharides (Kato et al. 1983), the adsorbing ability of these gels is lower than that of Sephadex LH-20.

5.1 Isolation of Oligosaccharides Containing Esterified Phenolic Acids

The phenolic acid-containing LCC dissolved in 0.1 M sodium acetate buffer, pH 4.6, was treated with cellulase "Onozuka RS" from *Trichoderma viride*, which was purified by gel filtration on Sephadex G-50 at 5° C using distilled water as an eluent. The absence of hydroxycinnamic acid ester hydrolase activity in the enzyme was checked by using chlorogenic acid as a substrate (Okamura and Watanabe 1982). The substrate and enzyme concentrations were 1% (w/v) and 0.1% (w/v), respectively. After incubation for 48 h at 37° C with a few drops of toluene, the precipitate formed was removed by centrifugation for 15 min at 8000 g and washed with the same buffer, followed by distilled water. The supernatant and washings were combined, passed through a column of Dowex 50 × 8 (H^+ form) and extracted twice with chloroform. The aqueous layer was concentrated to a small volume and poured into 5 vol of ethanol. The precipitate formed was centrifuged and washed twice with ethanol. The supernatant and washings were combined and concentrated to dryness. The dried material was solubilized in a small amount of distilled water and lyophilized.

The phenolic acid-containing oligosaccharides were separated from mixtures of simple sugars by adsorption chromatography on Sephadex LH-20 (57.0 × 6.0 cm) equilibrated with distilled water. The ethanol-soluble fraction from the enzymatic hydrolysate (1 g) was applied on the column and eluted with 5 l of distilled water. The eluent was then changed into 50% aqueous dioxane to elute the adsorbed materials. Fractions (20 ml) were collected at a flow rate of 100 ml/h. Each fraction was measured at absorbances of 280 nm for lignin and 325 nm for ferulic acid, respectively. The carbohydrate content of each fraction was monitored by measuring the absorbance at 480 nm due to the phenol-sulfuric acid assay (Dubois et al. 1956). The fractions having absorbances at 325 nm and 480 nm were pooled and lyophilized. Purification of the crude phenolic acid-containing oligosaccharides was carried out by HPLC on ODS column (1.9 × 15 cm) of μBondapak C18 (Waters, Millipore) using water–methyl alcohol (1 : 1, v/v) as an eluting solvent.

5.2 Structural Determination of Phenolic Acid-Containing Oligosaccharides

5.2.1 Molecular Weight Determination

Fast atom bombardment mass spectrometry (FAB-MS) and secondary ion mass spectrometry (SIMS) using double-focusing spectrometer are especially suitable for the analysis of carbohydrates (Kambara and Hishida 1981 a, b; Kambara 1982; Dell and Ballou 1983; Dell et al. 1983 a, b; Bosso et al. 1984; Dell 1987). This technique has been applied to determine the molecular weights of phenolic acid-containing tri- and tetra-saccharides (Mueller-Harvey et al. 1986; Kato et al. 1987 a). Usually, samples were suspended in glycerol (Glycerol matrix) and Xenon atoms were used for the bombardment. The signals of phenolic acid-containing oligosaccharides appeared as protonated $(M+H)^+$ and cationized (M+

Na)$^+$ and (M+K)$^+$ species in the positive recording mode, but appeared as deprotonated (M−H)$^-$ species in the negative recording mode.

5.2.2 Structural Determination of Phenolic Acid Moïety

Esterified phenolic acids were saponified and analyzed by GLC and HPLC (Sect. 3.4).

5.2.3 Structural Determination of Sugar Moiety

The phenolic acid-containing oligosaccharides were hydrolyzed with 2 M trifluoroacetic acid in a sealed tube for 1 h at 121° C (Albersheim et al. 1967). The hydrolysate was evaporated to dryness, dissolved in water and extracted with ethyl ether. The monosaccharides in the aqueous layer were converted into alditol acetates and analyzed by GLC (Sect. 3.1). The linkage analysis was carried out by methylation. Although complete methylation could be achieved by the combined methods of Hakomori (1964) and Kuhn (Wallenfels et al. 1963), ester linkages were splitted during the procedure. This defect was overcome by repeated methylation with methyl trifuluoromethanesulfonate for 3 h at 50° C according to the procedure of Prehm (1980).

5.2.4 UV and IR Spectroscopic Analyses

Phenolic acid content and composition in the phenolic acid containing-oligosaccharides were analyzed by measuring UV in 80% aqueous 1,4-dioxane with use of λmax values at 312 nm (*p*-coumaric acid) and 325 nm (ferulic acid). In the case of saponification of the feruloylated oligosaccharides, λmax value at 325 nm was shifted to 370 nm immediately after addition of 1 M sodium hydroxide due to formation of sodium phenolate, and gradually attained 347 nm by formation of sodium ferulate.

The mode of linkage between phenolic acids and carbohydrates was deduced to be ester linkage from IR band (potassium bromide discs) at 1710 cm^{-1} (C=O). Phenolic acid-containing oligosaccharides also had characteristic IR bands at 1620 cm^{-1} (conjugated aliphatic C=C), 1600 cm^{-1} and 1510 cm^{-1} (aromatic C=C) together with bands due to glycosidic linkages.

5.2.5 NMR Spectroscopic Analysis

NMR spectroscopic analysis has been shown to be particulary useful for structural determination of oligosaccharides (Bock et al. 1984; Bradbury and Jenkins 1984). This method has an advantage in its nondestructive procedure. The isolated phenolic acid-containing oligosaccharides and de-esterified oligosaccharides were dissolved in deuterium oxide and their one-dimensional ^{1}H- and ^{13}C-NMR spectra (Sect. 3.1.4) were measured (Kato et al. 1983, 1987a).

Recently, new techniques of two-dimensional (2D) NMR spectroscopy have been developed to analyze the structure of complex biomolecules by using various pulse sequences. Among the 2D-NMR techniques, *J*-resolved spectroscopy (Nagayama et al. 1977), ^{1}H,^{1}H-COSY (^{1}H-relayed ^{1}H-^{1}H correlation spectroscopy) (Bax and Freeman 1981), ^{1}H,^{13}C-COSY (^{13}C-relayed ^{1}H-^{13}C correlation spectroscopy) (Bax 1982), H,C-COLOC (^1H-^{13}C correlation spectroscopy via long-range couplings) (Kessler et al. 1984), HOHAHA (homonuclear Hartmann-Hann) (Summers et al. 1986) and RELAY-COSY (relayed coherence-transfer chemical shift-correlation spectroscopy) (Eich et al. 1982) were successfully applied to determining the primary structures of carbohydrates (Hall and Sukumar 1979; Appleton et al. 1986; Ikura and Hikichi 1987; Goux and Unkefer 1987; Lerner and Bax 1987; Byrd et al. 1987). All of their NMR signals could be assigned to give unambiguous information about their structures. Mueller-Harvey et al. (1986) have already applied ^1H,^1H-COSY technique to determine the structures of feruloylated and *p*-coumaroylated trisaccharides.

6 Concluding Remarks

The structure of LCC was difficult to elucidate and the information we have is still limited and fragmentary. However, recent progress in the development of new fractionation and analytical techniques enabled us to arrive at the conclusive result that carbohydrates are coupled to quinone methides produced as intermediates in the biosynthesis of lignin to form benzyl ether and ester linkages (Fig. 3), confirming Freudenberg's idea (Freudenberg 1968). This may imply that the lignin-carbohydrate linkages are formed at limited stages of growth and maturation of plant cells. Based on the model experiments (Hemmingson et al. 1978; Hemmingson 1979; Leary et al. 1983 a, b), it is however also suggested that the formation of lignin-carbohydrate linkages continue after lignification has stopped. Further investigation is also required to clarify the number and frequency of lignin-carbohydrate bonds in the actual plant cell walls.

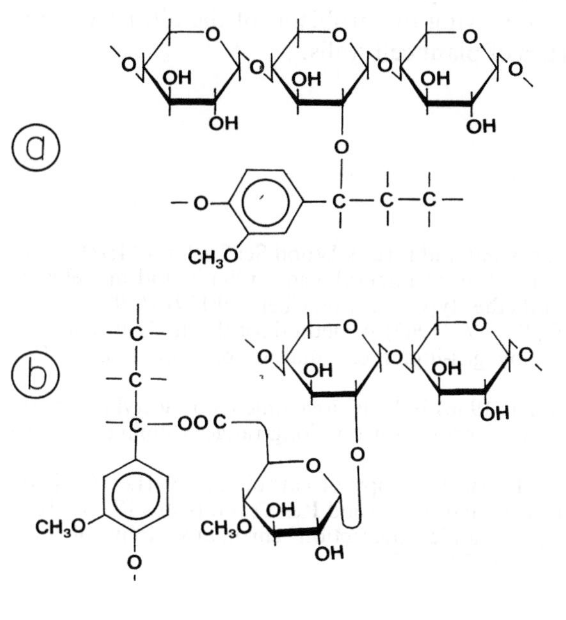

Fig. 3 a, b. Proposed types of linkages between lignin and carbohydrates. Benzyl ether linkage: Cpd. **a**. Benzyl ester linkage: Cpd. **b**

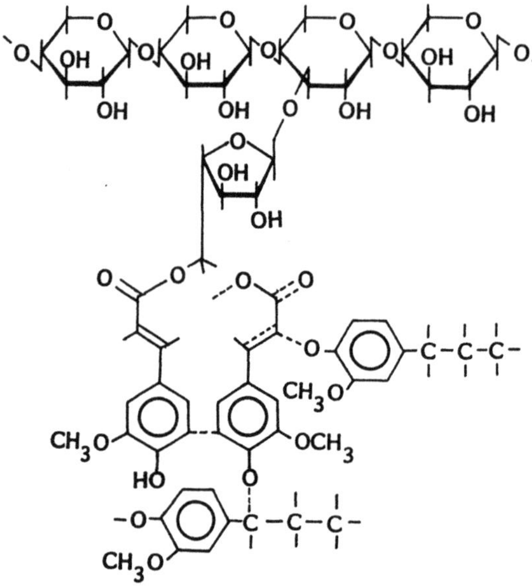

Fig. 4. A cross-linking model among lignin, carbohydrate and phenolic acid

Based on the present results, not only hemicellulosic polysaccharides and pectin but also cellulose was suggested to link to lignin (Pew and Weyna 1962; Eriksson et al. 1980). Ultrastructural models of LCC were proposed based on electron microscopic observations (Fengel 1976; Košiková et al. 1978). In the gramineous plant cell walls, it has been hypothesized that the ferulic acids in the feruloylated carbohydrates are coupled to form diferulate-polysaccharide cross-links by catalysis of peroxidase (Markwalder and Neukom 1976). Phenolic acids are also ether-linked to lignin (Scalbert et al. 1985, 1986). Such cross-linking of polysaccharides, lignin and phenolic acids (Fig. 4) together with the amphipathic property of LCC, presents a key for solving the problems of the ultrastructural arrangement and physical properties of plant cell walls.

References

Adler E (1977) Lignin chemistry-past, present and future. Wood Sci Technol 11:169–218

Adler E, Pepper JM, Eriksoo E (1957) Action of mineral acid on lignin and model substances of guaiacylglycerol-beta-aryl ether type. Ind Eng Chem 49:1391–1392

Albersheim P, Nevins DJ, English PD, Karr A (1967) A method for the analysis of sugars in plant cell-wall polysaccharides by gas-liquid chromatography. Carbohydr Res 5:340–345

Appleton ML, Cottrell CE, Behrman EJ (1986) N.M.R. assignments of acetyl and trityl groups in derivatized carbohydrates via proton-carbon long-range couplings. Carbohydr Res 158:227–235

Azuma J, Koshijima T (1983) ^{13}C-NMR spectroscopy of carbohydrates II. ^{13}C-NMR spectroscopy of xylan and related carbohydrate. Wood Res Techn Notes 17:132–169

Azuma J, Koshijima T (1985) Hydrophobic interaction chromatography of pine Björkman-LCC. Mokuzai Gakkaishi 31:383–387

Azuma J, Koshijima T (1988) Lignin-carbohydrate complexes from various sources. Meth Enzymol 161B:12–18

Azuma J, Takahashi N, Koshijima T (1981) Isolation and characterization of lignin-carbohydrate complexes from the milled-wood lignin fraction of *Pinus densiflora* Sieb. et Zucc. Carbohydr Res 93:91–104

Azuma J, Nomura T, Koshijima T (1985a) Lignin-carbohydrate complexes containing phenolic acids isolated from the culms of bamboo. Agric Biol Chem 49:2661–2669

Azuma J, Takahashi N, Isaka M, Koshijima T (1985b) Lignin-carbohydrate complexes extracted with aqueous dioxane from beech wood. Mokuzai Gakkaishi 31:587–594

Barker SA, Bourne EJ, Whiffen DH (1956) Use of infrared analysis in the determination of carbohydrate structure. Meth Biochem Anal III:213–245

Bax A (1982) Two-dimentional nuclear magnetic resonance in liquids. Reidel, Amsterdam, Neth

Bax A, Freeman R (1981) Investigations of complex networks of spin-spin coupling by two-dimensional NMR. J Magn Reson 44:542–561

Bebault GM, Berry JM, Choy YM, Dutton GGS, Funnell N, Hayward LD, Stephen AM (1973) Semimicro determination of sugar configuration by measurement of circular dichroism of alditol acetates. Can J Chem 51:324–326

Björkman A (1957a) Studies on finely divided wood Part 3. Extraction of lignin-carbohydrate complexes with neutral solvents. Sven Papperstidn 60:243–251

Björkman A (1957b) Lignin and lignin-carbohydrate complexes: extraction from wood meal with neutral solvents. Ind Eng Chem 49:1395–1398

Bock K, Pedersen C, Pedersen H (1984) Carbon-13 nuclear magnetic resonance data for oligosaccharides. Adv Carbohydr Chem Biochem 42:193–225

Bolker HI (1963) A lignin carbohydrate bond as revealed by infra-red spectroscopy. Nature 197:489–490

Bosso C, Defaye J, Heyraud A, Ulrich J (1984) Fast-atom bombardment-mass spectrometry for the characterization of cello- and malto-oligosaccharides. Carbohydr Res 125:309–317

Bradbury JH, Jenkins GA (1984) Determination of the structures of trisaccharides by ^{13}C-NMR spectroscopy. Carbohydr Res 126:125–156

Brice RE, Morrison IM (1982) The degradation of isolated hemicelluloses and lignin-hemicellulose complexes by cell-free, rumen hemicellulases. Carbohydr Res 101:93–100

Brownell HH (1971) Stability of the Lignin-carbohydrate complex. Tappi 54:66–71

Byrd BA, Egan W, Summers MF, Bax A (1987) New N.M.R.-spectroscopic approaches for structural studies of polysaccharides: application to the *Haemophilus influenzae* type a capsular polysaccharide. Carbohydr Res 166:47–58

Deferrari JO, Gros EG, Thiel IME (1972) Methylation with diazomethane-boron trifluoride etherate. Meth Carbohydr Chem IV:365–372

Dell A (1987) F.A.B.-mass spectrometry of carbohydrates. Adv Carbohydr Chem Biochem 45:19–72

Dell A, Ballou CE (1983) Fast-atom-bombardment, negative-ion mass spectrometry of the mycobacterial *O*-methyl-D-glucose polysaccharide and lipopolysaccharides. Carbohydr Res 120:95–111

Dell A, Morris HR, Egge H, Nicolai HV, Strecker G (1983a) Fast-atom-bombardment mass-spectrometry for carbohydrate-structure determination. Carbohydr Res 115:41–52

Dell A, York WS, McNeil M, Darvill AG, Albersheim P (1983b) The cyclic structure of β-D-(1$\rightarrow$2)-linked D-glucans secreted by *Rhizobia* and *Agrobacteria*. Carbohydr Res 117:185–200

Dubois M, Gilles KA, Hamilton JK, Rebers PA, Smith F (1956) Colorimetric method for determination of sugars and related substances. Anal Chem 28:350–356

Dudman WF, Bishop CT (1968) Electrophoresis of dyed polysaccharides on cellulose acetate. Can J Chem 46:3079–3084

Eich G, Bodenhausen G, Ernst RR (1982) Exploring nuclear spin systems by relayed magnetization transfer. J Am Chem Soc 104:3731–3732

Enoki A, Yaku F, Koshijima T (1983) Synthesis of LCC model compounds and their chemical and enzymatic stabilities. Holzforschung 37:135–141
Eriksson Ö, Lindgren BO (1977) About the linkage between lignin and hemicelluloses in wood. Sven Papperstidn 80:59–63
Eriksson Ö, Goring DAI, Lindgren BO (1980) Structural studies on the chemical bonds between lignins and carbohydrates in spruce wood. Wood Sci Technol 14:267–279
Fenckl VJ, Fengel D (1982) Fraktionierung von Lignin-Polysaccharid-Komplexen Teil 2: Chromatographische Trennung einer ligninreichen Fraktion des Alkali-Extraktes aus Fichten-Holocellulose. Holzforschung 36:269–273
Fengel D (1976) Studies on the supermolecular structure of cell wall components. Part 5. Ultrastructure of the fractions from ion exchange chromatography and of isolated lignin. Sven Papperstidn 79:24–28
Fengel D, Wegener G (1984) Lignin-polysaccharide complexes. In: Wood: chemistry, ultrastructure, reactions. Walter de Gruyter, Berlin pp 167–181
Ford CW (1986) Comparative structural studies of lignin-carbohydrate complexes from *Digitaria decumbens* (pangola grass) before and after chlorite delignification. Carbohydr Res 147:101–117
Fox A, Morgan SL, Hudson JR, Zhu ZT, Lau PY (1983) Capillary gas chromatographic analysis of alditol acetates of neutral and amino sugars in bacterial cell walls. J Chromatogr 256:429–438
Freudenberg K, Grion G (1959) Beitrag zum Bindungsmechanismus des Lignins und der Lignin-Kohlenhydrat-Bindung. Chem Ber 92:1355–1363
Freudenberg K, Harkin JM (1960) Modelle für die Bindung des Lignins an die Kohlenhydrate. Chem Ber 93:2814–2819
Freudenberg K (1968) The constitution and biosynthesis of lignin. In: Freudenberg K, Neish AC (eds) Constitution and biosynthesis of lignin. Springer, Berlin Heidelberg New York, pp 47–122 (Molecular Biology Biochemistry and Biophysics, vol 2)
Fry SC (1982) Phenolic components of the primary cell wall: feruloylated disaccharides of D-galactose and L-arabinose from spinach polysaccharide. Biochem J 203:493–504
Fry SC (1983 a) Feruloylated pectins from the primary cell wall: their structure and possible functions. Planta 157:111–123
Fry SC (1983 b) Incorporation of [^{14}C]cinnamate into hydrolase-resistant components of the primary cell wall of spinach. Phytochemistry, 23:59–64
Fukuda T, Kanda T (1976) Studies on tissue culture of tree-cambium. III. Separation of lignin-carbohydrate-protein complex from poplar callus tissue. Mokuzai Gakkaishi 22:112–118
Fukuda T, Tomimura Y (1976) Studies on tissue culture of tree-cambium. V. Enzymic hydrolysis of lignin-carbohydrate-protein complex in poplar cambium tissue. Mokuzai Gakkaishi 22:570–574
Galambos JT (1967) The reaction of carbazole with carbohydrates I. Effect of borate and sulfamate on the carbazole color of sugars. Anal Biochem 19:119–132
Gerwig GJ, Kamerling JP, Vliegenthart JFG (1978) Determination of the D and L configuration of neutral monosaccharides by high-resolution capillary G.L.C. Carbohydr Res 62:349–357
Gerwig GJ, Kamerling JP, Vliegenthart JFG (1979) Determination of the absolute configuration of monosaccharides in complex carbohydrates by capillary G.L.C. Carbohydr Res 77:1–7
Glinski AJ, Nicholls GA (1977) Peroxyacetic acid delignification of white birch and new evidence for lignin-carbohydrate bonds. Pap Puu 59:745–760
Goldschmid O (1971) Ultraviolet spectra. In: Sarkanen KV, Ludwig CH (eds) Lignins. Springer, Berlin Heidelberg New York, pp 241–266
Gorin PAJ (1981) Carbon-13 nuclear magnetic resonance spectroscopy of polysaccharides. Adv Carbohydr Chem Biochem 38:13–104
Goux WJ, Unkefer CJ (1987) The assignment of carbonyl resonances in ^{13}C-N.M.R. spectra of peracetylated mono- and oligo-saccharides containing D-glucose and D-mannose: an alternative method for structural determination of complex carbohydrates. Carbohydr Res 159:191–210

Grushnikov OP, Shorygina NN (1970) The present state of the problem of lignin-carbohydrate bonds in plant tissues. Russ Chem Rev 39:684–694
Gubler F, Ashford AE, Bacic A, Blakeney AB, Stone BA (1985) Release of ferulic acid esters from barley aleurone. II. Characterization of the feruloyl compounds released in response to GA_3. Aust J Plant Physiol 12:307–317
Hakomori S (1964) A rapid permethylation of glycolipid, and polysaccharide catalyzed by methylsulfinyl carbanion in dimethly sulfoxide. J Biochem 55:205–208
Hall LD, Sukumar S (1979) Two-dimensional *J* spectroscopy: ^{1}H NMR. spectra of mono- and di-saccharides. J Chem Soc Chem Comm 292–294
Harris PJ, Hartley RD (1980) Phenolic constituents of the cell walls of monocotyledons. Biochem Syst Ecol 8:153–160
Hartley RD (1981) Chemical constitution, properties and processing of lignocellulosic wastes in relation to nutritional quality for animals. Agric Environ 6:91–113
Hartley RD (1987) HPLC for the separation and determination of phenolic compounds in plant cell walls. In: Linskens HF, Jackson JF (eds) Modern methods of plant analysis, New series, vol 5. Springer, Berlin Heidelberg New York, pp 92–103
Hartley RD, Harris PJ (1981) Phenolic constituents of the cell walls of dicotyledons. Biochem Syst Ecol 9:189–203
Hartley RD, Jones EC (1975) Effect of ultraviolet light on substituted cinnamic acids and estimation of their *cis* and *trans* isomers by gas chromatography. J Chromatogr 107:213–218
Hartley RD, Jones EC (1977) Phenolic components and degradability of cell walls of grass and legume species. Phytochemistry 16:1531–1534
Hayashi A (1961) Studies on chemical combination between lignin and carbohydrate. Part X – On the existence of phenol-β-glycoside bond. J Agri Chem Soc Jpn 35:83–86
Hayashi A, Tachi I (1958) Studies on the chemical combination between lignin and carbohydrate: the existence of a phenolglycosidic bond in wheat straw lignin. Tappi 41:173A–178A
Hemmingson JA (1979) A new way of forming lignin-carbohydrate bonds. Etherification of model benzyl alcohols in alcohol / water mixtures. Aust J Chem 32:225–229
Hemmingson JA, Leary GJ, Miller IJ, Thomas WA, Woodhouse AD (1978) A new class of linkage between carbohydrates and lignin model compounds. J Chem Soc Chem Comm 92–93
Hergert HL (1971) Infrared spectra. In: Sarkanen KV, Ludwig CH (eds) Lignins. Springer, Berlin Heidelberg New York, pp 267–297
Himmelsbach DS, Barton II FE (1980) ^{13}C Nuclear magnetic resonance of grass lignins. J Agric Food Chem 28:1203–1209
Hjertén S (1976) Hydrophobic interaction chromatography of proteins on neutral adsorbents. Meth Protein Sep 2:233–243
Hofstee BHJ (1976) Hydrophobic adsorption chromatography of proteins. Meth Protein Sep 2:245–278
Hough L, Jones JKN, Wadman WH (1950) Quantitative analysis of mixtures of sugars by the method of partition chromatography. Part V. Improved methods for separation and detection of the sugars and their methylated derivatives on the paper chromatogram. J Chem Soc 1702–1706
Ikura M, Hikichi K (1987) Two-dimentional ^{1}H-N.M.R. studies of cello-oligosaccharides: the utility of multiple-relay chemical-shift-correlated spectroscopy. Carbohydr Res 163:1–8
Iversen T (1985) Lignin-carbohydrate bonds in a lignin-carbohydrate complex isolated from spruce. Wood Sci Technol 19:243–251
Iversen T, Wännström S (1986) Lignin-carbohydrate bonds in a residual lignin isolated from pine kraft pulp. Holzforschung 40:19–22
Iversen T, Westermark U (1985) Lignin-carbohydrate bonds in pine lignins dissolved during kraft pulping. Cellul Chem Technol 19:531–536
Iversen T, Westermark U, Samuelsson B (1987) Some comments on the isolation of galactose-containing lignin-carbohydrate complexes. Holzforschung 41:119–121

Jansson PE, Kenne L, Liedgren H, Lindberg B, Lönngren J (1976) A practical guide to the methylation analysis of carbohydrates. Chem Commun (Univ. of Stockholm) 8:1–74

Joseleau JP, Gancet C (1981) Selective degradations of the lignin-carbohydrate complex from aspen wood. Sven Papperstidn 84:R123–R127

Joseleau JP, Kesraoui R (1986) Glycosidic bonds between lignin and carbohydrates. Holzforschung 40:163–168

Kambara H (1982) Comparative study of $[Xe]^+$ and $[Ar]^+$ bombardment in secondary ion mass spectrometry for bioorganic compounds. Org Mass Spectrom 17:29–33

Kambara H, Hishida S (1981 a) Secondary ion mass spectra of non-volatile bioorganic compounds. Anal Chem 53:2340–2344

Kambara H, Hishida S (1981 b) Secondary ion mass spectra of oligosaccharides. Org Mass Spectrom 16:167–170

Katayama Y, Morohoshi N, Haraguchi T (1980) Formation of lignin-carbohydrate complex in enzymatic dehydrogenation of coniferyl alcohol II. Isolation of lignin-carbohydrate complex with guaiacylglycerol-β-coniferyl ether or other dimeric compound as lignin moiety. Mokuzai Gakkaishi 26:414–420

Kato A (1986) Lignin-carbohydrate complexes containing phenolic acids in Gramineae. Thesis of Doctor of Agriculture, Kyoto University, Kyoto, pp 29

Kato Y, Nevins DJ (1985) Isolation and identification of *O*-(5-*O*-feruloyl-α-L-arabinofuranosyl)-(1→3)-*O*-β-D-xylopyranosyl-(1→4)-D-xylopyranose as a component of *Zea* shoot cell-walls. Carbohydr Res 137:139–150

Kato A, Azuma J, Koshijima T (1983) A new feruloylated trisaccharide from bagasse. Chem Lett 137–140

Kato A, Azuma J, Koshijima T (1984) Lignin-carbohydrate complexes and phenolic acids in bagasse. Holzforschung 38:141–149

Kato A, Azuma J, Koshijima T (1987 a) Isolation and identification of a new feruloylated tetrasaccharide from bagasse lignin-carbohydrate complex containing phenolic acid. Agric Biol Chem 51:1691–1693

Kato A, Azuma J, Koshijima T (1987 b) Björkman LCC from sugar-cane bagasse. Mokuzai Gakkaishi 33:487–494

Kawamura I, Higuchi T (1954) Association of lignin and carbohydrates in wood XI. Nature of linkage between lignin and carbohydrate. J Soc Textile Cellul Ind Jpn 10:15–17

Kessler H, Griesinger C, Zarbock J, Loosli HR (1984) Assignment of carbonyl carbons and sequence analysis in peptides by heteronuclear shift correlation via small coupling constants with broadband decoupling in t_1 (COLOC). J Mag Reson 57:331–336

König WA, Benecke I, Bretting H (1981) Gas chromatographic separation of carbohydrate enantiomers on a new chiral stationary phase. Angew Chem Int Ed 20:693–694

Koshijima T, Taniguchi T, Tanaka R (1972) Lignin carbohydrate complex Pt. I. The influences of milling of wood upon the Björkman LCC. Holzforschung 26:211–217

Koshijima T, Yaku F, Tanaka R (1976) Fractionation of Björkman LCC from *Pinus densiflora.* Appl Polym Symp 28:1025–1039

Koshijima T, Watanabe T, Azuma J (1984) Existence of benzylated carbohydrate moiety in lignin-carbohydrate complex from pine wood. Chem Lett 1737–1740

Košíková B, Polčin J, Dandárová-Vašátková M, Joniak D (1969) Untersuchung des Bindungscharakters des Lignin-Saccharid-Komplexes, isoliert aus methyliertem Holz. Holzforschung 23:43–47

Košíková B, Polčin J, Joniak D (1973) Isolation and properties of the lignin-saccharidic complex in acetylated beech wood. Cellul Chem Technol 7:605–611

Košíková B, Polčin J, Joniak D (1977) NMR studies on the lignin-saccharidic complexes II. Hydrolysis of the lignin-saccharidic complex isolated from acetylated beechwood. Holzforschung 31:191–193

Košíková B, Zákutná L, Joniak D (1978) Investigation of the lignin-saccharidic complex by electron microscopy. Holzforschung 32:15–18

Košíková B, Joniak D, Kosáková (1979) On the properties of benzyl ether bonds in the lignin-saccharidic complex isolated from spruce. Holzforschung 33:11–14

Košíková B, Joniak D, Polčin J (1984) Behaviour of lignin-saccharidic spruce wood complex in thermal mechanical pulping. I TMP process. Cellul Chem Technol 18:63–71
Kreicberga Z, Aronchik BM, Sergeeva VN (1974) Enzymically destroyed wood. XII. Effect of the length of beating of birch wood and the degree of its biological degradation on the yield and chemical composition of lignocarbohydrate complexes. Khim Drev 74–77
Kringstad KP, Cheng CW (1969) Studies on lignin-hemicellulose complexes. Tappi 52:2382–2385
Kusakabe I, Kamiyama Y, Yasui T (1977) Methyl-β-D-glucoside as a satisfactory internal standard in gas-liquid chromatographic quantitative-analysis of alditolacetates derived from monosaccharides. Nippon Nogeikagaku Kaishi 51:167–169
Lai YZ, Sarkanen KV (1971) Isolation and structural studies. In: Sarkanen KV, Ludwig CH (eds) Lignins. Springer, Berlin Heidelberg New York, pp 165–240
Lapierre C, Lallemand JY, Monties B (1982) Evidence of poplar lignin heterogeneiity by combination of ^{13}C and ^{1}H NMR spectroscopy. Holzforschung 36:275–282
Leary GJ, Sawtell DA, Wong H (1983 a) The formation of model lignin-carbohydrate compounds in aqueous solution. Holzforschung 37:11–16
Leary GJ, Sawtell DA, Wong H (1983 b) NMR spectra of benzyl non-cyclic alkyl ether lignin model compounds. Holzforschung 37:213–215
Leontein K, Lingberg B, Lönngren J (1978) Assignment of absolute configuration of sugars by G.L.C. of their acetylated glycosides formed from chiral alcohols. Carbohydr Res 62:359–362
Leopold B (1952) Lignin. III. Oxidation of wood from *Picea abies* with nitrobenzene and alkali. Acta Chem Scand 6:39–48
Lerner L, Bax A (1987) Aplication of new, high-sensitivity, ^{1}H-^{13}C-N.M.R.-spectral techniques to the study of oligosaccharides. Carbohydr Res 166:35–46
Lindberg B (1972) Methylation analysis of polysaccharides. Meth Enzymol 28:178–195
Linnell WS, Thompson NS, Swenson HA (1966) Structure of the black spruce glucomannan II. Lignin-glucomannan relationships. Tappi 49:491–493
Little MR (1982) Separation, by G.L.C., of enantiomeric sugars as diastereo-isomeric dithioacetals. Carbohydr Res 105:1–8
Lomax JA, Gordon AH, Chesson A (1985) A multiple-column approach to the methylation analysis of plant cell-walls. Carbohydr Res 138:177–188
Ludwig CH (1971) Magnetic resonance spectra. In: Sarkanen KV, Ludwig CH (eds) Lignins. Springer, Berlin Heidelberg New York, pp 299–344
Lüdemann HD, Nimz H (1974) ^{13}C-Kernresonanzspektren von Ligninen, 2. Buchen- und Fichten-Björkman-Lignin. Makromol Chem 175:2409–2422
Lundquist K (1970) Acid degradation of lignin II. Separation and identification of low molecular weight phenols. Acta Chem Scand 24:889–907
Lundquist K, Lundgren R (1972) Acid degradation of lignin part VII. The cleavage of ether bonds. Acta Chem Scand 26:2005–2023
Lundquist K, Simonson R (1975) Lignin preparations with very low carbohydrate content. Sven Papperstidn 78:390
Lundquist K, Simonson R, Tingsvik K (1979) On the occurrence of carbohydrates in milled wood lignin preparations. Svensk Papperstidn 82:272–275
Lundquist K, Simonson R, Tingsvik K (1980) Studies on lignin carbohydrate linkages in milled wood lignin preparations. Sven Papperstidn 83:452–454
Markwalder HU, Neukom H (1976) Diferulic acid as a possible crosslink in hemicelluloses from wheat germ. Phytochemistry 15:836–837
McNeil M, Darvill AG, Fry SC, Albersheim P (1984) Structure and function of the primary cell walls of plants. Annu Rev Biochem 53:625–663
Merewether JWT (1957) A lignin-carbohydrate complex in wood. Holzforschung 11:65–80
Merewether JWT, Samsuzzaman LAM (1972) Studies of a lignin-carbohydrate complex Pt.I. The isolation of a water soluble lignin-carbohydrate complex from alkali extraction of *Eucalyptus obliqua*. Holzforschung 26:11–18

Meshitsuka G, Lee ZZ, Nakano J, Eda S (1982) Studies on the nature of lignin-carbohydrate bonding. J Wood Chem Technol 2:251–267
Minor JL (1982) Chemical linkage of pine polysaccharides to lignin. J Wood Chem Technol 2:1–16
Minor JL (1986) Chemical linkage of polysaccharides to residual lignin in loblolly pine kraft pulps. J Wood Chem Technol 6:185–201
Morohoshi N (1973) Chemical structure of lignin. Analysis of milled wood lignin from ezomatsu by NMR spectroscopy. Bull Exp For Tokyo Univ Agric Technol 10:183–194
Morrison IM (1973) Isolation and analysis of lignin-carbohydrate complexes from *Lolium multiflorum*. Phytochemistry 12:2979–2984
Mouttet PB, Robert A (1978) Mise en evidence de complexes lignine-polysaccharides dans la membrane de tiges de *Sequoia sempervirens* cultivées „in vitro". Holzforschung 32:173–177
Mueller-Harvey IM, Hartley RD, Harris PJ, Curzon EH (1986) Linkage of *p*-coumaryl and feruloyl groups to cell wall polysaccharides of barley straw. Carbohydr Res 148:71–85
Mukoyoshi S, Azuma J, Koshijima T (1981) Lignin-carbohydrate complexes from compression wood of *Pinus densiflora* Sieb et Zucc. Holzforschung 35:233–240
Nagayama K, Wüthrich K, Bachmann P, Ernst RR (1977) Two-dimensional *J*-resolved ^{1}H N.M.R. spectroscopy for studies of biological macromolecules. Biochem Biophys Res Commun 78:99–105
Nakamura Y, Higuchi T (1976) Ester linkage of *p*-coumaric acid in Bamboo lignin. Holzforschung 30:187–191
Neilson MJ, Richards GN (1982) Chemical structures in a lignin-carbohydrate complex isolated from the bovine rumen. Carbohydr Res 104:121–138
Nimz HH, Lüdemann HD (1976) Kohlenstoff-13-NMR-Spektren von Ligninen, 6. Lignin- und DHP-Acetate. Holzforschung 30:33–40
Nimz HH, Robert D, Faix O, Nemr M (1981) Carbon-13 NMR spectra of lignins, 8. Structural differences between lignins of hardwoods, softwoods, grasses and compression wood. Holzforschung 35:16–26
Obst JR (1982) Frequency and alkali resistance of lignin-carbohydrate bonds in wood. Tappi 65:109–112
Ohara S, Hosoya S, Nakano J (1980) Studies on the formation of lignin-carbohydrate complex during alkaline pulping processes. Mokuzai Gakkaishi 26:408–413
Oikawa Y, Toshioka T, Yonemitsu O (1982) Specific removal of O-methoxybenzyl protection by DDQ oxidation. Tetrahedron Lett 23:885–888
Okamura S, Watanabe M (1982) Purification and properties of hydroxycinnamic acid ester hydrolase from *Aspergillus japonicus*. Agric Biol Chem 46:1839–1848
Oshima R, Kumanotani J (1981) Determination of configuration of monosaccharides by HPLC on diastereomeric 1-deoxy-1-(N-acetyl-α-methylbenzylamino)alditol acetates. Chem Lett 943–946
Oshima R, Kumanotani J, Watanabe C (1983) Gas-liquid chromatographic resolution of sugar enantiomers as diastereoisomeric methylbenzylaminoalditols. J Chromatogr 259:159–163
Pew JC, Weyna P (1962) Fine grinding, enzyme digestion, and the lignin-cellulose bond in wood. Tappi 45:247–256
Prehm P (1980) Methylatioin of carbohydrates by methyl trifluoromethanesulfonate in trimethyl phosphate. Carbohydr Res 78:372–374
Reicher F, Corrêa JBC, Gorin PAJ (1982) Location of *O*-acetyl groups in the acidic D-Xylan of *Mimosa scabrella (bracatinga)*. A study of *O*-acetyl group migration. Carbohydr Res 135:129–140
Scalbert A, Monties B, Lallemand JY, Guittet E, Rolando C (1985) Ether linkage between phenolic acids and lignin fractions from wheat straw. Phytochemistry 24:1359–1362
Scalbert A, Monties B, Rolando C, Sierra-Escudero A (1986) Formation of ether linkage between phenolic acids and gramineae lignin: a possible mechanism involving quinone methides. Holzforschung 40:191–195
Schweer H (1982) Gas chromatographic separation of carbohydrate enantiomers as (−)-menthyloxime pertrifluoroacetates on silicone OV-225. J Chromatogr 243:149–152

Sergeeva LL, Semenovskii AV, Mal'tseva Al, Shorygina NN (1979) Isolation of a lignin-carbohydrate complex in the reductive decomposition of sprucewood by metallic sodium in liquid ammonia. Khim Drev 35–40
Shaltiel S (1974) Hydrophobic chromatography. Meth Enzymol 34:126–140
Shimada M, Fukuzuka T, Higuchi T (1971) Ester linkages of *p*-coumaric acid in bamboo and grass lignins. Tappi 54:72–78
Simonson R (1971 a) The himicellulose in the sulfate pulping process. Part 6. Hemicellulose-lignin compounds present in birch sulfate cooking liquors. Svensk Papperstidn 74:153–165
Simonson R (1971 b) The hemicellulose in the sulfate pulping process. Part 7. Crystallized xylan-lignin compounds. Sven Papperstidn 74:268–270
Sloneker JJ (1972) Gas-liquid chromatography of alditol acetates. Meth Carbohydr Chem IV:20–24
Smelstorius JA (1974) Chemical composition of wood of Australian-grown *Pinus radiata* D. Don III. Lignin-polysaccharide complexes. Holzforschung 28:99–101
Smith MM, Hartly RD (1983) Occurrence and nature of ferulic acid substitution of cell-wall polysaccharides in graminaceous plants. Carbohydr Res 118:65–80
Sone Y, Kakuta M, Misaki A (1978) Isolation and characterization of polysaccharides of "Kikurage," fruit body of *Auricularia auricula-judae*. Agric Biol Chem 42:417–425
Sudo K, Shimizu K, Sakurai K (1985) Characterization of steamed wood lignin from beech wood. Holzforschung 39:281–288
Summers MF, Marzilli LG, Bax A (1986) Complete ^{1}H and ^{13}C assignments of coenzyme B_{12} through the use of new two-dimensional NMR experiments. J Am Chem Soc 108:4285–4294
Takahashi N, Azuma J, Koshijima T (1982) Fractionation of lignin-carbohydrate complexes by hydrophobic-interaction chromatography. Carbohydr Res 107:161–168
Tanabe H, Kobayashi Y (1987) Effect of lignin-carbohydrate complex on marceration of mitsumata (*Edgeworthia papyrifera* Sieb. et Zucc.) bast by pectinolytic enzymes from *Erwinia carotovora*. Holzforschung 41:395–399
Tanaka R, Koshijima T (1972) Fractionation of Björkman LCC and separation of acetyl glucomannan from the LCC. Mokuzai Gakkaishi 18:403–408
Tanaka K, Nakatsubo F, Higuchi T (1976) Reactions of guaiacylglycerol-β-guaiacyl ether with several sugars. I. Reaction of quinonemethide with D-glucuronic acid. Mokuzai Gakkaishi 22:589–590
Tanaka K, Nakatsubo F, Higuchi T (1979) Reactions of guaiacylglycerol-β-guaiacyl ether with several sugars. II. Reactions of quinonemethide with pyranohexoses. Mokuzai Gakkaishi 25:653–659
Tomoda M, Kaneko S, Nakatsuka S (1975) Plant musilages. X. Isolation and characterization of a mucous polysaccharide, "Lilium-A glucomannan", from the bulbs of *Lilium auratum*. Chem Pharm Bull 23:430–436
Wallenfels K, Bechtler G, Kuhn R, Trischmann H, Egge H (1963) Permethylation of oligomeric and polymeric carbohydrates and quantitative analysis of the cleavage products. Angew Chem Int Ed 2:513–523
Wardi AH, Allen WS, Varma R (1974) A simple method for the detection and quantitative determination of hexuronic acids and pentoses. Anal Biochem 57:268–273
Watanabe T, Azuma J, Koshijima T (1985) Isolation of lignin-carbohydrate complex fragments by adsorption chromatography. Mokuzai Gakkaishi 31:52–53
Watanabe T, Kaizu S, Koshijima T (1986) Binding sites of carbohydrate moieties toward lignin in "lignin-carbohydrate complex" from *Pinus densiflora* wood. Chem Lett 1871–1874
Watanabe T, Azuma J, Koshijima T (1987) A convenient method for preparing lignin-carbohydrate complex from *Pinus densiflora* wood. Mokuzai Gakkaishi 33:798–803
Wegener G (1976) Beitrag zur Charakterisierung der Natriumchlorit-Delignifizierung von Fichtenholz. III. Lignin-Polysaccharid-Komplexe in Chlorit-Aufschlußlösungen. Papier 30:177–185
Wegener G, Fengel D (1979) Rapid ultrasonic isolation of milled wood lignins: Fractionation and degradation experiments. Tappi 62:97–100

Whitmore FW (1978) Lignin-carbohydrate complex formed in isolated cell walls of callus. Phytochemistry 17:421–425

Yaku F, Koshijima T (1971) Lignin-acetyl glucomannan interaction in Björkman LCC. Mokuzai Gakkaishi 17:267–268

Yaku F, Yamada Y, Koshijima T (1976) Lignin carbohydrate complex Pt. II. Enzymic degradation of acidic polysaccharide in Björkman LCC. Holzforschung 30:148–156

Yaku F, Tsuji S, Koshijima T (1979) Lignin carbohydrate complex Pt. III. Formation of micells in the aqueous solution of acidic lignin carbohydrate complex. Holzforschung 33:54–59

Yaku F, Tanaka R, Koshijima T (1981) Lignin carbohydrate complex Pt. IV. Lignin as side chain of the carbohydrate in Björkman LCC. Holzforschung 35:177–181

Yphantis DA (1964) Equilibrium ultracentrifugation of dilute solutions. Biochemistry 3:297–317

Fluorometric Measurement of Callose and Other 1,3-β-Glucans

H. Kauss

1 Introduction

1.1 Nature and Function of Callose

The term *callose* originates from cytochemical observations. Over the last hundred years many examples have been reported where staining and fluorescence techniques reveal that certain cell-wall appositions appear to consist of substances distinct from the materials which usually constitute the cell walls of higher plants (for a review on the early microscopical work see Eschrich 1956). Later it was shown that these staining properties are due to the presence of essentially linear 1,3-β-glucans (for citations see Fincher and Stone 1981; Evans et al. 1984), although some other linkages may be present (Hinch and Clarke 1982). The term callose is now used synonymously for linear 1,3-β-glucans exhibiting a degree of polymerization high enough to render the polymer water-insoluble.

Callose deposition may occur in the course of regular developmental steps, such as the elongation of pollen tubes, or during pollen maturation. More spectacular is an often rather rapid and precisely localized callose formation which is believed to seal cells and tissues after stress situations of various kinds. Examples are the closing of connections that unite protoplasts of contiguous cells (such as plasmadesmata or sieve pores), the formation of papillae at the sites of attempted invasion of host cells by fungal hyphae, and the rejection of self-incompatible pollen (for citations see Eschrich 1956; Fincher and Stone 1981; Kauss 1985, 1987).

1.2 Regulation of Callose Synthesis

The well-documented participation of callose deposition in defence reactions raises questions about the biochemical mechanisms governing its biosynthesis. Knowledge pertinent to these questions has been recently reviewed (Kauss 1985, 1987). One major aspect in the regulation of callose deposition relates to the activation by Ca^{2+} of the plasma-membrane-located 1,3-β-glucan synthase, which appears to be an integral membrane enzyme exhibiting Ca^{2+}-binding sites at its cytoplasmic side (Fink et al. 1987). The 1,3-β-glucan synthase is present with a high activity in the plasma membranes of all higher plant cells studied so far, but remains latent as long as the cells are not disturbed.

In suspension-cultured cells a massive callose deposition can be induced by certain substances which perturb the membrane permeability and it has been shown that this is accompanied by a pronounced increase in net Ca^{2+} uptake (Kauss 1987; Waldmann et al. 1988). The latter event is presumed to lead to an elevation of the cytoplasmic Ca^{2+} concentration, which in turn can activate in an allosteric-like manner the plasma membrane 1,3-β-glucan synthase. This chain of events directly leads to the deposition of callose in the area of the cell wall where the plasma membrane was perturbed, and explains why callose deposition can often occur rather rapidly and in a precisely localized fashion. Additional regulatory factors might be superimposed on the Ca^{2+} regulation of callose formation (Kauss 1987).

1.3 Advantages and Limits of Fluorometric Callose Measurement

The work cited on the regulation of callose deposition was facilitated greatly by the availability of a quantitative callose assay based on the cytochemical demonstration of callose with decolorized aniline blue which allows to visualize this polymer by a bright yellow fluorescence. The fluorescence is due to the presence in commercial aniline blue preparations of a fluorophore called Sirofluor (4,4′-[carbonylbis(benzene-4,1-diyl)bis(imino)]bisbenzenesulfonic acid, Evans et al. 1984). This substance appears to interact in a relatively specific manner with linear 1,3-β-glucans. The type of interaction of the fluorophore with these polymers is not absolutely clear. With 1,3-β-glucan oligomers an interaction through H-bonding appears to occur (Thistlethwaite et al. 1986). In addition, 1,3-β-glucans of a higher DP that adopt a triple-helix structure both in solution and in a dry state may also be capable of hydrophobic interaction with the fluorophore (Evans et al. 1984). As described by these authors and also evident from Table 1, fluoroescence is most pronounced with polymers exhibiting a linear backbone of 1,3-β-glucan and increases with DP (compare for instance laminarin / chrysolaminarin / pachyman). Chemically modified derivatives (O-carboxymethyl-pachyman) or 1,3-β-glucans with substituents at position 6 (scleroglucan, Phythophthora glucan) show less fluorescence, while β-glucans with both 1,3- and 1,4-linkages in a linear chain (barley glucan) exhibit essentially none.

The specificity of the aniline blue fluorescence procedure for the demonstration of callose has been questioned (e.g., Smith and McCully 1977). Thus regular cell walls always, and without obvious callose appositions, show some background fluorescence, although often with a slightly different color. This is especially pronounced at some specialized wall structures, e.g., at the cell walls of guard cells in barley leaves. The apparent lack of specificity of the cytochemical aniline blue fluorescence technique may partially result from the fact that fluorescence intensity depends on pH and ionic strength, and also from the presence in commercial aniline blue samples of several fluorophores in varying abundance. Accordingly, Stone et al. (1984) showed that less background fluorescence and a considerably improved contrast of callose-containing appositions are observed when the pure Sirofluor is used for histochemistry. The high price of Sirofluor prohibited, however, its routine use in the quantitative fluorometric callose assay described here. The problems due to lack of specificity have been overcome by making use of a solubilization step to isolate the callose, in combination with ad-

Table 1. Fluorescence efficiency of various polysaccharides in the aniline blue assay. Fluorescence was determined in a Kontron SMF 25 (sensitivity set to 357 V, exc. 393 nm, em. 479 nm) and is given in % relative to 4 μg of pachyman per 200 μl sample (total volume 1.4 ml)

Polymer used	Origin	Sol.[a]	Ref.	Relative fluorescence (%)		
				4 μg	40 μg	400 μg
Pachyman, DP 690-914	*Poria cocos*	A	b	*100*	–	–
Pachyman, DP 690-914	*Poria cocos*	D	b	44	–	–
Pachyman, 0-carboxymethyl	Derivative of pachyman	W	b	14	–	–
Paramylon, DP 710-790	*Euglena gracilis*	A	b	115	–	–
Curdlan, DP 200-620	*Alcaligenes* faecalis	A	c	121	–	–
Scleroglucan	*Sclerotium rolfsii*	A	d	36	–	–
Laminarin, DP ~ 20	*Laminaria digitata*	W	e	< 1	11	–
Laminarin	*Laminaria hyperborea*	W	f	< 1	9	–
Chrysolaminarin, DP ~ 34	*Poterioochromonas malhamensis*	W	g	13	98	–
Chrysolaminarin	*Navicula pelliculosa* (diatoms)	W	h	< 1	19	–
Barley 1,3/1,4-β-glucan	Endosperm cell walls	W	c	0	0	< 1
Hemicellulose A	Corn cobs	A	i	0	2.0	12
Hemicellulose B	Corn cobs	A	i	0	2.4	18
Zymosan A	*Saccharomyces cerevisiae*	A	j	< 1	26	–
Glucan A1	*Phythophtora parasitica*	W	k	0	1	7
Amylopectin	Corn endosperm	W	l	0	1	6
Glycogen	Rabbit liver	W	m	0	< 1	3
Pectin	Citrus fruits	W	l	0	< 1	5

[a] Solvent: W = solubilized in hot water, A = solubilized by heating to 80° in 1 N NaOH for 15 min, D = autoclaved at 121° for 15 min in dimethylsulfoxide and directly applied in 50 μl of this solvent (see text, section 2.5).
[b] Sample from B.A. Stone, Bundoora (Australia), DP according to Yamaguchi and Makino (1977).
[c] Sample from B.A. Stone, produced with *A. faecalis var. myxogenes,* strain 10C3K-V; DP according to Ogawa and Tsurugi (1973).
[d] Sample provided by A. Domard, Grenoble.
[e] Sigma, DP according to Evans et al. (1984).
[f] Sample from Ferak (Berlin).
[g] Chrysomonadales, Kauss and Kriebitzch (1969).
[h] Sample provided by B.E. Volcani, La Jolla (USA).
[i] Whistler and Feather (1965).
[j] Sigma.
[k] Provided by G. Franz, Regensburg, see Bruneteau et al. (1988).
[l] Sigma.
[m] Serva, Heidelberg (FRG).

equately buffered solutions. In addition, the use of a spectrofluorometer, in contrast to a fluorescence microscope, allows selection of the exact wavelengths for excitation and emission. In this procedure the use of Sirofluor instead of commercial aniline blue for the quantitative fluorometric callose determination appears to offer no real advantage, as illustrated by the following observation. Tobacco leaf discs were infiltrated with digitonin to induce callose formation. In respective extracts the same amounts of callose were determined both with decolorized aniline blue and with Sirofluor (Hildebrandt 1987).

The major advantage of the fluorometric callose assay described is that it properly reflects the situation as observed in the fluorescence microscope and exhibits a certain specificity for linear 1,3-β-glucans (Table 1). In addition, many samples can be handled in a comparatively short time, using small amounts of plant material. This allows extended time-course studies and the variation of many other parameters during physiological experiments. In contrast, permethylation analysis is not only more time-consuming but would also monitor for instance glucans of the barley type where both 1,3- and 1,4-β-linked glucose molecules occur in the same chain.

There are also some drawbacks in the fluorometric method for callose determination described here. The first of these results from the fact that in certain plants or tissues a considerable percentage of the callose cannot be readily solubilized in dilute alkali. The following two examples illustrate that the solubility properties of callose cannot be predicted. Most of our published work (Köhle et al. 1985; Waldmann et al. 1988) was performed with a soybean strain producing callose fully soluble in 1 N NaOH. These cells were originally provided by J. Ebel (Freiburg). Five years later a new subculture of the presumably same strain was again brought to our laboratory and found to form callose, but most of it was now insoluble in 1 N NaOH. Obviously, the original cell suspension culture had developed in the two laboratories into different lines. Similarly, digitonin-induced callose in tobacco leaf discs is partly alkali-insoluble (Hildebrandt 1987), whereas protoplasts from the same source can be induced by digitonin to produce fully alkali-soluble callose (unpublished). As outlined below, the problems related to the solubility of callose may, however, be overcome by partial acid hydrolysis with 2 N H_2SO_4 or 80% formic acid (see below), which presumably results in a decrease of the DP. Alternatively, dimethlysulfoxide might be used as a solvent (Sect. 2.5).

Other serious problems arise when soluble autofluorescent substances are present or are formed on treatment of the plant material with alkali. This problem may be avoided by selectively solubilizing them prior to the extraction of callose with alkali. If this is not possible, a correction may be achieved to a certain degree by the substraction of separately determined values for autofluorescence (see below). In one instance, both corrective manipulations were, however, only partially successful, namely for the determination of callose formed in local lesion-forming tobacco leaves infected with tobacco mosaic virus (Hildebrandt 1987). During the hypersensitive reaction various phenolic polymers appear to be formed, which are difficult to remove by simple washing procedures, and which also appear to interfere to some extent with the fluorescence efficiency of the callose / Sirofluor complex.

Another major limitation of the fluorometric callose determination stems from the fact that the magnitude of the fluorescence signal varies greatly with the degree of polymerization of the 1,3-β-glucans (Table 1, Evans et al. 1984). The absolute amount of callose measured depends, therefore, greatly on the polymer sample used to establish the calibration curve; using laminarin it may apparently be up to 100-fold higher than with pachyman (Table 1). Although absolute values of callose are not needed for the evaluation of the results in most physiological studies, it is, nevertheless, informative to obtain an estimate. With chitosan-treated soybean cells, for instance, fluorometry employing a pachyman calibration curve gave a callose value which was only about a fifth of that estimated from permethylation studies; presumably the callose produced in vivo contained at least some polymer chains which are shorter than pachyman (Köhle et al. 1985). The same authors found, in contrast, that the amount of callose produced in vitro from radioactive UDP-glucose by soybean microsomes was about the same, regardless whether it was determined fluorometrically and against a pachyman calibration curve or was calculated from the radiotracer content. On prolonged incubation at high substrate and optimal effector concentration the plasma-membrane-located 1,3-β-glucan synthase is obviously able to produce callose which approaches pachyman in chain length and, therefore, exhibits comparable fluorescence efficiency.

It should also be stressed that the quantitative fluorometric callose determination, as described here, allows a complete interpretation of the physiological events studied only when they are accompanied by fluorescence microscopy. The microscopic image is often more interpretable than quantitative measurements, and the eye appears to be surprisingly sensitive to single fluorescent callose spots, compared to a spectrofluorometer where the same amount of callose would be diluted over the entire measuring cuvette.

2 Procedures

The fluorometric callose procedure has evolved from experiments on the induction of callose deposition in suspension-cultured plant cells (Köhle et al. 1985; Waldmann et al. 1988). The method will, therefore, be first described for this material. Adaptation of the procedure for other plant tissues and for 1,3-β-glucans other than callose will then be discussed, in the hope of providing a guideline for the application of the method to new problems.

2.1 Preparation of Plant Material

Suspension-cultured cells were derived from *Glycine max, Catharanthus roseus, Petroselium hortense, Phaseolus vulgaris* and *Nicotiana tabacum*. The concentration of digitonin necessary to optimally induce callose formation varies from 15 to 30 μM, depending on the cell type, suspension medium, and physiological condition used. Callose formation may also be induced with other amphipathic compounds (Kauss 1987) or with 0.5 to 1.0 mg chitosan/g cell fresh weight (Köhle et al. 1985).

Before treatment, cells are washed on a funnel with 10 mM Tes / NaOH (pH 7.0) or 10 mM Mes / Bis-Tris buffer (pH 5.8) containing 2% (w/v) sucrose and 4% (v/v) of the respective growth medium. The cells (300 mg fresh weight /

5 ml) are then suspended in the same buffer. Callose induction in full-strength growth medium is also possible but generally requires about twofold higher concentrations of inducers. Five milliliters of cell suspension are pipetted to a 20 ml-vial, the inducer added in about 0.1 ml of water and the vial gently rolled on a roller mixer (42 rpm) for 2 to 4 h. The cells are than allowed to settle and about 3 ml of the supernatant is removed to allow, e.g., the determination of K^+-efflux. The vials are filled with ethanol (final conc 80% to 90%) and allowed to stand for at least 1 h in order to kill the cells and solubilize some autofluorescent material. If necessary, the cells can be kept in the ethanol overnight, or longer.

The cells are dried by suction on a 2.5-cm polyester net with about 17 µm apertures (e.g., Scrynel PE 17 HD from Zurich Bolting Cloth Mfg. Co, 8803 Rüschlikon, Switzerland), which can be used repeatedly. Several milliliters of 2 N H_2SO_4 are added to the cells on the filter which are then kept for 15 min at room temperature to allow a partial acid hydrolysis of the alkali-insoluble fraction of callose. The acid is then washed off with two portions of about 10 ml each of distilled water.

The acid step should be omitted if all the callose is readily soluble in alkali (see Sect. 2.4), as found for instance with cells of *C. roseus* and certain *G. max* strains. Under these conditions ordinary cellulose filters or Whatman glass microfiber filters (GF/A) instead of the polyester net may be used to collect the cells.

2.2 Callose Extraction and Assay

After washing, the cells are transferred with a spatula into a tightly fitting ground-glass or Teflon-glass Potter and homogenized in 5 ml of 1 N NaOH. Commonly, about 20 to 30 strokes are necessary to disintegrate all the cells. This is essential for the callose extraction step which follows and should be controlled microscopically (see below). Alternatively, desintegration of the cells in 3 ml of NaOH might be performed by sonification, e.g. for 30 sec with *C. roseus*. The suspension of broken cells is transferred to a centrifuge tube and heated at 80° for 15 min to solubilize the callose. After centrifugation (5 min, 380 g) the supernatant is used for the callose assay.

It was found that after pachyman in 1 N NaOH had stood at room temperature for up to 20 h there was no decrease in fluorescence efficiency. Similarly, after heating for 15 min at 80° the fluorescence efficiency decreased by only about 3%. Obviously, this 1,3-β-glucan is stable in 1 N NaOH to such a degree that the results are not greatly affected. It appears, however, possible that other 1,3-β-glucans are less stable under alkaline conditions. For this reason, the fluorometric assays should be completed on the day of cell extraction. An aliquot of the same 1 N NaOH cell extract may also be used to determine the protein content according to Lowry et al. (1951) to provide a correction factor if unequal amounts of cells or tissue were used within a given set of experiments, or if a considerable part of the cells were not homogenized.

Aliquots (200 µl) from the alkali extract (or 200 µl of 1 N NaOH for a blank) are mixed with 400 µl of 0.1% (w/v) aniline blue in water. This solution can be kept for several months at 4°. Good results have been obtained with aniline blue

WS from Merck (Darmstadt). Alternatively, a solution containing 5 μg/ml of pure Sirofluor (Biosupplies, P.O. Box 835, Parkville, Victoria 3052, Australia) may be used, although there was no advantage in terms of the assay (Hildebrandt 1987). When aniline blue is used the assay now exhibits a violet-red color. As a next step, 210 μl of 1 N HCl are added to cause a neutral to slightly acidic pH-value which is indicated by a change to a deep blue color. The final pH value is adjusted by addition of 590 μl of 1 M glycine / NaOH, pH 9.5, and the tubes are vigorously mixed. When the assay is adapted for a particular system, the solutions of aniline blue, acid and buffer might be combined and pipetted (1.2 ml) in one step.

After incubation for 20 min at 50° and a further 30 min at room temperature the aniline blue solution is essentially decolorized and the sample is ready for fluorescence measurement. Any slight turbidity should *not* be removed by centrifugation: About 20% of the fluorescence of water-insoluble 1,3-β-glucans, such as pachyman or callose, are lost on centrifugation for 10 min in an Eppendorf bench-top centrifuge.

2.3 Instrumentation and Calibration

The fluorescence of the sample may be read in any fluorometer allowing an excitation around 400 nm. The instrument is set to zero with a blank, containing all above additions except 1,3-β-glucan, and the fluorescence of the samples monitored. We routinely use a Kontron SFM 25 spectrofluorometer with excitation 393 nm and emission 479 nm (Table 1). A Jobin-Yvon JY3D spectrofluorometer (excitation 400 nm, emission 510 nm) and an filter fluorometer Eppendorf 1100+1030 (excitation filter 405+436 nm, cut-off filter 470 nm) have also been used. The assay volume given above is sufficient for a 0.4 × 1.0 cm half-micro cuvette but could be doubled to allow measurement in a standard cuvette. In any case, cuvettes must always be placed with the same side toward the excitation beam and should be fully translucent. It is recommended that only water be used for cuvette rinsing as acetone may lead to smearing on the surface.

At the Kontron SFM 25 the instrument sensitivity is determined by setting the high voltage. At 357 V the calibration curve is linear up to 4 μg pachyman, the highest value being set as 100% relative fluorescence (Table 1). At a higher sensitivity (400 V) 2 μg of pachyman represents the upper limit of the calibration curve. With laminarin from *Laminaria digitata* (Table 1) and a sensitivity setting to 400 V the calibration curve was linear up to 200 μg. When this or other water-soluble polymers are used, the solubilization with NaOH and the neutralization with HCl can be omitted.

2.4 How to Optimize for New Problems

If the fluorometric procedure described above has to be adapted for a particular tissue and a new physiological problem, several parameters have to be systematically altered in order to optimize the method and to circumvent as far as possible the drawbacks indicated in Sect. 1.3. To facilitate such an optimization, callose formation can be artificially induced by digitonin or chitosan, not only in suspension cells but also in cut-out leaf discs or stripes. A solution of 30 to 60 μM digitonin is vacuum-infiltrated for a few minutes. The samples are then allowed to stand for 3 to 6 h in order to induce massive callose formation.

During method optimization it is essential that the callose induction as well as the efficiency of its extraction by NaOH are controlled by fluorescence microscopy. For this purpose, the cells or tissues should be examined microscopically before and after homogenization. A 0.1% (w/v) solution of aniline blue WS in 1 M glycine/NaOH buffer, pH 9.5, can be used for this purpose. This solution can be kept for months at room temperature although its blue color fades rapidly. The staining solution is added to the tissue or cell samples on the microscope slide. The plant material can be examined either without any pretreatment, or after removal of chlorophyll by boiling in ethanol or standing in ethanol for a prolonged period at room temperature. Without extraction by ethanol details of the cell structure can easily be studied by additional observation under phase contrast, but a less clear fluorescence image is obtained. In contrast, after ethanol extraction fluorescence of callose spots is more intense and gives a clear image of wall details. If cells or cell debris after NaOH extraction have to be stained, they should be washed once by water to remove the alkali. A yellow fluorescence of the callose deposits is observed, for instance, using the Zeiss filter set 18 (exc. 390–420 nm, color splitter 425 nm, secondary light filter 450 nm). With green, living cells the Zeiss filter set 02 may be used (exc. 365 nm, color splitter 395 nm, secondary filter 420 nm) which suppresses the red chlorophyll fluorescence and provides a blue-white callose fluorescence.

When callose induction is clearly visible in the cells but the NaOH extracts show little or no fluorescence in the spectrofluorometer, it is likely that the pellet containing the walls will still fluoresce under the microscope, indicating the presence of NaOH-insoluble callose. In these cases the polymer might be rendered soluble using the partial acid hydrolysis step (2 N H_2SO_4 at room temperature) performed as described above for the suspension cells, followed by washing away of the acid. With small amounts of cells or with glucan samples this step could be performed by addition of 100 μl of 2 N H_2SO_4, followed – after a suitable time at room temperature – by neutralization with 100 μl of 2 N NaOH, and a further 200 μl of 2 N NaOH for the extraction at 80° for 15 min. Another possibility for partial acid hydrolysis of callose is to heat the sample at 80° with 80% (v/v) formic acid followed by drying the sample overnight in a stream of air (Ogawa and Tsurugi 1973). The time required for optimal hydrolysis has to be determined as an extended hydrolysis leads to a considerable decrease in fluorescence. A 20-min period was optimal for ethanol-extracted tobacco leaves homogenized directly in the 80% formic acid (Hildebrandt 1987).

The problem of autofluorescence may be fully or partly overcome by further extraction of the ethanol-treated sample. The appropriate solvent has to be determined, 20% ethanol or a buffer at pH 9.5 should be tried. In the case of protoplasts from barley leaves washing the ethanol-extracted material simply with water was successful. In any case, autofluorescence of the 1 N NaOH-extract must be monitored by running a parallel sample without aniline blue but containing all other assay constituents. If the values measured without aniline blue are relatively low, they can be subtracted from the values for the callose sample. In performing such a procedure it is, however, necessary to ensure by recovery experiments that little or no fluorescence quenching occurs. For this purpose a 1,3-β-glucan is added to the cell extract. The highest amount used for the calibration curve is appropriate (e.g., 4 μg of pachyman or 200 μg of laminarin per 200 μl extract). The wavelengths for maximal excitation and emission should also be controlled.

2.5 Callose Extraction with Dimethylsulfoxide

As an alternative to the extraction with dilute alkali, callose might also be solubilized by autoclaving in dimethylsulfoxide, as suggested by Pillonel et al. (1980). We have recently used this procedure with protoplasts from barley leaves (unpublished). The fluorometric callose assay can be used with the direct addition of up to 50 μl of this solvent, as outlined below.

Protoplasts (6×10^5) were suspended in 3 ml of 25 mM Tes / NaOH, pH 7.0, containing 5 mM KCl, 0.12 mM $CaCl_2$, 0.6 M glucose and 2% (w/v) sucrose and incubated for 10 to 18 h in the presence of 15 to 25 μM digitonin. After chlorophyll extraction with about 85% ethanol the pellet was washed twice with water to remove autofluorescent material. The material was transferred into 3-ml heavy-walled round-bottom Sovirel tubes, dried by a stream of air for about 1 h and 0.2 ml of dimethylsulfoxide added. The tubes were sealed with a Teflon-lined screw-cap and autoclaved for 30 min at 121°. After cooling to room temperature the tubes were centrifuged in a bench-top centrifuge and 40 μl of the supernatant mixed with 360 μl of water, 400 μl of 0.1% (w/v) aniline blue WS in water and 600 μl of 1 M glycine / NaOH, pH 9.5. After heating for 20 min at 50° and a further 30 min at room temperature the fluorescence was read as outlined in Sect. 2.2.

It is evident from Table 1 that pachyman autoclaved in dimethylsulfoxide and directly added into the assay in 50 μl of this solvent gives a lower fluorescence, compared to a solution in alkali (line 1). This appears not to be due to a partial decomposition of the polymer, as a sample solubilized in the same solvent by heating for 15 min to 80° gave identical fluorescence values. Addition of the pachyman in more than 50 μl of dimethylsulfoxide per 1.4 ml standard assay resulted in a still more drastically decreased fluorescence, indicating that it is the solvent itself which affects the fluorescence efficiency of the complex between polymer and fluorophore (see also Evans et al. 1984). Nevertheless, with increased spectrofluorometer sensitivity (390 V instead of the 357 V used for Table 1) a calibration curve could be established with 0 to 4 μg of pachyman added in 40 μl of dimethylsulfoxide. Again, only releative amounts of callose can be determined using such a calibration curve, but this rapid procedure was sufficient to work out many physiological parameters involved in callose formation by barley protoplasts.

3 Application to Other 1,3-β-Glucans

The assay procedure described above for callose could also be used for the determination of other 1,3-β-glucans in biochemical or physiological experiments.

As first suggested by Evans et al. (1984) and at present worked out in detail by A. Bonhoff and H. Grisebach (Freiburg, pers. comm.), the assay may be used to determine the activity of 1,3-β-glucanases. At the spectrofluorometer sensitivity used for Table 1 up to 400 μg of laminarin in 200 μl aliquots may serve as a substrate. These aliquots may be taken over time, heated for a few minutes and the decrease in fluorescence with aniline blue determined according to Sect. 2.3, but without the use of 1 N NaOH and HCl for its neutralization.

In addition, studies on the biosynthesis or degradation of laminarin or chrysolaminarin appear to be timely. These reserve glucans are produced in tremendous amounts by various algae – especially marine Chromophyta – but few details of the respective metabolism are known (for a 1,3-β-glucan phosphorylase see Kauss and Kriebitzsch 1969).

Several 1,3-β-glucans have striking effects on the immune system in mammals and this might, in part, be due to the fact that they can trigger the alternate pathway of the complement system. One important impact of this phenomenon might be the antitumor action of some plant and fungal branched 1,3-β/1,6-glucans (see Franz and Kraus 1987). Similarly, there are also reports on the activation of the prophenoloxidase activating system in arthropods by fungal 1,3-β-glucans, a defence system exhibiting some similarities with the complement system in mammals (Söderhall and Hall 1984). Studies on the biochemical and cellular mechanisms underlying these various pharmacological effects might be greatly facilitated by the rapid and sensitive method for the determination of 1,3-β-glucans which is described here.

Acknowledgments. The work in my laboratory was financially supported by grants from the „Deutsche Forschungsgemeinschaft" and from the „Fonds der chemischen Industrie". The present report is the summary of the experience made over several years by the following co-workers: H. Köhle, W. Jeblick, Th. Waldmann, E. Hildebrandt, A. Deutsch and B. Pranke. The generous supply of substances and the helpful discussions of W. Blaschek (Regensburg), A. Domard (Grenoble), G. Franz (Regensburg), H. Meier (Fribourg), and especially of B. A. Stone (Bundoora) are also acknowledged.

References

Bruneteau MJ, Fabre J, Perret J, Michel S, Ricci P, Joseleau JP, Kraus J, Schneider M, Blaschek W, Franz G (1988). Antitumor active β-D-glucans from *Phytophthora parasitica*. Carbohydr Res 175:137–143

Eschrich W (1956) Kallose. Protoplasma 47:487–530

Evans NA, Hoyne PA, Stone BA (1984) Characteristics and specificity of the interaction of a fluorochrome from aniline blue (Sirofluor) with polysaccharides. Carbohydr Polymers 4:215–230

Fincher GB, Stone BA (1981) Metabolism of noncellulosic polysaccharides. In: Tanner W, Loewus FA (eds) Encyclopedia of plant physiology, vol 13B. Springer, Berlin Heidelberg New York, pp 68–132

Fink J, Jeblick W, Blaschek W, Kauss H (1987) Calcium ions and polyamines activate the plasma membrane-located 1,3-β-glucan synthase. Planta 171:130–135

Franz G, Kraus J (1987) Pflanzliche Polysaccharide mit Antitumorwirkung. Z Phytother 8:114–118

Hildebrandt E (1987) Untersuchungen zur Callosebildung während der Läsionsentwicklung bei der Tabaksorte *N. tabaccum* var. Samsun NN. nach der Infektion mit Tabakmosaikvirus. Thesis, University of Kaiserslautern

Hinch JM, Clarke AE (1982) Callose formation in *Zea mays* as a response to infection with *Phytophthora cinnamomi*. Physiol Plant Pathol 21:113–124

Kauss H (1985) Callose biosynthesis as a Ca^{2+}-regulated process and possible relations to the induction of other metabolic changes. J Cell Sci Suppl 2:89–103

Kauss H (1987) Some aspects of calcium-dependent regulation in plant metabolism. Annu Rev Plant Physiol 38:47–72

Kauss H, Kriebitzsch C (1969) Demonstration and partial purification of a β-(1—→3)-glucan phosphorylase. Biochem Biophys Res Commun 6:926–930

Köhle H, Jeblick W, Poten F, Blaschek W, Kauss H (1985) Chitosan-elicited callose synthesis in soybean cells as a Ca^{2+}-dependent process. Plant Physiol 77:544–551

Lowry OH, Rosebrough NH, Farr AL, Randall RJ (1951) Protein measurement with the folin phenol reagent. J Biol Chem 193:265–275

Ogawa K, Tsurugi J (1973) The dependence of the conformation of a (1–3)-β-D-glucan on chain-length in alkaline solution. Carbohydr Res 29:397–403

Pillonel C, Buchala AJ, Meier H (1980) Glucan synthesis by intact cotton fibres fed with different precursors at the stages of primary and secondary wall formation. Planta 149:306–312

Smith MM, McCully ME (1977) A critical evaluation of the specificity of aniline blue-induced fluorescence. Protoplasma 95:229–254

Söderhall K, Hall L (1984) Lipopolysaccharide-induced activation of prophenoloxidase activating system in crayfish haemocyte lysate. Biochim Biophys Acta 797:99–104

Stone BA, Evans NA, Bonig I, Clarke AE (1984) The application of Sirofluor, a chemically defined fluorochrome from aniline blue for the histochemical detection of callose. Protoplasma 122:191–195

Thistlethwaite P, Porter I, Evans N (1986) Photophysics of the aniline blue fluorophore: a fluorescent probe showing specificity toward (1—→3)β-D-Glucans. J Phys Chem 90:5058–5063

Waldmann T, Jeblick W, Kauss H (1988) Induced net Ca^{2+}-uptake and callose biosynthesis in suspension-cultured plant cells. Planta 173:88–95

Whistler RL, Feather MS (1965) Hemicellulose. Extraction from annual plants with alkaline solutions. Methods Carbohydr Chem 5:144–145

Yamaguchi H, Makino K (1977) Studies on a microchemical method for determination of the degree of polymerization of neutral oligo- and polysaccharides. J Biochem 81:563–569

Measuring β-Glucan Deposition in Plant Cell Walls

T. Hayashi

1 Introduction

Four types of β-glucans are known to occur in the cell walls of higher plants. These are mixed-linkage glucans (only in monocotyledons),1,3-β-glucan, cellulose, and xyloglucan. The mixed-linkage glucans contain 1,3- and 1,4-linkages distributed in a typical sequence in most species (Bacic and Stone 1980; Kato et al. 1981 a, 1982; Staudte et al. 1983), but at random in maize (Kato and Nevins 1984). The mixed-linkage glucans are characteristically found in monocotyledonous coleoptiles (Wada and Ray 1978), leaves (Kato et al. 1981 a), stems (Wilkie 1979), and seeds (Nevins et al. 1978; Stinard and Nevins 1980). In endosperm and aleurone cell walls, the polysaccharides exist in several forms and can be extracted with water and with alkali. The matrix of these glucans appears to be a gel-like structure in the middle lamella of cell walls.

1,3-β-Glucan (callose) is generally considered to be a wound-response glucan (Currier 1957), but in some unusual cases it is synthesized by nonperturbed systems such as developing cotton fibers (Huwyler et al. 1978; Maltby et al. 1979) and elongating pollen tube (Vithanage et al. 1980). The backbone of cotton fiber 1,3-β-glucan has been found to contain a small proportion of branches at C-*O*-6 (Hayashi et al. 1987 b). In many studies on cultured protoplasts, the first and major polysaccharide generated is callose (Blaschek et al. 1981; Klein et al. 1981; Takeuchi and Komamine 1981; Hayashi et al. 1986). Wounded or stressed plants often screte masses of this glucan into periplasmic spaces (Drake et al. 1978) and it is always encountered in products formed by plasma membranes from UDP-glucose in vitro (R. L. Anderson and Ray 1978; Mueller and Maclachlan 1983). This enzyme activity (Glucan Synthetase II) is therefore commonly used as a plasma membrane marker (Ray 1979). This enzyme may be involved in defence mechanisms (Aist 1983) in higher plants. Callose synthases, as solubilized by digitonin, are commonly of high molecular weight (> 500,000) and present a doughnut-shaped structure similar in size to the rosettes and terminal globules obserbed in freeze-fracture studies (Hayashi et al. 1987 b). It has been suggested (Delmer 1977) that the enzymes may be altered forms of cellulose synthase.

Recently, a small quantity of 1,3-β-glucan (Kato and Nevins 1985) has been isolated from a hot-water extract of *Zea mays* shoot cell-wall preparations which consist mostly of mixed-linkage glucan. In the biosynthesis of *Lolium multiflorum* mixed-linkage glucan (Smith and Stone 1973), the proportion of 1,3-and 1,4-linkages in the total products has been shown to vary with the concentration of UDP-glucose. At 1 μM UDP-glucose, 90% of the residues in the total product are 1,4-linked, whereas the proportion falls to 49% at 1 mM UDP-glucose. Therefore, the ratio of 1,3-linkages to 1,4-linkages in the enzyme system is probably dependent on the concentration of UDP-glucose. However, it is uncertain whether the formation of these two β-glucans (1,3-β-glucan and mixed-linkage glucan) are regulated in vivo by concentrations of UDP-glucose.

Cellulose is a simple polymer composed of glucose residues joined together by 1,4-β-linkages, which has a strong tendency to self-associate into microfibrils. Xyloglucan occurs widely in the primary walls where it is found in close association with cellulose microfibrils as part of the basic wall structure. Xyloglucan possesses a cellulose-like 1,4-β-glucan backbone substituted with 1,6-α-xylosyl residues. Species-specific differences occur as to the distribution of additional side-chains containing fucosyl-galactosyl residues in dicots, and galactosyl residues in monocots.

Xyloglucan isolated from plant cell walls can be bound to cellulose in vitro (Hayashi et al. 1987a). Such reconstitution probably creates an organized xyloglucan : cellulose complex. The macromolecular complex of xyloglucan and cellulose has been isolated from elongating regions of etiolated pea epicotyls (Hayshi and Maclachlan 1984). The association between the two polysaccharides is very strong because concentrated alkali (24% KOH) is needed to dissolve the complex. Standard reagents for breaking weak hydrogen bonds such as 8 M urea are ineffective, and DMSO extracts only a small part of the xyloglucan. However, the xyloglucan component of this complex is much more accessible to endoglucanase hydrolysis than cellulose microfibrils (Hayashi et al. 1984). Xyloglucan probably functions in the primary cell walls as a cementing matrix materials which contributes cross-links and rigidity to the cellulose framework. The integrity of xyloglucan could then control the ability of microfibrils to loosen and the whole cell to expand during growth. The biosynthesis and biodegradation of the two 1,4-β-glucans (cellulose and xyloglucan) is regulated by different processes. Cellulose is made at the plasma membrane (Shore and Maclachlan 1975; Mueller and Brown 1980; Herth and Weber 1984) and its assembly is probably regulated by xyloglucan to prevent the fasciation of microfibrils and to spread the microfibril network over the cell surface (Hayashi et al. 1987a), while xyloglucan is made in the Golgi (Ray 1980; Hayashi and Matsuda 1981a; Camirand et al. 1987; Hayashi et al. 1988) by concurrent transfer of glucose and xylose (Hayashi and Matsuda 1981b) and its turnover is controlled by auxin-induced endo-1,4-β-glucanases (Hayashi et al. 1984).

This chapter describes the chemical and fragmentation analysis of β-glucans in higher plants. The chemical analysis includes the mass fragmentation patterns of the alditol acetates obtained from permethylated β-glucans. A part of the fragmentation analysis deals with the enzymic degradation of β-glucans. In addition, the visualization of β-glucans in plant tissues by microscopy is described.

2 Chemical Analysis

Several procedures for the various steps of chemical analysis and chromatography have been reported previously. In the following paragraphs, only the procedures currently used in our laboratory are described in detail.

2.1 Colorimetry

β-Glucans can be measured quantitatively according to the methods described below. The phenol-sulfuric acid method which measures total sugar contents (Dubois et al. 1956) is not specific for the determination of β-glucans. The Nelson-Somogyi method which measures reducing sugars (Somogyi 1952) is mostly used to determine the hydrolysis rate of polysaccharides through detection of new reducing ends. The iodine-sodium sulfate method (Kooiman 1960) is a specific assay for xyloglucan.

A. Phenol-sulfuric Acid Method. Neutral sugars, oligosaccharides, polysaccharides and their derivatives including the methyl ethers with free or potentially free reducing groups give an orange-yellow color when treated with phenol and concentrated sulfuric acid (Dubois et al. 1956). The reaction is sensitive (1–20 μg) and the color is stable for 24 h. A sugar solution (0.2 ml) containing 1–20 μg of sugars and 0.2 ml of 5% phenol (w/w) are mixed and 1.0 ml of concentrated sulfuric acid (96.9%) is added rapidly, the stream of acid being directed against the liquid surface rather than against the side of the test tube. The mixture is vortexed gently to obtain good mixing. The tubes are allowed to stand 30 min and the optical density at 490 nm is measured. A linear standard curve is obtained using glucose for β-glucan estimation. An OD of 1 corresponds to approximately 22 μg per 0.2 ml of sugar solution for glucose.

An important application of this technique is the quantitative determination of cellulose. This cellulose assay first requires the pretreatment of biological materials to remove other carbohydrates such as pectins and hemicelluloses by alkali (24% KOH or 17.5% NaOH). The resulting insoluble materials is washed with water and 1 M acetic acid, and solubilized with ice-cold 72% sulfuric acid for 1 h. A portion of the mixture is analyzed by the phenol/sulfuric acid method as described above. Alternatively, plant materials can be extracted with acetic-nitric reagent (80% acetic acid/conc nitric acid, 10:1) in a boiling water bath for 30 min (Updegraff 1969). The resulting insoluble material is washed with water and solubilized with ice-cold 72% sulfuric acid for 1 to 12 h, and the amount of cellulose is determined by phenol/sulfuric acid method.

B. Nelson-Somogyi Method. This method (Somogyi 1952) measures the reducing power of sugars. To 0.5 ml of sugar solution containing 2–40 μg of reducing sugar is added 0.5 ml of reagent A into a borosilicate tube. The tube is put in a boiling water bath for 20 min and cooled in cold water bath. Then, 0.5 ml of reagent B is added, and the mixture is vortexed to remove forming CO_2. After 30 min, the absorbance is measured at 500 nm. A linear standard curve is obtained with a standard glucose solution: an OD of 1 corresponds to approximately 34 μg per 0.5 ml of sugar solution for glucose.

Reagent A. (a) Dissolve 12 g of sodium carbonate and 6 g of sodium potassium tartrate in 125 ml of distilled water. Add 20 ml of 10% copper sulfate solution, and then dissolve 8 g of sodium bicarbonate in the mixture. (b) Dissolve 90 g of sodium sulfate in 250 ml of boiling distilled water and boil the mixture for

5 min. (c) Mix solutions a and b in a graduated cylinder and bring to 500 ml with distilled water. Store this reagent in a dark bottle.

Reagent B. Dissolve 25 g of ammonium molybdate in 450 ml of distilled water and add to the solution 21 ml of concentrated sulfuric acid. In a separate beaker, dissolve 3 g of sodium arsenate in 20 ml of distilled water. Add this to the ammonium molybdate solution and bring volume to 500 ml. Incubate at 37° C for 24 h and store this reagent (reagent B) at room temperature.

C. Iodine-sodium Sulfate Method. This method is a specific assay for xyloglucan (Kooiman 1960). To 0.2 ml of a solution containing 3–60 µg of xyloglucan are added 0.1 ml of a solution of 0.5% I_2 in 1% KI and 1.0 ml of 20% Na_2SO_4. The mixture is vortexed and the tube is allowed to stand 60 min at 4° C in the dark. The absorbance is measured at 640 nm. A linear standard curve is obtained with xyloglucan and an OD of 1 corresponds to approximately 60 µg per 0.2 ml of sugar solution for pea xyloglucan. Since soluble starch and glycogen are also stained with the iodine reagent, the sample should be pretreated with salivary α-amylase. It should be noted that the staining of xyloglucan with iodine is dependent on the molecular weight of the polysaccharide; poor coloring was observed using low-molecular-weight pea xyloglucan (less than MW 10,000).

2.2 Chromatography

The fragment oligosaccharides from fragmented β-glucans are measured by gel filtration and/or paper chromatography and the composition in monosaccharides is established by gas chromatography. Both gel filtration and paper chromatography are conducted effectively at elevated temperatures. In our laboratory, a range of 50°–80° C has been found to be optimal for gel filtration and paper chromatography: high temperatures improve the resolution and reduce the separation time. We show the procedure of high-temperature paper chromatog-

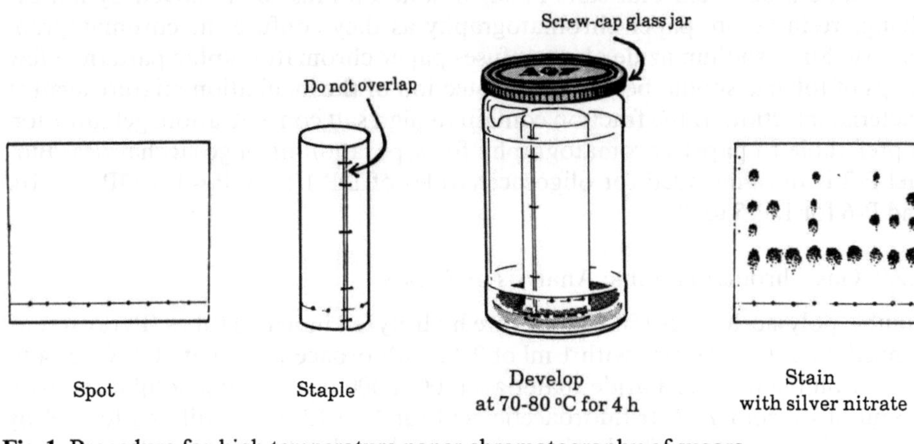

Fig. 1. Procedure for high temperature paper chromatography of sugars

raphy in detail (Fig. 1) because this would give easier and very powerful for sugar analysis.

2.2.1 High-Temperature Paper Chromatography of Sugars

Paper chromatography is most useful for characterization of enzymic hydrolyzates. High-temperature paper chromatography (Fig. 1) is performed at 70°–80° C on Whatman 3MM filter paper (18 × 22 cm) with 1-propanol/ethyl acetate/water, 7:1:2 or 3:3:1 (Hayashi et al. 1980). The 3:2:1 solvent is used to characterize disaccharides, i.e., isoprimeverose (R_{glc} 0.76), cellobiose (R_{glc} 0.61), laminaribiose (R_{glc} 0.76) and gentiobiose (R_{glc} 0.50), and the 7:1:2 is for high-molecular-weight oligosaccharides. Samples are applied 2–3 cm apart along a line 2 cm from the bottom of the paper; the paper is curled into a cylinder, stapled, and placed in a glass jar (kitchen jar or instant coffee bottle) containing the solvent. The jar is tightly sealed with a screw-cap and placed in a 70°–80° C oven for 4 h. High temperatures shorten the time of development and give better separation of sugars.

Sugars on the chromatogram are detected by the alkaline silver nitrate method (Robyt and French 1963). The dried paper is developed by rapidly dipping it into saturated silver nitrate/acetone, 1:1000 (v/v). The paper is air-dried and dipped into 40% sodium hydroxide/methanol, 1:100 (v/v). Sugar spots develop immediately and the paper is rapidly dipped into Kodak film fixer (or 20% $Na_2S_2O_3$) to whiten the background of the chromatogram. The paper is washed with tap water for 30 min and air-dried. The chromatogram obtained is stable without decolorization. This method of chromatography is very convenient and gives reproducible results.

For quantitative determination of sugars on paper chromatogram, the unstained paper is cut into 5 mm strips (2 cm wide) from origin to top and each strip is then eluted with 1 ml of water for 30 min in a test tube without agitation. Aliquots of the sugar solution are transferred into borosilicate tubes and the sugar in the solutions is determined by the phenol/sulfuric acid method (Dubois et al. 1956), as described earlier.

It should be noted that salts in sugar solution must be removed by ion-exchange resin before paper chromatography as they confuse the chromatogram pattern. Since sodium azide also confuses paper chromatographic pattern, a few drops of toluene should be used for protection of the incubation mixture against bacterial infection. If the fraction contains a high salt concentration, gel filtration is preferable to paper chromatography for separation of oligosaccharides. Bio-Gel P-2 is recommended for oligosaccharides of DP 1 to 6, P-4 for DP 3 to 10, and P-6 for DP 5 to 30.

2.2.2 Gas Chromatographic Analysis of Sugars

Soluble polysaccharides (50–500 µg) are hydrolyzed in sealed tubes (Pyrex screw-capped tube, 1.3 × 10 cm) with 1 ml of 2 M trifluoroacetic acid at 100° C for 4 to 6 h. Insoluble polysaccharide materials (100–1 000 µg) are first solubilized with 0.3 ml of ice-cold 70% trifluoroacetic acid for 1 to 12 h, and diluted to 2 M by the addition of water (0.7 ml). Following this, the solubilized materials are hydro-

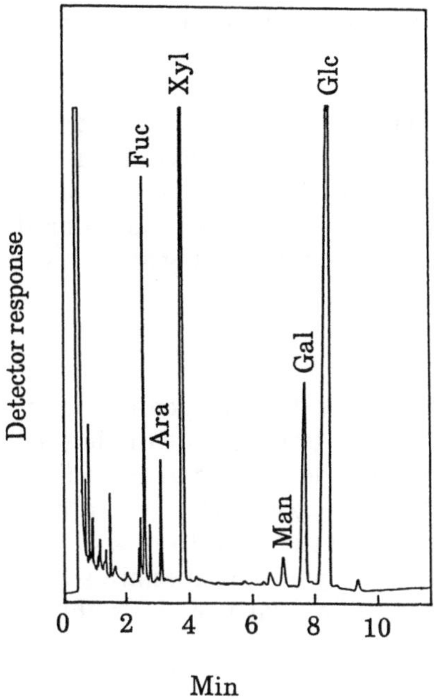

Fig. 2. Gas chromatographic pattern of alditol acetates from partially purified pea xyloglucan. The polysaccharide fraction was obtained by extraction of the 4% KOH-insoluble material of pea epicotyls with 24% KOH

lyzed at 100° C as described for soluble polysaccharides, and the hydrolyzates are evaporated to dryness.

The preparation of alditol acetates from monosaccharides is carried out according to the method described by Blakeney et al. (1983). The hydrolyzates are reduced with 1 ml of 0.5 M $NaBH_4$ in 2 M NH_4OH at 60° C for 1 h. The mixture is again concentrated to dryness and 0.1 ml of acetic acid, 0.5 ml of ethyl acetate and 1.5 ml of acetic anhydride are added and mixed well. Fifty µl of perchloric acid is added into the mixture. After 5 min, the tube is cooled on ice, and 4 ml of ice-cold water and 0.1 ml of 1-methyl-imidazole are added and the mixture kept on ice for 5 min. One ml of dichloromethane is then added into the mixture and vortexed well. The resulting lower dichloromethane layer is transferred into another tube and the dichloromethane is concentrated to dryness under nitrogen gas. The resulting alditol acetates are dissolved in 1.5 ml of dichloromethane and passed through filter paper to remove insoluble materials. The alditol acetates are examined by gas-liquid chromatography at 200° C on a glass capillary column (0.25 mm × 15 m) of DB-225 (0.25 µm) obtained from J & W Scientific Inc., California, USA. The chromatographic pattern of partially purified pea xyloglucan is shown in Fig. 2.

2.3 Methylation Analysis

Methylation analysis is one of the most important procedures for the study of linkages in the polysaccharide. The procedures in this section include permethylation of all hydroxyl groups in the polysaccharide, acid hydrolysis of the permethylated polysaccharide to give a mixture of partially methylated sugars, reduction of the sugars to partially methylated alditols, acetylation to partially methylated alditol acetates, and qualitative and quantitative analysis of the alditol acetates by GC-MS. The free hydroxyls formed after hydrolysis of the permethylated polysaccharide show the positions in which the sugar residues are substituted. Therefore, one of the most important criteria in the procedures is the completeness of the methylation of the polysaccharide. Methylation was originally described by Hakomori (1964) and modified by Sandford and Conrad (1966). The methylated samples are converted into the corresponding alditol acetates and analyzed by the method of Lindberg (1972).

2.3.1 Preparation of Methyl Sulfinyl Carbanion

All procedures should be conducted under a stream of dry nitrogen or argon. Sodium hydride in mineral oil (2 g) is carefully washed four times with dry hexane to remove mineral oil. Twenty ml of dry dimethyl sulfoxide is added to the sodium hydride and the mixture is stirred at 52°–55° C (never exceed 60° C) on a heat plate. After 1 h, a green solution is obtained and transferred into tubes (5 ml each). Since the solution should be stored under nitrogen, the tubes are sealed with a rubber septum and a screw-cap. The methyl sulfinyl carbanion is stable at 4° C for at least 2 years.

2.3.2 Methylation of Polysaccharides

All procedures should be conducted under dry nitrogen or argon. The polysaccharide (50–1 000 µg) in opened Pyrex screw-cap tube (1.3 × 10 cm) is kept in a dessicator with phosphorus pentaoxide overnight, and then dissolved in dry 0.5 ml of DMSO and sealed under nitrogen. The tube is agitated in the ultrasonic bath for 30 min to dissolve the sample, and methyl sulfinyl carbanion (0.5 ml) is added. The mixture is sealed under dry nitrogen and agitated in an ultrasonic bath at 45° to 55° C for 4 h. Methyl iodide (0.5 ml) is then added dropwise with external cooling in ice water. The first few drops should be very carefully added into the mixture with shaking to allow the complete methylation of sugar. The reaction mixture is again agitated in the ultrasonic bath for 30 min. Two ml of water are then added and the mixture is dialyzed against tap water overnight. The dialyzed polysaccharide solution is concentrated to dryness under a stream of air in a Pyrex tube. Most polysaccharides require at least two successive repeats of the methylation procedure described above for complete methylation. The levels of methylation can be checked by the decrease in –OH absorption (3 200–3 700 cm^{-1}) of the sample in chloroform by IR analysis. After methylation is completed, the polysaccharide is extracted twice with 2 ml of chloroform, and concentrated to dryness in a Pyrex tube.

Methylation of oligosaccharides is accomplished by a single methylation procedure. The methylated sugars are solubilized in 2 ml of chloroform and washed ten times with water to remove salts and dimethyl sulfoxide.

2.3.3 Preparation of Partially Methylated Alditol Acetates

The methylated polysaccharide is dissolved in 1.5 ml of 2 M trifluoroacetic acid and hydrolyzed for 1 h at 120° C in a sealed tube on a heat block. The mixture is concentrated to dryness and 1 ml of 0.5 M $NaBH_4$ in 2 M NH_4OH is added to the tube. The partially methylated aldoses are reduced in the sealed tube at 60° C for 1 h. The resulting alditols are then acetylated according to the method described earlier (Blakeney et al. 1983). It should be noted that partially methylated xylitol acetates are especially volatile and may be lost during concentration of solutions. Therefore, methylation analysis of xyloglucan sometimes gives relatively low levels of terminal xylose residues.

2.3.4 Gas Chromatographic Analysis of Partially Methylated Alditol Acetates

Complete separation of partially methylated alditol acetates is required for the qualitative and quantitative analyses. The best separation and quantitation of partially methylated alditol acetates is obtained from gas-liquid chromatography using a glass capillary column with helium gas as a carrier. Combined gas chromatography–mass spectrometry (GC-MS) is a preferable technique for identification of the acetates. The recommended column is DB-225 (0.25 mm × 15 m) which can also be used for the determination of sugar composition (Fig. 2). The column oven temperature is raised linearly from 140° to 190° C at a rate of 2° C/min. The gas-liquid chromatographic pattern of cotton 1,3-β-glucan is shown in Fig. 3 A and that of cotton xyloglucan in Fig. 3 B.

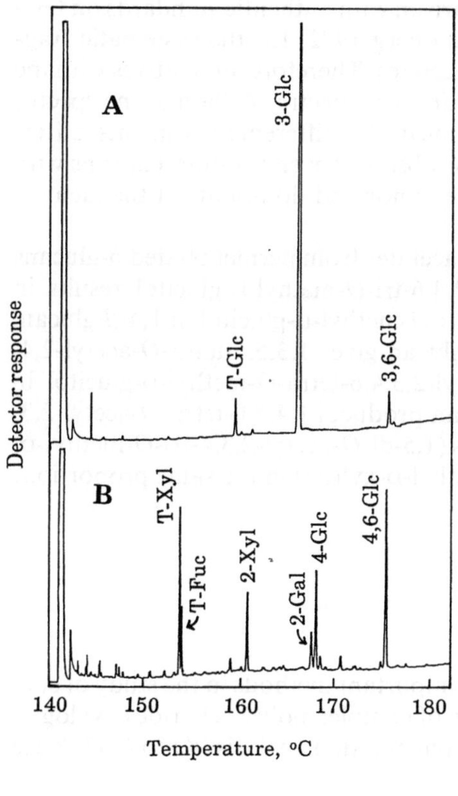

Fig. 3 A, B. Gas chromatographic patterns of partially methylated alditol acetates from cotton 1,3-β-glucan (**A**) and xyloglucan (**B**)

Table 1. Detector response factors for different alditol acetates from permethylated β-glucans

Partially methylated alditol acetates	Carbon content (numbers)	Response factor
Trimethyl glucitol	15	1
Tetramethyl glucitol	14	0.93
Dimethyl glucitol	16	1.07
Trimethyl xylitol	12	0.80
Dimethyl xylitol	14	0.93
Trimethyl fucositol	14	0.93

When a flame-ionization detector (FID) is used, there is a linear relationship between detector response and carbon amount. Therefore, as shown in Table 1, the molar response must be calibrated by the response factors of carbon content for different partially methylated alditol acetates.

2.3.5 GC-MS Analysis of Partially Methylated Alditol Acetates

The partially methylated alditol acetates are separated on the chromatography column and the mass spectrum of each compound is interpreted. Nonsugar components from impurities in the original sample or in the solvent sometimes give strong peaks, which may cause confusion. These extra peaks are distinguished from those of the alditol acetates by comparison with authentic standards on GC-MS. There are certain generalizations (Lindberg 1972) for the systematic fragmentation of partially methylated alditol acetates. Therefore, derivatives with the same substitution pattern show very similar mass spectra. Although the spectra should be essentially the same when determined with different instruments, differences in relative intensities of peaks may be observed even for different runs with the same instrument. These differences are minor and do not affect the identifications.

The mass spectra of the typical alditol acetates from permethylated β-glucans are shown in Fig. 4. 1,3,5-Tri-*O*-acetyl-2,4,6-tri-*O*-methyl-D-glucitol results in 1,3-β-glucan and 1,4,5-tri-*O*-acetyl-2,3,6-tri-*O*-methyl-D-glucitol in 1,4-β-glucan. Methylation analysis of branched 1,3-β-glucan gives 1,3,5,6-tetra-*O*-acetyl-2,4-di-*O*-methyl-D-glucitol and 1,5-di-*O*-acetyl-2,3,4,6-tetra-*O*-methyl-D-glucitol in the ratio of 1:1. Methylation of xyloglucan produces 1,4,5,6-tetra-*O*-acetyl-2,3-di-*O*-methyl-D-glucitol and methyl xylitols (1,5-di-*O*-acetyl-2,3,4-tri-*O*-methyl-D-xylitol plus 1,2,5-tri-*O*-acetyl-3,4-di-*O*-methyl-D-xylitol) in the same proportion.

3 Fragmentation Analysis

Fragmentation analysis is one of the most important methods in the study of subunit structure, sequence, and fine structure of complex polysaccharides. Xyloglucan is well suited for characterization by fragmentation analysis (Kato 1978; Kato and Matsuda 1980; Hayashi et al. 1980; Hayashi and Maclachlan 1984) because the polysaccharide is composed of subunit structures (Hayashi et al. 1981; Hayashi and Matsuda 1981c). In the study of cotton 1,3-β-glucan, a small proportion of branches at C-*O*-6 was discovered (Hayashi et al. 1987b) by degradation of the polysaccharide with *Basidiomycete* exo-1,3-β-glucanase and identification of gentiobiose (6-*O*-β-D-glucopyranosyl-D-glucopyranose). Table 2 shows the action mode of β-D-glucanases for use in the fragmentation analysis of β-glucans. The fragments obtained by enzymic hydrolysis are also useful for the identification of polymers synthesized in vitro by plant enzyme preparations.

3.1 Mixed-Linkage Glucan

Lichenase does not hydrolyze simple 1,4-β-glucan 1,3-β-glucan, and xyloglucan, but degrades 1,3- and 1,4-mixed glucan (M.A. Anderson et al. 1976; Huber and Nevins 1977). The enzyme recognizes successive (1→3)-β-glucosyl-(1→4)-β-glucosyl residues but hydrolyzes only the 1,4-β-glucosidic linkage. The degradation by lichenase is very specific and the affinity for the glucan is very high. Therefore, the identification and quantitation of mixed-linkage glucan has been commonly carried out by measuring the characteristic oligosaccharide fragments (3-*O*-β-cellobiosyl-D-glucose and 3-*O*-β-cellotriosyl-D-glucose) released by lichenase (EC 3.2.1.73) prepared from *Bacillus subtilis* (Anderson et al. 1976; Huber and Nevins 1977) or germinating barley (Woodward and Fincher 1982). Lichenase preparation from *Bacillus subtilis* (usually obtained as α-amylase preparation of Sigma) does not contain other endo-β-glucanase activities and is good enough to use directly for hydrolysis.

3.2 1,3-β-Glucan and Cellulose

Most of 1,3-β-glucans contain a small number of β(1→6) branches in the glucan backbone, as observed in cotton fiber 1,3-β-glucan (Hayashi et al. 1987b), or internal backbone β(1→6) linkages, as observed in laminarin. Methylation analysis is required to determine whether gentiobiose is resulting from branched residues or from internal linkages. Cotton fiber 1,3-β-glucan is 48% digested by *Rhizopus arrhizus* endo-1,3-β-glucanase and completely digested by *Basidiomycete* exo-1,3-β-glucanase. Hydrolysis products with the endo-enzyme were glucose, laminaribiose, and a small amount of laminaritriose. Degradation with the exo-enzyme yielded glucose and gentiobiose in a molar ratio of about 20:1. Since this enzyme acts on the nonreducing end of chains, and methylation analysis of the polysaccharide yields terminal glucose, 3-linked glucose and 3,6-linked glucose in the ratio of 1:23:1 (Fig. 3A), it can be concluded that the glucan backbone of cotton fiber 1,3-β-glucan is substitued by branches of 1,6-linked glucose residues.

Digestion of cellulose with endo-1,4-β-glucanase (cellulase) yields mainly cellobiose, and cellodextrins and glucose. The problem is the limited level of endohydrolysis due to product inhibition by glucose and cellobiose, and the low affin-

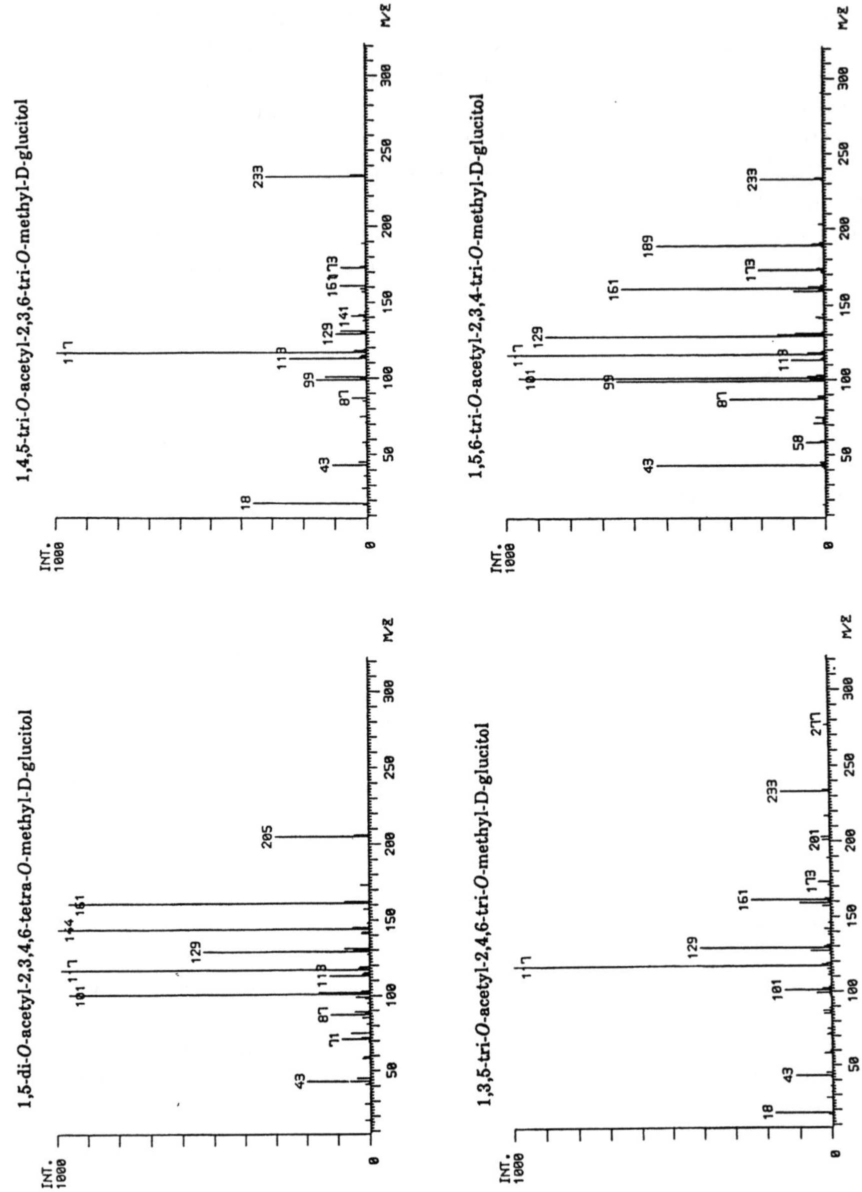
1,5-di-O-acetyl-2,3,4,6-tetra-O-methyl-D-glucitol
1,4,5-tri-O-acetyl-2,3,6-tri-O-methyl-D-glucitol
1,3,5-tri-O-acetyl-2,4,6-tri-O-methyl-D-glucitol
1,5,6-tri-O-acetyl-2,3,4-tri-O-methyl-D-glucitol
INT. 1000
M/Z

Fig. 4

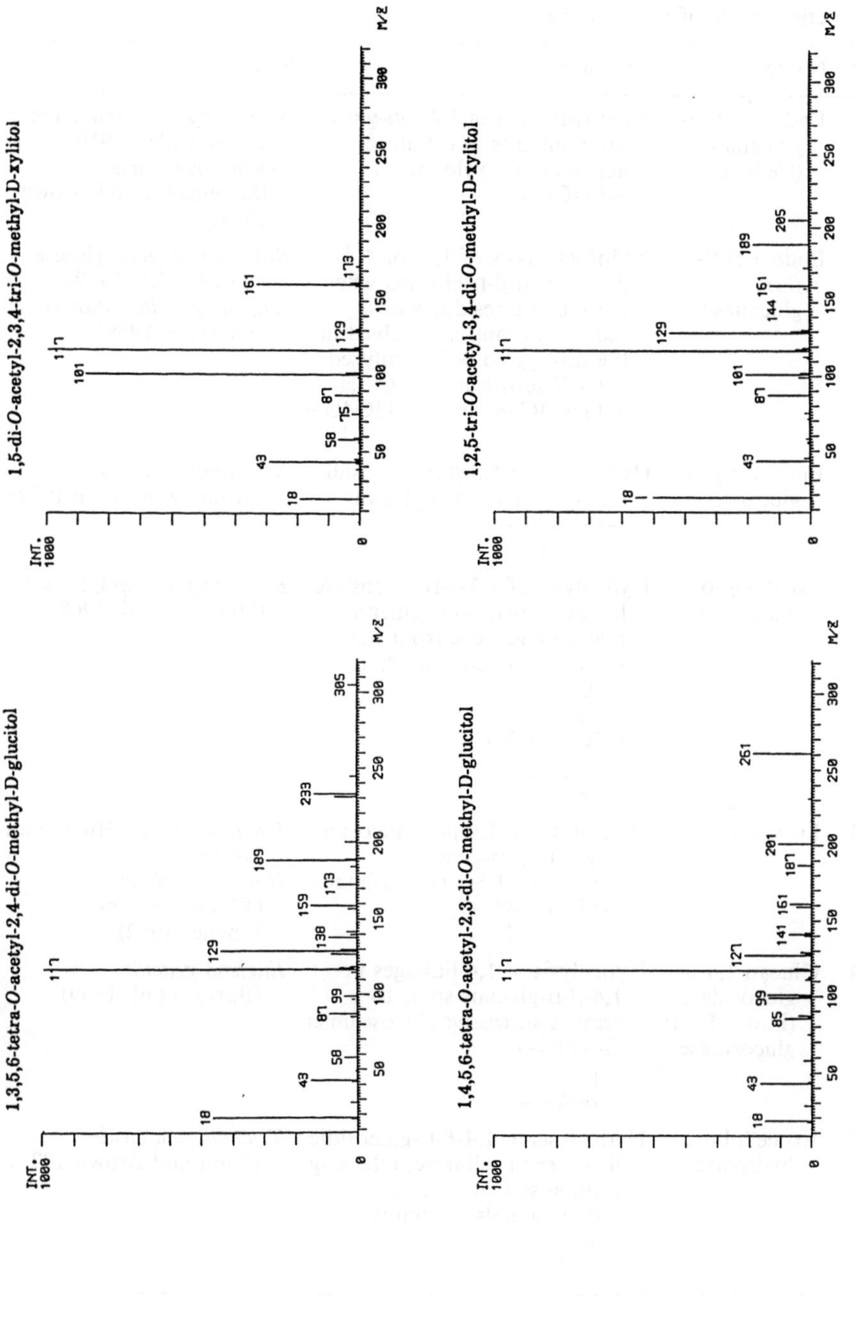

Fig. 4. Mass fragmentation patterns of the alditol acetates from permethylated β-glucans

Table 2. Action mode of β-D-glucanases

Number	Name	Reaction	Source
3.2.1.4	Endo-1,4-β-D-glucanase (Cellulase)	Endohydrolysis of 1,4-β-D-glucosidic linkages in cellulose, lichenan and xyloglucan →G4G4G→ ↑	*Streptomyces griseus* (Reese and Mandels 1959) *Trichoderma viride* (Shoemaker and Brown 1978)
3.2.1.6	Endo-1,3(4)-β-D-glucanase	Endohydrolysis of 1,3- or 1,4-linkages in β-D-glucans when the glucose residue whose reducing group is involved in the linkage to be hydrolyzed is itself substituted at C-3. →G3G3G→ or →G3G4G→ ↑ ↑	*Rhizopus arrhizus* (Reese and Mandels 1959) *Euglena gracilis* (Barras and Stone 1969)
3.2.1.39	Endo-1,3-β-D-glucanase	Hydrolysis of 1,3-β-D-glucosidic linkages in 1,3-β-D-glucans →G3G3G→ ↑	*Nicotiana glutinosa* (Moore and Stone 1972)
3.2.1.58	Exo-1,3-β-D-glucanase	Hydrolysis of 1,3-β-D-glucosidic linkages in 1,3-β-D-glucans, releasing glucose from the nonreducing ends of chains G 6 G3G3G3G→ ↑ ↑ ↑ *order*→	*Basidiomycete sp* QM 806 (Huotari et al. 1968)
3.2.1.73	Lichenase	Hydrolysis of 1,4-β-D-glucosidic linkages in β-D-glucans containing 1,3- and 1,4-bonds →G4G3G4G→ ↑	*Bacillus subtilis* (Huber and Nevins 1977) *Hordeum vulgare* (Woodward and Fincher 1982)
3.2.1.74	Glucan 1,4-β-glucosidase (Exo-1,4-β-D-glucosidase)	Hydrolysis of 1,4-linkages in 1,4-β-D-glucans so as to remove successive glucose units G4G4G→ ↑ ↑ *order*→	*Euglena gracilis* (Barras et al. 1969)
3.2.1.91	Exocellobiohydrolase	Hydrolysis of 1,4-β-D-glucosidic linkages in cellulose, releasing cellobiose from the non reducing ends of chains G4G4G→ ↑	*Trichoderma viride* (Gum and Brown 1977)

ity of crystalline cellulose to attack. Exocellobiohydrolase from *Trichoderma viride* (Gum and Brown 1977) has a higher affinity for crystalline cellulose and produces cellobiose at high rates (Chanzy et al. 1983). When both glucanases are provided together, one may enhance the action of the other. It should be noted that 1,3-β-glucan and mixed-linkage glucan are also hydrolyzable with the exocellobiohydrolase to produce glucose.

3.3 Xyloglucan

3.3.1 Hydrolysis by *Aspergillus oryzae* Enzymes

Structural studies on xyloglucans of plant cell walls have indicated that these polysaccharides possess a backbone of cellulose-like 1,4-β-glucan substituted with xylosyl residues in a regular pattern. Therefore, hydrolysis of xyloglucan with the carbohydrate hydrolases from an *Aspergillus oryzae* enzyme preparation yields monosaccharides (glucose, galactose and fucose) and the disaccharide isoprimeverose, in which all xylosyl residues of the polysaccharide are recovered (Hayashi et al. 1981). The isoprimeverose, 6-*O*-α-D-xylopyranosyl-D-glucopyranose, is a smallest common structural subunit. It is obtained by hydrolysis of crude cell-wall polysaccharides because the hydrolases are capable to degrade all sugar linkages except α-xyloside.

Enzyme hydrolysis is carried out in 1 ml of 50 mM sodium acetate buffer, pH 5.0, containing 1 mg crude *A. oryzae* enzyme preparation, polysaccharide, and covered with a few drops of toluene. Hydrolysis proceeds for 12 h at 40° C. Isoprimeverose does not undergo hydrolysis, even when the reaction time is prolonged to 24 h or when higher enzyme concentrations are used. The activity of the xyloglucan hydrolases is maximal between pH 4.0 and 6.0. The reaction is stopped by heating in a boiling water bath for 5 min. The enzymic hydrolyzates are treated with Dowex 50 W (H^+ type) to remove sodium ions and then evaporated to remove acetic acid. Identification of isoprimeverose is carried out by paper chromatography with *n*-propanol / ethyl acetate / water, 3 : 2 : 1, as shown in Fig. 1.

3.3.2 Systematic Degradation

A. Degradation of Xyloglucan with Endo-1,4-β-glucanase (Cellulase). Xyloglucan is highly branched and the substitution pattern of this glucan is such that the 1,4-linked glucose backbone is only hydrolyzable by endo-1,4-β-glucanases (EC 3.2.1.4) at unsubstituted glucose units (every fourth residue).

Hydrolysis of pea xyloglucan (Hayashi and Maclachlan 1984) with *Streptomyces griseus* endo-1,4-β-glucanase yields a nonasaccharide (glucose / xylose / galactose / fucose, 4 : 3 : 1 : 1) and a heptasaccharide (glucose / xylose, 4 : 3) in a molar ratio close to 1 : 1. The hydrolyzate of cotton fiber xyloglucan (Hayashi and Delmer 1988) contains four kinds of oligosaccharide units: decasaccharide (glucose / xylose / galactose / fucose, 4 : 3 : 2 : 1), nonasaccharide (glucose / xylose / galactose / fucose, 4 : 3 : 1 : 1), octasaccharide (glucose / xylose / galactose, 4 : 3 : 1),

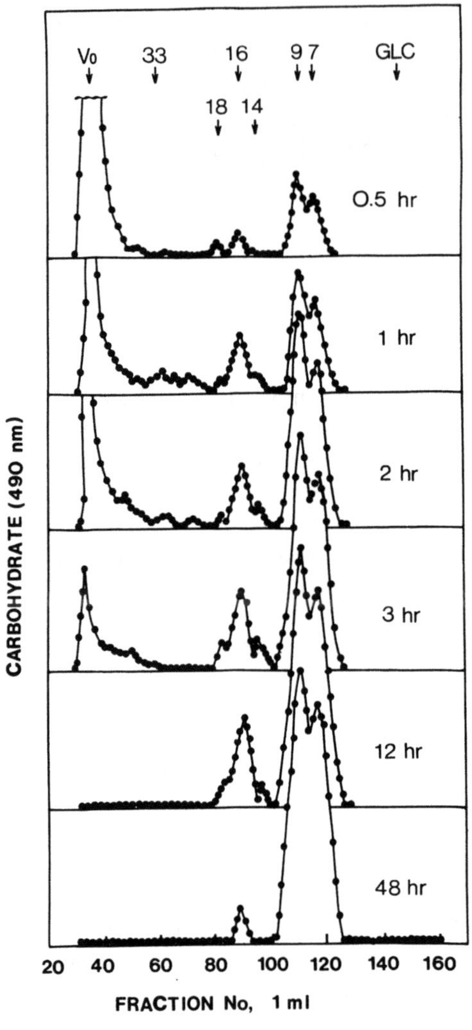

Fig. 5. Fractionation on Bio-Gel P-6 columns of products formed from pea xyloglucan following treatment with *Streptomyces* endoglucanase. The degree of polymerization of standards is indicated, i.e., *33* laminarin; *9* xyloglucan-derived nonasaccharide; *7* heptasaccharide. The identity of higher oligosaccharides, i.e., *18* (9→9), *16* (9→7), *14* (7→7), was confirmed by elution, exhaustive endoglucanase hydrolysis, and rechromatography on a Bio-Gel P-4 gel. (Reproduced from Hayashi and Maclachlan 1984)

and heptasaccharide (glucose / xylose, 4:3). Hydrolysis of monocot xyloglucan (Kato et al. 1981 b; Kato et al. 1982) often produces a hexasaccharide (glucose / xylose, 4:2), a pentasaccharide (glucose / xylose, 3:2), cellobiose and glucose but no heptasaccharide. Partial endohydrolysis of pea xyloglucan gives dimers of the oligosaccharide units: mostly 16 saccharides, plus small amounts of 18 and 14 saccharides (Hayashi and Maclachlan 1984) (Fig. 5). The structure of the dimer containing 16 saccharides is nonasaccharide→heptasaccharide. Since the nona- and heptasaccharides accumulated in a molar ratio of 1:1 at all stages of hydrolysis, and the main dimer contained the two oligosaccharides, it has been proposed that pea xyloglucan contains these units distributed primarily in alternating sequence.

Hydrolysis of xyloglucan with endoglucanase is carried out as follows. Xyloglucan (1 to 10 mg) is dissolved in 0.5 ml of 50 mM acetate buffer (pH 5.0) and incubated at 40° C with 0.1 mg of *Streptomyces* endo-1,4-β-glucanase. The reaction is stopped by heating the mixture in a boiling water bath for 5 min. The hydrolyzates collected at intervals are fractionated on a Bio-Gel P-4(200–400 mesh) column (1.5 × 100 cm) at 60° C to separate nona- and heptasaccharides. The oligosaccharide trimers or dimers are fractionated on a Bio-Gel P-6 (−400 mesh) column (1.5 × 100 cm) at 60° C (Fig. 8.5). Crude endoglucanase preparations from *Streptomyces griseus* do not contain other β-glucanase and glycosidase activities. It should also be noted that the endo-1,4-β-glucanase from *Trichoderma viride* often over-hydrolyzes the oligosaccharides, i.e., the heptasaccharide to tetrasaccharide (glucose / xylose, 2:2) and trisaccharide (glucose / xylose, 2:1) (Hayashi et al. 1980) by hydrolyzing substituted 1,4-linked glucose at every second glucose unit.

B: Degradation of Oligosaccharides with Aspergillus Glycosidases. When the nona- and heptasaccharides obtained from pea xyloglucan are isolated and treated with *Aspergillus* glycosidases, the paper chromatography of these second hydrolyzates demonstrates the presence of monosaccharides and of isoprimeverose. A β-glucosidase partially purified by DEAE-Sephadex A-50 ion-exchange chromatography and Sephadex G-200 gel filtration (Kato et al. 1985) from *Aspergillus oryzae* produces isoprimeverose and the pentasaccharide L-fucosyl-D-galactosyl-D-xylosyl-D-glucosyl-D-glucose from the nonasaccharide (Kato 1978). The specificity of this enzyme is completely different from that of Emulsin β-glucosidase (EC 3.2.1.21).

Hydrolysis of the nonasaccharide with bovine kidney α-fucosidase gives fucose and octasaccharide (glucose / xylose / galactose, 4:3:1) (Camirand and Maclachlan 1986). The octasaccharide is not sensitive to digestion with β-galactosidase from *Aspergillus niger* or from *Charonia lampas.*

C. Precipitation Test with Lectins. For the two-dimensional double-diffusion method, a Petri dish is layered with 1% Ionagar in saline containing 0.02% NaN_3. Wells are made in the set agar with an Ouchterlony technique. Polysaccharides (20 μg) or lectins (50 μg) are added to the center basin and/or the six peripheral wells. The plate is then incubated in a water-saturated atmosphere at room temperature.

Xyloglucans which contain terminal α-L-fucosyl-(1→2)-β-D-galactosyl side-chains will complex with the α-L-fucose-binding lectins from *Ulex europaeus* (Pereira et al. 1978). Xyloglucans which contain terminal β-D-galactosyl residues as side-chains will complex with β-D-galactose-binding lectin from *Ricinus communis* (Van Wauwe et al. 1974). In Ouchterlony-style precipitation tests (Fig. 6), pea and soybean xyloglucans clearly react with the fucose-binding lectin. Partillay degraded pea xyloglucan (lower than 20000 molecular weight) slightly reacted with the fucose-binding lectin but, after mild acid treatment, lost this property along with fucose and became reactive with the galactose-binding lectin. *Tamarindus* seed xyloglucan and monocot xyloglucan which contain no fucose also react with the galactose-binding lectin. This precipitation test is useful for the determination of terminal fine structure of xyloglucans (Hayashi and Maclachlan 1984).

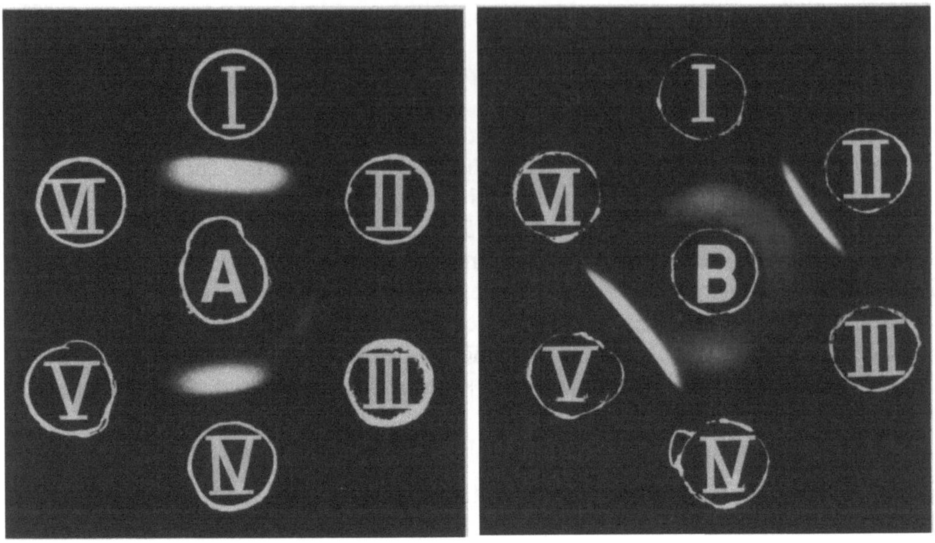

Fig. 6 A, B. Precipitation test of xyloglucans with lectins from *U. europaeus* (**A**) and *R. communis* (**B**). *I* pea xyloglucan; *II* mild acid-treated pea xyloglucan; *III* partially endoglucanase-treated pea xyloglucan; *IV* soybean xyloglucan; *V Tamarindus* seed xyloglucan; *VI* carboxymethylcellulose. (Reproduced from Hayashi and Maclachlan 1984)

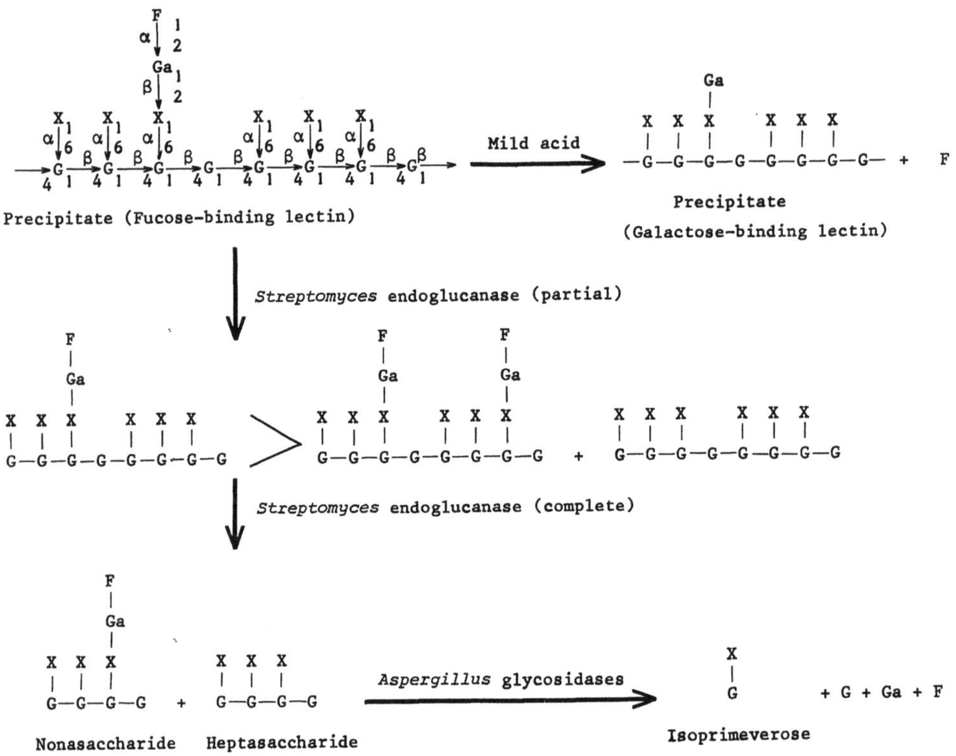

Fig. 7. Systematic degradation of pea xyloglucan. *F* fucose; *Ga* galactose; *G* glucose; *X* xylose. (Reproduced from Hayashi and Maclachlan 1984)

The scheme of systematic degradation of pea xyloglucan is shown in Fig. 7. This degradation involves the enzymic hydrolyses and mild acid hydrolysis of the polysaccharide.

4 Visualization

Since β-glucans have great biological significance for plant cells, the visualization of their deposition on the cell walls is required for biochemical study. In this section, we introduce recent attempts at β-glucan observation using antibodies, lectins, enzymes and specific dyes. These techniques are being constantly improved in our laboratory.

4.1 1,3-β-Glucan and Mixed-Linkage Glucan

Aniline blue is known to induce fluorescence in dilute alkaline solutions when complexed with callose (Wood and Fulcher 1984). However, commercial aniline blue preparations contain several fluorochromes whose abundance varies from batch to batch. Evans and Hoyne (1982) have obtained a pure and chemically defined aniline blue fluorochrome called Sirofluor. Sirofluor complexes specifically with $(1\rightarrow3)$-β-glucan and the fluorescence induced is independent of pH over a wide range (Stone at al. 1984; Stone 1984). The advantage of this fluorochrome is the specific detection of 1,3-β-glucans in the walls of living cells.

Purified *Bacillus* lichenase conjugated to fluorescein isothiocyanate (Stone 1984) has been used as a specific stain for mixed-linkage glucan.

4.2 Cellulose

4.2.1 Fluorescence-Labeled Xyloglucan

Since xyloglucan binds to cellulose, the fluorescein-labeled xyloglucan is used to stain cellulose microfibrils in the plant cell walls (Hayashi et al. 1987a). A time-course study shows that xyloglucan binding continued for at least 4 h at 40° C. Binding is inhibited above pH 6. Binding capacity varies for cellulose of different origins and is directly related to the relative surface area of the microfibrils. The binding of xyloglucan to cellulose is very specific and is not affected by the presence of a tenfold excess of $(1\rightarrow2)$-β-glucan, $(1\rightarrow3)$-β-glucan, $(1\rightarrow6)$-β-glucan, $(1\rightarrow3,1\rightarrow4)$-$\beta$-glucan, arabinogalactan, or pectin (Table 3).

Preparation of fluorescence-labeled xyloglucan was done according to the method of Glabe et al. (1983). Twenty mg of CNBr are dissolved in 0.2 ml of water and added to a solution containing 40 mg of pea xyloglucan in 1 ml of water. The mixture is adjusted to pH 11 and maintained at that pH for 5 min by addition of 0.2 M NaOH. The activated polysaccharide is desalted by gel filtra-

Table 3. Relative binding activity of xyloglucan to pea cellulose in the presence of various polysaccharides. Each polysaccharide (50 μg) was added to 5 μg of radioactive xyloglucan and 100 μg of pea cellulose suspended in 25 mM acetate buffer (pH 5.0). At equilibrium, approximately 5 μg of xyloglucan can bind to 100 μg of pea cellulose in the standard mixture. (Reproduced from Hayashi et al. 1987a)

Polysaccharide	Relative binding activity %
None	100
1,2-β-glucan	103.4
Laminarine	99.3
Pachyman	100.7
Pustulan	99.2
Lichenan	92.0
Xylan	47.0
Arabinogalactan	106.7
Pectin	107.9

tion on a 1 × 20 cm column of Sephadex G-50 in 0.2 M sodium borate at pH 8.0. The polysaccharide fractions (void volume) are pooled and incubated overnight at room temperature with 4 mg of fluoresceinamine. The fluorescein-labeled xyloglucan is separated from the untreated fluoresceinamine by gel filtration on Sephadex G-50 (1 × 20 cm) equilibrated in 50 mM phosphate-buffered saline (pH 7.0). Fractions (1.0 ml) are collected and absorbance at 440 nm is measured. The polysaccharide fractions are measured by the phenol / sulfuric acid method (Dubois et al. 1956), and pooled. The fluorescence-labeled xyloglucan obtained is used directly for binding assay or for staining cellulose microfibrils.

A conjugate of cellulase with fluorescein isothiocyanate was also used to stain cellulose (Seibert et al. 1978). However, the enzyme probably has some affinity for xyloglucan.

4.3 Xyloglucan

4.3.1 Fluorescence-Labeled Lectins

Xyloglucans from dicotyledons contain many side-chains terminated with α-D-fucosyl residues; those from monocotyledons contain several side-chains with terminal β-D-galactosyl residues. The latter can complex with β-galactose-binding lectins and the former with fucose-binding lectins as shown in Fig. 6 (Hayashi and Maclachlan 1984). Pea xyloglucan was examined by fluorescence microscopy using a fluorescein fucose-binding lectin from *Ulex europaeus* as probe. Since plant cell walls contain other fucosyl residues such as fucosylated glycoproteins, the labeled lectin can be applied to use only for cell-wall ghosts which are composed of xyloglucan and cellulose.

4.3.2 Immuno-Gold Localization

Localization of xyloglucan on thin sections of suspension-cultured sycamore cells has been carried out by using a polyclonal antibody against the polysaccharide (Moore et al. 1986). Immunogold visualization shows that xyloglucan is localized throughout the entire wall and middle lamella. The antibody has been prepared as follows: purified xyloglucan is coupled to the protein carrier ovalbumin, and the xyloglucan-ovalbumin antigen is emulsified in Freud's adjuvant and injected into rabbits. Antibodies against ovalbumin are removed by preabsorbing antisera to ovalbumin before use.

5 Concluding Remarks

Studies on three-dimensional structures of plant β-glucans are presently in progress in several laboratories. Our understanding of the analysis of β-glucan is focused on the spatial conformation of xyloglucan in order to understand the association between this polysaccharide and cellulose microfibrils by hydrogen bonds, and to study the cross-links of the microfibrils with xyloglucan in the primary walls. Visualization of β-glucans is now becoming important in showing the formation, secretion, deposition, and degradation of β-glucans in plant cell walls, with an ultimate goal of explaining some of the phenomena associated with cell growth, differentiation, regeneration and defence mechanisms in plants.

Acknowledgment. I am grateful to Dr. Anne Camirand, Départment Biochimie, Université de Montréal, Canada, for her critical reading of the manuscript.

References

Aist JR (1983) Structural responses as resistance mechanisms. In: Bailey JA, Deverall BJ (eds) The dynamic of host defence. Academic Press, Australia, pp 33–70

Anderson MA, Cook J, Stone BA (1976) Enzymatic determination of 1→3;1→4-β-glucans in barley grain and other cereals. J Inst Brew 84:233–239

Anderson RL, Ray PM (1978) Labeling of the plasma membrane of pea cells by a surface-localised glucan synthetase. Plant Physiol 61:723–730

Barras DR, Stone BA (1969) β-1,3-Glucan hydrolases from *Euglena gracilis*. II. Purification and properties of the β-1,3-glucan exo-hydrolase. Biochim Biophys Acta 191:342–353

Barras DR, Moore AE, Stone BA (1969) Enzyme-substrate relations among β-glucan hydrolases. Adv Chem Ser 95:105–138

Bacic A, Stone BA (1980) A (1→3)-and(1→4)-linked β-D-glucan in the endosperm cell walls of wheat. Carbohydr Res 82:372–377

Blakeney AB, Harris PJ, Henry RJ, Stone BA (1983) Simple and rapid preparation of alditol acetates for monosaccharide analysis. Carbohydr Res 113:291–299

Blaschek W, Haass D, Koehler H, Franz G (1981) Cell wall regeneration by *Nicotiana tabacum* protoplasts: chemical and biochemical aspects. Plant Sci Lett 22:47–57

Camirand A, Maclachlan G (1986) Biosynthesis of the fucose-containing xyloglucan nonasaccharide by pea microsomal membranes. Plant Physiol 82:379–383

Camirand A, Brummell D, Maclachlan G (1987) Fucosylation of xyloglucan: localization of the transferase in dictyosomes of pea stem cells. Plant Physiol 84:753–756

Chanzy H, Henrissat B, Vuong R, Schülein M (1983) The action of 1,4-β-D-glucan cellobiohydrolase on *Valonia* cellulose microcrystals. A electron microscopic study. FEBS Lett 153:113–118

Currier HB (1957) Callose substance in plant cells. Am J Bot 44:478–488

Delmer DP (1977) The biosynthesis of cellulose and other plant cell wall polysaccharides. Rec Adv Phytochem 11:45–77

Drake GA, Carr DJ, Anderson WP (1978) Plasmolysis, plasmodesmata and electrical coupling of oat coleoptile cells. J Exp Bot 29:1205–1214

Dubois M, Gilles KA, Hamilton JK, Rebers PA, Smith F (1956) Colorimetric method for determination of sugars and related substances. Anal Chem 28:350–356

Evans NA, Hoyne PA (1982) A fluorochrome from aniline blue: structure, synthesis and fluorescence properties. Aust J Chem 35:2571–2575

Glabe CG, Harty PK, Rosen SD (1983) Preparation and properties of fluorescent polysaccharides. Anal Biochem 130:287–294

Gum EK Jr, Brown RD Jr (1977) Comparison of four purified extracellular 1,4-β-D-glucan cellobiohydrolase enzymes from *Trichoderma viride*. Biochim Biophys Acta 492:225–231

Hakomori S (1964) A rapid permethylation of glycolipid, and polysaccharide catalyzed by methylsulfinyl carbanion in dimethyl sulfoxide. J Biochem 55:205–208

Hayashi T, Delmer DP (1988) Xyloglucan in the cell walls of the cotton fibre. Carbohydr Res 181:273–277

Hayashi T, Maclachlan G (1984) Pea xyloglucan and cellulose. I. Macromolecular organization. Plant Physiol 75:596–604

Hayashi T, Matsuda K (1981 a) Biosynthesis of xyloglucan in suspension-cultured soybean cells. Evidence that the enzyme system of xyloglucan synthesis does not contain β-1,4-glucan 4-β-D-glucosyltransferase activity (EC 2.4.1.12). Plant Cell Physiol 22:1571–1584

Hayashi T, Matsuda K (1981 b) Biosynthesis of xyloglucan in suspension-cultured soybean cells. Occurrence and some properties of xyloglucan 4-β-D-glucosyltransferase and 6-α-D-xylosyltransferase. J Biol Chem 256:11117–11122

Hayashi T, Matsuda K (1981 c) Biosynthesis of xyloglucan in suspension-cultured soybean cells. Synthesis of xyloglucan from UDP-glucose and UDP-xylose in the cell-free system. Plant Cell Physiol 22:517–523

Hayashi T, Kato Y, Matsuda K (1980) Xyloglucan from suspension-cultured soybean cells. Plant Cell Physiol 21:1405–1418

Hayashi T, Kato Y, Matsuda K (1981) Biosynthesis of xyloglucan in suspension-cultured soybean cells. An assay method for xyloglucan xylosyltransferase and attempted synthesis of xyloglucan from UDP-D-xylose. J Biochem 89:325–328

Hayashi T, Wong YS, Maclachlan G (1984) Pea xyloglucan and cellulose. II. Hydrolysis by pea endo-1,4-β-glucanses. Plant Physiol 75:605–610

Hayashi T, Polonenko DR, Camirand A, Maclachlan G (1986) Pea xyloglucan and cellulose. IV. Assembly of β-D-glucans by pea protoplasts. Plant Physiol 82:301–306

Hayashi T, Marsden MPF, Delmer DP (1987 a) Pea xyloglucan and cellulose. V. Xyloglucan-cellulose interactions in vitro and in vivo. Plant Physiol 83:384–389

Hayashi T, Read SM, Bussell J, Thelen M, Lin FC, Brown RM Jr, Delmer DP (1987 b) UDP-glucose: (1→3)-β-glucan synthases from mung bean and cotton. Differential effects of Ca^{2+} and Mg^{2+} on enzyme properties and on macromolecular structure of the glucan product. Plant Physiol 83:1054–1062

Hayashi T, Koyama T, Matsuda K (1988) Formation of UDP-xylose and xyloglucan in soybean Golgi membranes. Plant Physiol 87:341–345

Herth W, Weber G (1984) Occurrence of the putative cellulose-synthesizing "rosettes" in the plasma membrane of *Glycine max* suspension culture cells. Naturwissenschaften 71:153–154

Huber DJ, Nevins DJ (1977) Preparation and properties of a β-D-glucanase for the specific hydrolysis of β-D-glucans. Plant Physiol 60:300–304

Huotari FI, Nelson TE, Smith F, Kirkwood S (1968) Purification of an exo-β-D-(1→3)-glucanase from *Basidiomycete* species QM 806. J Biol Chem 243:952–956
Huwyler HR, Franz G, Meier H (1978) β-1,3-glucans in the cell walls of cotton fibres (*Gossypium arboreum* L.). Plant Sci Lett 12:55–62
Kato Y (1978) Studies on the structure and functional role of xyloglucan in immature Leguminosae. Ph D thesis, Tohoku University, Sendai, Japan
Kato Y, Matsuda K (1980) Structure of oligosaccharides obtained by hydrolysis of mung bean xyloglucan with *Trichoderma viride* cellulase. Agric Biol Chem 44:1759–1766
Kato Y, Nevins DJ (1984) Enzymic dissociation of *Zea* shoot cell wall polysaccharides. II. Dissociation of (1→3),(1→4)-β-D-glucan by purified (1→3),(1→4)-β-D-glucan 4-glucanohydrolase from *Bacillus subtilis*. Plant Physiol 75:745–752
Kato Y, Nevins DJ (1985) A (1→3)-β-D-glucan isolated from *zea* shoot cell wall preparations. Plant Physiol 78:20–24
Kato Y, Iki K, Matsuda K (1981 a) Cell-wall polysaccharides of immature barley plants. I. Isolation and characterization of a β-D-glucan. Agric Biol Chem 45:2737–2744
Kato Y, Iki K, Matsuda K (1981 b) Cell-wall polysaccharides of immature barley plants. II. Characterization of a xyloglucan. Agric Biol Chem 45:2745–2753
Kato Y, Shiozawa R, Takeda S, Ito S, Matsuda K (1982) Structural investigation of a β-D-glucan and a xyloglucan from bamboo-shoot cell walls. Carbohydr Res 109:233–248
Kato Y, Matsushita J, Kubodera T, Matsuda K (1985) An novel enzyme producing isoprimeverose from oligoxyloglucans of *Aspergillus oryzae*. J Biochem 97:801–810
Klein AS, Montezinos D, Delmer DP (1981) Cellulose and 1,3-glucan synthesis during the early stages of wall regeneration in soybean protoplasts. Planta 152:105–114
Kooiman P (1960) A method for the determination of amyloid in plant seeds. Recl Trav Chim Pays-Bas 79:675–678
Lindberg B (1972) Methylation analysis of polysaccharides. Methods Enzymol 28:178–195
Maltby D, Carpita NC, Montezinos D, Kulow C, Delmer DP (1979) β-1,3-Glucan in developing cotton fibers. Structure, localization, and relationship of synthesis to that of secondary wall cellulose. Plant Physiol 63:1158–1164
Moore AE, Stone BA (1972) A β-1,3-Glucan hydrolase from *Nicotiana glutinosa*. I. Extraction, purification and physical properties. Biochim Biophys Acta 258:238–247
Moore PJ, Darvill AG, Albersheim P, Staehelin LA (1986) Immunogold localization of xyloglucan and rhamnogalacturonan I in the cell walls of suspension-cultured sycamore cells. Plant Physiol 82:787–794
Mueller SC, Brown RM Jr (1980) Evidence for an intramembrane component associated with a cellulose microfibril synthesizing complex in higher plants. J Cell Biol 84:315–326
Mueller SC, Maclachlan G (1983) Radio-autographic visualization of β-glucan formed by pea membranes from UDP-glucose. Can J Bot 61:1266–1275
Nevins DJ, Yamamoto R, Huber DJ (1978) Cell wall β-D-glucans of five grass species. Phytochemistry 17:1503–1505
Pereira MEA, Kisailus EC, Gruezo F, Kabat EA (1978) Immunochemical studies on the combining site of the blood group H-specific lectin 1 from *Ulex europaeus* seeds. Arch Biochem Biophys 185:108–115
Ray PM (1979) Separation of maize coleoptile cellular membranes that bear different types of glucan synthetase activity. In: Reid E (ed) Plant organelles. Horwood, Chichester, pp 135–146
Ray PM (1980) Cooperative action of β-glucan synthetase and UDP-xylose xylosyltransferase of Golgi membranes in the synthesis of xyloglucan-like polysaccharide. Biochim Biophys Acta 629:431–444
Reese ET, Mandels M (1959) β-D-1,3 Glucanases in fungi. Can J Microbiol 5:173–184
Robyt J, French D (1963) Action pattern and specificity of an amylase from *Bacillus subtilis*. Arch Biochem Biophys 100:451–467
Sandford PA, Conrad HE (1966) The structure of the *Aerobacter aerogenes* A3 (S1) polysaccharide. I. A reexamination using improved procedures for methylation analysis. Biochemistry 5:1508–1517

Seibert GR, Benjaminson MA, Hoffman H (1978) A conjugate of cellulase with fluorescein isothiocyanate: a specific stain for cellulose. Stain Technol 53:103–106
Shoemaker SP, Brown RD Jr (1978) Enzymic activities of endo-1,4-β-D-glucanases purified from *Trichoderma viride*. Biochim Biophys Acta 523:133–146
Shore G, Maclachlan GA (1975) The site of cellulose synthesis. Hormone treatment alters the intracellular location of alkali-insoluble β-1,4-glucan (cellulose) synthetase activities. J Cell Biol 64:557–571
Smith MM, Stone BA (1973) β-glucan synthesis by cell-free extracts from *Lolium multiflorum* endosperm. Biochim Biophys Acta 313:72–94
Somogyi M (1952) Notes on sugar determination. J Biol Chem 195:19–23
Staudte RG, Woodward JR, Fincher GB, Stone BA (1983) Water-soluble (1→3),(1→4)-β-D-glucans from barley (*Hordeum vulgare*) endosperm. III. Distribution of cellotriosyl and cellotetraosyl residues. Carbohydr Polymers 3:299–312
Stinard PS, Nevins DJ (1980) Distribution of noncellulosic β-D-glucans in grasses and other monocots. Phytochemistry 19:1467–1468
Stone BA (1984) Noncellulosic β-glucans in cell walls. In: Dugger WM, Bartnicki-Garcia S (eds) Structure, function, and biosynthesis of plant cell walls. Waverly, Baltimore, pp 52–74
Stone BA, Evans NA, Bonig I, Clarke AE (1984) The application of Sirofluor, a chemically defined fluorochrome from aniline blue for the histochemical detection of callose. Protoplasma 122:191–195
Takeuchi Y, Komamine A (1981) Glucans in the cell walls regenerated from *Vinca rosea* protoplasts. Plant Cell Physiol 22:1585–1594
Updegraff DM (1969) Semimicro determination of cellulose in biological materials. Anal Chem 32:420–424
Van Wauwe JP, Loontiens FG, De Bruyne CK (1974) On the interaction of *Ricinus communis* lectin with plant amyloids. Biochim Biophys Acta 354:148–151
Vithanage HIMV, Gleeson PA, Clarke AE (1980) The nature of callose produced during self-pollination in *Secale cereale*. Planta 148:498–509
Wada S, Ray PM (1978) Matrix polysaccharides of oat coleoptile cell walls. Phytochemistry 17:923–931
Wilkie KB (1979) The hemicelluloses of grasses and cereals. Adv Carbohydr Chem Biochem 36:215–264
Wood PJ, Fulcher RG (1984) Specific interaction of aniline blue with (1→3)-β-D-glucan. Carbohydr Polymers 4:49–72
Woodward JR, Fincher GB (1982) Substrate specificities and kinetic properties of two 1,3;1,4-β-glucan endohydrolases from germinating barley (*Hordeum vulgare*). Carbohydr Res 106:111–122

Methods Used in the Chemistry of Lignin Biodegradation

T. UMEZAWA and T. HIGUCHI

1 Introduction

Lignin is one of the main constituents of xylem tissue and has been thought of as a major renewable resource. Hence, understanding lignin biodegradation is important not only from the viewpoint of pure biochemistry and the physiology of wood-degrading microorganisms but also with regard to the utilization of lignocelluloses. Significant advances in the elucidation of the chemistry and biochemistry of lignin biodegradation were made during the last decade (Kirk and Farrell 1987). Lignin is a complex and heterogeneous aromatic polymer formed via dehydrogenative polymerization of *p*-hydroxycinnamyl alcohols, and its chemical structure is known only in the form of a statistical average of several characteristic intermonomer linkages (substructures, Fig. 1) which involve several different C–C and C–O–C (ether) bonds (Adler 1977). Furthermore, unaltered original lignin associated with plant tissue, which is referred to as "protolignin", has not been isolated successfully. Therefore, "lignin preparations" used as sub-

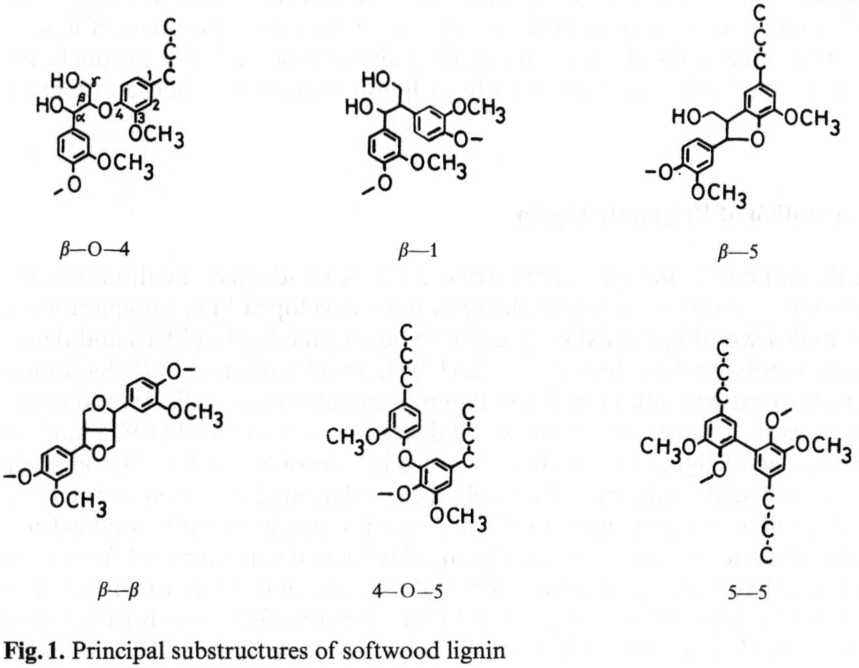

Fig. 1. Principal substructures of softwood lignin

strates in biodegradation of lignin have been limited by the inherent properties of lignin. As the next best solution, two types of "lignin preparations" have been used as substrates of microbial degradation: One of them is polymeric lignins including wood itself, milled wood lignin (MWL), cellulolytic enzyme lignin (CEL) and synthetic lignin (DHP) (Chen and Chang 1985), and the other is low-molecular-weight model compounds with lignin substructure (Higuchi 1985). The lignin-substructure model compounds are indispensable for the elucidation of mechanisms of specific degradative reactions. On the other hand, the degradation of polymeric lignin by microorganisms provides a general view of the chemistry of lignin biodegradation. The two approaches are complementary, and have contributed definitively to the elucidation of the chemistry of lignin biodegradation.

Methods for the study of microbial degradation of polymeric lignin were reviewed by Crawford (1981). Hence, this paper will focus on the methods used in the chemistry of microbial degradation of lignin-substructure model compounds. The literature cited is mainly concerned with lignin degradation by white-rot basidiomycetes.

2 Degradation of Polymeric Lignin

Degradation fractions of polymeric lignin by microorganisms are composed of the polymeric degraded lignin and the low-molecular-weight degradation products. Both fractions can be separated by gel-filtration or solvent extraction. The low-molecular-weight degradation products are analyzed and identified by methods similar to those used in the analysis of degradation products of lignin-substructure models. On the other hand, the polymeric degradation products are analyzed by methods employed usually in lignin chemistry (Chen and Chang 1985).

2.1 Preparation of Polymeric Lignin

One of the difficulties in lignin chemistry is the lack of adequate methods for lignin isolation. Nevertheless, lignin chemists have developed lignin preparations, such as milled wood lignin (MWL), cellulolytic enzyme lignin (CEL) and dehydrogenation polymer (synthetic lignin, DHP) (Lai and Sarkanen 1971; Sarkanen 1971), as the next best solution. These lignin preparations, as well as wood itself, have been used as substrates of microbial degradation. Crawford (1981) made a detailed review of lignin preparations for use in microbial studies. Wood itself should be used as the substrate of microbial degradation after solvent extraction, which eliminates contamination of the degradation products with wood extractives. Milled wood lignin (Björkman lignin, MWL), which is extracted from ball-milled wood tissues using dioxane-water (9:1), is known to be as a relatively unchanged lignin preparation. The yield of MWL from conifer woods is usually less than 50% (Björkman 1956). Yield of cellulolytic enzyme lignin (CEL), which is extracted with dioxane-water from cellulase-treated ball-milled wood, is greater than that of MWL (Chang et al. 1975). Accordingly, CEL may be more representative, although it contains still 12%–14% of associated carbohydrate, which is a drawback for the examination of the effects of carbohydrates on lignin degradation. Thus, MWL and CEL are the best lignin preparations extracted from natural lignocelluloses. On the other hand, dehydrogenation polymers of *p*-hydroxycinnamyl alcohols (such as coniferyl, sinapyl and *p*-coumaryl alcohol) (DHP, synthetic lignin) (Sarkanen 1971) have no associated polysaccharide. Both stable and radio isotopes-labeled polymeric lignins have been prepared (Kirk et al. 1975; Haider and Trojanowski 1975; Ellwardt et al. 1981; Crawford 1981) and applied successfully in lignin biodegradation research, as described in Sect. 2.2.

2.2 Analysis of Polymeric Lignin Degradation Products

2.2.1 Quantification of Lignin

The Klason method (Lai and Sarkanen 1971) is the most accurate method for the determination of lignin in wood, although it requires a relatively large amount of sample. This method has been used for the quantification of residual lignin during microbial degradation of wood (Kirk and Chang 1974). UV-absorption methods, including the acetyl bromide method, are also used for the determination of lignin content (Goldschmid 1971). These methods require smaller amounts of sample than the Klason method, and are often used for the detection of lignin degradation products eluted from a chromatographic column, such as gel permeation chromatography or gel filtration. However, care should be taken with the quantitative interpretation of the chromatogram monitored by the methods, since an accurate estimation of the extinction coefficient of lignin degradation products is difficult. Radioisotopic methods which use ^{14}C-labeled polymeric lignin can avoid this difficulty, since they measure the number of labeled carbon atoms, and are not affected by the structural modification of lignin during the microbial degradation, such as the formation of carbonyl groups conjugated with an aromatic ring, which influences the quantification of lignin by an UV-absorption method. The radioisotopic method, which was reviewed by Crawford (1981), has allowed unequivocal and highly sensitive assays of lignin biodegradation (Kirk et al. 1975; Haider and Trojanowski 1975). For example, optimum culture parameters for lignin degradation by a white-rot fungus *Phanerochaete chrysosporium* have been determined for the first time by using this method (Kirk et al. 1978).

2.2.2 Chemical and Spectrometric Analysis of Polymeric Lignin Degradation Products

Polymeric fractions of the lignin biodegradation products can be separated from all other degradation products by solvent extraction (Chua et al. 1982) or gel filtration (Kirk and Chang 1974), and are analyzed by comparison with intact lignin preparations, employing methods used in usual lignin analyses, such as elemental analysis, functional group analysis, analysis of lignin substructures, and gel filtration (Ishikawa et al. 1963; Hata 1966; Kirk and Chang 1975; Chua et al. 1982; Tai et al. 1983; Chen and Chang 1985). A general view of lignin biodegradation has been constructed based on the results of the analyses. These usual methods of lignin analysis were described extensively by Browning (1967), Sarkanen and Ludwig (1971), and Tanahashi and Higuchi (1988): The methoxyl group, phenolic and aliphatic hydroxyl groups, and carbonyl and carboxyl groups are detected and estimated by various chemical and spectrometric analyses. $KMnO_4$ oxidation, alkaline nitrobenzene oxidation and acidolysis provide important information on both condensed and noncondensed aromatic rings, the molecular ratio of the syringyl unit to the guaiacyl unit (S/V ratio), and the amount of β-O-4 substructure, respectively. [^{1}H]- and [^{13}C]-NMR provide information about the framework of lignin skeletons. Molecular weight distribution of lignin degradation products was estimated by using gel permeation chromatography (GPC) or, more conveniently, gel filtration (Kirk and Chang 1974; Connors et al. 1978; Chua et al. 1982; Tai et al. 1983). The possible association of lignin components in organic solvents may seriously affect gel filtration profiles of lignin samples. This effect can be eliminated when LiCl is added to *N,N*-dimethylformamide (DMF) as an eluent (Connors et al. 1980).

Alternatively, DHP labeled with ^{13}C in specific sites, such as [ring-^{13}C]DHP, [side-chain-^{13}C]DHP and [methoxyl-^{13}C]DHP, were prepared, incubated with microorganisms and the degradation products were analyzed by [^{13}C]-NMR. Since [^{13}C]-NMR signals of the labeled carbon atoms are intensified relative to those of the unlabeled ones, specific alterations in the chemical environment of the labeled carbon atom can be followed by a decrease of the original signals and by the appearance of new prominent resonance (Ellwardt et al. 1981; Haider et al. 1985). Recently, solid state CP-MAS [^{13}C]-NMR spectroscopy has been developed, and it can analyze solid state samples, such as insoluble high polymer of lignin, while conventional NMR fails to do so. The CP-MAS [^{13}C]-NMR was applied to the analysis of kraft lignin degraded by *Coriolus versicolor* (Schaefer et al. 1981).

2.3 Analysis of Low-Molecular-Weight Degradation Products

The identification of low-molecular-weight degradation products isolated from woods decayed by microorganisms suggests possible reactions involved in lignin biodegradation. These products can be identified by the usual analytical methods used in the identification of fungal degradation products of lignin-substructure models (Sect. 3). Comprehensive investigations of the low-molecular-weight fraction of degraded lignin were reported (Chen et al. 1982, 1983; Tai et al. 1983; Chen and Chang 1985). Methyl alcohol extracts of spruce and birch woods decayed by the white-rot fungus *Phanerochaete chrysosporium* were further fractionated into acidic, phenolic and neutral fractions. The acidic fraction was again extracted with chloroform so as to separate the low-molecular-weight components. The low-molecular-weight, chloroform-soluble components were analyzed by GC-MS after acetylation and methylation (Sect. 2.2), and degradation products, such as vanillic acid and syringic acid, were identified.

3 Degradation of Lignin-Substructure Model Compounds

Not only in the study of lignin biodegradation, but also in general lignin chemistry, one of the important features is that lignin-substructure model compounds (Fig. 2) have always been employed as substrates of reactions to be investigated, especially in the elucidation of mechanisms for the specific chemical reactions, such as sulfonation and acidolysis of lignin. Otherwise, mechanisms for specific reactions of lignin would not be elucidated.

In the substructure model approach, the lignin-substructure model compounds should be synthesized. The degradation products of lignin-substructure model compounds are purified by chromatography, and identified by spectrometry. Then the identified products (degradation intermediates) are synthesized and used as substrates for further degradation study. Based on the chemical structure of the degradation products, mechanisms for the reactions involved are presumed, and confirmed by experiments using stable isotopes (Higuchi 1985, 1986). Some of the degradative reactions of the substructure models can be used to isolate ligninolytic enzymes (Tien and Kirk 1983; Glenn et al. 1983).

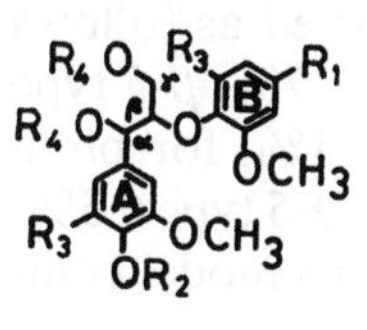

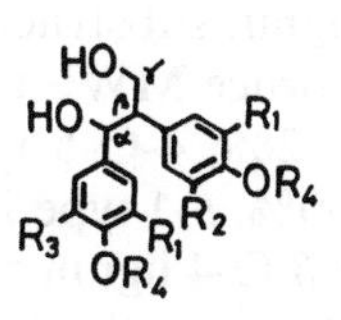

COMPOUND	R_1	R_2	R_3	R_4
I	$CH=CHCH_2OH$	H	H	H
II	CHO	H	H	H
III	H	H	H	H
IV	H	CH_3	H	H
V	H	CH_2CH_3	H	H
VI	CH_2OH	H	OCH_3	H
VII	H	CH_2CH_3	H	CH_2CH_3

COMPOUND	R_1	R_2	R_3	R_4
VIII	OCH_3	H	H	H
IX	OCH_3	H	OCH_3	H
X	H	H	H	CH_3

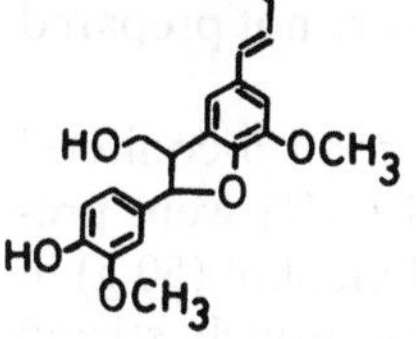

XI

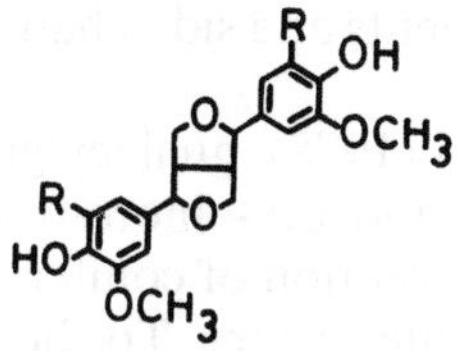

XII R:H
XIII R:OCH_3

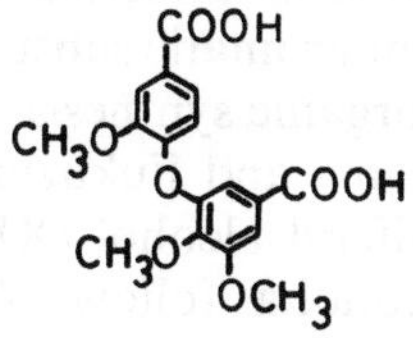

XIV

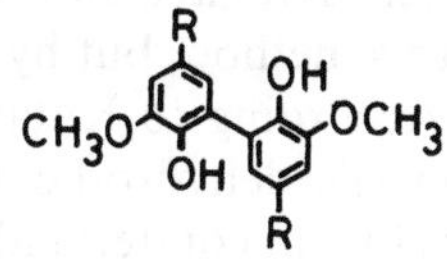

XV R:CHO
XVI R:COOH
XVII R:$CH=CHCH_2OH$

Fig. 2. Lignin-substructure model compounds

Care should be taken in the interpretation of results obtained by using the lignin substructure models. The substructure model approach is based on the assumption that reactivities of specific functional groups of the substructure model compounds toward lignin-degrading enzymes (i.e., specific reaction mechanisms) are similar to those of the functional groups of protolignin. However, the actual reactivity of protolignin may be influenced by factors other than the reactivity of specific functional groups, since molecular weights of the lignin-substructure model compounds are smaller than those of protolignin, and since protolignin constitutes a cell-wall matrix intimately associated to hemicelluloses. For a white-rot basidiomycete *Phanerochaete chrysosporium*, in the beginnings of substructure model studies, Weinstein et al. (1980) demonstrated that several of the culture parameters which affect lignin degradation have similar effects on the degradation of β-O-4 lignin-substructure model dimers.

In the following sections, syntheses of lignin substructure models and analysis of their degradation products by microorganisms are described.

3.1 Preparation of Lignin-Substructure Model Compounds

The frequency of lignin substructures in lignin molecules is reported as follows (Adler 1977): For spruce MWL (Fig. 1), β-O-4 type 48%, β-1 type 7%, β-5 type 9%–12%, β-β type 2%, 4-O-5 type 3.5%–4%, 5-5 type 9.5%–11%; for birch MWL, β-O-4 type 60%, β-1 type 7%, β-5 type 6%, β-β type 3%, 4-O-5 type 6.5%, 5-5 type 4.5%. The β-O-4 lignin substructure is predominant and its model compounds were, therefore, synthesized and used the most frequently.

3.1.1 Dehydrogenative Dimerization of *p*-Hydroxycinnamyl Alcohols

Synthetic lignins (DHP) are prepared by dehydrogenative polymerization of *p*-hydroxycinnamyl alcohols catalyzed by the horseradish peroxidase / H_2O_2 system. When the dehydrogenation reaction is stopped in the early stage, certain lignin-substructure model dimers can be prepared (Freudenberg 1965; Nakatsubo and Higuchi 1975a; Katayama and Fukuzumi 1978; Kamaya et al. 1980). Contained among them are pinoresinol (XII), syringaresinol (XIII), dehydrodiconiferyl alcohol (XI) and guaiacylglycerol-β-coniferyl ether (I). This is an easy method, and important substructure models, such as β-O-4, β-5, and β-β substructures, can be prepared in a one-pot reaction. However, lignin-substructure models that have different aromatic substituents and side-chains are not prepared by this method, but by organic synthesis.

According to Katayama and Fukuzumi (1978), guaiacylglycerol-β-coniferyl ether (I), dehydrodiconiferyl alcohol (XI) and D,L-pinoresinol (XII) were prepared from coniferyl alcohol as follows. A solution of coniferyl alcohol (50 g) in acetone (200 ml) was added into 5 l of desalted water. To the vigorously stirred solution an aqueous solution (200 ml) of peroxidase (40 mg, 58.8 units/mg, Miles-Seravac. Inc.) and 3% H_2O_2 (200 ml) were added all together. After coniferyl alcohol had disappeared (checked by TLC), the reaction mixture saturated with NaCl was extracted with methylene chloride. The methylene chloride layer was dried over anhydrous Na_2SO_4, and the solvent was evaporated. The reaction products thus obtained were chromatographed with silicic acid gel (110 × 9 cm, eluent: methylene chloride / methyl alcohol = 20 : 1) to give D,L-pinoresinol (XII), dehydrodiconiferyl alcohol (XI) and guaiacylglycerol-β-coniferyl ether (I) (10 g, yield: 20%). D,L-Pinoresinol (XII) was further purified by silica gel column chromatography (70 × 5 cm, eluent: methylene chloride / methyl alcohol = 20 : 1) and subsequently recrystallized from ligroin to give white crystals (5 g, yield: 10%). Dehydrodiconiferyl alcohol (XI) was recrystallized from methylene chloride-petroleum ether (white crystals, 12 g, yield: 24%).

3.1.2 Chemical Synthesis of Lignin-Substructure Model Compounds

Several lignin-substructure model compounds were synthesized (Nakatsubo 1981) and used as substrates for microbial degradation (Higuchi 1985, 1986). In particular, β-O-4 and β-1 lignin substructure models have been used most frequently. Hence, syntheses of both models are mainly described in this section. Furthermore, syntheses of many authentic samples of microbial degradation products of lignin-substructure models have been reported by many investigators (Higuchi 1985, 1986), which would be instructive for the synthesis of new model substrates.

β-O-4 Lignin-Substructure Model Dimers. A β-O-4 lignin-substructure model compound was first synthesized by Adler et al. (1952); it was 1,3-dihydroxy-1-(3,4-dimethoxyphenyl)-2-(2-methoxyphenoxy)propane (veratrylglycerol-β-guaiacyl ether) (IV). This compound has no phenolic hydroxyl group on the A-ring and no side-chain on the B-ring (Fig. 2). A free phenolic analog, 1,3-dihydroxy-1-(4-hydroxy-3-methoxyphenyl)-2-(2-methoxyphenoxy)propane (guaiacylglycerol-β-guaiacyl ether) (III), was subsequently synthesized by Adler and Eriksoo (1955). Both compounds were also synthesized by similar methods (Kratzl et al. 1959; Hosoya et al. 1980; Ralph and Young 1981; Landucci et al. 1981) and by completely different methods (Pearl and Gratzl 1962; Miksche et al. 1966; Nakatsubo et al. 1975; Lundquist and Remmerth 1975; Ahvonen et al. 1983).

β-O-4 model compounds which have side-chains on the B-ring represent more accurately the β-O-4 lignin substructure than the β-O-4 models without side-chains on the B-ring, such as β-guaiacyl ethers. As this type of compounds, the following were synthesized: Guaiacylglycerol-β-coniferyl ether (I) (Freudenberg and Eisenhut 1955; Nakatsubo and Higuchi 1980a; Katayama et al. 1981), 1,3-dihydroxy-2-(4-formyl-2-methoxyphenoxy)-1-(4-hydroxy-3-methoxyphenyl) propane (guaiacylglycerol-β-vanillin ether) (II) (Nimz 1967; Katayama et al. 1981), 1,3-dihydroxy-2-(2,6-dimethoxy-4-hydroxymethylphenoxy)-1-(3,5-dimethoxy-4-hydroxyphenyl)propane (VI) (Miksche 1973).

In our laboratory, β-O-4 lignin-substructure models are synthesized by two different methods. One is the method of Nakatsubo et al. (Nakatsubo et al. 1975; Nakatsubo and Higuchi 1980a; Katayama et al. 1981). The experimental procedures were recently described in detail by Nakatsubo (1988) and should be referred to. The other is the modified method of Adler et al. (1952): a nonphenolic β-O-4 lignin-substructure model, 1,3-dihydroxy-1-(4-ethoxy-3-methoxyphenyl)-

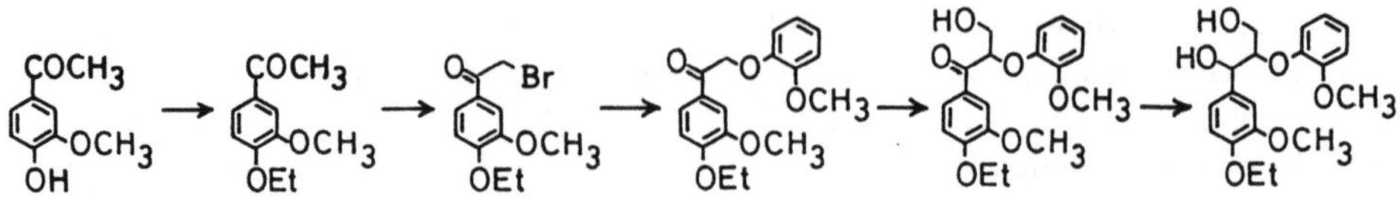

(XVIII) (XIX) (XX) (XXI) (XXII) (V)

Fig. 3. Synthetic route for 1,3-dihydroxy-1-(4-ethoxy-3-methoxyphenyl)-2-(2-methoxyphenoxy)propane (V)

2-(2-methoxyphenoxy)propane (V), was prepared from acetovanillone (XVIII) via the following five steps (Fig. 3) (Umezawa and Higuchi 1985 b). 4′-Ethoxy-3′-methoxyacetophenone (XIX): To a stirred solution of acetovanillone (XVIII) (5 g, 30.1 mmol) in DMF (10 ml), ethyl iodide (2.88 ml, 36 mmol) and subsequently finely powdered K_2CO_3 (5 g, 36 mmol) were added. The stirring was continued at room temperature until the starting material disappeared (about 3 h, checked by TLC). The reaction mixture was poured into 200 ml of diethyl ether, which was washed with 10 ml of 1 N NaOH, then with distilled water until the washings became neutral, and finally with 10 ml of a saturated NaCl solution. The diethyl ether solution was dried over anhydrous Na_2SO_4, and evaporated in vacuo to give yellow crystals (5.8 g, 99% yield). 2-Bromo-4′-ethoxy-3′-methoxyacetophenone (XX): 4′-Ethoxy-3′-methoxyacetophenone (XIX) was brominated by a modification of the method reported by King and Ostrum (1964). To a stirred solution of (XIX) (2.0 g, 10.3 mmol) in ethyl acetate (30 ml), were added $CuBr_2$ (4.7 g, 21 mmol) and then trace amounts of 47% HBr by a glass capillary. The reaction suspension was refluxed for 3 h, and poured into 100 ml of ethyl acetate. The ethyl acetate solution was washed several times with distilled water until the washings became neutral, and then with a saturated NaCl solution (10 ml), dried over anhydrous Na_2SO_4, and subsequently evaporated in vacuo to give brownish black residue. The residue was recrystallized from ethyl acetate / *n*-hexane (2.0 g, 71% yield). 4′-Ethoxy-3′-methoxy-2-(2-methoxyphenoxy)acetophenone (XXI): To a stirred solution of 2-bromo-4′-ethoxy-3′-methoxyacetophenone (XX) (153 mg, 0.56 mmol) and guaiacol (69 mg, 0.56 mmol) in DMF (1 ml), finely powdered anhydrous K_2CO_3 (147 mg, 1.07 mmol) was added. The stirring was continued at room temperature until the starting materials disappeared (2.5–7 h, checked by TLC). The reaction suspension was poured into 10 ml of distilled water, which was extracted twice with diethyl ether (50 ml and 20 ml). The combined diethyl ether was washed with a saturated NaCl solution, dried over anhydrous Na_2SO_4 and evaporated in vacuo. The residue was purified by silica gel TLC (developing solvent: ethyl acetate / *n*-hexane = 1 : 3, by volume) to yield the pure expected compound (158 mg, 89% yield). 1-(4-Ethoxy-3-methoxyphenyl)-3-hydroxy-2-(2-methoxyphenoxy)propane-1-one (XXII): To a stirred solution of 4′-ethoxy-3′-methoxy-2-(2-methoxyphenoxy)acetophenone (XXI) (338 mg, 1.07 mmol) in dimethylsulfoxide (DMSO) (3 ml), paraformaldehyde (32 mg, 1.07 mmol) and finely powdered K_2CO_3 (41.4 mg, 0.3 mmol) were added. The reaction suspension was stirred at room temperature until the starting material disappeared (0.5–1 h, checked by TLC). The reaction suspension was poured into distilled water (20 ml), and was extracted with ethyl acetate (50 ml).

The ethyl acetate solution was washed four times with a saturated NaCl solution (5 ml × 4), dried over anhydrous Na_2SO_4 and evaporated in vacuo to give pale yellow crystals, which were recrystallized from ethyl acetate to give white crystals (246 mg, 66%). The mother liquor (130 mg), which consisted mainly of the expected compound, was not further purified. 1,3-Dihydroxy-1-(4-ethoxy-3-methoxyphenyl)-2-(2-methoxyphenoxy)propane (V): To a stirred solution of 1-(4-ethoxy-3-methoxyphenyl)-3-hydroxy-2-(2-methoxyphenoxy)propane-1-one (XXII) (246 mg, 0.71 mmol) in a mixture (2 ml) of dioxane and methyl alcohol (1 : 1, by volume), $NaBH_4$ (57 mg, 1.5 mmol) was added at 0° C. After stirring at the same temperature was continued for 10 min, the reaction mixture was poured into distilled water, which was extracted twice with ethyl acetate (100 ml and 30 ml). The combined ethyl acetate solution was washed with a saturated NaCl solution, dried over anhydrous Na_2SO_4 and evaporated in vacuo to give the expected compound (colorless oil) in quantitative yield.

β-1 Lignin-Substructure Model Dimers. This type of compound was first synthesized by Lundquist and Miksche (1965): 1,2-Bis(4-hydroxy-3-methoxyphenyl)-1,3-propanediol (VIII) was synthesized. Preparation of β-1 lignin-substructure model compounds which have two differently substituted aromatic rings by the method is, however, difficult. Nakatsubo and Higuchi (1975b) reported a synthetic method for the preparation of β-1 type lignin model compounds which does not suffer from this disadvantage. Subsequently, different methods of β-1 substructure model synthesis were reported (Brežný and Pufflerová 1978; Kristersson and Lundquist 1980; Nonni and Dence 1982). Kirk and Nakatsubo (1983) reported a convenient synthesis of a β-1 type lignin model, 1,2-bis(4-methoxyphenyl)-1,3-propanediol (X), from commercially available desoxyanisoin, although this model compound, having only one methoxyl group on each aromatic ring, is not a typical model of woody lignin, but a model for glass lignin.

In our laboratory, β-1 lignin-substructure models were synthesized by the method of Nakatsubo and Higuchi (1975b). The experimental procedures, which were described in detail by Nakatsubo (1988), are outlined as follows (Fig. 4). 4′-Benzyloxy-3′-methoxyacetophenone (XXIII): To a stirred solution of acetovanillone (XVIII) (6.64 g, 0.04 mol) and benzyl chloride (5.1 ml, 0.048 mol) in 100 ml of tetrahydrofuran (THF), finely powdered anhydrous K_2CO_3 (8.3 g, 0.06 mol) and tetra-*n*-butylammonium iodide (1.5 g, 0.004 mol) were added and the reaction mixture was refluxed for 2 h with vigorous stirring. After cooling the reaction mixture to room temperature, the suspension was filtered and the precipitate was washed three times with 40 ml of ethyl acetate. The combined filtrate and washings were evaporated to dryness in vacuo to give a yellow oil which was dissolved in 100 ml of ethyl acetate. The ethyl acetate solution was washed with 40 ml of a saturated NaCl solution, dried over anhydrous Na_2SO_4 and evaporated in vacuo to give slightly yellow crystals. The product was recrystallized from ethyl alcohol and *n*-hexane to give needles (9.7 g, 95% yield). Methyl benzylhomovanillate (XXIV): 4′-Benzyloxy-3′-methoxyacetophenone (XXIII) (5.1 g, 0.02 mol) was added slowly to a solution of thallium(III) nitrate [$Tl(NO_3)_3 \cdot 3H_2O$, 9.3 g, 0.021 mol) and 75% perchloric acid (10 ml) in 50 ml of methyl alcohol. The reaction mixture was stirred for 4 h at room temperature. The thallium(I) nitrate pre-

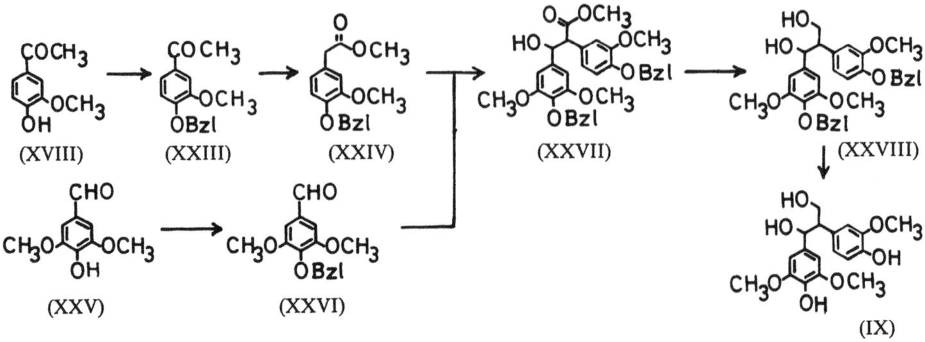

Fig. 4. Synthetic route for 1-(4-hydroxy-3,5-dimethoxyphenyl)-2-(4-hydroxy-3-methoxyphenyl)-1,3-propanediol (IX)

cipitate was filtered off and washed three times with 30 ml of ethyl acetate. The combined filtrate and washings were diluted with 200 ml of ethyl acetate and the solution was washed with a saturated $NaHCO_3$ solution until the washings became neutral. This solution was washed once with a saturated NaCl solution, dried over anhydrous Na_2SO_4 and evaporated in vacuo to give a brown oil. This oil was then dissolved in about 20 ml of methylene chloride and passed through a silica gel column to obtain a slightly yellow oil, which was crystallized from ethyl alcohol and *n*-hexane to afford colorless crystals (4.8 g, 84% yield). Benzyl syringaldehyde (XXVI): Syringaldehyde (XXV) (5.46 g, 0.03 mol) and benzyl chloride (4.14 ml, 0.036 mol) were dissolved in 100 ml of THF. Finely powdered K_2CO_3 (6.22 g, 0.045 mol) and tetra-*n*-butyl-ammonium iodide (1.1 g, 0.003 mol) were added and the reaction mixture was refluxed for 14 h with vigorous stirring. After subjecting the reaction mixture to the same treatment as had been used in the above preparation of 4′-benzyloxy-3′-methoxyacetophenone (XXIII), a slightly orange-colored oil was obtained, which was crystallized from ethyl alcohol and *n*-hexane to give needles (7.5 g, 92% yield). β-Hydroxy ester (XXVII): Diisopropylamine (2.02 g, 0.02 mol, freshly distilled from Na metal) was dissolved in 70 ml of anhydrous THF in a 200-ml 3-necked flask equipped with a dropping funnel, a drying tube containing anhydrous calcium sulfate, and a rubber septum. The flask was cooled in a dry-ice / acetone bath (−10° C) for 30 min. To the solution 13 ml of *n*-butyl lithium (0.02 mol) in *n*-hexane (1.57 N) was added dropwise from a syringe over a period of 30 min. The stirring was then continued for an additional 30 min at the same temperature. After cooling the reaction mixture below −70° C, methyl benzylhomovanillate (XXIV) (4.29 g, 0.015 mol) dissolved in 20 ml of anhydrous THF was added dropwise over a period of 20 min. The reaction mixture was stirred for an additional 30 min. Benzyl syringaldehyde (XXVI) (4.08 g, 0.015 mol) dissolved in 20 ml of anhydrous THF was added dropwise over a period of 30 min. After addition of the solution, the bath temperature was gradually raised to −50° C over a period of 80 min. The reaction mixture was neutralized by the addition of dry ice, diluted with 200 ml of ethyl acetate. The ethyl acetate solution was washed once with water (100 ml) and three times with a saturated NaCl solution (50 ml), dried over anhydrous Na_2SO_4 and evaporated in vacuo to yield an oil. Crystallization from ethyl alcohol (100 ml) gave colorless crystals consisting of both the *erythro* and *threo* isomers (roughly 1:1 ratio) (7 g, 84% yield). 1-(4-Hydroxy-3,5-dimethoxyphenyl)-2-(4-hydroxy-3-methoxyphenyl)-1,3-propanediol (IX): To a stirred solution of β-hydroxy ester (XXVII) (5.58 g, 0.01 mol) in 40 ml of anhydrous THF, triethylamine (8.4 ml, 0.06 mol, freshly distilled from Na metal) was added dropwise under nitrogen. Chlorotrimethylsilane (4.35 ml, 0.03 mol) was then added dropwise over a period of 10 min at 0° C. The reaction mixture was brought to room temperature and stirred for 30 min to complete the formation of the trimethylsilyl ether derivative. Lithium aluminum hydride (2.3 g, 0.06 mol) was placed in a 500-ml 3-necked flask equipped with a dropping funnel, a reflux condenser with a drying tube containing anhydrous calcium sulfate, and a rubber septum. Anhydrous THF (50 ml) was slowly added to the flask under nitrogen. The THF solution of the trimethylsilyl ether derivative was transferred to the dropping funnel. This solution was added dropwise over a period of 30 min under nitrogen with stirring and gentle reflux at 50° C. After the addition, the stirring was continued for an additional 30 min. After the reaction mixture had been cooled in an ice-water bath, 2.16 ml of water (0.12 mol) diluted with 10 ml of THF was carefully added dropwise to the solution. About 16 ml of water was then added dropwise to the vigorously stirred solution until finely precipitated white inorganic salts appeared. The precipitate was filtered off and washed three times with 40 ml of ethyl acetate. The combined filtrate and washings were washed successively with a 2 N HCl, a saturated NaCl solution, a saturated $NaHCO_3$ solution and a saturated NaCl solution, then dried over anhydrous Na_2SO_4. The solvents were evaporated in vacuo to give the compound (XXVIII) (a colorless oil, 5.3 g, 100% yield). To a stirred solution of compound (XXVIII) in 100 ml of methyl alcohol, 10% palladium charcoal (1 g) was added and the reaction mixture was stirred for 4 h under hydrogen. The palladium charcoal was filtered off and washed three times with 30 ml of methyl alcohol. The combined filtrate and washings were evaporated in vacuo to give a colorless oil which was dissolved in 20 ml of methyl alcohol / methylene chloride = 2:8, by volume. The solution was applied to a silica gel column (silica gel: 10 g) and the compound was eluted with the same solvent. The eluent was evaporated in vacuo to give a colorless oil. The product was crystallized from 40 ml of ethyl acetate to afford the expected compound (IX) [colorless crystals, 3.22 g, 92% overall yield from (XXVII)].

β-5 Lignin-Substructure Model Dimers. β-5 lignin-substructure model compounds, dehydrodiconiferyl alcohol (XI) and its syringyl and *p*-hydroxyphenyl analogs, were synthesized by Nakatsubo and Higuchi (1979).

β-β Lignin-Substructure Model Dimers. A β-β lignin-substructure model compound, pinoresinol (XII), was prepared by Kratzl and Miksche (1963) and by Fujimoto et al. (1982).

Other Lignin-Substructure Model Dimers. The following compounds can be used for 5-5 (biphenyl) substructure model compounds: dehydrodivanillin (XV), dehydrodivanillic acid (XVI), 2,2′-dihydroxy-3,3′-dimethoxy-5,5′-di(3-hydroxy-1-propenyl)biphenyl (XVII) which were synthesized by Elbs and Lerch (1916), Booth and Saunders (1956), Freudenberg et al. (1961), Pew (1963), and Freudenberg and Renner (1965). As for 4-O-5 (diphenyl ether) substructure models, many compounds involving diphenyl ether bond, such as 4′,5-dicarboxy-2,2′,3-trimethoxy-diphenyl ether (XIV), were prepared (Freudenberg and Renner 1965; Larsson and Miksche 1969; Kuroda and Inoue 1986).

Oligolignols. Synthesis of the below listed lignin-substructure model trimers and tetramers was also reported. The synthetic methods of the oligomers are essentially the same as those for the above mentioned dimers. (β-5)-(β-O-4) trimer (Nakatsubo and Higuchi 1980b; Ralph et al. 1986), (β-5)-(β-1) trimer (Nakatsubo and Higuchi 1980c), (β-O-4)-(β-1) trimer (Namba et al. 1980), (β-O-4)-(β-β) trimer (Kamaya et al. 1980), (β-O-4)-(α-O-γ) trimer (Umezawa et al. 1983), (β-O-4)-(α-O-4) trimer (Katayama et al. 1987), (β-O-4)-(β-O-4) trimer (Ralph et al. 1986), (β-O-4)-(β-β)-(β-O-4) tetramer (Kamaya et al. 1980), (β-O-4)-(β-O-4)-(β-O-4) tetramer (Hyatt 1987), (β-O-4)-(5-5)-(β-O-4) tetramer (Hyatt 1987).

3.1.3 Lignin-Substructure Model Compounds as Substrates of Microbial Degradation

Lignin-substructure model compounds thus prepared are submitted to microbial degradation after purification. Special care should be taken for the impurity present in the substrate, because the amounts of degradation products are usually very small. For example, when 1 μmol (348 μg) of 1,3-dihydroxy-1-(4-ethoxy-3-methoxyphenyl)-2-(2-methoxyphenoxy)propane (V) was degraded by lignin peroxidase of *P. chrysosporium*, the yield of degradation products were in the range of 1.1–130 nmol (Umezawa et al. 1986a). The purity of the lignin-substructure model should be checked before use by melting point measurement or by at least two different chromatographic conditions. In our laboratory, the model compound is analyzed by TLC with two different solvent systems, since different compounds may sometimes give practically the same Rf value (Section 3.2.2).

3.2 Analysis of Degradation Products

Structure determination of products is fundamental for lignin biodegradation study. First of all, a mixture of microbial or enzymatic degradation products of a lignin-substructure model are recovered from the microbial culture or enzymatic reaction mixture by extraction. To determine the structure of each product, the mixture is separated by chromatography and each product is identified by spectrometry.

3.2.1 Recovery of Degradation Products

The progress of degradation of substrates was followed by extracting one of the replicate cultures and examining the extract by TLC (Nakatsubo et al. 1981), or by UV spectrometric examination of the culture filtrate (Ohta et al. 1979).

Most of microbial or enzymatic degradation products of lignin-substructure model compounds can be extracted with organic solvents such as ethyl acetate. For example, the cultures of *Phanerochaete chrysosporium* after incubation with the model compounds were filtered, and both the mycelial mat ground in sand and the filtrate were extracted (Enoki et al. 1981). Alternatively, the whole cultures of the fungus were extracted, after the mycelial mat was ruptured by vigorous shaking (Umezawa et al. 1982). For acidic substances, the cultures or the enzymatic reaction mixtures are acidified to pH 2 with 1 N HCl, and extracted. Some hydrophilic compounds, however, cannot be extracted from an aqueous solution by the method. These compounds can be recovered by extraction of evaporation residue or lyophilized residue. For example, 1-(4-hydroxy-3-methoxyphenyl)-1,2,3-propanetriol was isolated from the water-soluble and ethyl acetate-insoluble fractions of the dehydrogenation products of coniferyl alcohol (Higuchi et al. 1974).

3.2.2 Separation and Purification of Degradation Products

Thin layer chromatography (TLC) and column chromatography. TLC is indispensable in the purification of lignin-substructure model compounds as substrates of fungal degradation, and in the separation of their degradation products. Recently, experimental procedures of preparative and analytical TLC of fungal degradation products of lignin and lignin model compounds were described (Umezawa and Higuchi 1988).

Microbial and enzymatic degradation products are usually separated successfully with silica gel TLC containing a fluorescent indicator. For co-chromatography, developing duplicate samples with at least two different solvent systems is recommended, since different compounds sometimes give practically identical Rf values. In our laboratory, ethyl acetate / *n*-hexane system (2:1–1:8, by volume) and methyl alcohol / methylene chloride / *n*-hexane system (5:95:0–0:100:0, and 0:2:1) are usually used as developing solvents for the silica gel TLC (Kieselgel 60 F_{254}, thickness: 0.25, 0.5, 2 mm) analysis of microbial and enzymatic degradation products of lignin-substructure models (Higuchi 1986). Lignin-related phenolic acids, such as protocatechuic acid, vanillic acid, ferulic acid and caffeic acid, are separated on a silica gel layer with benzene / dioxane / acetic acid (90:25:4, by volume) and benzene / methyl alcohol / acetic acid (90:16:8, by volume) (Pastuska 1961). Separated substances on the fluorescent layer can be detected under UV light (254 nm), because the lignin-related aromatic compounds absorb UV light. This method is not destructive, and therefore quite useful for preparative TLC. Color-producing reagents (Table 1) are also useful for detection of the separated samples, especially in analytical TLC.

For analytical TLC, precoated glass plates (20 × 20 cm, thickness 0.25 mm), usually cut into small plates (1.5 × 5 cm), are used. After preliminary analysis by analytical TLC, preparative separations are usually accomplished by means of TLC or column chromatography. Column chromatography is usually used for the separation or purification of relatively large amounts of substances, while TLC is used for fine separation of small amounts of substances. Precoated silica gel plates (20 × 20 cm, thickness 0.25, 0.5 and 2 mm) are used for preparative sep-

Table 1. Color-producing reagents

Compound	Reagent	Procedure	Results
Aldehydes ketones	2,4-Dinitro-phenyl hydrazine	Dissolve 1 g reagent in 1000 ml ethyl alcohol and 10 ml conc. HCl; spray	Yellow to orange spots on pale orange-yellow background
Phenols	$FeCl_3$/ $K_3Fe(CN)_6$	(a) 2% $FeCl_3$ in H_2O, (b) 2% $K_3Fe(CN)_6$ in H_2O, mix 1 ml (a) and 1 ml (b) before use; spray	Blue to purple spots
p-Hydroxy-benzyl alcohols	2,6-Dichloro-quinone-4-chloroimide	(a) Dissolve 1 g reagent in 100 ml ethyl alcohol, (b) 1 N NaOH; spray with (a), then (b)	Blue spots
Cinnam aldehydes	Phloroglucinol/ HCl (Wiesner reagent)	(a) Dissolve 1 g reagent in 50 ml ethyl alcohol, (b) conc. HCl; mix 2 ml (a) and 1 ml (b) before use; spray	Red purple spots
Organic compounds	H_2SO_4/HCHO	Mix 90 parts conc. H_2SO_4 with 10 parts 35% solution of HCHO in water; spray and heat at 170° C on hot plate for 30 sec	Various colors
Organic compounds	Phosphomolybdic acid	Dissolve 0.5 g reagent in 10 ml ethyl alcohol; dip the plate into the solution, dry and then heat at 170° C on hot plate for 30 sec	Various colors
Organic compounds	I_2	Place the plate in a closed tank containing a few iodine crystals	Brown spots on yellow background

aration of samples. Applicable amounts of samples, dependent upon their nature, are as follows: Thickness 0.25 mm, <10–15 mg; 0.5 mm, <20–30 mg; 2 mm, <100–200 mg. Separated zones are visualized under UV light, then scraped or scratched off. The silica gel is put into a cotton-plugged column, and the substances are eluted from the gel with methyl alcohol / methylene chloride (20:80, by volume), and recovered by solvent evaporation. The process from detection to evaporation should be accomplished as soon as possible in order to avoid undesirable decomposition.

Column chromatography is used for preparative separation of relatively large amounts of compounds. As in TLC, silica gel is most widely used for the separation of lignin-related low-molecular-weight aromatic compounds. The weight of packing is usually 20–100 times the amount of the sample to be chromatographed. The size of the column is chosen to give roughly a 10:1 ratio of height to diameter for the amount of adsorbent required. Elution solvents (moving phase) are almost the same as those for silica gel TLC. Analytical TLC using the similar stationary phase is quite useful for the determination of the elution solvent for column chromatography. The detection of bands of materials being eluted can be made by analytical TLC.

High performance liquid chromatography (HPLC). Wider ranges of samples can be analyzed by HPLC; especially many samples which are not to be analyzed by GC, such as thermolabile compounds, can be analyzed by HPLC without derivatization. HPLC is a chromatography which does not give direct information on the chemical structure of substances, and is used mainly for preparative separation and quantitative analysis of lignin degradation products, although direct coupling of HPLC and MS (LC-MS) is being developed. For stationary phase, octadecylsilane is often used for the analysis of degradation products of lignin and lignin-substructure models by microorganisms and their enzymes (Kirk et al. 1986; Chen et al. 1982).

Gas chromatography (GC). Gas chromatography (GC) is used for the quantitative analysis and separation step of GC-MS analysis of microbial degradation products of lignin and lignin-substructure model compounds. Columns packed with less polar methyl silicone polymer and its modification (OV-1, OV-101, OV-17) were often employed for this purpose (Enoki et al. 1980; Chen et al. 1982; Umezawa and Higuchi 1985a, b; Umezawa et al. 1986a, b). Recently, fused silica, bonded phase open tubular capillary column, which has a much better resolution than the conventional packed column, has been developed and is used increasingly in lignin biodegradation studies (Umezawa et al. 1986a; Kirk et al. 1986; Miki et al. 1986). Internal diameter of the usual capillary column is 0.1–0.33 mm, whereas wide-bore (or megabore) capillary columns, which can be used with most packed column detectors, have internal diameters of 0.53–0.75 mm. Wide-bore columns are less efficient than narrow-bore capillary columns, but they have larger sample capacities. They are, therefore, invaluable for quantifying widely disproportionate sample components, where one component is abundant and the amount of another component is very small. In such cases, preliminary analysis of the sample is conducted by use of a wide-bore column. If the resolution is not sufficient, the sample can be analyzed with a narrow-bore capillary column after the removal of large-amount components by using proper methods, such as TLC.

Derivatization of less volatile compounds to volatile compounds is essential for GC (or GC-MS) analysis. Acetylation, methylation (ethylation), trimethylsilylation and their combination (acetylation and subsequent methylation) are often employed for the analysis of fungal degradation products of lignin and lignin-substructure model compounds (Enoki et al. 1981; Chen et al. 1982; Umezawa and Higuchi 1984).

Phenolic and alcoholic hydroxyl groups are acetylated by acetic anhydride / pyridine. To ethyl acetate solution of the sample, acetic anhydride and pyridine are added (ethyl acetate / acetic anhydride / pyridine = 1 : 1 : 1, by volume) and stirred for 10–24 h at room temperature. Subsequently, the solvent and excess reagents are evaporated off under high vacuum without heating (Umezawa and Higuchi 1984; Umezawa et al. 1986a, b). For this purpose a rotary vacuum pump and a dry ice-methyl alcohol trap ($< -70°$ C) are used.

Carboxylic acids are not acetylated, but carboxylic acids and phenolic hydroxyl groups are converted to methyl (or ethyl) esters and methyl (or ethyl) ethers, respectively, by using diazomethane (or diazoethane). Alcoholic hydroxyl groups are not affected by the treatment. Samples are dissolved into methyl alcohol, and then an ether solution of diazomethane (or diazoethane) (Fieser and Fieser 1967; Sandler and Karo 1983) is added dropwise. Reactions of carboxylic acids finish immediately with the evolution of nitrogen gas and decolorization of the diazoalkane (diazomethane: yellow, diazoethane: orange). As for phenolic hydroxyl groups, longer reaction periods are required (10 min–3 h). After desired periods of reaction, the solvent and excess reagent are evaporated off. Thus, the alkylation by use of diazoalkanes is very simple and easy. However, since diazoalkanes are toxic and explosive, operations should be conducted in a draft chamber behind a safety screen. Care should be taken for possible side reactions: Aldehydes can react with diazoalkanes to give the corresponding ketone. 2-(2-Benzyloxy-5-formyl-3-methoxyphenyl)acetic acid treated with excess diazoethane gives its ethyl ester and ethyl 2-(2-benzyloxy-3-methoxy-5-propionylphenyl)acetate (Umezawa, Nakatsubo and Higuchi unpublished). *N,N*-Dimethylformamide dialkyl acetals are also used as esterification reagents of carboxylic acids (Thenot et al. 1972).

Another important derivatization is trimethylsilylation (TMS). Both phenolic and alcoholic hydroxyl groups and carboxylic acids of lignin degradation products can be silylated by *N,O*-bis(trimethylsilyl)trifluoroacetamide (BSTFA). This derivatization method is very useful, especially for routine analysis. This method was applied to GC-MS analysis of fungal and enzymatic degradation products of β-O-4 and β-1 lignin-substructure model compounds (Enoki et al. 1981; Glenn et al. 1983; Kirk and Nakatsubo 1983). Trimethylsilylated samples, however, cannot be stored for a long period of time, while the acetates of hydroxyl groups and methyl (or ethyl) esters and ethers of carboxylic acids and phenols are usually stable during long-period storage and thin-layer and column chromatographic separation. The acetylation and the alkylation may, therefore, be effective for the analysis of unknowns which require time-consuming repeated purification by TLC or column chromatography.

3.2.3 Structure Determination

The structures of purified microbial degradation products can be determined by use of spectroscopy such as MS, [^{1}H]-NMR, [^{13}C]-NMR and IR. Alternatively, the partially separated or unpurified degradation products are often analyzed by GC-MS after proper derivatization. If authentic samples of possible degradation products are available, structures of the degradation products are determined by comparison of their spectral data, chromatographic behavior (retention time in GC or HPLC, and Rf value in TLC) and melting points with those of the authentic samples. Accordance of Rf value in TLC and/or retention time in GC or HPLC between the unknown and the authentic sample is never a sufficient criterion but a necessary condition for identification. On the other hand, if the authentic samples are not available, purified products can be identified by interpretation of the spectral data as much as possible. The determined structure should be confirmed by organic synthesis of the corresponding authentic sample.

[^{1}H]- and [^{13}C]-NMR Spectrometry. NMR provides direct information about numbers, types of proton or carbon atoms, and proton-carbon framework of an

organic molecule, which are very important for structure determination, although the amounts of samples required are relatively large ([^{1}H]-NMR: >0.5 mg, [^{13}C]-NMR: >20 mg). Comparison of NMR spectra with those of related compounds is of great importance for the interpretation of the unknown spectra. Many [^{1}H]- and [^{13}C]-NMR spectra of lignin-related compounds and microbial degradation products of lignin-substructure models have been reported (Ludwig 1971; Nakatsubo 1981; Lüdemann and Nimz 1973, 1974; Nimz and Lüdemann 1974, 1976; see also references cited in Janshekar and Fiechter 1983; Higuchi 1985, 1986).

Mass Spectrometry (MS). Since mass spectrometry (MS) can be coupled with GC, and since the amounts of samples required for MS are very small (<0.1 μg), MS (especially GC-MS) is used most frequently in the analysis of degradation products of lignin-substructure models by microorganisms and their enzymes, which are usually complex mixtures of a small amount of compounds.

An important piece of information provided by MS is a kind of molecular fingerprint. In electron impact (EI)-MS, the chance that two different compounds will have identical mass spectra is small. Degradation products of lignin-substructure models are very often identified by matching their mass spectra as well as retention times in GC and [^{1}H]-NMR spectra with those of the authentic samples (Higuchi 1985, 1986).

MS provides information about molecular weight, structure (by interpretation of the fragmentation pattern) and elemental composition (by high-resolution MS). Although EI-MS has been used most widely for the identification of organic compounds involving lignin-related compounds, molecular ions of some compounds are missing by the ionization method. Whereas chemical ionization (CI)-MS gives the information about molecular weight in many cases where EI-MS failed to do so. Furthermore, stable isotopes have been often used in the study of the mechanisms for chemical reactions involved in lignin biodegradation, where isotope-labeled products ought to be analyzed by MS (see Sect. 3.2.4).

GC-MS is also effective for quantification of products. The stable isotope dilution method by use of selected ion monitoring is an excellent method for quantification (Baillie 1981), which can compensate imperfect resolution of GC and error caused by loss in extraction, derivatization and purification. This method was recently employed in the quantification of fungal and enzymatic degradation products of β-O-4 lignin models (Umezawa et al. 1986a; Yokota et al. 1988).

Infrared Spectroscopy (IR). Since IR spectrum gives a fingerprint of a compound, it has been used for identification of compounds. If the two samples have identical IR spectra, the compounds are almost certainly identical. Russell et al. (1961) used for the first time a lignin-substructure model compound in the study of the microbial degradation of lignin, where IR was used to identify its degradation product. Another important piece of information obtained from IR spectra is the functional groups of the sample molecule, such as carbonyl and carboxyl groups.

Examples for the Identification of Degradation Products. Many degradation products have been identified by matching the above mentioned data of the spectrom-

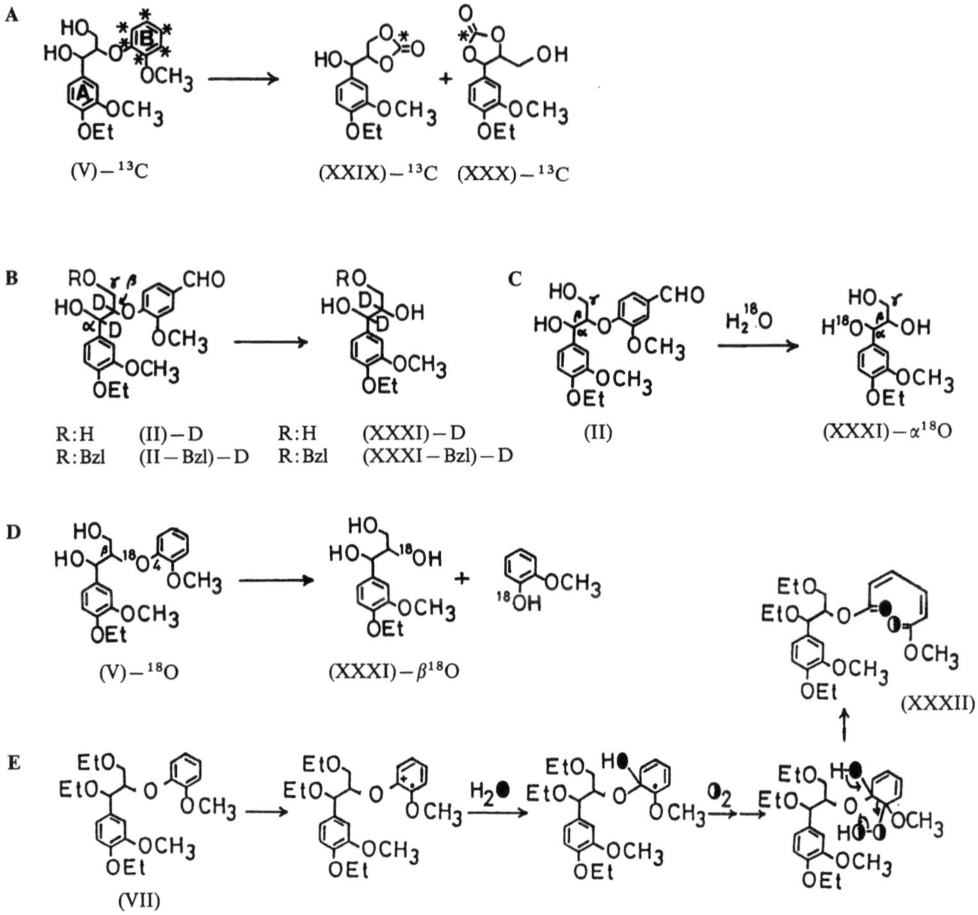

Fig. 5 A–E. Examples of stable isotope experiments in microbial and enzymatic degradation of β-O-4 lignin substructure model compounds. **A–D** Degradation by intact cells of *Phanerochaete chrysosporium*; **E** Degradation by lignin peroxidase of *Phanerochaete chrysosporium*; * ^{13}C; *D* ^{2}H, ● ^{18}O of $H_2{}^{18}O$; ◐ ^{18}O of $^{18}O_2$; *Et* CH_2CH_3; *Bzl* $CH_2C_6H_5$

etry as well as the retention time in GC and HPLC, and Rf value in TLC with those of the authentic sample (Higuchi 1985, 1986).

Example: Identification of an aromatic ring cleavage product (Fig. 5 A) (Umezawa and Higuchi 1985 b; Umezawa et al. 1986 b). Ligninolytic cultures (20 ml / 300 ml Erlenmeyer flask) of *Phanerochaete chrysosporium* Burds. (ME-446) were grown at 39° C without agitation in a nitrogen-limiting, glucose, dilute mineral salts medium (Kirk et al. 1978; Nakatsubo et al. 1981). 1,3-Dihydroxy-1-(4-ethoxy-3-methoxyphenyl)-2-(2-methoxyphenoxy)propane (V) was added to the 6-day-old cultures as a solution of DMF (3.44 mg in 0.1 ml of DMF / 20 ml culture, 5 cultures) without sterilization (Umezawa et al. 1982). No contamination occurred on the addition of substrates. The final amount of DMF was 0.1 ml per 20 ml culture, at which concentration growth or lignin decomposition was not influenced (Kirk et al. 1978). All the cultures were flushed with sterilized 100% O_2 immediately after the addition of the model compound, and the stationary incubation was continued at 39° C for further 93 h. Subsequently, the mycelial mats were ruptured by shaking vigorously, all the cultures were combined and extracted three times with ethyl acetate (100 ml × 3). The combined ethyl acetate was washed with a saturated NaCl solution (10 ml), dried by passing through Na_2SO_4 (25 g) column, and the solvent was evaporated off in vacuo. The residue was dissolved in ethyl acetate (1 ml), and acetic anhydride (1 ml) and pyridine (1 ml) were added with stirring. After the stirring had been continued for 24 h at room temperature, the solvent and excess reagents were evaporated. The acetylated degradation products thus obtained were separated twice by silica gel TLC as follows. Precoated silica gel plates containing fluorescent indicator [Kieselgel 60 F_{254}, Merck, Art. 5744 (20 × 20 cm, layer thickness 0.5 mm) and Art. 5715 (20 × 20 cm, layer thickness 0.25 mm)] were developed with ethyl acetate before sample application, then dried, cut into proper size and used for sample separation. The acetylated degradation products were separated by the predeveloped plate (20 × 10 cm, thickness 0.5 mm) (developing solvent: methylene chloride / *n*-hexane = 2 : 1, by volume, developed once; then methylene chloride, developed four times). Separated zones were visualized under UV light (254 nm) (more than ten zones were detected), and each zone was recovered separately. Two fractions among them [apparent Rf value after repeated developments: 0.2 (Fraction 1) and 0.3 (Fraction 2)] were again separated by TLC to give pure compounds X_1 and X_2, respectively (Fraction 1: 20 × 5 cm, thickness 0.25 mm, developing solvent, ethyl acetate / *n*-hexane = 1 : 2, by volume, developed once; Fraction 2: 20 × 5 cm, thickness 0.25 mm, developing solvent, ethyl acetate / *n*-hexane = 1 : 4, by volume, developed twice). [^{1}H]-NMR and MS indicated that compounds X_1 and X_2 are 1-(4-ethoxy-3-methoxyphenyl)-1,2,3-propanetriol-2,3-cyclic carbonate (XXIX) and 1-(4-ethoxy-3-methoxyphenyl)-1,2,3-propanetriol-1,2-cyclic carbonate (XXX), respectively. Furthermore, the structures were confirmed by the fact that their [^{1}H]-NMR spectra, mass spectra, retention times in GC, Rf values in TLC were identical to those of synthesized authentic samples.

Subsequently, [^{13}C]-tracer experiment showed that both the compounds are aromatic ring cleavage products. 1,3-Dihydroxy-1-(4-ethoxy-3-methoxyphenyl)-2-[U-ring-^{13}C](2-methoxyphenoxy)propane (V)-^{13}C was synthesized and degraded by the same condition as above. Degradation products were extracted and acetylated as above, and the acetylated degradation products were analyzed by GC-MS (glass column, 1 m × 2.6 mm i.d.; 1.5% OV-17 on Chromosorb W (AW-DMCS); column temperature: 170–240° C, 5° C/min). Mass spectra of the cyclic carbonates showed that their molecular ion peaks were m/z 311, one mass unit higher than those of unlabeled authentic samples, indicating that they are the B-ring cleavage products.

3.2.4 Mechanisms for Lignin Degradation and Stable Isotope Experiments

Many stable isotope experiments have been applied successfully to the examination of the mechanisms for microbial degradation of lignin-substructure model compounds (Higuchi 1986). Figure 5 shows the examples of the stable isotope experiments.

Example 1: As for the formation of arylglycerol from β-O-4 lignin-substructure model compound by white-rot fungi [e.g., 1-(4-ethoxy-3-methoxyphenyl)-1,2,3-propanetriol (XXXI) from 1,3-dihydroxy-1-(4-ethoxy-3-methoxyphenyl)-2-(2-methoxyphenoxy)propane (V)], earlier papers suggested possible involvement of hydrolysis of the β-O-4 bond and/or β-hydroxylation of the β-O-4 substructure. Tracer experiments showed, however, that both are not the case at least for the degradation by *Phanerochaete chrysosporium*, and that the arylglycerol was formed via O-C_4 cleavage of the β-O-4 bond: β-O-4 model compounds, doubly deuterated at the Cα and Cβ positions, (II)-D and (II-Bzl)-D, were incubated with the fungus. Mass spectra showed that the arylglycerols (XXXI) and (XXXI-Bzl) formed during the degradation retained the deuterium atoms at both the Cα and Cβ positions, indicating no involvement of the β-hydroxylation (Fig. 5B) (Umezawa et al. 1983; Umezawa and Higuchi 1985c). Subsequently, 1,3-dihydroxy-1-(4-ethoxy-3-methoxyphenyl)-2-(4-formyl-2-methoxyphenoxy)-propane (II) was incubated with the fungus under $H_2{}^{18}O$. Mass spectrum showed that ^{18}O was not incorporated into the Cβ position of the formed arylglycerol (XXXI), although small amounts of ^{18}O were incorporated into the Cα position, indicating no involvement of the hydrolysis (Fig. 5C) (Umezawa and Higuchi 1984).

[2-^{18}O]-1,3-dihydroxy-1-(4-ethoxy-3-methoxyphenyl)-2-(2-methoxyphenoxy)-propane (V)-^{18}O was synthesized and degraded by *P. chrysosporium* to yield the arylglycerol (XXXI) and guaiacol. Mass spectrometric analysis showed that β-oxygen of (XXXI) and phenolic oxygen of guaiacol were quantitatively derived from the β-ethereal oxygen of the substrate (Fig. 5D). This result showed that both products are formed via different pathways; the arylglycerol (XXXI) was formed via O-C_4 cleavage of the β-O-4 bond, whereas guaiacol was formed via Cβ-O cleavage of the β-O-4 bond (Umezawa and Higuchi 1985a).

Example 2: 1,3-Diethoxy-1-(4-ethoxy-3-methoxyphenyl)-2-(2-methoxyphenoxy)propane (VII) was synthesized and incubated with lignin peroxidase of *P. chrysosporium* / H_2O_2 system under $^{18}O_2$ / $H_2{}^{16}O$ and $^{16}O_2$ / $H_2{}^{18}O$. GC-MS analysis of the aromatic ring cleavage product, methyl muconate of arylglycerol (XXXII), showed that one oxygen atom of the ester carbonyls was derived from $H_2{}^{18}O$ and the other from $^{18}O_2$. Based on the results, the mechanism involving nucleophilic attack of water and radical coupling with dioxygen was proposed for the aromatic ring cleavage of β-O-4 lignin-substructure model compounds by the enzyme (Fig. 5E) (Umezawa and Higuchi 1987).

References

Adler E (1977) Lignin chemistry – past, present and future. Wood Sci Technol 11:169–218

Adler E, Eriksoo E (1955) Guaiacylglycerol and its β-guaiacyl ether. Acta Chem Scand 9:341–342

Adler E, Lindgren BO, Saedlén U (1952) The beta-guaiacyl ether of alpha-veratrylglycerol as a lignin model. Sven Papperstidn 55:245–254

Ahvonen T, Brunow G, Kristersson P, Lundquist K (1983) Stereoselective syntheses of lignin model compounds of the β-O-4 and β-1 types. Acta Chem Scand B37:845–849
Baillie TA (1981) The use of stable isotopes in pharmacological research. Pharmacol Rev 33:81–132
Björkman A (1956) Finely divided wood. I. Extraction of lignin with neutral solvents. Sven Papperstidn 59:477–485
Booth H, Saunders BC (1956) Studies in peroxidase action. Part X. The oxidation of phenols. J Chem Soc 940–948
Brežný R, Pufflerová A (1978) Synthesis of 1,2-bis(4-hydroxy-3-methoxyphenyl)- and 1,2-bis(3,4-dimethoxyphenyl)-1,3-propanediol. Collect Czech Chem Commun 43:3263–3267
Browning BL (1967) Methods of wood chemistry, vol II. John Wiley, New York
Chang H-m, Cowling EB, Brown W, Adler E, Miksche G (1975) Comparative studies on cellulolytic enzyme lignin and milled wood lignin of sweetgum and spruce. Holzforschung 29:153–159
Chen C-L, Chang H-m (1985) Chemistry of lignin biodegradation. In: Higuchi T (ed) Biosynthesis and biodegradation of wood components. Academic Press, Orland FL, p 535
Chen C-L, Chang H-m, Kirk TK (1982) Aromatic acids produced during degradation of lignin in spruce wood by *Phanerochaete chrysosporium.* Holzforschung 36:3–9
Chen C-L, Chang H-m, Kirk TK (1983) Carboxylic acids produced through oxidative cleavage of aromatic rings during degradation of lignin in spruce wood by *Phanerochaete chrysosporium.* J Wood Chem Technol 3:35–57
Chua MGS, Chen C-L, Chang H-m, Kirk TK (1982) ^{13}C-NMR spectroscopic study of spruce lignin degraded by *Phanerochaete chrysosporium* I. New structures. Holzforschung 36:165–172
Connors WJ, Lorenz LF, Kirk TK (1978) Chromatographic separation of lignin models by molecular weight using Sephadex LH-20. Holzforschung 32:106–108
Connors WJ, Sarkanen S, McCarthy JL (1980) Gel chromatography and association complexes of lignin. Holzforschung 34:80–85
Crawford RL (1981) Lignin biodegradation and transformation. John Wiley, New York
Elbs K, Lerch H (1916) Über Dehydrodivanillin. J Prakt Chem 93:1–9
Ellwardt P-C, Haider K, Ernst L (1981) Untersuchungen des mikrobiellen Ligninabbaues durch ^{13}C-NMR-Spektroskopie an spezifisch ^{13}C-angereichertem DHP-Lignin aus Coniferylalkohol. Holzforschung 35:103–109
Enoki A, Goldsby GP, Gold MH (1980) Metabolism of the lignin model compounds veratrylglycerol-β-guaiacyl ether and 4-ethoxy-3-methoxyphenylglycerol-β-guaiacyl ether by *Phanerochaete chrysosporium.* Arch Microbiol 125:227–232
Enoki A, Goldsby GP, Gold MH (1981) β-Ether cleavage of the lignin model compound 4-ethoxy-3-methoxyphenylglycerol-β-guaiacyl ether and derivatives by *Phanerochaete chrysosporium.* Arch Microbiol 129:141–145
Fieser LF, Fieser M (1967) Reagents for organic synthesis. John Wiley, New York
Freudenberg K (1965) Lignin: its constitution and formation from p-hydroxycinnamyl alcohols. Science 148:595–600
Freudenberg K, Eisenhut W (1955) Synthese des Guajacylglycerin-β-coniferyläthers und weitere Versuche im Zusammenhang mit dem Lignin. Chem Ber 88:626–633
Freudenberg K, Renner K-C (1965) Über Biphenyle und Diaryläther unter den Vorstufen des Lignins. Chem Ber 98:1879–1892
Freudenberg K, Jovanović V, Topfmeier F (1961) Versuche mit deuteriertem Coniferylalkohol zur Bestimmung der Substitution am C-Atom 5 der Coniferyleinheit des Lignins. Chem Ber 94:3227–3238
Fujimoto H, Nakatsubo F, Higuchi T (1982) Synthesis of pinoresinol. Mokuzai Gakkaishi 28:555–562
Glenn JK, Morgan MA, Mayfield MB, Kuwahara M, Gold MH (1983) An extracellular H_2O_2-requiring enzyme preparation involved in lignin biodegradation by the white rot basidiomycete *Phanerochaete chrysosporium.* Biochem Biophys Res Commun 114:1077–1083

Goldschmid O (1971) Ultraviolet spectra. In: Sarkanen KV, Ludwig CH (eds) Lignins: occurrence, formation, structure and reactions. John Wiley, New York, p 241
Haider K, Trojanowski J (1975) Decomposition of specifically ^{14}C-labelled phenols and dehydropolymers of coniferyl alcohol as models for lignin degradation by soft and white rot fungi. Arch Microbiol 105:33–42
Haider K, Kern HW, Ernst L (1985) Intermediate steps of microbial lignin degradation as elucidated by ^{13}C NMR spectroscopy of specifically ^{13}C-enriched DHP-lignins. Holzforschung 39:23–32
Hata K (1966) Investigations on lignins and lignification XXXIII. Studies on lignins isolated from spruce wood decayed by *Poria subacida* BII. Holzforschung 20:142–147
Higuchi T (1985) Degradative pathways of lignin model compounds. In: Higuchi T (ed) Biosynthesis and biodegradation of wood components. Academic Press, Orland FL, p 557
Higuchi T (1986) Catabolic pathways and role of ligninases for the degradation of lignin substructure models by white-rot fungi. Wood Res 73:58–81
Higuchi T, Nakatsubo F, Ikeda Y (1974) Enzymic formation of arylglycerols from *p*-hydroxycinnamyl alcohols. Holzforschung 28:189–192
Hosoya S, Kanazawa K, Kaneko H, Nakano J (1980) Synthesis of guaiacylglycerol-β-guaiacyl ether. Mokuzai Gakkaishi 26:118–121
Hyatt JA (1987) Synthesis of some tetrameric lignin model compounds containing β-O-4 and 5,5′-interunit linkages. Holzforschung 41:363–370
Ishikawa H, Schubert WJ, Nord FF (1963) Investigations on lignins and lignification. XXVII. The enzymic degradation of softwood lignin by white-rot fungi. Arch Biochem Biophys 100:131–139
Janshekar H, Fiechter A (1983) Lignin: biosynthesis, application, and biodegradation. In: Fiechter A (ed) Advances in biochemical engineering / biotechnology 27. Springer, Berlin Heidelberg New York, p 119
Kamaya Y, Nakatsubo F, Higuchi T (1980) Synthesis of a trimeric lignin model compound composed of β-O-4 and syringaresinol substructures. Mokuzai Gakkaishi 26:471–475
Katayama T, Nakatsubo F, Higuchi T (1981) Syntheses of arylglycerol-β-aryl ethers. Mokuzai Gakkaishi 27:223–230
Katayama T, Kawai S, Sogo M, Higuchi T (1987) Degradation of guaiacylglycerol-α-vanillyl alcohol-β-vanillin diether, a lignin substructure model containing a non-cyclic benzyl aryl ether, by *Fusarium solani* M-13-1. Mokuzai Gakkaishi 33:503–510
Katayama Y, Fukuzumi T (1978) Enzymatic synthesis of three lignin-related dimers by an improved peroxidase-hydrogen peroxide system. Mokuzai Gakkaishi 24:664–667
King LC, Ostrum GK (1964) Selective bromination with copper(II) bromide. J Org Chem 29:3459–3461
Kirk TK, Chang H-m (1974) Decomposition of lignin by white-rot fungi I. Isolation of heavily degraded lignins from decayed spruce. Holzforschung 28:217–222
Kirk TK, Chang H-m (1975) Decomposition of lignin by white-rot fungi II. Characterization of heavily degraded lignins from decayed spruce. Holzforschung 29:56–64
Kirk TK, Farrell RL (1987) Enzymatic "combustion": the microbial degradation of lignin. Annu Rev Microbiol 41:465–505
Kirk TK, Nakatsubo F (1983) Chemical mechanism of an important cleavage reaction in the fungal degradation of lignin. Biochim Biophys Acta 756:376–384
Kirk TK, Connors WJ, Bleam RD, Hackett WF, Zeikus JG (1975) Preparation and microbial decomposition of synthetic [^{14}C]lignins. Proc Natl Acad Sci USA 72:2515–2519
Kirk TK, Schultz E, Connors WJ, Lorenz LF, Zeikus JG (1978) Influence of culture parameters on lignin metabolism by *Phanerochaete chrysosporium*. Arch Microbiol 117:277–285
Kirk TK, Tien M, Kersten PJ, Mozuch MD, Kalyanaraman B (1986) Ligninase of *Phanerochaete chrysosporium*. Mechanism of its degradation of the non-phenolic arylglycerol β-aryl ether substructure of lignin. Biochem J 236:279–287
Kratzl K, Miksche GE (1963) Zur Synthese des DL-Pinoresinols. Eine neue Darstellungsweise des 2,3-Di-α-hydroxyvanillylbutandiols-(1,4). Monatsh Chem 94:434–439

Kratzl K, Kisser W, Gratzl J, Silbernagel H (1959) Der β-Guajacyläther des Guajacylglycerins, seine Umwandlung in Coniferylaldehyd und verschiedene andere Arylpropanderivate (Modellversuche zum hydrolytischen Abbau des Coniferenlignins). Monatsh Chem 90:771–782

Kristersson P, Lundquist K (1980) A new synthetic route to lignin model compouds of the 1,2-diaryl-1,3-propanediol type. Acta Chem Scand B34:213–215

Kuroda K, Inoue Y (1986) Synthesis of diphenyl ethers composed of 4-*O*-5 structures. Mokuzai Gakkaishi 32:285–288

Lai YZ, Sarkanen KV (1971) Isolation and structural studies. In: Sarkanen KV; Ludwig CH (eds) Lignins: occurrence, formation, structure and reactions. John Wiley, New York, p 165

Landucci LL, Geddes SA, Kirk TK (1981) Synthesis of ^{14}C labeled 3-methoxy-4-hydroxy-α-(2-methoxyphenoxy)-β-hydroxypropiophenone, a lignin model compound. Holzforschung 35:66–69

Larsson S, Miksche GE (1969) Gaschromatographische Analyse von Ligninoxydationsprodukten III. Oxydativer Abbau von methyliertem Björkman-Lignin (Fichte). Acta Chem Scand 23:3337–3351

Lüdemann H-D, Nimz H (1973) Carbon-13 nuclear magnetic resonance spectra of lignins. Biochem Biophys Res Commun 52:1162–1169

Lüdemann H-D, Nimz H (1974) ^{13}C-Kernresonanzspektren von Ligninen, 2. Buchen- und Fichten-Björkman-Lignin. Makromol Chem 175:2409–2422

Ludwig CH (1971) Magnetic resonance spectra. In: Sarkanen KV, Ludwig CH (eds) Lignins: occurrence, formation, structure and reactions. John Wiley, New York, p 299

Lundquist K, Miksche GE (1965) Nachweis eines neuen Verknüpfungsprinzips von Guajacylpropaneinheiten im Fichtenlignin. Tetra Lett 2131–2136

Lundquist K, Remmerth S (1975) New synthetic routes to lignin model compounds of the arylglycerol-β-aryl ether type. Acta Chem Scand B29:276–278

Miki K, Renganathan V, Gold MH (1986) Mechanism of β-aryl ether dimeric lignin model compound oxidation by lignin peroxidase of *Phanerochaete chrysosporium*. Biochemistry 25:4790–4796

Miksche GE (1973) Zum alkalischen Abbau von Arylglycerin-β-(2,6-dimethoxy-4-alkylaryl)-ätherstrukturen. Acta Chem Scand 27:1355–1368

Miksche GE, Gratzl J, Fried-Matzka M (1966) Zur Synthese der beiden diastereomeren Formen des Guajacylglycerin-β-(2-methoxyphenyl)-äthers und des Guajacylglycerins. Acta Chem Scand 20:1038–1043

Nakatsubo F (1981) Enzymic dehydrogenation of *p*-coumaryl alcohol and syntheses of oligolignols. Wood Res 67:59–118

Nakatsubo F (1988) Synthesis of lignols and related compounds. In: Wood WA, Kellogg ST (eds) Methods Enzymol, Vol 161. Academic Press, San Diego, p 57

Nakatsubo F, Higuchi T (1975a) Enzymic dehydrogenation of *p*-coumaryl alcohol I. Structure of dimeric compounds. Holzforschung 29:64–68

Nakatsubo F, Higuchi T (1975b) Synthesis of 1,2-diarylpropane-1,3-diols and determination of their configurations. Holzforschung 29:193–198

Nakatsubo F, Higuchi T (1979) Syntheses of phenylcoumarans. Mokuzai Gakkaishi 25:735–742

Nakatsubo F, Higuchi T (1980a) Synthesis of guaiacylglycerol-β-coniferyl and β-coniferyl aldehyde ethers. Wood Res 66:23–29

Nakatsubo F, Higuchi T (1980b) Synthesis of trimeric lignin model compounds composed of phenylcoumaran and β-O-4 structures. Mokuzai Gakkaishi 26:31–36

Nakatsubo F, Higuchi T (1980c) Synthesis of trimeric lignin model compound composed of phenylcoumaran and β-1 structures. Mokuzai Gakkaishi 26:107–111

Nakatsubo F, Sato K, Higuchi T (1975) Synthesis of guaiacylglycerol-β-guaiacyl ether. Holzforschung 29:165–168

Nakatsubo F, Kirk TK, Shimada M, Higuchi T (1981) Metabolism of a phenylcoumaran substructure lignin model compound in ligninolytic cultures of *Phanerochaete chrysosporium*. Arch Microbiol 128:416–420

Namba H, Nakatsubo F, Higuchi T (1980) Synthesis of trimeric lignin model compound composed of β-O-4 and β-1 structures. Mokuzai Gakkaishi 26:426–431

Nimz H (1967) Über zwei aldehydische Dilignole aus Fichtenlignin. Chem Ber 100:2633–2639

Nimz H, Lüdemann H-D (1974) ^{13}C-Kernresonanzspektren von Ligninen, 5. Oligomere Ligninmodellsubstanzen. Makromol Chem 175:2577–2583

Nimz HH, Lüdemann H-D (1976) Kohlenstoff-13-NMR-Spektren von Ligninen, 6. Lignin- und DHP-Acetate. Holzforschung 30:33–40

Nonni AJ, Dence CW (1982) The synthesis of 1,2-diarylpropane-1,3-diols via the Ivanoff reaction. J Wood Chem Technol 2:161–168

Ohta M, Higuchi T, Iwahara S (1979) Microbial degradation of dehydrodiconiferyl alcohol, a lignin substructure model. Arch Microbiol 121:23–28

Pastuska G (1961) Silica gel layer chromatography of phenols and phenolcarboxylic acids I. Z Anal Chem 179:355–358

Pearl IA, Gratzl J (1962) Studies on lignin and related products. XVI. Syntheses of lignin model compounds of the phenylglycerol β-ether and related series. J Org Chem 27:2111–2114

Pew JC (1963) Evidence of a biphenyl group in lignin. J Org Chem 28:1048–1054

Ralph J, Young RA (1981) Synthesis of the lignin model compounds *threo*-guaiacylglycerol-β-guaiacyl ether and *threo*-veratrylglycerol-β-guaiacyl ether. Holzforschung 35:39–41

Ralph J, Ede RM, Wilkins AL (1986) Synthesis of trimeric lignin model compounds composed of β-aryl ether and phenylcoumaran structures. Holzforschung 40:23–30

Russell JD, Henderson MEK, Farmer VC (1961) Metabolism of lignin model compounds by *Polystictus versicolor*. Biochim Biophys Acta 52:565–570

Sandler SR, Karo W (1983) Organic functional group preparations, 2nd edn, Vol 1. Academic Press, New York

Sarkanen KV (1971) Precursors and their polymerization. In: Sarkanen KV, Ludwig CH (eds) Lignins: occurrence, formation, structure and reactions. John Wiley, New York, p 95

Sarkanen KV, Ludwig CH (eds) (1971) Lignins: occurrence, formation, structure, and reactions. John Wiley, New York

Schaefer J, Sefcik MD, Stejskal EO, McKay RA; Hall PL (1981) Characterization of the catabolic transformation of lignin in culture using magic-angle carbon-13 nuclear magnetic resonance. Macromolecules 14:557–559

Tai D, Terazawa M, Chen C-L, Chang H-m, Kirk TK (1983) Biodegradation of guaiacyl and guaiacyl-syringyl lignins in wood by *Phanerochaete chrysosporium*. In: Higuchi T, Chang H-m, Kirk TK (eds) Recent advances in lignin biodegradation research. Uni Publishers, Tokyo, p 44

Tanahashi M, Higuchi T (1988) Chemical degradation methods for characterization of lignins. In: Wood WA, Kellogg ST (eds) Methods Enzymol, Vol 161. Academic Press, San Diego, p 101

Thenot J-P, Horning EC, Stafford M, Horning MG (1972) Fatty acid esterification with N,N-dimethylformamide dialkyl acetals for GC analysis. Anal Lett 5:217–223

Tien M, Kirk TK (1983) Lignin-degrading enzyme from the Hymenomycete *Phanerochaete chrysosporium* Burds. Science 221:661–663

Umezawa T, Higuchi T (1984) Incorporation of $H_2{}^{18}O$ into the Cα but not the Cβ position in degradation of a β-O-4 lignin substructure model by *Phanerochaete chrysosporium*. Agric Biol Chem 48:1917–1921

Umezawa T, Higuchi T (1985a) Role of guaiacol in the degradation of arylglycerol-β-guaiacyl ether by *Phanerochaete chrysosporium*. FEMS Microbiol Lett 26:123–126

Umezawa T, Higuchi T (1985b) Aromatic ring cleavage in degradation of β-O-4 lignin substructure by *Phanerochaete chrysosporium*. FEBS Lett 182:257–259

Umezawa T, Higuchi T (1985c) Arylglycerol formation in degradation of β-O-4 lignin substructures with and without α-O-γ bond by *Phanerochaete chrysosporium*. Wood Res 71:25–31

Umezawa T, Higuchi T (1987) Mechanism of aromatic ring cleavage of β-O-4 lignin substructure models by lignin peroxidase. FEBS Lett 218:255–260

Umezawa T, Higuchi T (1988) Analysis of lignin degradation intermediates by thin layer chromatography and gas chromatography / mass spectrometry. In: Wood WA, Kellogg ST (eds) Methods Enzymol, Vol 161. Academic Press, San Diego, p 200

Umezawa T, Nakatsubo F, Higuchi T (1982) Lignin degradation by *Phanerochaete chrysosporium*: metabolism of a phenolic phenylcoumaran substructure model compound. Arch Microbiol 131:124–128

Umezawa T, Nakatsubo F, Higuchi T (1983) Degradation pathway of arylglycerol-β-aryl ethers by *Phanerochaete chrysosporium*. Agric Biol Chem 47:2677–2681

Umezawa T, Shimada M, Higuchi T, Kusai K (1986a) Aromatic ring cleavage of β-O-4 lignin substructure model dimers by lignin peroxidase of *Phanerochaete chrysosporium*. FEBS Lett 205:287–292

Umezawa T, Kawai S, Yokota S, Higuchi T (1986b) Aromatic ring cleavage of various β-O-4 lignin model dimers by *Phanerochaete chrysosporium*. Wood Res 73:8–17

Weinstein DA, Krisnangkura K, Mayfield MD, Gold MH (1980) Metabolism of radiolabeled β-guaiacyl ether-linked lignin dimeric compounds by *Phanerochaete chrysosporium*. Appl Environ Microbiol 39:535–540

Yokota S, Umezawa T, Higuchi T (1988) Influence of alkoxyl ring substituents on the degradability of β-O-4 lignin substructure model dimers by *Phanerochaete chrysosporium*. Mokuzai Gakkaishi 34:65–74

Measuring Lignin Degradation

M. Kuwahara

1 Introduction

Lignin is a complex high-molecular compound in which the phenylpropane units are linked to each other by a variety of carbon-to-carbon and ether linkages. Three cinnamyl alcohol precursors, p-coumaryl, coniferyl, and sinapyl alcohols are subjected to dehydrogenative polymerization to form lignin. The relative proportion of these three precursor units varies with the plant species and the location of the lignin in the tissue as well as in the cell wall. Lignin is accumulated at the highest concentration in the middle lamella and most abundantly in the secondary wall of the plant cell.

Degradation of lignin aims at the clarification of the lignin structure. Another purpose of lignin degradation comes from the industrial importance; pulping and other potential utilization of cellulosic materials for the production of food, fodder, and chemicals.

The measurement of lignin degradation is essentially achieved by characterization of degradation products and/or comparison of the chemical structure of decayed lignin with that of sound lignin, using a variety of chemical and physical techniques. The identification of degradation products formed by chemical reactions has been indispensable for deducing and reconstituting the implicate structure of lignin. On the other hand, the isolation and characterization of low molecular aromatic compounds which are produced by microbial attack have yielded significant items of information on the mechanism of biochemical depolymerization of lignin. The isolation and identification of degradation products by white-rot fungi were reviewed by Chen and Chang (1985).

This chapter deals with the methods for estimating the change of lignin structure or lignin content, mainly caused by biological attack, in plant materials.

2 Lignin Preparations as Substrates for Lignin Degradation Studies

Wood samples (blocks, chips and powders), agricultural wastes and other plant materials have been used as substrates for the study of lignin degradation. Experiments using these nonartificial substrates give important information on the conversion and degradation of lignin in natural environments and on delignification treatments for the utilization of this plant biomass. On the other hand, it is preferable to use isolated lignin preparations for the study of the mechanism or process of lignin degradation. However, a method which does not result in change or modification of the structure of lignin during the isolating procedure has not been found yet. The molecular weight of lignins extracted even under moderate conditions is thought to be much lower than that of "protolignin" as it exists in the plant tissues ("native lignin" refers to a fraction of the protolignin, which can be extracted with a neutral solvent such as ethanol; defined as "Brauns' native lignin"). It is also unavoidable that lignin preparations are contaminated with sugars from cellulose and hemicellulose closely connected with lignin in the plant cell wall. The same as lignin, preparations extracted from plants, and synthetic lignins, especially labeled with radioisotopes, are excellent substrates for elucidating the degradation mechanism of lignin under controlled conditions.

Milled wood lignin (MWL) is probably the best preparation known so far for biodegradation studies, although it is not considered to represent whole lignin. The procedure of preparation of lignin was originally reported by Björkman (1956) and improved methods have been proposed (Pearl and Busche 1960; Lundquist and Simonson 1975). Air-dried wood chips are destroyed mechanically and further milled in the presence or absence of toluene, using a ball mill. The wood powder obtained is extracted with dioxane-water for a few days or longer (up to a few months). The extract is evaporated to give as residue a MWL preparation. The purified MWL so obtained is a cream-colored powder representing a yield of approximately 50% of the total lignin, containing 2%–8% of sugars. The molecular weight of the preparation varies according to the conditions of extraction; it lies between 8 000–11 000, which corresponds to 40–60 phenylpropane units.

The idea of removing cellular polysaccharides by the cellulolytic action of wood-rotting fungi so as to leave lignin was originally introduced by Schubert and Nord (1950). Sawdust was inoculated with the brown-rot fungi *Poria vaillantii* or *Lenzites sepiaria* and incubated for a few months. The treated materials were extracted with ethanol and the residues after evaporation was purified by re-precipitation in an appropriate solvent. To improve this laborious procedure, a cellulolytic enzyme preparation is used to dissolve cellulose and hemicellulose (cellulase lignin) (Pew and Weyna 1962). If spruce wood is first ground in a vibratory mill and then treated with 2 N NaOH, at most 95% of the polysaccharides can be removed by the digestion with the cellulase preparation of *Trichoderma viride*, leaving a lignin that is little soluble in organic solvents or alkali. Cellulase lignin is thought to be the most representative of the bulk of a plant's lignin, since a cellulolytic enzyme does not essentially affect the chemical structure of lignin.

Kraft lignin, lignosulfonate and other lignins isolated during the pulping process, and Klason lignin obtained by sulfuric acid hydrolysis of the cellulose fraction of the plant, can be used as substrates for lignin degradation. However, it is necessary to take into account that lignin is modified and fragmented during the isolation procedure.

The application of synthetic lignins (dehydrogenation polymers, DHPs) and model compounds to lignin degradation studies has brought useful results, especially for the detection of the degradation activity on the specific linkage or substructure (β-O-4, β-1, β-5, β-β and others). The idea of the synthesis of lignin arose out of the last step in lignin biosynthesis which consists in the radical polymerization of phenylpropane units (p-coumaryl, coniferyl or sinapyl alcohol) by peroxidase. Enzymic dehydrogenation of these aromatic alcohols by the use of laccase or perxoidase leads to artificial lignins. In practice, the substrate alcohols are polymerized in a buffer solution with purified horseradish peroxidase in the presence of hydrogen peroxide. Several procedures have been developed for the preparation of DHPs. In the original method described by Freudenberg (1965), a solution of coniferyl alcohol and hydrogen peroxide was added in equal parts ("Zulauf" process) to result in the "bulk polymer" composed of mainly β-5, β-β and β-O-4 structures. The procedure in which the alcohol solution is added to hydrogen peroxide dropwise ("Zutropf" process) yields "end-wise polymer" abundant in the β-O-4 linkages (Sarkanen 1971). The molecular weight of DHP obtained by the latter process was found to be as high as that of spruce MWL (11000) and is higher than that obtained by the former method. Tanahashi and Higuchi (1981) proposed the dialysis membrane method to obtain higher-molecular-weight (83000) DHP. In this procedure, polymerization is carried out in a peroxidase-containing cellophane tube soaked in the mixed solution of coniferyl alcohol and peroxide. Besides the homopolymer of coniferyl alcohol (guaiacyl lignin), which is used as a model of gymnosperm lignin, the copolymer of sinapyl and coniferyl alcohols (syringyl / guaiacyl lignin) is also synthesized to represent angiosperm lignin (Faix et al. 1985). High molecular fractions of synthesized DHP are used as the substrate after exclusion of the low molecular product by gel chromatography, using a LH-type Sephadex column. Various types of oligolignols have been synthesized and used as substrates for degradation studies (Higuchi and Nakatsubo 1980; Glenn et al. 1983).

3 Methods for Measuring the Degradation of Lignin

Lignin can be determined either directly or indirectly. However, because lignin has a complex and heteropolymeric structure, no specific method for the quantitative determination of lignin has been developed and most of the assay methods have insuperable limitations.

Conventionally, gravimetric analysis based on acid-insolubility of lignin has been mostly employed for measuring the change of lignin content in plant samples. Recent advances in spectroscopy have provided sensitive assay methods for detecting structural changes of lignin. Some of these methods are applicable for quantitative assays.

3.1 Objective Evaluation of Lignin Degradation

The degration of lignin and the delignification of plant tissue are macroscopically estimated by decolorization of the wood samples. Lignin degradation by fungi results in white-rot, forming white-mottled and white pocket areas on woods. These areas can be distinguished from the intact area by transmission and scanning electron microscopy and also by conventional low-magnification microscopy.

The degradation of lignin has been shown to cause drastic changes in micromorphological cell-wall structures (Ruel and Barnoud 1985). The patterns of lignin degradation of the wood cell wall vary with the fungal species and with the plant species attacked by rot. When predominantly white-rot fungi attack the lignin of the cell wall, the middle lamella composed of deposited lignin is degraded, the cells are separated from one another and the delignified secondary remaining cell wall has a fibrillar appearance (Blanchette 1984; Otjen et al. 1987). Histochemical techniques have been shown to be helpful in the estimation of lignin degradation either by chemical or by biological processes. Early observations using fluorescence microscopy showed that the contrast between the area attacked by fungi and the sound zone was remarkable. Yamashita et al. (1978) found that cell wall decayed by white-rot fungi showed varying behavior against staining with acrydine orange; the original red color changed to green on fungal attack. They defined this area as the "degrading zone" where ligninolytic enzymes were acting. They also reported that this staining property was applicable to the oxidative delignification of wood by chlorite.

3.2 Chemical Analysis

3.2.1 Gravimetry

The generally adopted method for lignin determination is gravimetric analysis of 72%-sulfuric acid-insoluble lignin (Klason lignin or sulfuric acid lignin). In this procedure polysaccharides are hydrolyzed and solubilized by the strong acid. Although standardized by Tappi (1983) and modified properly (Effland 1977), this method suffers from some defects. The lignin residue obtained by this procedure is not identical with protolignin, because the lignin is being damaged and chemically modified during acid treatment. Additional treatment is required to remove protein, polyphenol and other components insoluble in sulfuric acid. Determining lignin in this manner does not take into account the lignin that has been dissolved in the sulfuric acid (acid-soluble lignin). Acid-soluble lignin is analyzed by measuring absorbance at 205 nm – which is the most characteristic absorption maximum of lignin – of the filtrate from acid-insoluble lignin determination (Schöning and Johansson 1965). In hardwood the soluble lignin may be as much as 3% to 5%, which corresponds to 9% to 16% of the total lignin (Klason lignin + acid-soluble lignin). Annual and perennial plants, such as wheat and bamboo, contain 1.5 to 2% acid-soluble lignin.

Change of Klason lignin content is determined by comparing the lignin content of treated or decayed materials with that of sound materials. This method has been employed for estimating the effect of mechanical, chemical and biological pretreatments of agricultural residues and plant materials whose delignification has for aim the utilization of the cellulosic component (Chua and Wayman 1979; Levonen-Munoz et al. 1983; Avgerinos and Wang 1983; Gharpuray et al. 1983; Tanahashi and Higuchi 1983; Gould 1984; Sawada et al. 1987). This method is also useful for specifying fungi which delignify woods and other plants (Kirk and Highley 1973; Blanchette 1984; Otjen et al. 1987).

3.2.2 Elemental and Functional Group Analyses and Degradative Characterization

A comparison of the elemental composition of decayed lignin with that of sound lignin provides information on the chemical change of lignin. In most cases, lignin degradation by fungi results in the concomitant increase in oxygen content and decrease in hydrogen content (Tai et al. 1983). The elemental composition has been usualy expressed as C_9-unit formulae (Kirk and Chang 1974; Kirk and Fenn 1982), based on the fundamental phenylpropane (C_6–C_3) structure of lignin. However, lignin is composed of only three elements, C, O, and H, and analytical data do not always reveal the structural change of lignin. Analyses of functional groups, such as methoxyl, carbonyl, carboxyl, and phenolic and aliphatic hydroxyls provide more useful criteria of lignin decay. Especially the analysis of methoxyl content is crucial for the standard of degradation of lignin, because a drastic decrease in methoxyl caused by demethylation or demothoxylation is usually observed during lignin decay. The change of functional group contents during lignin degradation is demonstrated by Kirk and Chang (1975) and summarized by Kirk and Fenn (1982), as shown in Table 1.

Table 1. Property changes in lignins caused by white-rot fungi (Kirk and Fenn 1982; Janshekar and Fiechter 1983)[a]

Property	Change[b]		Method of analysis[c]
	Increase	Decrease	
Carboxyl content	+		C, UV, IR, NMR
Hydroxyl content			
Phenolic	+[d]	+	C, UV, NMR
Aliphatic		+	C, NMR
Carbonyl	+		UV, IR, NMR
Hydrogen/carbon	+		C
Oxygen/carbon		+	C
Methoxyl/carbon		+	C
Yield of methoxylated aromatic acids on oxidative degradation after methylation[e]		+	C
Yield of principal acidolysis products[f]		+	C
Yield of major products on nitrobenzene oxidation[g]		+	C

[a] These results are mainly based on Higuchi et al. (1955), Hata (1966), Kirk and Chang (1974, 1975), and Chua et al. (1982).

[b] Degraded lignins were compared to sound lignins.

[c] C = various chemical procedures; UV, IR and NMR = ultraviolet, infrared and nuclear magnetic resonance spectrometry, respectively.

[d] Hata (1966) reported an increase in phenolic hydroxyl content and variable results with aliphatic hydroxyl content.

[e] The major product is 3,4-dimethoxybenzoic (veratric) acid from gymnosperm lignin. From angiosperms, veratric acid and tri-O-methyl gallic acid are dominant.

[f] The major product from sound lignin is 3-hydroxy-1-(4-hydroxy-3-methoxyphenyl)-2-propanone. Content of the ketol in the degradation products from rotted lignin is low, whereas vanillic acid is dominant.

[g] Vanillin (gymnosperms) and vanillin + syringaldehyde (angiosperms) are the major products.

Oxidative degradation reactions which have been applied to the elucidation of the chemical structure of lignin provide also useful evidence for the change of structure brought about by fungal decay (Kirk and Fenn 1982). Acidolysis is carried out in dioxane / water in the presence of added HCl. Methylated and ethylated lignins are oxidatively degraded by permanganate. The degradation products are subjected to gas chromatography. Degradation of decayed lignin resulted in less methoxylated benzoic acid content than in the case of sound lignin. Acidolysis gave a markedly lower yield of principal single-ring products from the degraded lignin, whereas a higher yield of vanillic acid was observed (Kirk and Cheng 1975).

3.2.3 UV Spectrophotometry

Lignin in the plant cell wall can be solubilized by warm acetyl bromide and subjected to measuring absorbance at 280 nm (Johnson et al. 1961). Kawase (1963) reported that when this method was applied to the wood decayed by white-rot fungi, lignin content was almost identical with that obtained by Klason's method, while this method gave higher lignin content in the case of brown-rot type decay.

3.3 Spectroscopy

3.3.1 UV Spectroscopy

UV spectrometry gives information on the change in the aromatic ring structure and the functional groups of lignin. Although lignin does not show a sharp absorption spectrum, intense absorbance is observed around 205 and 280 nm, due to mainly to aromatic moieties (Schöning and Johansson 1965). The position and number of functional groups (hydroxyl, methoxyl and side-chain structures) and the pH of the solution affect the intensity of absorbance and absorption maxima. Generally speaking, drastic change brought about by decay results in decrease of absorbance at 205 and 280 nm. The UV spectrum of spruce MWL decayed by some white-rot fungi had greatly increased absorbance around 260 and 300 nm, which reflected an increase in aromatic carboxyl and α-carbonyl, respectively (Kirk and Chang 1974). Difference spectra obtained by subtracting the spectra of reduced lignin from those of the nonreduced lignin showed the increase in absorbance at 310 nm, indicating the increase in α-carbonyl by wood decay (Kirk and Chang 1975). Decrease in absorbance at 350 nm of alkali-neutral difference spectra indicates the presence of p-hydroxyphenyl content by the fungal action (Kirk and Chang 1975).

3.3.2 IR Spectroscopy

IR spectrometry has been used extensively to analyze the structure of lignin and IR peaks assigned to functional groups and chemical linkages based on observation using model compounds. Although quantitative analysis has been undertaken, it has been limited (Sarkanen et al. 1967 a, b, c; Obst 1982). IR analysis was applied to estimate the change of functional groups in heavily degraded lignins from decayed spruce, and it showed an increase in conjugated carboxyl and a decrease in aromatic ring content (Kirk and Chang 1974). Diffuse reflectance Fourier transform IR spectrometry enables a rapid determination of lignin (Schultz et al. 1985; Schultz and Glasser 1986). This spectrometry was found to give more precise quantitative structural information than conventional methods. Methoxyl content, aromatic hydrogen content, phenolic hydroxyl content, guaiacyl / syringyl ratio, and "hydrolysis" and "condensation" ratios were found to be significantly correlated with peak intensities. New IR methodology is expected to give further information on fungal decay of lignin.

3.3.3 NMR Spectroscopy

Although ^{1}H-NMR spectrometry has been used for the analysis of lignin structure, recent progress in ^{13}C-NMR technique enabled the analysis of native and decayed lignins as well as degradation products of lignin. Nimz et al. (1984) described ^{13}C-NMR spectra of structural units in spruce and beech MWL and DHP from coniferyl alcohol. Quantitative measurements of syringyl / guaiacyl, aromatic / methoxyl, aromatic / aliphatic, methoxyl / aryl ratios have been presented (Obst and Ralph 1983; Landucci 1985; Lapierre et al. 1985; Obst and Landucci 1986). Two-dimensional ^{13}C-J[^{1}H]-NMR[2D-^{13}C-NMR] technique was introduced in the analysis of lignin structure by Nimz et al. (1984). ^{13}C-enriched lignin obtained from Aspen wood grown in $^{13}CO_2$ was analyzed by solid-state cross polarization and magic-angle spinning (CP/MAS) NMR and 2D-INADEQUATE NMR (Bardet et al. 1986). The syringyl / guaiacyl ratio in hardwood was also measured by CP/MAS (Manders 1987).

The structures of biodegraded spruce and birch lignins, as well as those of low-molecular degradation products, were analyzed by ^{13}NMR (Chua et al. 1982; Tai et al. 1983). Structural changes in lignin during Kraft cooking shown by ^{13}C-NMR (Robert et al. 1984).

3.4 Gel Permeation Chromatography

Gel permeation chromatography and high-performance liquid chromatography (HPLC), both of them being size-exclusion types, have been extensively used to estimate the molecular weight distribution of lignin preparations and the change of molecular weight during chemical treatments of lignin. The degradation process of lignin by microbiological attack can be also monitored by gel chromatography. Gel chromatography is useful for the fractionation of lignin based on molecular size. However, because of polydispersity of lignin, gel chromatography always suffered from a lack of proper molecular weight standards. As in the case of other polymers, fractionated polystyrene has been used for the standards of molecular weight distribution of lignin (Connors et al. 1980; Pellinen and Salkinoja-Salonen 1985; Dordick et al. 1986). In the system using Sephadexes with aqueous eluents, protein markers have been adopted (references; Table 2). However, these standards are thought to reflect neither the average polymerization number nor the average molecular weight of lignin. Apart from these experiment, Wagner et al. (1986) suggested that polystyrene sulfonate could be employed as intermediate standard in low-pressure size-exclusion chromatography. Another problem in gel chromatography is the formation of artifacts due to adsorption on the gel, ionic interaction, and intermolecular association. The association complexes are suggested to be formed by hydrogen bonding or hydrophobic interactions between lignin molecules to affect the elution profiles. Association can be avoided by using a strongly polar mobile phase, such as dimethyl sulfoxide (DMSO) or dimethylformamide (DMF) (Kristersson et al. 1983). Bimodal elution profile observed in Sephadex LH chromatography changed to a monomodal profile by the addition of LiCl to the eluent (Connors et al. 1980).

HPLC provides a rapid and convenient method to estimate the molecular weight distribution of lignin. Polystyrene and styrene-divinyl benzene copolymer gel columns have been used with a polar organic solvent such as DMF or tetrahydrofuran (THF) as a mobile phase. Pellinen and Salkinoja-Salonen (1985) described the separation of explosion wood lignin, Kraft lignin (pine) and Indulin AT-R (Kraft lignin preparation) by HPLC, with THF as a solvent. They suggested that synthetic lignin model compounds of various molecular weights from 154 to 638 showed neither significant adsorption on the gel nor formation of association complexes, and gave a relevant calibration curve for molecular weight distribution of lignin. Molecular weight distributions of MWL (Faix et al. 1981 a), DHP (Faix et al. 1981 b), and Kraft lignin (Walsh and Campbell 1986) were also analyzed by HPLC.

Gel chromatography has been used for detecting depolymerization of lignin by ligninolytic microorganisms and enzymes. The application of gel chromatography, listed in references, are summarized in Table 2.

3.5 Radioisotopic Methods

Because of their high sensitivity for detection, radioisotopic methods have been successfully used as tools for measuring lignin degradation. The evolution of $^{14}CO_2$ from labeled lignin is an indication of lignin degradation. A change in molecular weight distribution in gel chromatography, as determined by radioactivity, also indicates the depolymerization of lignin. Exclusively ^{14}C-labeled lignocellulose, or lignin preparations obtained from it, and ^{14}C-DHPs have been employed as substrates for studies on lignin degradation.

^{14}C-labeled lignin is synthesized by growing plants fed with radio-labeled precursors, such as ferulic, cinnamic and p-coumaric acids and phenylalanine. Radio-labeled lignocellulose can be obtained by simply administering the precursor through a cut stem. To avoid physiological change in lignin synthesis caused by injury of the plant, syntheses by growing seedling, shoots or twigs have been carried out. The precursor is incorporated preferentially into lignin and labeled lignocellulose, or lignin is isolated after the process for the removal of low-molecular contaminants. R. L. Crawford and Crawford (1978) proposed the protocol for the preparation of ^{14}C-[lignin]-lignocellulose. ^{14}C-MWL is prepared from

Table 2. Assay of depolymerized lignins by gel permeation chromatography

Lignin	Column	Organism	Reference
MWL (spruce)	Sephadex G-25	*Polyporus versicolor*	Kirk and Lundquist (1970)
Extractive lignin (spruce)	Sephadex G-50	*Polyporus anceps, Coriolus versicolor*	Kirk and Chang (1974)
Kraft lignin, lignosulfonate from ^{14}C DHP	Sephadex G-25	*Phanerochaete chrysosporium*	Lundquist et al. (1977)
Kraft lignin, MWL (spruce)	Sephadex CL-6B	*Fomes anosus*	Hüttermann et al. (1980)
Kraft lignin	Sephadex G-75, LH-60, S-200	*Phanerochaete chrysosporium*	Janshekar et al. (1982)
Kraft lignin	Sephadex G-75	*Phanerochaete chrysosporium*	Leisola et al. (1983)
Kraft lignin, lignosulfonate, dioxane straw lignin, alkali straw lignin	Sephadex G-75	*Phanerochaete chrysosporium*	Ulmer et al. (1983)
DHP (^{14}C)	Sephadex LH-20	*Phanerochaete chrysosporium*	Chua et al. (1983)
DHP (^{14}C) organosolve lignin (^{14}C)	HPLC (lipophilic alkyl ammonium chloride)	*Polyporus pinsitus, P. platensis*	Trojanowski et al. (1987)
Kraft lignin, MWL (pine), exploded lignin, DHP, OBLL[a]	HPLC (TSK4000PW, TSK3000PW, ultrastyragel)	Mixed culture of bacteria	Pellinen et al. (1987)
Lignosulfonate	Sephadex G-50	Peroxidase *(Trametes versicolor)*	Lobarzewski et al. (1982)
Extractive lignin (^{14}C methyl) (spruce)	Sephadex LH-20	Ligninase *(Phanerochaete chrysosporium)*	Tien and Kirk (1983)
MWL (spruce)	Sephadex LH-20	Laccase *(Coriolus versicolor)*	Evans (1985)
Lignosulfonate	Sephadex G-50	Laccase and glucose oxidase *(Trametes versicolor)*	Szklarz and Leonowicz (1986)
Thioglycolic lignin	Sephadex G-50	Laccase *(Rigidoporus lignosus)*	Geiger et al. (1986)
MWL (spruce), alkali straw lignin	Sephadex G-75	Ligninase *(Phanerochaete chrysosporium)*	Haemmerli et al. (1986)
MWL (ezomatsu), Brauns native lignin	Sephadex G-10, LH-20	Laccase *(Coriolus versicolor)*	Morohoshi et al. (1987)
DHP, MWL, Kraft lignin[b]	Sephadex LH-20	Peroxidase (Horseradish)	Dordick et al. (1986)
DHP[b]	Sephadex LH-20	Peroxidase (Horseradish)	Lewis et al. (1987)

[a] Oxygen bleaching liquor lignin.
[b] Reaction was carried out in a organic medium (dioxane/aqueous buffer).

^{14}C-lignocellulose plant tissue (R. L. Crawford et al. 1980) with a procedure described earlier. Specifically labeled ^{14}C-[lignin]-lignocellulose can be synthesized by feeding with the specifically labeled precursors. Feeding with ^{14}C-[side-chain]-, ^{14}C-[ring]-, or ^{14}C-[methoxyl]-labeled ferulic acid may yield side-chain-, ring- or methoxyl-labeled lignin, respectively. These lignins were partly prepared by R. L. Crawford et al. (1980).

^{3}H-labeled lignins were prepared by chemical reduction of MWLs (Faix et al. 1985). MWL of spruce (guaiacyl lignin) has been oxidized with 2,3-dichloro-5,6-dicyanobenzo-quinone (DDQ), and subsequent reduction with $NaB^{3}H_4$ yielded tritiated MWL. Tritiation of birch lignin (syringyl / guaiacyl lignin) is achieved by a similar procedure. Faix et al. (1985) used ^{3}H-MWL as reference lignins for DHP degradation studies.

^{14}C-labeled lignin preparations have been used for studies on mineralization of lignin by soil and microflora (D. L. Crawford et al. 1977; Benner et al. 1984), screening of lignin-degrading bacteria (Odier et al. 1981) and fungi (Agosin et al. 1985), and lignin degradation by white-rot fungi (Reid and Seifert 1982; McCarthy et al. 1984; Reid 1985). ^{14}C-lignocellulose is preferable to DHP because lignocellulose combined with carbohydrate provides natural growth conditions for microorganisms. Degradation of ^{14}C-lignocellulose or lignin permits to evolve $^{14}CO_2$ which is trapped and counted by a scintilation counter.

^{14}C-DHPs are used for lignin degradation studies more extensively than radio-labeled lignocellulose. Procedures of synthesis of radioactive DHPs are basically the same as those of cold DHPs, except that radioactive phenolic alcohols are used as precursors. ^{14}C-coniferyl or ^{14}C-coumaryl alcohol is preferably used as a substrate for polymerization. DHP specifically labeled on the aromatic ring, phenylpropanoide side-chain or methoxyl can be synthesized from the phenolic alcohol labeled on the appropriate carbon (Kirk et al. 1975; Haider et al. 1977). Kirk et al. (1975) described the synthesis of labeled DHP from specifically labeled coniferyl alcohol. [Side-chain-^{14}C]coniferyl alcohol was synthesized by condensation of [1,2-^{14}C]acetaldehyde with vanillin methoxymethyl ether. The reaction product is converted to conferaldehyde, to be reduced to give side-chain labeled coniferyl alcohol. For the synthesis of [ring-^{14}C]coniferyl alcohol uniformly labeled phenol was used as a starting material. Phenol is converted successively to guaiacol, vanillin, and finally to coniferyl alcohol. For the synthesis of [methoxy-^{14}C]coniferyl alcohol, 2-(2-hydroxyphenoxy)-5-nitrobenzophenone was first methylated by diazo[^{14}C]methane prepared from N-[^{14}C]methyl-N-nitroso-p-toluene-sulfonamide. The reaction product is converted to coniferyl alcohol via ethyl ferulate. Specifically labeled coniferyl alcohol is polymerized using horseradish peroxidase in the presence of hydrogen peroxide in the phosphate buffer. The molecular weight number average ($\overline{Mn}$) for DHP was 1490–1600, which corresponded to an average number of 8.1 units per polymer molecule ($\overline{Xn}$) assuming a unit weight of 185.

Radioactive DHP composed of copolymer of coniferyl and synapyl alcohols was also synthesized using the analogous reaction system (Faix et al. 1985). The copolymer represents syringyl / guaiacyl lignin which comprises angiosperm lignin, whereas homopolymer of coniferyl alcohol represents guaiacyl lignin of the gymnosperm.

^{14}C-DHPs are thought to be the best model compounds of lignin to be used extensively for lignin degradation studies which involve the location of the chemical linkage attacked by fungi (Kirk et al. 1975), clarifying culture conditions for the expresion of ligninolytic activity (Kirk et al. 1978; Leatham and Kirk 1983; Ulmer et al. 1984; Faison and Kirk 1985; Commanday and Macy 1985), and the production of ligninolytic enzymes (Jäger et al. 1985; Asther et al. 1987). ^{14}C-DHPs were shown to be useful substrates for the screening of lignin-degrading microorganisms (Trojanowski and Hüttermann 1987). When DHP is used as the substrate, usually it is incubated with growing microorganisms, and evolved $^{14}CO_2$ is trapped in alkaline fluid (KOH or ethanolamine) for measurement of radioactivity by a liquid scintillation counter.

3.6 Microscopy and Related Techniques – Estimation of Lignin in Situ

3.6.1 UV Microscopy

The UV microscopic method makes it possible to estimate the lignin content of various regions of the wood cell wall both qualitatively and quantitatively. Nearly all UV microscopic studies on wood and pulp are carried out on ultra-thin cross sections. Measurements of absorbance around 280 nm caused by aromatic structure are made directly on a microscope equipped with a microscope photometer. It is also possible to differentiate distribution of syringyl and guaiacyl units based on their absorption maxima, that is, guaiacyl lignin has a sharp peak at 280 nm, while syringyl lignin has a broad peak around 270 nm (Fergus and Goring 1970a). By assuming the molecular coefficients of guaiacyl and syringyl lignins at 280 nm to be 15.6 cm^{-1} and 4.5 cm^{-1}, respectively, the content of these lignins in the cell wall was determined (Fergus and Goring 1970b). By these experiments it was demonstrated that vessels in birch wood contained predominantly guaiacyl lignin, whereas the fiber and ray parenchyma secondary walls contained a large proportion of syringyl lignin. On the other hand, middle lamellae around the fibers and ray cells contained a guaiacyl-syringyl lignin mixture or copolymer. A flake-like middle lamella, instead of a cross section, in mechanical pulp can be directly subjected to UV spectrometric analysis to give lignin weight concentration and the content of free phenolic hydroxyl groups in the middle lamella, based on the absorbances at 280 and 300 nm (Boutelje and Eriksson 1982).

The quantitation of change in lignin content during fungal decay by using UV spectrometry is limited. Hiroi and Tamai (1983) estimated guaiacyl / syringyl distribution in the whole area of a hardwood cell by measuring differential absorbance values between 270 and 280 nm, using a UV microscopic photometer equipped with a scanning stage. The values were numerically treated and its distribution was printed out. The area that was of negative differential values corresponded to the area that had higher absorbance at 280 nm, namely, the guaiacyl-rich region. By contrast, the area of positive values represented the syringyl-rich region. This technique may be a useful one for estimating the change in lignin content.

A combination of TEM and UV microscopic techniques was reported to be available for the estimation of fungal decay (Nyhlen and Nilsson 1987). The delignified area of wood lacked UV absorption at 280 nm, although this area was still visible by phase-contrast microscopy. At 300 nm, a reacting zone (Yamashita et al. 1978), where ligninolytic enzymes were acting, was revealed adjacent to the unaffected lignified wall layers. The higher absorbance of the reacting zone was thought to indicate a modification of the lignin, and the increased absorbance was consistent with demethylation. By TEM observation of stained thin cross sections, the reacting zone could be precisely located.

3.6.2 Scanning Electron Microscopy Coupled with Energy-Dispersive X-Ray Analysis (SEM-EDXA)

A quantitative assay of lignin distribution in wood cell walls can be carried out by SEM-EDXA technique (Kolar et al. 1982; Saka and Thomas 1982). This technique involves selective bromination of lignin in a nonaqueous system and subsequent analysis of bromine concentrations in different morphological regions by the determination of Br X-ray intensities emitted from various morphological regions as a fine electron beam impinges upon this sections (0.5 μm) of sample wood by an energy-dispersive X-ray analyzer. Bromine distribution curves are determined along a line over a cross section. Results obtained by this technique supported the view on the distribution of syringyl and guyaiacyl lignins in cell-wall regions obtained by UV spectrometry. The results also suggested that during the neutral sulfite delignification of spruce wood, the sulfonated lignin seems to be dissolved more rapidly from the middle lamella (Kolar et al. 1982).

3.7 Calorimetry

Differential scanning calorimetry (DSC) enables direct analysis of sound and degraded wood samples (Reh et al. 1986, 1987). In the calorimeter the specimen and an inert reference sample are subjected to increasing temperatures over the range of 30° to 600° C at a specific heating rate. Thermograms obtained are evaluated with respect to the number of peaks, the temperatures at the individual maxima, the quantity of absorbed or released energy, and the change of the heat capacity at each phase transition.

Thermograms of most of the tested sound woods showed two distinct peaks with maxima at 480° C and 350° C and a shoulder at about 330° C. These peaks are attributed to cellulose, lignin and hemicellulose, respectively. The area under each peak represents the energy of combustion of the substance, and this corresponds to its mass. During wood decay by lignin-degrading fungi, the combustion temperature of lignin changed, indicating an alteration of the native lignin structure. DSC data were thought to show that the first step of lignin degradation by a white-rot *Fomes fomentarius* was the lowering of the combustion temperature from 480° C to 430° C. By analyzing thermograms, the sequence of degradation of wood components during wood decay by white- and brown-rots can be followed.

This method is thought to offer several advantages over chemical techniques; (1) no pretreatment of the wood are required, (2) only small amounts are required for each assay, (3) local differences in wood, such as "mottled rots", can be analyzed, and (4) quantity and quality of all main wood components can be estimated at once in a single DSC-thermogram.

References

Agosin A, Daudin J-J, Odier E (1985) Screening of white-rot fungi on (^{14}C)lignin-labelled and (^{14}C)whole-labelled wheat straw. Appl Microbiol Biotechnol 22:132–138

Asther M, Corrieu G, Drapron R, Odier E (1987) Effect of Tween 80 and oleic acid on ligninase production by *Phanerochaete chrysosporium* INA-12. Enzyme Microb Technol 9:245–249

Avgerinos GC, Wang DIC (1983) Selective solvent delignification for fermentation enhancement. Biotechnol Bioeng 25:67–83

Bardet M, Gagnaire D, Nardin R, Robert D, Vincendon M (1986) Use of ^{13}C enriched wood for structural NMR investigation of wood and wood components, cellulose and lignin, in solid and in solution. Holzforschung 40:Suppl 17–24

Benner R, MacCubbin AE, Hodson RE (1984) Anaerobic biodegradation of the lignin and polysaccharide components of lignocellulose and synthetic lignin by sediment microflora. Appl Environ Microbiol 47:998–1004

Björkman A (1956) Studies on finely divided wood. Part I. Extraction of lignin with neutral solvents. Sven Papperstidn 59:477–485

Blanchette RB (1984) Screening wood decayed by white rot fungi for preferential lignin degradation. Appl Environ Microbiol 48:647–653

Boutelje JB, Eriksson I (1982) An UV-microscopy study of lignin in middle lamella fragments for fibers of mechanical pulp of spruce. Sven Papperstidn 85:R39–42

Chen C-L, Chang H-M (1985) Chemistry of lignin biodegradation. In: Higuchi H (ed) Biosynthesis and biodegradation of wood components. Academic Press, Orlando, p 535

Chua MGS, Wayman M (1979) Characterization of autohydrolysis aspen (*P. tremuloides*) lignin. Part 1. Compostion and molecular weight distribution of extracted autohydrolysis lignin. Can J Chem 57:1141–1149

Chua MGS, Chen C-L, Chang H-M (1982) ^{13}C NMR spectroscopic study of spruce lignin degraded by *Phanerochaete chrysosporium*. I. New structures. Holzforschung 36:165–172

Chua MGS, Choi S, Kirk TK (1983) Mycelium binding and depolymerization of synthetic ^{14}C-labeled lignin during decomposition by *Phanerochaete chrysosporium*. Holzforschung 37:55–61

Commanday F, Macy JM (1985) Effect of substrate nitrogen on lignin degradation by *Pleurotus ostreatus*. Arch Microbiol 142:61–65

Connors W, Sarkanen S, McCarthy LJ (1980) Gel chromatography and association complexes of lignin. Holzforschung 34:80–85

Crawford DL, Crawford RL, Pometto AL (1977) Preparation of specifically labelled ^{14}C-(lignin)- and ^{14}C-(cellulose)-lignocelluloses and their decomposition by the microflora of soil. Appl Environ Microbiol 33:1247–1251

Crawford RL, Crawford DL (1978) Radioisotopic methods for the study of lignin biodegradation. Dev Ind Microbiol 19:35–49

Crawford RL, Robinson LE, Cheh AM (1980) ^{14}C-labeled lignins as substrates for the study of lignin biodegradation and transformation. In: Lignin biodegradation: microbiology, chemistry, and applications, vol I. CRC Press, Boca Raton, p 61

Dordick JS, Marletta MA, Klibanov AM (1986) Peroxidases depolymerize lignin in organic media but not in water. Proc Natl Acad Sci USA 83:6255–6257

Effland MJ (1977) Modified procedure to determine acid-insoluble lignin in wood and pulp. Tappi 60:143–144

Evans CS (1985) Laccase activity in lignin degradation by *Coriolus versicolor* in vivo and in vitro studies. FEMS Microbiol Lett 27:339–343

Faison BD, Kirk TK (1985) Factors involved in the regulation of ligninase activity in *Phanerochaete chrysosporium*. Appl Environ Microbiol 49:299–304

Faix O, Lange W, Besold G (1981 a) Molecular weight determination of DHP's from mixtures of precursors by steric exclusion chromatography (HPLC). Holzforschung 35:137–140

Faix O, Lange W, Salud EC (1981 b) The use of HPLC for the determination of average molecular weights and molecular weight distributions of milled wood lignins from *Shorea polysperma* (Blco.). Holzforschung 35:3–9

Faix O, Mozuch MD, Kirk TK (1985) Degradation of gymnosperm (guaiacyl) vs. angiosperm (syringyl / guaiacyl) lignins by *Phanerochaete chrysosporium*. Holzforschung 39:203–208

Fergus BJ, Goring DAI (1970 a) The location of guaiacyl and syringyl lignins in birch xylem tissue. Holzforschung 24:113–117

Fergus BJ, Goring DAI (1970 b) The distribution of lignin in birch wood as determined by ultraviolet microscopy. Holzforschung 24:118–124

Freudenberg K (1965) Lignin: its constitution and formation from p-hydroxycinnamyl alcohols. Science 148:595–600

Geiger JP, Huguenin B, Nicole M, Nandris D (1986) Laccase of *Rigidoporus lignosus* and *Phellinus noxius* II. Effect of *R. lignosus* laccase L1 on thioglycolic lignin of *Hevea*. Appl Biochem Biotechnol 13:97–110

Gharpuray MM, Lee Y-H, Fan LT (1983) Structure modification of lignocellulosics by pretreatments to enhance enzymatic hydrolysis. Biotechnol Bioeng 25:157–172

Glenn JK, Morgan MA, Mayfield MB, Kuwahara M, Gold MH (1983) An extracellular H_2O_2-requiring enzyme preparation involved in lignin biodegradation by the white rot basidiomycete *Phenerochaete chrysosporium*. Biochem Biophys Res Commun 114:1077–1083

Gould JM (1984) Alkaline peroxide delignification of agricultural residues to enhance enzymatic saccharification. Biotechnol Bioeng 26:46–52

Haemmerli SD, Leisola MSA, Fiechter A (1986) Polymerisation of lignins by ligninases from *Phanerochaete chrysosporium*. FEMS Microbiol Lett 35:33–36

Haider K, Martin JP, Rietz E (1977) Decomposition in soil of ^{14}C-labeled coumaryl alcohols; free and linked into dehydropolymer and plant lignins and model humic acids. Soil Sci Soc Am J 41:556–562

Hata K (1966) Investigations of lignins and lignification. XXXIII Studies on lignins isolated from spruce wood decayed by *Poria subacida* BII. Holzforschung 20:142–147

Higuchi T, Nakatsubo F (1980) Synthesis and biodegradation of oligolignols. Kemia-Kemi No. 9:481–488

Higuchi T, Kawamura I, Kawamura H (1955) Properties of the lignin in decayed wood. J Jpn For Soc 37:298–302

Hiroi T, Tamai A (1983) Degradation of beech wood components by a white-rot fungus, *Grifora frondosa*. In: Higuchi H, Chang H-M, Kirk TK (eds) Recent advances in lignin biodegradation research. Uni Publ, Tokyo, p 34

Hüttermann A, Herche C, Haars A (1980) Polymerization of water-insoluble lignins by *Fomes annosus*. Holzforschung 34:64–66

Jäger A, Croan S, Kirk TK (1985) Production of ligninases and degradation of lignin in agitated submerged cultures of *Phanerochaete chrysosporium*. Appl Environ Microbiol 50:1274–1278

Janshekar H, Fiechter A (1983) Lignin: biosynthesis, application, and biodegradation. In: Fiechter A, Jeffries (eds) Advances in biochemical engineering / biotechnology, vol 27. Springer, Berlin Heidelberg New York Tokyo, p 119

Janshekar H, Brown C, Haltmeier T, Leisola M, Fiechter A (1982) Bioalteration of Kraft pine lignin by *Phanerochaete chrysosporium*. Arch Microbiol 132:14–21

Johnson DB, Moore WE, Zank LC (1961) The spectrophotometric determination of lignin in small wood samples. Tappi 44:793–798

Kawase K (1963) Spectrophotometric determination of lignin in decayed wood (Japanese). Mokuzai Gakkaishi 9:72–77

Kirk TK, Chang H-M (1974) Decomposition of lignin by white-rot fungi I. Isolation of heavily degraded lingins from decayed spruce. Holzforschung 28:217–222

Kirk TK, Chang H-M (1975) Decomposition of lignin by white-rot fungi II. Characterization of leavity degaded lignics from decayed spruce. Holzforschung 29:56–64

Kirk TK, Fenn P (1982) Formation and action of the ligninolytic system in basidiomycetes. In: Franklad, Hedger, Swift (eds) Decomposer basidiomycetes, Cambridge University Press, London, p 67

Kirk TK, Highley TL (1973) Quantitative changes in structural components of conifer woods during decay by white- and brown-rot fungi. Phytopathology 63:1338–1342

Kirk TK, Lundquist K (1970) Comparison of sound and white-rotted sapwood of sweetgum with respect to properties of the lignin and composition of extractives. Sven Papperstidn 73:294–306

Kirk TK, Connors WJ, Bleam RD, Hackett WF, Zeikus JG (1975) Preparation and microbial decomposition of synthetic [^{14}C]lignins. Proc Natl Acad Sci USA 72:2515–2519

Kirk TK, Schultz E, Connors WJ, Lorenz LF, Zeikus JG (1978) Influence of culture parameters on lignin metabolism by *Phanerochaete chrysosporium.* Arch Microbiol 117:277–285

Kolar JJ, Lindgren BO, Treiber T (1982) The distribution of lignin between fiber wall and middle lamella. Sven Papperstidn 85:R21–26

Kristersson P, Lundquist K, Simonson R, Tingsvik K (1983) Gel permeation chromatography of lignin carbohydrate compounds. Holzforschung 37:51–53

Lapierre C, Monties B, Guittet E, Lallemand J-Y (1985) The quantitative measurements in hardwood lignin ^{13}C NMR spectra. Holzforschung 39:367–368

Leatham GF, Kirk TK (1983) Regulation of ligninolytic activity by nutrient nitrogen in white-rot basidiomycetes. FEMS Microbiol Lett 16:65–67

Leisola M, Ulmer D, Haltmeier T, Fiechter A (1983) Rapid solubilization and depolymerization of purified Kraft lignin by thin layers of *Phanerochaete chrysosporium.* Appl Microbiol Biotechnol 17:117–120

Levonen-Munoz E, Bone DH, Daugulis AJ (1983) Solid state fermentation and fractionation of oat straw by basidiomycetes. Eur J Appl Microbiol Biotechnol 18:120–123

Lewis NG, Razal RA, Yamamoto E (1987) Lignin degradation by peroxidase in organic media: a reassessment. Proc Natl Acad Sci USA 84:7925–7927

Lundquist K, Simonson R (1975) Lignin preparations with very low carbohydrate content. Sven Papperstidn 78:390

Lundquist K, Kirk TK, Connors WJ (1977) Fungal degradation of Kraft lignin and lignin sulfonates prepared from synthetic ^{14}C-lignin. Arch Microbiol 112:291–296

Lunducci LL (1985) Quantitative ^{13}C NMR characterization of lignin 1. A methodology for high precision. Holzforschung 39:355–359

Lobarzewski J, Trojanowski J, Mojtas-Wasileska M (1982) The effects of fungal peroxidase on Na-lignosulfonates. Holzforschung 36:173–176

Manders WF (1987) Solid-state ^{13}C NMR determination of the syringyl / guaiacyl ratio in hardwoods. Holzforschung 41:13–18

McCarthy AJ, MacDonald MJ, Paterson A, Broda P (1984) Degradation of [^{14}C]lignin-labelled wheat lignocellulose by white-rot fungi. J Gen Microbiol 130:1023–1030

Morohoshi N, Fujita K, Wariishi H, Haraguchi T (1987) Degradation of lignin by the extracellular enzymes of *Coriolus versicolor* V. Degradation of isolated lignin fractions by laccases I, II, and III. Mokuzai Gakkaishi 33:310–315

Nimz HH, Tschirner U, Stähle M, Lehmann R, Schlosser M (1984) Carbon-13 NMR spectra of lignins, 10. Comparison of structural units in spruce and beech lignin. J Wood Chem Technol 4:265–284

Nyhlen L, Nilsson T (1987) Combined T.E.M. and UV-microscopy on delignification of pine wood by *Phlebia radiata* and four other white rotters. In: Odier (ed) Lignin enzymic and microbial degradation, INRA Publications, Paris, p 277

Obst JR (1982) Guaiacyl and syringyl lignin composition in hardwood cell components. Holzforschung 36:143–152
Obst JR, Landucci LL (1986) Quantitative ^{13}C NMR of lignin-methoxyl: aryl ratio. Holzforschung 40:Suppl 87–92
Obst JR, Ralph J (1983) Characterization of hardwood lignin: investigation of syringyl / guaiacyl composition by ^{13}C nuclear magnetic resonance spectroscopy. Holzforschung 37:297–302
Odier E, Janin G, Monties B (1981) Poplar lignin decomposition by Gram-negative aerobic bacteria. Appl Environ Microbiol 41:337–341
Otjen L, Blanchett R, Effland M, Leatham G (1987) Assessment of 30 white rot basidiomycetes for selective lignin degradation. Holzforschung 41:343–349
Pearl IA, Busche LR (1960) Studies on the chemistry of aspenwood XII. Studies on the preparation of milled wood lignin from quaking aspenwood. Tappi 43:970–974
Pellinen J, Salkinoja-Salonen M (1985) High-performance size-exclusion chromatography of lignin and its derivatives. J Chromatogr 328:299–308
Pellinen J, Jokela J, Salkinoja-Salonen MS (1987) Degradability of different lignins by bacteria. Holzforschung 41:271–276
Pew JC, Weyna P (1962) Fine grinding, enzyme digestion, and the lignin-cellulose bond in wood. Tappi 45:247–256
Reh U, Kraepelin G, Lamprecht I (1986) Use of differential scanning calorimetry for structural analysis of fungally degraded wood. Appl Environ Microbiol 52:1101–1106
Reh U, Kraepelin G, Schulz U (1987) Differential scanning calorimetry in wood biodegradation studies. Sequence of white-rot and brown-rot decay. In: Odier E (ed) Lignin enzymic and microbial degradation. INRA Publications, Paris, p 291
Reid ID (1985) Biological delignification of aspen wood by solid-state fermentation with the white-rot fungus *Merulius tremellosus*. Appl Environ Microbiol 50:133–139
Reid ID, Seifert KA (1982) Effect of an atmosphere of oxygen on growth, respiration, and lignin degradation by white-rot fungi. Can J Bot 60:252–260
Robert DR, Bardet M, Gellerstedt G, Lindfors EL (1984) Structural changes in lignin during Kraft cooking, Part 3. On the structure of dissolved lignins. J Wood Chem Technol 4:239–263
Ruel R, Barnoud F (1985) Degradation of wood by microorganism. In: Higuchi T (ed) Biosynthesis and biodegradation of wood components. Academic Press, Orland, p 441
Saka S, Thomas RJ (1982) Evaluation of quantitative assay of lignin distribution by SEM-EDXA technique. Wood Sci Technol 16:1–18
Sarkanen KV (1971) Precursors and their polymerization. In: Sarkanen KV, Ludwig CH (eds) Lignins. John Wiley, New York, p 95
Sarkanen KV, Chang H-M, Allan GG (1967 b) Species variation in lignins II. Conifer lignins. Tappi 50:583–587
Sarkanen KV, Chang H-M, Allan GG (1967 a) Species variation in lignins II. Hardwood lignins. Tappi 50:587–591
Sarkanen KV, Chang H-M, Ericsson B (1967 c) Species variation in lignins I. Infrared spectra of guaiacyl and syringyl models. Tappi 50:572–575
Sawada T, Kuwahara M, Nakamura Y, Suda H (1987) Effect of a process of explosion for the effective utilization of biomass. Int Chem Eng 27:686–693
Schöning AG, Johansson G (1965) Absorptiometric determination of acid-soluble lignin in semichemical bisulfite pulps and in some woods and plants. Sven Papperstidn 68:607–613
Schubert WJ, Nord FF (1950) Investigations of lignin and lignification II. The characterization of enzymatically liberated lignin. J Am Chem Soc 72:3835–3838
Schultz TP, Glasser WG (1986) Quantitative structural analysis of lignin by diffuse reflectance Fourier transform infrared spectrometry. Holzforschung 40:Suppl 37–44
Schultz TP, Templeton MC, McGinnis GD (1985) Rapid determination of lignocellulose by diffuse reflectance Fourier transform infrared spectrometry. Anal Chem 57:2867–2869
Szklarz G, Leonowicz A (1986) Cooperation between fungal laccase and glucose oxidase in the degradation of lignin derivatives. Phytochemistry 25:2537–2539

Tai D, Terazawa M, Chen CL, Chang H-M, Kirk TK (1983) Biodegradation of guaiacyl and guaiacyl-syringyl lignins in wood by *Phanerochaete chrysosporium*. In: Higuchi T, Chang HM, Kirk TK (eds) Recent advances in lignin biodegradation research. Uni Publ, Tokoy, p 44

Tanahashi M, Higuchi T (1981) Dehydrogenative polymerization of monolignols by peroxidase and H_2O_2 in a dialysis tube I. Preparation of highly polymerized DHPs. Wood Res No 67:29–42

Tanahashi M, Higuchi T (1983) Conversion and total utilization of forest-biomass by explosion process (Japanese). Wood Res Tech Notes No 19:34–65

TAPPI Standard T 222 om-83

Tien M, Kirk TK (1983) Lignin-degrading enzyme from the Hymenomycete *Phanerochaete chrysosporium* Burds. Science 221:661–663

Trojanowski J, Hüttermann A (1987) Screening of wood inhabiting fungi for their capacity to degrade and to solubilize ^{14}C-labelled lignin. Microbios 50:91–97

Trojanowski A, Milstein O, Majcherczyk A, Haars A, Hüttermann A (1987) Solubilization and polymerization of lignin by several woodinhabiting fungi. In: Odier E (ed) Lignin enzymic and microbial degradation. INRA Publication, Paris, p 223

Ulmer DC, Leisola MSA, Schmidt BH, Fiechter A (1983) Rapid degradation of isolated lignins by *Phanerochaete chrysosporium*. Appl Environ Microbiol 45:1795–1801

Ulmer DC, Leisola MSA, Fiechter A (1984) Possible induction of ligninolytic system of *Phanerochaete chrysosporium*. J Biotechnol 1:13–24

Wagner BA, To T, Teller DE, McCarthy JL (1986) Estimation of the distribution of molecular weights of lignins by use of low pressure size exclusion chromatography. Holzforschung 40:Suppl 67–73

Walsh AR, Campbell AG (1986) HPSEC analysis of Kraft lignin on a Bondagel column. Holzforschung 40:263–266

Yamashita Y, Fukuzawa K, Ishida S (1978) Histochemical observation of wood attacked by white rot fungi, *Coriolus versicolor* and *Cryptoderma yamanoi*. Res Bull Coll Exp For 35:109–121

Methods for Studying the Plant Cytoskeleton

R. E. Williamson, F. Grolig, U. A. Hurley,
P. P. Jablonski, D. W. McCurdy, and G. O. Wasteneys

1 Introduction

The cytoskeleton is composed of organized arrays of polymeric proteins of which actin and tubulin are the most abundant. The assembled proteins provide a structural framework which supports motility and imparts order and directionality to cytoplasmic processes. This subcellular order is magnified to the level of the tissue and organ by the cytoskeleton's role in orienting cell division and cellulose deposition and hence in defining the direction in which cells, tissues and organs will expand. Cytoskeletal function has been recently reviewed (Lloyd 1987); this chapter focuses on some important methods. We describe:

i) Methods of two-dimensional immunoblotting by which isoforms of actin and tubulin can be resolved and detected with commercially available antibodies.
ii) Methods which use the same antibodies to localize actin and tubulin by immunofluorescence.
iii) Methods by which giant algal cells can be exploited to raise monoclonal antibodies that bind to cytoskeletal components.

2 Immunoblotting with Commercial Antibodies to Identify Isoforms of Actin and Tubulin Separated by Two-Dimensional Gel Electrophoresis

Actin and tubulin from plant sources can now be identified by immunoblotting using commercially available antibodies (Table 1). The antibodies recognize epitopes that are widely conserved in plants, animals and fungi. Methods for one-dimensional separation of proteins are now in common use and have been described elsewhere (Hames and Rickwood 1981). Two-dimensional separations (see Celis and Bravo 1984) by isoelectric point and M_r resolve the products of some but not necessarily all of the various genes commonly encoding isoforms of actin and tubulin. Some important post-translational modifications (e.g., phosphorylation, tyrosination, acetylation) can also be detected by using metabolic labelling or antibodies that are not commercially available.

Our method follows O'Farrell (1975) with modifications that reduce some of the problems frequently seen with plants. For detecting all proteins in an extract, silver staining (Fig. 1 a) is commonly preferred to fluorography of proteins labelled in vivo; detection limits are broadly comparable.

Table 1. Monoclonal antibodies reacting with some major proteins of the plant cytoskeleton

Protein	Immunogen	Comments	Source
Actin	Chicken gizzard cytoskeleton	Also binds a 58 kDa plant protein[a]	Amersham
	Chicken gizzard actin		J. Lessard[b]
Myosin[c]	3T3 cytoskeleton		Amersham (anti-pan myosin)
α-tubulin	Chick brain microtubules		Amersham
	Yeast tubulin	Requires Tyr at C-terminus of α-tubulin[d]	Sera Lab (clone YL1/2)
β-tubulin	Chick brain microtubules		Amersham

[a] McCurdy and Williamson (1987).
[b] Dr. J. Lessard, Children's Hospital Research Foundation, Cincinnati, Ohio, 45229 has made this available for a small charge to cover costs.
[c] Myosin is usually studied on one-dimensional gels since high M_r proteins are not usually well focused on two-dimensional gels.
[d] Wehland et al. (1983).

2.1 Sample Preparation

Proteins must be efficiently extracted, protected from modification by proteases and other molecules released during homogenization and brought into a suitable buffer for isoelectric focusing. Solubilization can use almost any conventional buffer to which various protectants may be added.

Thioglycollate (10 mM) is a recognized inhibitor of phenolic oxidation reactions (Anderson 1968) and minimizes charge heterogeneity artefacts in, for example, pea chloroplast extracts (Fig. 1 b and c). These appear as lines of regularly spaced spots of identical M_r decreasing in intensity towards the acidic side of the gel (see O'Farrell 1975 for discussion). Each spot represents the gain by the parent protein (the most basic spot) of a single, extra negative charge generated by poorly understood reactions. Such spots may be confused with the products of different genes or the results of post-translational changes, such as phosphorylation, that also add negative charge. Proteins deliberately modified to show charge heterogeneity can be used to calibrate isoelectric focusing gels ("Carbamalytes", Pharmacia) and indicate whether suspected charge heterogeneity spots have the appropriate spacing.

CHAPS (3-[(3-cholamidopropyl) dimethylammonio] 1-propane sulphonate) is a zwitterionic detergent recommended (Perdew et al. 1983) for improving the resolution of hydrophobic proteins and for reducing the background in silver-staining. We find substantial improvements with crude plant extracts.

Protamine (Mayer et al. 1987) is a convenient and effective alternative to DNAase (O'Farrell 1975) to remove DNA which can cause the production of smears across the gel rather than discrete spots.

Proteolysis. A fluorimetric assay of the hydrolysis of casein labelled with fluorescein isothiocyanate (FTC-casein) conveniently indicates proteolytic activity in an extract and determines which protease inhibitors control it. It is more

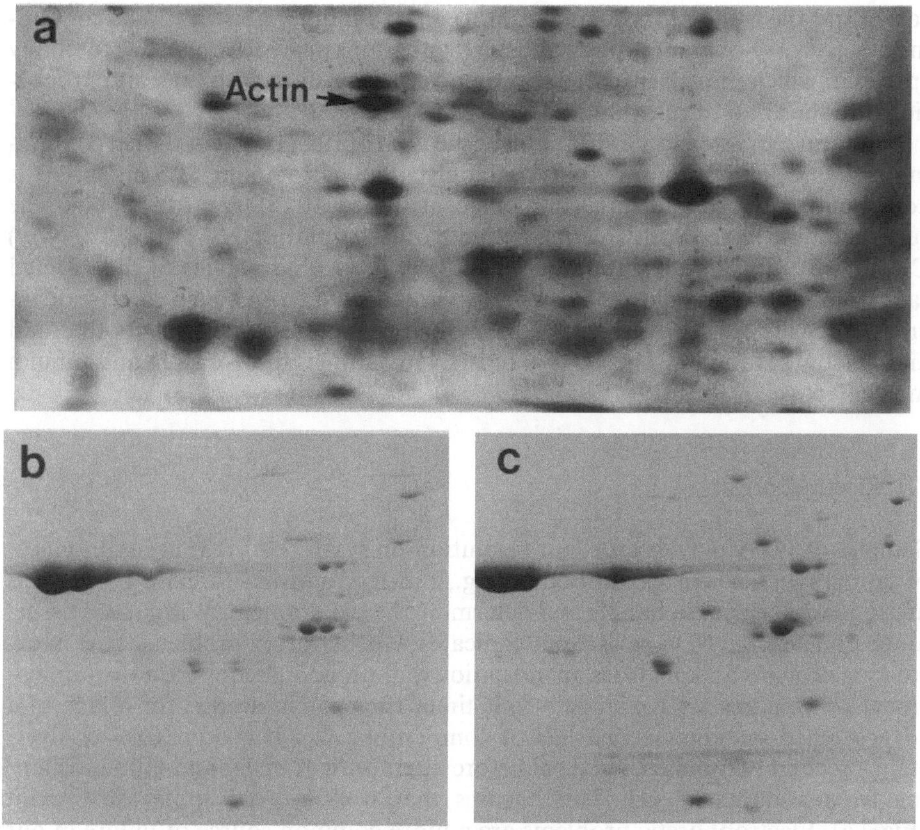

Fig. 1 a–c. Two-dimensional polyacrylamide gel electrophoresis of plant proteins. **a** Portion of a gel on which a crude *Chara* extract was focused on a pH 5–6 gradient and detected by silver-staining. Note the numerous, well-resolved spots that are detected by this method, even when using such a narrow pH range. Actin was identified by immunoblotting and selective extraction using LSS (Williamson et al. 1987); **b, c** Separation of pea chloroplast proteins showing the production of multiple spots at various places by charge heterogeneity (**b**) and its reduction by thioglycollate (**c**). Proteins were separated on pH 4–6.5 gradients and detected by Coomassie staining. Basic proteins are *on the left* of all gels

sensitive than colorimetric assays and cheaper than methods using iodination, although these may be used where a fluorimeter is not available.

Because no single protein substrate will be degraded by all proteases, such assays cannot measure total proteolysis in the extract. Nevertheless, since inhibitors are not completely specific for a particular class of proteases (Storey and Wagner 1986), a combination of inhibitors completely inhibiting FTC-casein degradation generally minimizes proteolysis of plant proteins. We have successfully used this approach to abolish the degradation of plant actin and tubulin that is visible as immunostained bands of lower than expected M_r. The effective concentrations and combination of inhibitors should be determined empirically in each case. Storey and Wagner (1986) give a comprehensive list of inhibitors for the various classes of plant proteases; Twining (1984) describes the preparation of FTC-casein and the assay of its degradation. Using Twining's assay, extracts containing 300 µg ml^{-1} of plant protein and 2 h incubations provide conditions for many species in which the relative rate of proteolysis is constant with time and proportional to the amount of extract added.

Tissue is extracted in a mortar and pestle with buffer (1 ml g^{-1} tissue) containing protease inhibitors and 10 mM thioglycollate. Protamine (1.5 mg ml^{-1}) is added and, after 10 min, the sample is centrifuged at 15000 g for 10 min. The supernatant is mixed with an equal volume of lysis solution [9 M urea, 2% (w/v) CHAPS, 5% (v/v) β-mercaptoethanol, 6.25% (v/v) Pharmalytes of the selected range], solid urea added to 9 M and a few grains of sucrose added to provide increased density during overlaying. Pharmalytes are available covering both broad and narrow pH ranges. The pH 5–6 or pH 4–6.5 ranges resolve actin and tubulin most effectively.

2.2 Electrophoresis

Samples can be run with excellent resolution on small slabs that are economical of samples and expensive antibodies (e.g., Bio-Rad's Mini Protean). Larger gels are somewhat easier to handle and determine the pH gradient. With a new tissue, a one-dimensional SDS gel quickly indicates whether gross problems exist in extracting proteins and provides an indication of the loading required; a two-dimensional gel requires approximately four times the protein needed for a lane of a silver-stained one-dimensional gel of comparable size. It is advisable to silver-stain a second two-dimensional gel before attempting transfer and immunostaining from a duplicate gel. This ensures that well-resolved spots are present (Fig. 1 a). Electrophoretic problems are a more common source of failure in our experience than the immunostaining itself.

Apparatus. Isoelectric focusing is commonly performed in rods of acrylamide that are subsequently laid on top of an SDS-slab gel for separation by M_r. The glass plates for the slab or the clamps holding them are usually bevelled on their internal top edge to assist seating of the focused rod. Apparatus for both steps is available commercially (e.g., Bio-Rad, Hoeffer) or can be built in a workshop (e.g., O'Farrell 1975; Hames and Rickwood 1981). Whatever system is used, it is highly desirable to run two or more slabs simultaneously to accomodate controls, replication, etc. By incorporating a thin glass divider plate in the middle of the slab, some systems allow the proteins from one rod to partition between two slabs, one of which is immunostained, the other silver-stained. If the nitrocellulose is cut accurately to the size of the slab, the position of the immunostained spots can be related to the pattern of silver-stained spots using fractional x, y coordinates to correct for size changes in the stained gel.

Isoelectric Focusing. Rods are cast in glass tubes (e.g., 120 mm × 4.5 mm internal diameter) that have been cleaned in alcoholic KOH and sealed at the base with Parafilm. The gel mix [9 M urea, 10% (w/v) CHAPS, 4% (w/v) acrylamide and 6.25% (v/v) Pharmalytes] is prepared, filtered and degassed for 1 h. Riboflavin

(0.14 mg ml^{-1}) in a 1% (v/v) TEMED solution is added with gentle swirling and the tubes filled to about 10 mm from the top. A syringe with a large gauge needle reaching to the bottom of the tubes helps to avoid air bubbles. Each gel is carefully overlaid with distilled water and polymerized for 2 h in front of fluorescent lights. The water is replaced with 30 µl of lysis solution and the Parafilm removed. After filling the lower reservoir of the electrophoresis tank with 0.1 M H_2SO_4, the rods are placed in the focusing apparatus and filled to the top with 0.1 M degassed NaOH. The upper chamber is filled with degassed NaOH and the gels prefocused for 15 min at 200 V, 20 min at 300 V and 25 min at 400 V with the negative electrode connected to the upper reservoir. When the NaOH has been siphoned out of the top reservoir and from each tube, the sample can be loaded on to the top of each gel rod using a microsyringe and then covered with overlay solution [9 M urea, 3% (v/v) Pharmalytes of the appropriate pH range]. After first refilling each tube and then the upper reservoir with NaOH, we routinely focus overnight at 400 V and for 3 h at 800 V. After removing the glass tubes from the apparatus, the gel rods are loosened by injecting water from a syringe needle between the rod and the glass tube and expelled by pressure applied from a syringe connected to the glass tube through plastic tubing. The pH gradient can be determined on a duplicate rod by equilibrating 5 mm gel pieces with 0.5 ml degassed distilled water for 2 h. If not required immediately, rods with focused proteins can be stored frozen after wrapping in plastic film.

Second Dimension. The separation by M_r involves running a conventional SDS gel modified to accept focused proteins from the rod gel rather than conventional sample wells. We find that different brands and grades of SDS greatly effect the quality of separation and silver-staining using the method of Wray et al. (1981). We routinely use Sigma SDS or the Pierce "SDS lauryl" product; slight differences in the relative mobility of various spots in the two detergents may be exploited to minimize overlapping of particular spots of interest.

Loading the Rod Gel. We equilibrate the rod gel in SDS sample buffer [62.5 mM Tris-HCl, pH 6.8, 5% (v/v) β-mercaptoethanol, 2.3% (w/v) SDS, 10% (v/v) glycerol] for 15–30 min in order to reduce streaking in the second dimension. A molten solution of 1% (w/v) agarose in SDS sample buffer is pipetted along the top of the stacking gel that has been cast to the top of the glass plates. Before the agarose has time to set the rod gel is laid out along a folded piece of Parafilm and carefully guided onto the top of the slab gel, without trapping air bubbles between the rod gel and the stacking gel. Extra molten agarose can then be pipetted over the rod gel to seal it in place. Plastic gloves should be worn during this operation to prevent contaminating skin proteins appearing in the slab gel. We routinely align the rod gel so that the basic end of the pH gradient appears on the left-hand side of the assembled two-dimensional gel. Molecular weight standards can be conveniently applied from an agarose cylinder formed by dissolving the standards in double strength SDS sample buffer and mixing with an equal volume of 1% (w/v) agarose solution. After setting in a suitable tube, the agarose cylinder is laid alongside the rod and secured with molten agarose. The upper and lower chambers are carefully filled with running buffer (25 mM Tris-HCl, 192 mM glycine, 0.1% (w/v) SDS, pH 8.3) and a few drops of bromophenol blue (2 mg ml^{-1}) added to the upper chamber to act as tracker dye.

2.3 Transfer

We use essentially the method of Towbin et al. (1979). Overnight transfer is usually convenient and effective, although shorter transfers (1 to 1.5 h) can be used with the thinner mini-slabs. For very high-molecular-weight proteins (e.g., dynein heavy chains), 24 h transfers are desirable. The common cytoskeletal proteins are not prone to wash off nitrocellulose, thus making it unnecessary to fix with glutaraldehyde in the way that is advantageous with proteins such as calmodulin. To facilitate interpretation, the gel is placed in a standard orientation with respect to the nitrocellulose (e.g., basic side to the left) and a known corner marked. Following transfer, it is convenient to transiently stain the transferred polypeptides with 0.2% (w/v) Ponceau S in 3% (w/v) TCA (e.g., Pratt et al. 1986) which will reveal the major spots only.

2.4 Immunostaining

Blocking the excess protein-binding capacity of the nitrocellulose is the crucial step for achieving specific antibody staining. Milk [5% (w/v) non-fat powder in TSB; 10 mM Tris, 150 mM NaCl, pH 7.25] usually provides efficient blocking. Gentle shaking for 1 h in a petri dish is adequate. Primary antibodies are diluted as recommended by the suppliers and applied as a thin film to cover the nitrocellulose filter laid face up on a flat sheet of Parafilm. After 60 min, the filter is washed in a petri dish with several changes of TBS-Tween [TBS with 0.05% (v/v) Tween 20 (Sigma)] using vigorous wrist shaking for about 15 s with each change. We routinely detect bound antibody with biotinylated anti-mouse (or anti-rat for YL1/2 – see Table 1) followed by streptavidin-peroxidase complex (Amersham)

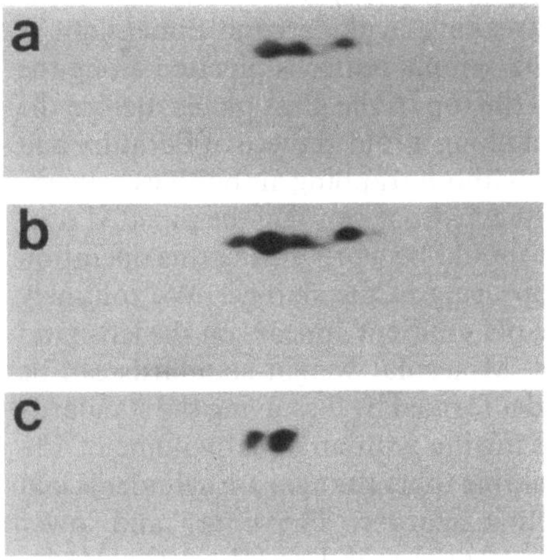

Fig. 2 a–c. Immunostains using Amersham anti-actin showing multiple actin isoforms in crude extracts of **a** cotyledons, **b** hypocotyl, and **c** roots from 7-day-old seedlings of *Arabidopsis thaliana*. A pH 5–6 gradient was used for each gel to resolve the isoforms. Note the slightly but consistently higher M_r shown by the most acidic isoform in cotyledons and hypocotyls. Prolonged development of the immunoblot indicates that each organ contains most, perhaps all, of the isoforms, so that differences in the degree of expression or of post-translational modifications occur in the different organs rather than truly organ-specific isoforms existing

using the recommended dilutions and TBS-Tween washes. Development uses 2×100 ml of 0.06% (w/v) chloro-1-naphthol dissolved in 40 ml of methanol and diluted to 200 ml with TBS. The blot is protected from light during development and photographed using Ilford Tech Pan and overhead lighting. Alkaline phosphatase (e.g., Sigma A-5153) has a greater sensitivity if required or if naturally biotinylated proteins or endogenous peroxidases are a problem. Identify these as spots on filters processed through all stages except exposure to primary antibody.

An example of the results obtained for actin is shown in Fig. 2. Seedlings of *Arabidopsis thaliana* show multiple actin spots whose relative intensity varies in different organs.

2.5 Specificity

Monoclonal antibodies detect only a single epitope on their antigen so that other proteins may be detected that are related only by the possession of that epitope in an otherwise unrelated sequence. Problems such as this need not preclude their use to detect tubulin and actin isoforms on blots if other reactive proteins are well separated from the authentic cytoskeletal proteins. Conversely, the products of particular members of a gene family encoding a protein may lack that particular epitope but possess the remainder of the sequence. Thus YL1/2, which recognizes the C-terminus of α-tubulin when this carries a terminal tyrosine residue (Wehland et al. 1983), fails to detect the product of a *Physarum* gene that encodes a protein lacking the C-terminal tyrosine (Monteiro and Cox 1987). This isoform can be recognized, however, by monoclonals directed to other epitopes of α-tubulin.

3 Immunofluorescence in Higher Plant Cells

The commercially available antibodies can be satisfactorily used in immunofluorescence, the most effective of the alternative antibodies often differing according to the cell under study. A method for tubulin immunofluorescence applicable to many higher plant cells was established several years ago (Wick et al. 1981). It has been applied to many species with only minor variations in, for example, time of exposure to wall-degrading enzymes, etc. The original method and some suggested improvements (e.g., Simmonds et al. 1985; Schroeder et al. 1985) should enable a new tissue to be tackled with reasonable confidence.

Phalloidin derivatives have frequently been preferred to antibodies for localizing actin in higher plants. These fungal products derivatized with rhodamine or fluorescein (Molecular Probes, Oregon, U.S.A.) bind to F-actin and their small size minimizes problems of penetration that may be experienced with antibodies. As a result, many spectacular pictures have been produced with extremely short and simple procedures (e.g., Traas et al. 1987). It is not possible, however, to assess the specificity of phalloidin by a method comparable to the immunoblotting routinely employed to establish the specificity of antibodies in immunofluorescence. The specificity of phalloidin is assumed from its tight binding to animal actin. The common control of showing that staining is abolished by incubating with an excess of nonderivatized phalloidin tests only that fluorescent derivatization has not modified the binding of the parent compound; it does not address the specificity of the parent compound itself. There is good reason to believe that the majority of the binding is to actin, but we suggest that immunofluorescence is a valuable adjunct to phalloidin localization, particularly where no other evidence (such as cytochalasin-sensitivity) implicates actin.

4 Methods for Giant Algal Cells

Characean algae produce internodal cells that reach lengths of several centimetres and contain spectacular arrays of microtubules and actin filaments (Fig. 3). The following methods permit reliable immunofluorescence localizations, reactivation of streaming, selective extraction or stabilization of cytoskeletal components and provide a rapid screening method to select monoclonal antibodies that bind to cytoskeletal components. Similar methods could probably be applied to many of the other giant algal and fungal cells that have rarely been exploited for cytoskeletal studies (see for example Menzel and Schliwa 1986) and the stabilization and extraction methods could probably be adapted for use with protoplasts from higher plants.

4.1 Intracellular Perfusion for Reactivation, Localization and Selective Extraction of Cytoskeletal Structures

Intracellular perfusion removes the tonoplast and much of the streaming endoplasm to leave the cortical cytoskeletal components of an open cell directly bathed in a solution of controlled composition and accessible to macromolecules, organelle suspensions, etc. We use a rapid flow of solution to quickly remove tonoplast and endoplasm; Tazawa and colleagues use slower perfusion with a low calcium solution to destabilize the tonoplast (Tazawa et al. 1976). Table 2 lists the composition of our standard solutions and their abbreviated names.

4.1.1 Perfusion Technique

The equipment needed for perfusion is shown in Fig. 4 and should be collected conveniently ready for use before starting to prepare a cell. Grease is applied to the base of the reservoirs, except over the grooves, and as two lines across the slide. Good quality vacuum grease seals well and can be dispensed from a disposable 2-ml syringe. Adjacent internodal and leaf cells are removed with scissors and the chosen internodal cell blotted lightly on tissue. The cell is positioned on the slide and pressed lightly into the grease lines. Grease is applied over the top

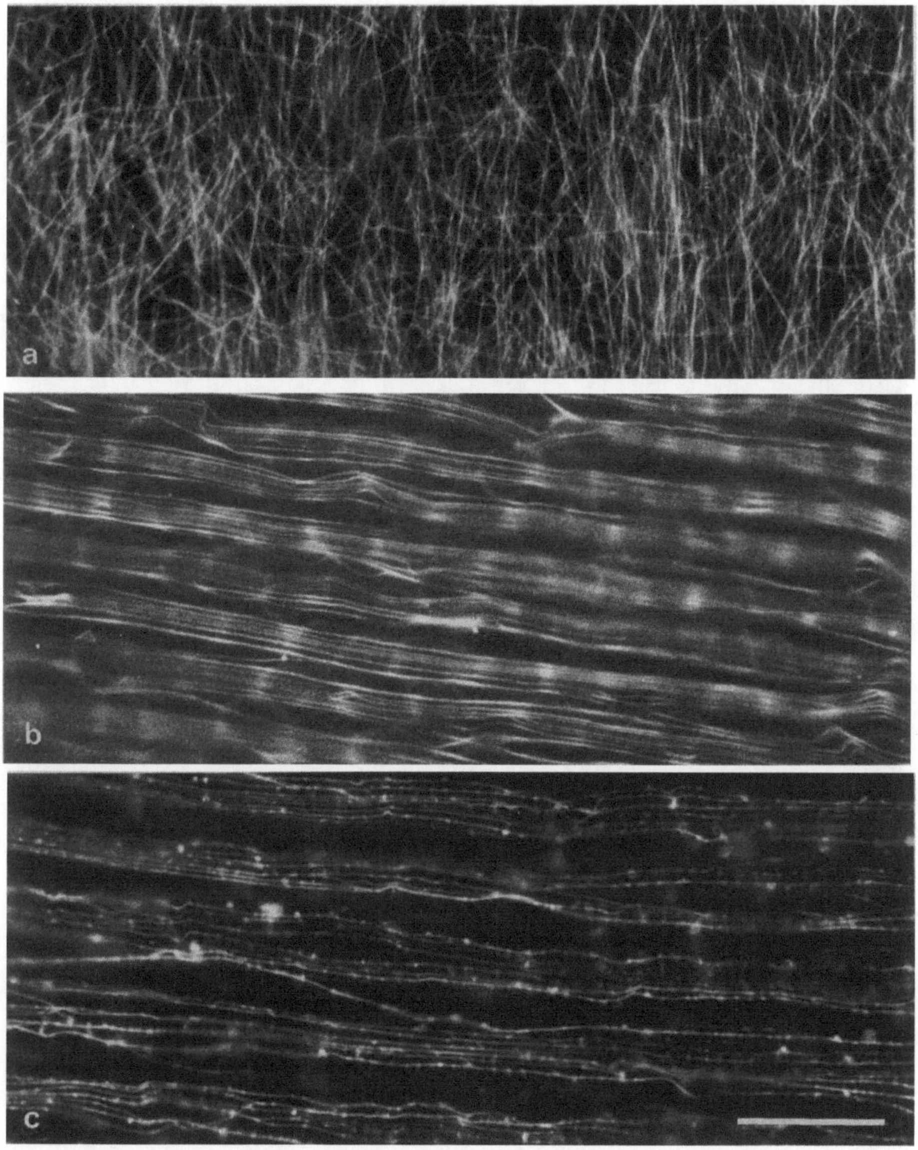

Fig. 3a–c. Cytoskeletal structures in three *Chara* cells seen by immunofluorescence. **a** Microtubules seen in a perfused cell lie above the chloroplasts. Glutaraldehyde fixation and anti-α-tubulin (YL1/2); **b** Actin bundles lie below the chloroplasts in an intact cell and their fluorescence is attenuated except where they pass gaps between the chloroplasts. Formaldehyde fixation and antibody CC6 (Williamson et al. 1986), a "structure-specific stain"; **c** The same structures are seen more clearly and without periodic attentuation when they lie above the chloroplasts after the cell has been cut open longitudinally. Glutaraldehyde fixation and Amersham's anti-myosin. Note the staining of both actin bundles and organelles immobilized by perfusion with AFPS. Bar line = 20 μm

Table 2. Composition of perfusion solutions in mM (Triton and glycerol are % v/v)

Solution	Abbreviation	Sucrose	KCl	$MgCl_2$	EGTA	$CaCl_2$	Pipes	ATP	Glycerol	Triton	pH	pCa
ATP-free perfusion solution	AFPS	200	70	4.49	5.0	1.48	10				7.0	ca7
ATP-containing perfusion solution	ACPS	200	70	5.48	5.0	1.48	10	1.0			7.0	ca7
Low-salt sucrose	LSS	330					2.0	1.0			7.0	
Cytoskeleton stabilizing buffer	CSB			5.00	2.0		100		14.6	1.0	6.9	8

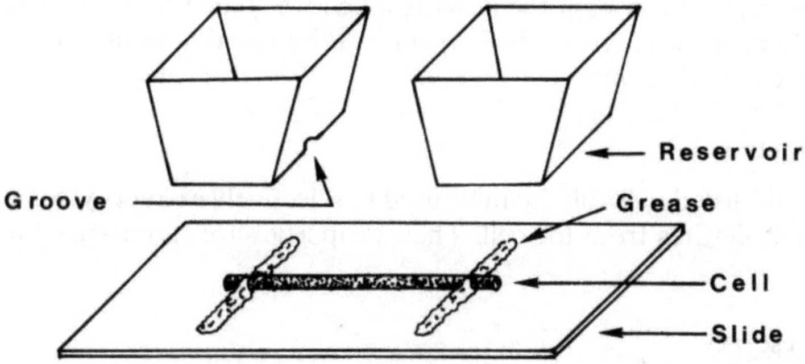

Fig. 4. Arrangements for intracellular perfusion of *Chara* internodal cells. The reservoirs are pressed down firmly onto the cell. Vacuum grease above and below the cell and on the base of the reservoir provides a watertight seal. The cell ends lying within the reservoirs are removed with scissors and solution passed through the cell interior under a small pressure difference

of the cell where it passes through the grease lines and the reservoirs pressed down with the grooves over the cell to make a good seal to the slide and the cell. Perfusion solution is added to the reservoirs and silicon fluid (Dow Corning) or liquid paraffin used to cover the cell where it runs between the reservoirs. (Aqueous bathing media isotonic with the perfusion solution can be used, but structural preservation is excellent with these nonaqueous media.) A coverslip cut to match the distance between the reservoirs is applied, supported at each corner by a small deposit of vacuum grease to avoid flattening the cell. The progress of the perfusion can be monitored microscopically, preferably using differential interference contrast optics. A difference in solution levels of 5 to 10 mm is established between the two reservoirs and the ends of the cell removed with small, sharp scissors. (If the cell is still turgid, cutting the cell causes a disruptive surge of the contents. Turgor can be reduced by leaving subsequent cells to dry in air for a few seconds after blotting.) The rapid solution flow through the open cell removes the tonoplast and the bulk of the endoplasm within 3 min. This can be assumed, provided liquid is being transferred between the reservoirs or verified microscopically when perfusing with AFPS, as the moment at which organelle movements near the actin bundles cease and the residual organelles bind tightly to the actin bundles (Williamson 1975). Some batches of cells are rather fragile and should be perfused more slowly. Cells can now be subject to immunofluorescence, reactivation or selective extraction. All further solution changes are also made by perfusion.

4.1.2 Reactivation

Initial perfusion with AFPS leaves organelles immobilized along the actin bundles. After removing AFPS from both wells, the organelles can be reactivated to move at in vivo velocities along the actin by a slow flow of ACPS (Williamson 1975). Addition of just two or three drops of ACPS to one well introduces sufficient ATP for reactivation to occur within about 10 s and the flow can then be halted by an equal addition to the other well. This prevents undue loss of the motile organelles whose association with the actin is much weakened by ATP. The velocity of the organelle movements declines from its initial peak and motile organelles are rarely seen after 30 min.

4.1.3 Selective Extraction

The perfusion solutions listed in Table 2 can be used to selectively extract cytoskeletal and associated structures from the cell. Their properties are summarized in Table 3.

4.1.4 Immunofluorescence

The cell is perfused with the perfusion solution appropriate to the experiment (Table 3) and the same solution or PBS used subsequently for dilution, etc. After initial perfusion to remove the tonoplast and bulk endoplasm, the cell is fixed for 20 min with 1% (v/v) glutaraldehyde. [1% (w/v) paraformaldehyde adequately preserves actin (Williamson et al. 1986) but not microtubules (Wasteneys and Williamson 1987).] Subsequent treatments required are: washing for 10 min; primary antibodies [diluted according to the supplier's instructions in solutions containing 1% (w/v) bovine serum albumin] gently perfused for 30 to 60 min; washing; secondary antibodies diluted and perfused as before for 30 to 60 min; final wash. Microtubules lying between the chloroplasts and plasma membrane are easily seen in intact cells (Fig. 3a) but the actomyosin system lying on the endoplasmic side of the chloroplasts is partially obscured by the light-absorbing chloroplasts (Fig. 3b). The actomyosin can be placed above the chloroplasts by bisecting the cell longitudinally with a razor blade, cutting the two halves into 10 to 15-mm-long pieces and placing them cytoplasmic face down in 20 μl of mounting medium on a coverslip. After transferring the coverslips to a slide, the chloroplasts lie beneath the actomyosin when viewed by epifluorescence (Fig. 3c). A mounting medium of 0.1% (w/v) phenylenediamine in Moviol 4–88 (Hoechst;

Table 3. Preservation (+) or extraction (−) of cytoplasmic components by the solutions defined in Table 2. (Entries left blank have not been determined)

Solution	1ary or[a] 2ary	Micro-tubules	Actin	Actin-assoc organelles	Myosin[b]	Endo-plasmic filaments[c]	Chloro-plasts
AFPS	1	+	+	+	+	+	+
ACPS	1,2	+	+	−	Trace	−	+
HC-AFPS[d]	1	+	+	+	Part	−	+
LSS[e]	1,2		−	−			+
CSB	2	+	+				−

[a] 1 indicates use for primary perfusion, 2 indicates use after initial perfusion with AFPS or ACPS.
[b] Seen by immunofluorescence with anti-myosin (unpublished observation).
[c] Putative myosin filaments seen by electron microscopy (Williamson 1979).
[d] HC-AFPS = high Ca^{2+} ATP-free perfusion solution. Identical to AFPS except that the $MgCl_2$, EGTA and $CaCl_2$ concentrations are 4.0, zero and 0.1 mM, respectively.
[e] Selective extraction of actin microfilaments (Williamson 1978).

Osborn and Weber 1982) or 50% (v/v) glycerol in PBS reduces fading with fluorescein-conjugated antibodies. Chloroplast autofluorescence is suppressed with a red-excluding barrier filter (e.g., KP580) to supplement the manufacturer's standard fluorescein filters.

We have used Lessard's anti-actin (Otey et al. 1986) successfully to detect the actin bundles of *Chara* with glutaraldehyde fixation. Interestingly, the Amersham anti-actin appears to require that the cells are treated with acetone, a requirement for partial denaturation, we suspect, rather than for permeabilization to secure antibody access (see Williamson et al. 1987).

4.2 Production of Monoclonal Antibodies Using Immunofluorescence Screening

The immunological aspects of producing monoclonal antibodies have been described in detail elsewhere (Campbell 1984). We will discuss the particular questions that arise when monoclonal antibodies to cytoskeletal proteins are produced using immunofluorescence to select clones secreting antibodies binding to cytoskeletal structures. This selection method permits immunization with crude extracts obviating the need to purify the immunogen. It potentially allows previously unknown cytoskeletal proteins to be identified by immunoblotting with antibodies first shown to bind to actin- or microtubule-containing structures. We find, however, that only a relatively small proportion of the selected antibodies can be employed successfully for immunoblotting but that many more antibodies can be of great value as structure-specific "stains" (e.g., Fig. 3 b; see Williamson et al. 1986).

4.2.1 Immunization

The proportion of useful antibodies should be increased by enriching the extract for cytoskeletal proteins before injection into mice. We originally used perfusion extraction to simply remove the endoplasm (Williamson et al. 1986) but more recently have used nonionic detergents in the presence of buffers that stabilize actin filaments and microtubules (CSB in Table 3). Protein injected into the mouse can be either native or denatured with 0.1% (w/v) SDS in the hope of increasing the proportion of antibodies that react with denatured proteins in immunoblots. Denaturation may, however, impair the ability of certain proteins either to elicit any antibodies or antibodies that can be detected by immunofluorescence. Antibodies to the actin bundles of *Chara*, for example, are readily obtained with native proteins but, in our experience, rarely with denatured proteins.

4.2.2 Screening

The chosen methods have to deal with 100 or more samples per day. We now screen all samples by immunofluorescence and all positive lines (binding to cytoskeletal and other sites) by immunoblotting. Once sufficient "structure-specific stains" (i.e., antibodies working in immunofluorescence) have been obtained, only those also working on blots are cloned and characterized. We carry out immunofluorescence on segments cut from large internodal cells of *Chara* and use a commercially available apparatus to blot large numbers of different antibodies.

Cell segments are prepared by immersing large internodal cells in AFPS and cutting them with scissors into 1 to 2 mm segments. A few minutes of gentle swirling in a petri dish washes out the tonoplast and most of the endoplasm leaving the cortical cytoplasm including actin and microtubules. Antibodies can now enter through the cut end of the segments. Glutaraldehyde fixation (1% in AFPS for 20 min) gives better preservation of microtubules than formaldehyde; both preserve actin. Segments are washed in the dish for about 10 min with AFPS. Subsequent processing is conducted in the wells of a microtitre plate that have been pierced with a needle to allow solutions to be exchanged. The holed plate is seated in another plate which has slightly larger wells and supernatant samples (75 μl approx.) added to each well. A segment is transferred to each well and immersed in the supernatant for 30 min. The holed plate carrying the cell segments is lifted out of the larger-welled plate and the bases of its wells touched to a pad of tissue to complete drainage of the supernatants. The holed plate is seated in a second plate with larger wells and the segments washed with two changes of AFPS added to cover the segments. Subsequent processing is the standard sequence of second antibody, final washing and mounting in AFPS (six segments to a slide). A sample of hybridoma culture medium should be routinely included as a negative control to see that no interfering molecules are present. Most products sold as "anti-mouse IgG" without preabsorption against other antibody classes satisfactorily detect those other antibody classes and offer adequately low background staining at low cost. The segments present an immunofluorescent image that is a disordered version of that produced by perfusion which should be used to define fully the binding sites of new antibodies. Preservation is adequate for screening and about 2.5 h are required to process a tray of segments and a similar period for the microscopy.

These immunofluorescence methods should be fairly directly applicable to other giant cells. The same strategy has been applied to fungal spores (Hardham et al. 1986) and fern spermatocytes (Marc et al. 1988). Commercial antibodies to cytoskeletal proteins can be used to assist in the development of screening methods and included as positive controls during screening.

Immunoblotting can be carried out on many samples using a commercially available device (Immunetics, Miniblotter II) and antibody samples as small as 120 μl that are obtainable from small hybridoma wells early in screening. A standard-sized slab gel run with a single sample well spanning its full width can be treated with 45 different antibodies, the nitrocellulose filter being removed from the device after incubation with the primary antibodies.

4.2.3 Comments

Antibodies working on blots appeared with a frequency of approx. 1 in 15 among over 200 antibodies judged positive on immunofluorescence (unpublished observations). This will reflect both the problem of antibodies elicited to conformational epitopes that are destroyed during denaturation by SDS and a problem of sensitivity, since some antigens will be present in indetectable quantities in our crude extracts (references in Williamson et al. 1986). The former problem is obviously significant, however, in that only one of nine antibodies obtained to purified, native calmodulin detects calmodulin fractionated on SDS gels even when large quantities of pure calmodulin are loaded. All nine, however, detect calmodulin transferred from native gels (unpublished observations). A very large number of antibodies must therefore be screened to have a reasonable chance of identifying and localizing a specific protein. Antibodies that provide valuable immunofluorescent "stains" for specific structures but not for specific proteins can be obtained much more readily.

5 Concluding Remarks

Antibodies reacting with the major proteins of the plant cytoskeleton are now readily available. The methods we have described should permit their use in immunofluorescence and two-dimensional immunoblotting to expand further our knowledge of their intracellular location and isoform complexity in plants. We hope that the methods for giant algal cells that we have described may be adaptable to other giant cells (see Fukui and Nagai 1985) and may encourage those working on organelle movements in other plant cells to exploit these algal cells as effectively as animal cell biologists have recently done (e.g., Adams and Pollard 1986).

References

Adams RJ, Pollard TD (1986) Propulsion of organelles isolated from *Acanthamoeba* along actin filaments by myosin-I. Nature 322:754–756

Anderson JW (1968) Extraction of enzymes and subcellular organelles from plant tissues. Phytochemistry 7:1973–1988

Campbell AM (1984) Monoclonal antibody technology. Elsevier, Amsterdam, pp 265

Celis JE, Bravo R (1984) Two-dimensional gel electrophoresis of proteins. Academic Press, Orlando, pp 487

Fukui S, Nagai R (1985) Reactivation of cytoplasmic streaming in a tonoplast-permeabilised cell model of *Acetabularia.* Plant Cell Physiol 26:737–744

Hames BD, Rickwood D (1981) Gel electrophoresis of proteins: a practical approach. IRL Press, London, pp 290

Hardham AR, Suzaki E, Perkin JL (1986) Monoclonal antibodies to isolate-, species-, and genus-specific components on the zoospores and cysts of the fungus *Phytophthera cinnamomi.* Can J Bot 64:311–321

Lloyd CW (1987) The plant cytoskeleton: the impact of fluorescence microscopy. Annu Rev Plant Physiol 38:119–139

Marc J, Gunning BES, Hardham AR, Perkin JL, Wick SM (1988) Monoclonal antibodies to surface and cytoskeletal components of the spermatozoid of *Pteridium aquilinum.* Protoplasma 142:5–14

Mayer JE, Hähne G, Palme K, Schell J (1987) A simple and general plant tissue extraction procedure for two-dimensional gel electrophoresis. Plant Cell Rep 6:77–81

McCurdy DW, Williamson RE (1987) An actin-related protein inside pea chloroplasts. J Cell Sci 87:449–456

Menzel D, Schliwa M (1986) Motility in the siphonous green alga *Bryopsis*. I. Spatial organisation of the cytoskeleton and organelle movements. Eur J Cell Biol 40:275–285
Monteiro MJ, Cox RA (1987) Primary structure of an α-tubulin gene of *Physarum polycephalum*. J Mol Biol 193:427–438
O'Farrell PH (1975) High resolution two-dimensional electrophoresis of proteins. J Biol Chem 250:4007–4021
Osborn M, Weber K (1982) Immunofluorescence and immunocytochemical procedures with affinity-purified antibodies: tubulin-containing structures. Methods Cell Biol 24A:97–132
Otey CA; Kalnoski NH, Lessard JL, Bulinski JC (1986) Immunolocalization of the gamma isoform of nonmuscle actin in cultured cells. J Cell Biol 102:1726–1737
Perdew GH, Schaup HW, Selivonchick DP (1983) The use of zwitterionic detergent in two-dimensional gel electrophoresis of trout liver microsomes. Anal Biochem 135:453–455
Pratt LH, McCurdy DW, Shimazaki Y, Cordonnier M-M (1986) Immunodetection of phytochrome: immunocytochemistry, immunoblotting and immunoquantitation. In: Linskens HF, Jackson JF (eds) Immunology in plant sciences. Springer, Berlin Heidelberg New York, pp 50–74 (Modern methods of plant analysis, New Series, vol 4)
Schroeder M, Wehland J, Weber K (1985) Immunofluorescence microscopy of microtubules in plant cells; stabilisation by dimethylsulfoxide. Eur J Cell Biol 38:211–218
Simmonds DH, Seagull RW, Setterfield G (1985) Evaluation of techniques for immunofluorescent staining of microtubules in cultured plant cells. J Histochem Cytochem 33:345–352
Storey RD, Wagner FW (1986) Plant proteases: a need for uniformity. Phytochemistry 25:2701–2709
Tazawa M, Kikuyama M, Shimmen T (1976) Electric characteristics and cytoplasmic streaming of Characeae cells lacking tonoplast. Cell Struct Funct 1:165–176
Towbin H, Staehelin T, Gordon J (1979) Electrophoretic transfer of proteins from polyacrylamide gels to nitrocellulose sheets: procedure and some applications. Proc Natl Acad Sci USA 76:4350–4354
Traas JA, Dornau JH, Rawlins DT, Shaw PJ, Watts J, Lloyd CW (1987) An actin network is present in the cytoplasm throughout the cell cycle of carrot cells and associates with the dividing nucleus. J Cell Biol 105:387–395
Twining SS (1984) Fluorescein isothiocyanate-labeled casein assay for proteolytic enzymes. Anal Biochem 143:30–34
Wasteneys GO, Williamson RE (1987) Microtubule orientation in developing cells of *Nitella*: a quantitative analysis. Eur J Cell Biol 43:14–22
Wehland J, Willingham MC, Sandoval IV (1983) A rat monoclonal antibody reacting specifically with the tyrosylated form of α-tubulin. I. Biochemical characterisation, effects on microtubule polymerisation in vitro, and microtubule polymerisation and organisation in vivo. J Cell Biol 97:1467–1475
Wick SM, Seagull RW, Osborn M, Weber K, Gunning BES (1981) Immunofluorescence microscopy of organized microtubule arrays in structurally stabilized meristematic plant cells. J Cell Biol 89:685–690
Williamson RE (1975) Cytoplasmic streaming in *Chara*: a cell model activated by ATP and inhibited by cytochalasin B. J Cell Sci 17:655–668
Williamson RE (1978) Cytochalasin B stabilises the sub-cortical actin bundles of *Chara* against a solution of low ionic strength. Cytobiologie 18:107–113
Williamson RE (1979) Filaments associated with the endoplasmic reticulum in the streaming cytoplasm of *Chara corallina*. Eur J Cell Biol 20:177–183
Williamson RE, Perkin JL, McCurdy DW, Craig S, Hurley UA (1986) Production and use of monoclonal antibodies to study the cytoskeleton and other components of the cortical cytoplasm of *Chara*. Eur J Cell Biol 41:1–8
Williamson RE, McCurdy DW, Hurley UA, Perkin JL (1987) Actin of *Chara* giant internodal cells: a single isoform in the subcortical filament bundles and a larger, immunologically related protein in the chloroplasts. Plant Physiol 85:268–272
Wray W, Boulikas T, Wray VP, Hancock R (1981) Silver staining of proteins in polyacrylamide gels. Anal Biochem 118:197–203

Analysis of Extensin Structure in Plant Cell Walls

K. M. M. Swords and L. A. Staehelin

1 Introduction

It has been realized for a long time that the primary cell wall of plants is organized into distinct cellulose-containing and middle lamella layers. However, only recently through the use of immunocytochemical studies (Moore and Staehelin 1988; Stafstrom and Staehelin 1988), has it been possible to establish unambiguously where specific carbohydrate and glycoprotein components reside in these layers, thereby providing crucial information for the development of cell wall models. For example, we now know that cellulose microfibrils surround the protoplast in a multilayered sheath (Roland and Vian 1979). Xyloglucan, the major hemicellulose of dicot walls, is hydrogen-bonded to the surface of cellulose (Valent and Albersheim 1974; Hayashi and MacLachlan 1984) and restricted to the cellulose layer of the wall (Moore and Staehelin 1988). Neighboring cells share a common middle lamella, where the acidic pectins rhamnogalacturonan I and homogalacturonans are exclusively localized (Albersheim et al. 1960; Selvendran 1985; Moore and Staehelin 1988). And interspersed throughout the cellulose-containing regions is the structural glycoprotein extensin (Stafstrom 1986; Stafstrom and Staehelin 1988).

The name "extensin" originally described an insoluble glycoprotein component that was distinguished by its high hydroxyproline content (Lamport 1965). Since that time, a number of hydroxyproline-rich glycoproteins (HRGPs) with extensin-like properties have been isolated from many sources, including potato tuber (Leach et al. 1982), soybean seed coats (Cassab et al. 1985), carrot root (Stuart and Varner 1980), tobacco callus (Mellon and Helgeson 1982), and tomato suspension cells (Smith et al. 1984). Extensin appears to be secreted to the cell wall as a monomer and covalently cross-linked into an inextensible, rigid matrix (Cooper and Varner 1984). This matrix forms during the later stages of cell elongation and is probably important in growth control (Lamport 1965; Sadava et al. 1973; van Holst et al. 1980) and in plant morphogenesis (Basile 1980). In mature tissues, extensin levels can be boosted further by stress-related stimuli, such as ethylene exposure (Toppan et al. 1982), mechanical wounding (Chrispeels 1969; Stuart and Varner 1980), cold temperature Weiser and Wallner *personal communication* and heat stress (Stermer and Hammerschmidt 1987). In general, these proteins strengthen the cell wall, as high extensin levels can protect cells from lysis in hypotonic medium (Lamport 1965) and can enhance resistance to plant pathogens (Esquerre-Tugaye et al. 1979; Hammerschmidt et al. 1984).

The wall-strengthening ability of extensin arises from the covalent cross-linking of monomers into a network (Wilson and Fry 1986) that is interwoven with the mesh of cellulose fibrils and matrix polysaccharides. However, the exact spatial arrangement of this network and its mechanism of assembly are still a matter of debate. Lamport and Epstein (1983) have proposed the "warp and weft" model of wall organization in which cellulose microfibrils are believed to be oriented parallel to the plane of the wall where they act as a "warp", while extensin is oriented perpendicular to the wall, behaving as a "weft". These two interwoven elements are then supended, in some yet undefined way, in a hydrophilic-pectin gel. There is no direct evidence supporting this theory to date. Moreover, the analysis of extensin in muro has been impeded by the extreme insolubility of the wall-bound network. As a result, researchers have turned their attention to salt-elutable precursors and extensin fragments released by trypsin or $NaClO_2$ treatments of cell walls, searching for features in primary sequence and physical structure that play a role in network assembly.

Both the amino acid sequence and pattern of glycosylation are important in determining the secondary structure of extensin molecules. The major extensin of dicots is about 35%–45% hydroxyproline, and rich in serine, lysine, tyrosine and histidine (Stuart and Varner 1980; Smith et al. 1984; Chen and Varner 1985). Most of the hydroxyproline is found within the sequence Ser-Hyp-Hyp-Hyp-Hyp (Chen and Varner 1985; Smith et al. 1986). This amino acid block is abundant in extensins from carrot and tomato and is probably characteristic of this class of proteins (Wilson and Fry 1986). The predicted amino acid sequence from a carrot extensin genomic cDNA shows that the Ser-Pro_4 sequence is dispersed throughout the peptide and represents over half of the amino acids (Chen and Varner 1985).

Carbohydrate makes up approximately two-thirds of the glycoproteins' mass (Stuart and Varner 1980) and the hydroxyproline residues are the key glycosylation sites. The bulk of the carbohydrate occurs as tetra- and tri-arabinosides bound to hydroxyproline in an O-glycosidic linkage, with smaller amounts of di- and mono-arabinosides (Lamport 1967; van Holst and Varner 1984). The terminal nonreducing arabinose is in an α-linkage, while the others are β-linked (Akiyama et al. 1980). Arabinose represents about 90% of the total sugar content; the balance is made up of single galactose residues O-linked to some serines (Lamport et al. 1973; van Holst and Varner 1984).

The carbohydrates may help stabilize the molecule in a rigid, polyproline II helical structure (van Holst and Varner 1984). Analysis by circular dichroism shows that the glycosylated form is 100% polyproline II, whereas extensin deglycosylated with anhydrous hydrogen fluoride (Mort and Lamport 1977) is only 50% polyproline II (van Holst and Varner 1984). This shape difference can be verified by rotary shadowing electron microscopy and gel filtration. With rotary shadowing, the native, carbohydrate-coated extensin appears to be a linear 84 nm rod with interspersed kinks, while the deglycosylated molecule appears to be much more globular (Stafstrom and Staehelin 1986a). Gel filtration measurements indicate that the glycosylated rigid-rod has a Stokes' radius of 89 Å, while deglycosylated extensin is more compact, with a Stokes' radius of 11 Å (Stafstrom and Staehelin 1986a).

Molecular modeling suggests that the β-linked tetra-arabinosides bound to hydroxyproline can follow or "nest" with the polyproline II helix with the same helical pitch (Traub and Shmueli 1963) and can form three hydrogen bonds with the carbonyl groups of the peptide backbone (Lamport 1980). In this way, the tri- and tetra-arabinosides may help stabilize the whole molecule without any particular sequence specificity (Stafstrom and Staehelin 1986 a). Tetra- and tri-arabinoside oligomers may be sufficiently long to cover the 2–15 amino acid long non-glycosylated regions (Chen and Varner 1985), stabilizing them into a helical structure. If this is the case, the more prevalent tetra-arabinosides could lend greater long-range rigidity than tri-arabinosides (Stafstrom and Staehelin 1986 a).

Glycosylation and the conformation it imposes on the peptide backbone may be important to the structural role that extensin plays in the cell wall. Inhibition of extensin hydroxylation and subsequent glycosylation with the iron chelator α,α'-dipyridyl does not block extensin synthesis, transport to the cell walls, or the cross-linking of monomers (Chrispeels 1970). The matrix formed, however, cannot inhibit cell elongation in either pea epicotyls or root tips as it would normally (Sadava and Chrispeels 1973; Vaughan 1973). Therefore, glycosylation appears to be important for extensin function, and may be necessary for proper orienting of the monomers in the wall prior to cross-linking (Stafstrom and Staehelin 1986 a). Immunolocalization studies have shown that the glycosylated extensin-1 of carrot is evenly distributed across the primary cell wall, but is absent from the middle lamella and cell corners (Stafstrom 1986; Stafstrom and Staehelin 1988). Extensin has also been shown to be expressed in a developmental and tissue-specific manner in soybean seed coats. In this case, only the walls of the palisade epidermal cells, the underlying hourglass cells and the vascular elements have high levels of HRGP when probed with specific antibodies (Cassab and Varner 1987).

As secreted extensin monomers are incorporated into the wall network, there is a concomitant increase in a new phenolic amino acid, isodityrosine (IDT). This oxidatively coupled dityrosine is the proposed intermolecular link between extensin molecules (Fry 1982) and its formation is probably catalyzed by a wall-bound peroxidase (Lamport 1986). A crude peroxidase preparation eluted from the cell walls of tomato suspension culture cells appears to catalyze the conversion of extensin monomers to oligomers in vitro (Everdeen et al. 1988). Isodityrosine has also been detected in isolated monomers as an intramolecular bridge within the sequence Tyr-Lys-Tyr (Epstein and Lamport 1984). Discontinuities or kinks have been seen in rotary shadowed extensin-1 monomers (Stafstrom and Staehelin 1986 b) whose positions may coincide with sites of potential internal IDT cross-linkages predicted from sequence data (Stafstrom and Staehelin 1986 b; Chen and Varner 1985).

A molecule of polyproline II helix requires a length of approximately 76 residues to wrap around a cellulose microfibril (Lamport and Epstein 1983). In order to create long, uninterrupted stretches of extensin capable of interweaving with cellulose microfibrils, some of the potential isodityrosine bridges must be formed intramolecularly. In fact, Lamport and Epstein (1983) predict that two to three times more intramolecular than intermolecular IDT linkages must be formed to establish the appropriate lengths of extensin. This correlates well with the values measured experimentally by Stafstrom and Staehelin (1986 b), where salt-extractable extensin-1 monomers from carrot already contain an estimated 5.3 half IDT linkages per molecule versus a predicted total of 7.75 half IDTs per monomer in immobilized networks (Stafstrom and Staehelin 1986 b).

These observations, together with the fact that about 60% of extensin-1 oligomers are end-linked, have lead Stafstrom and Staehelin (1986 b) to hypothesize that intramolecular bridging is important for "inactivating" internal tyrosines and for creating kinks, leaving non-participating tyrosines at the ends free for intermolecular cross-linking. Internal IDT sites have the Tyr-Lys-Tyr sequence (Epstein and Lamport 1984), whereas the tyrosines located at the ends of the molecule are typically found in different sequences (Chen and Varner 1985). If cross-linking can be controlled at these different sequences, the proportion of intra- and intermolecular IDTs may dictate the porosity and rigidity of the final extensin matrix.

In this chapter, we will describe how monomeric and oligomeric extensin can be isolated from cell walls and detail methods that have been important in their analysis. Techniques that have been used to visualize individual molecules and networks will also be reviewed. And finally, we will discuss how immunolocalization of HRGPs has augmented our understanding of extensin organization in the cell wall.

2 Isolation of Extensin Precursors

2.1 Elution of Precursors from Cell Walls

Extensin monomers and small oligomers can be eluted with high salt from the cell walls of excised carrot tap root (Stuart and Varner 1980), tomato suspension culture cells (Smith et al. 1984) and soybean seed coats (Cassab et al. 1985). These molecules are presumed to be precursor forms not yet cross-linked in the wall. In particular, tissue that is either mechanically wounded or in a "physical state of wounding," such as supension culture cells (Bolwell 1984), accumulate large pools of these precursors.

In the case of tomato tissue culture cells, three distinct hydroxyproline-rich proteins, P1a, P1b and P2, have been isolated by one of two methods (Smith et al. 1984). The first makes use of a cell column, which has the advantage of separating both wall-bound peroxidase and extensin. In this case, 10–20 ml of cultured cells are harvested 4–6 days after subculturing and packed into a small column (8 × 100 mm). After a brief washing with dH_2O, a linear 0–100 mM $CaCl_2$ gradient (nonbuffered, pH 6.0) is applied at a flow rate of 15–20 ml/h. Wall-bound peroxidase is eluted first at about 10 mM $CaCl_2$, whereas the HRGP-rich peak elutes later at 30 mM $CaCl_2$.

Alternatively, larger quantities of extensin can be recovered by gently agitating 650 ml of cells in 50 mM $CaCl_2$ for 5 min, and collecting the cells on a coarsely sintered funnel. The extensin-rich eluate is recovered, made 10% trichloroacetic acid (w/v), and precipitated overnight at 4° C. The basic charge and high level of glycosylation keeps the HRGPs soluble, while most other proteins are precipitated and can be removed via centrifugation at 15,000 g (15 min). The supernatant is dialyzed against water for 48 h (4° C) and lyophilized to give the crude extensin fraction.

We have used a protocol modified from several sources to extract extensin precursors from incubated carrot discs (Stuart and Varner 1980; Smith et al. 1984; Stafstrom and Staehelin 1986b). Fresh commercial carrots, preferably with tops intact, are cut into parenchyma strips and then sliced, by hand or food processor, into 1 mm discs. The cut tissue (150–200 g) is incubated in 25 mM potassium phosphate buffer, pH 6.0 and 50 mM chloramphenicol for 4 days (25° C) at 200 rpm on a rotary shaker. To retard bacterial growth, the discs are rinsed in dH_2O and the buffer is changed daily during the length of the incubation. The "aged" discs are rinsed several times in dH_2O and homogenized in the incubation buffer in a Virtis homogenizer (The Virtis Co., Inc., Gardiner, NY). All procedures are performed at 0°–4° C. The resulting slurry is washed six–eight times with large volumes of cold dH_2O and filtered through two layers of Miracloth (Calbiochem, La Jolla, CA). The cell remnants are extracted three times with 150 ml of cold 0.2 M $CaCl_2$ (10 min each) and the salt-released supernatants pooled. After concentration down to about 50 ml using an Amicon (Amicon Corp., Danvers, MA) ultrafiltration device (PM-10 filter), the supernatant is made 10% TCA (w/v) and treated as described for the tomato crude extract.

The seed coat of soybean contains an HRGP which differs in its amino acid content and glycosylation from the extensin-1 of carrot (Cassab et al. 1985). To isolate the soybean cell wall glycoprotein (Cassab et al. 1985), about 5 g of fresh seed coats (*Glycine max* var. Provar) are ground in 0.1% potassium acetate buffer (pH 5.0) with 4 mM $Na_2S_2O_5$, 1% insoluble polyvinyl pyrrolidone (PVP) in a glass homogenizer. The resulting cell walls are washed once with 0.5% Nonident P-40 detergent with 2 mM $Na_2S_2O_5$, resuspended and washed ten times with 50 ml of cold 2 mM $Na_2S_2O_5$. The clean cell wall fraction is then extracted for 12 h with 10 ml of 0.2 M $CaCl_2$, 4 mM $Na_2S_2O_5$ at 5° C to yield an HRGP-rich supernatant.

2.2 Carboxymethyl Cellulose Ion Exchange Chromatography

Cation exchange chromatography is an important step in the purification of plant extensins. These proteins typically have an isoelectric point between 10 and 12 (Cooper et al. 1987) and will bind to CM-Cellulose or other similar matrix, while the bulk of contaminating proteins are eluted in the void volume. It has been demonstrated in both the tomato and carrot systems that the different HRGPs elute from the column at different ionic strengths, allowing separation of the various extensins from one another (Smith et al. 1984; Stafstrom and Staehelin 1986b). As is the case with many of these techniques, a variety of columns and buffer systems have been used with success by different groups. Here, we will describe only three possible approaches.

In our hands, a modification of the Lamport protocol (Smith et al. 1984) has worked well for the isolation of two extensins from carrot (Stafstrom and Staehelin 1986b, 1987). Crude extensin is resuspended in a small volume of 25 mM Tris-HCl, pH 8.0 buffer, loaded on a CM-Cellulose (microgranular/ Sigma Corp., St. Louis, MO) cation exchange column equilibrated with the same buffer, and run at a flow rate of 12 ml/h. After elution of the void volume, a linear 0–

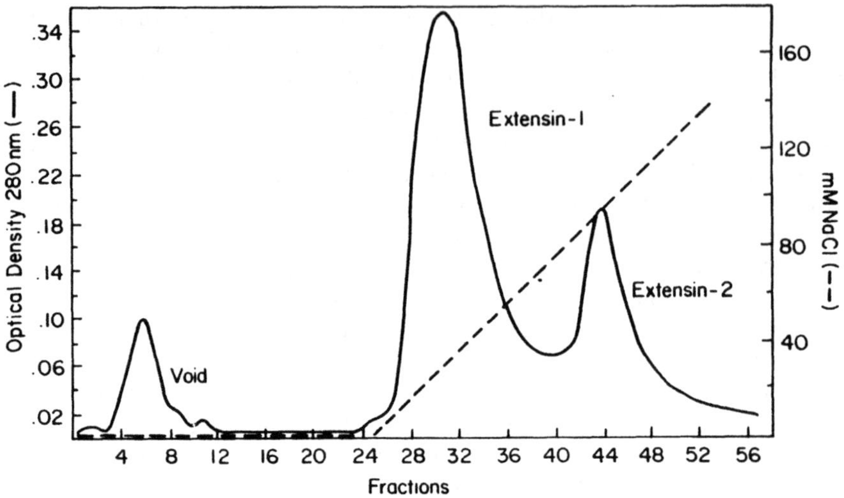

Fig. 1. Elution profile (see text for details) of carrot extensin-1 and extensin-2 from a carboxymethyl cellulose ion exchange column

400 mM NaCl gradient is applied. Under these conditions, the major extensin-1 (Ext-1) peak elutes at 25 mM NaCl and contains both monomers and oligomers, while a second hydroxyproline-rich protein extensin-2 (Ext-2) contaminated with several low MW proteins elutes later at about 90 mM salt (Fig. 1). The Ext-1 and Ext-2 peaks can be desalted and rechromatographed to insure separation from one another.

Precursors from tomato suspension culture elutes can be separated on a 90 × 1.5 cm BioRex 70 (100–200 mesh) column (Smith et al. 1986) in 30 mM Na-Phosphate buffer run at a flow rate of 60 ml/h. In this case, a linear increasing salt gradient (0–1 M NaCl) is superimposed on a decreasing pH gradient (pH 7.6–6.1) to afford separation of the HRGPs into distinct P1a, P1b and P2 peaks.

Soybean HRGP can be isolated using a CM-Sepharose CL-6B (Sigma) column (30 × 2.5 cm) equilibrated with 20 mM Tris-HCl, pH 8.0 (Cassab et al. 1985). The sample is loaded, washed with buffer and a linear 20–500 mM Tris-HCl gradient is applied. The hydroxyproline-containing fractions (Drozdz et al. 1976) are then pooled and made 1.4 g/cc final density with CsCl and ultracentrifuged at 250,000 g for 72 h (Beckman SW-65 Ti rotor). The tubes are fractionated and the HRGP-rich fractions recovered and dialyzed against 100 mM Na-phosphate overnight, and against dH_2O for 2 days.

2.3 Gel Filtration Chromatography

Protein eluted from the ion exchange column can be further purified using gel filtration chromatography. In this step, molecules are separated on the basis of apparent size, or Stokes' radius, enabling the recovery of pure extensin-2 and the separation of extensin-1 oligomers and monomers from carrot preparations

(Stafstrom and Staehelin 1986 b). We typically use a Sephacryl S-400 gel filtration column (1.1 × 48 cm), equilibrated with 150 mM Tris-HCl, pH 8.0 buffer and run at a flow rate of 9 ml/h. When the extensin-1 fraction is sized, oligomers run as a fast-eluting shoulder preceding the main peak of Ext-1 monomers. Sizing resolves the extensin-2 fraction into a leading peak of Ext-2 and trailing peaks of lower MW contaminants. The recovered Ext-1 and Ext-2 peaks are electrophoretically pure and have been used in EM analysis, biochemical studies and as antigens in the production of antibodies (Stafstrom and Staehelin 1986 b, 1987, 1988).

3 Deglycosylation of Extensin

Both extensin-1 and extensin-2 of carrot are highly glycosylated, with about 60% of their total weight represented by carbohydrate (Stuart and Varner 1980; Stafstrom and Staehelin 1987). In order to study how glycosylation contributes to the configuration of the molecule, it is necessary to remove the sugar side-chains. Mort and Lamport (1977) pioneered the use of anhydrous hydrogen fluoride, which by controlling the temperature and duration of reaction, can be used to cleave N- or O-linked oligosaccharide side-chains without disrupting the peptide backbone. Since extensin glycosylation is entirely O-linked arabinosyl and galactosyl residues, treatment with this cleavage agent should yield a deglycosylated molecule. A modification of the Mort and Lamport method is used to deglycosylate carrot extensin-1 (van Holst and Varner 1984; Stafstrom and Staehelin 1986 a). Less than 1.0 mg of purified extensin-1 is lyophilized in a plastic microfuge tube and dried further under reduced pressure with P_2O_5 overnight. The thoroughly dried protein is treated with 30 μl of dry methanol and 270 μl anhydrous hydrogen fluoride in pyridine for 90 min at room temperature, then quenched with 1.0 ml of dH_2O. The resulting solution is dialyzed overnight against water and the deglycosylated molecules used in rotary shadowing and biochemical studies.

4 Visualization of Isolated Molecules

The glycerol-spray technique of Tyler and Branton affords an easy and reliable means of spreading isolated molecules onto a mica surface. Purified extensin is suspended at a concentration of 1 μg/ml in 50% glycerol, and 25–50 μl of this suspension is sprayed onto freshly cleaved discs (3–6 mm diam) of ruby mica using a modified airbrush (Tyler and Branton 1980; Stafstrom and Staehelin 1986 a, b). The samples are dried in vacuo in a Balzers freeze-etch apparatus, rotary shadowed with platinum at a 5°–6° angle and stabilized with carbon at a 90° angle. Replicas are floated in dH_2O and mounted on 300 mesh grids. The glycerol droplets dry with a particular morphology, appearing as spots of "dirt" at low magnification (Tyler and Branton 1980). On close inspection, however, these regions

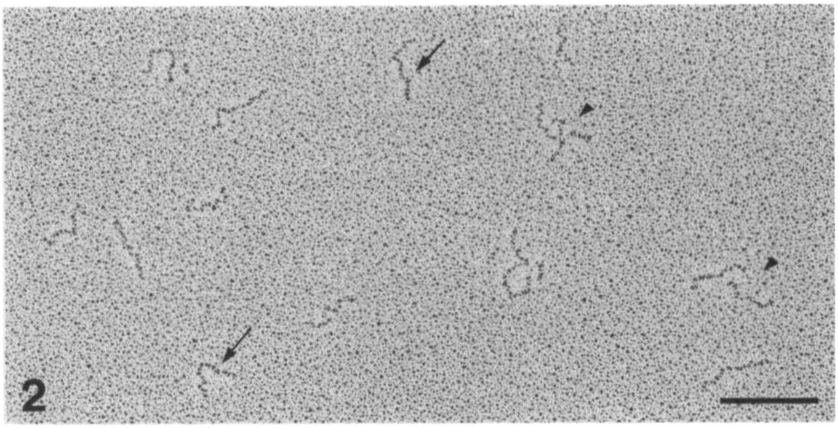

Fig. 2. Rotary shadowed field of purified carrot extensin-1 containing both monomer (*arrows*) and oligomer (*arrowheads*) forms of the molecule. (Micrograph kindly provided by Dr. J. P. Stafstrom). Bar line = 0.1 µm

always contained extensin molecules (Stafstrom and Staehelin 1986a, b). This technique works for both monomeric and oligomeric carrot extensin and appears to have little shear effect on the linear molecules (Fig. 2). An alternative method is described by Mould et al. (1985), where a small volume of suspended molecules is spread between two sheets of mica. This mica sandwich technique has prevented shearing of very long and fragile molecules, such as actin microfilaments, and may have some application in examination of cross-linked extensin complexes.

One potential drawback of these mica mounting techniques is that ionic interactions between the positively charged protein and negatively charged mica substrate may influence the proteins' morphology. Heckman et al. (1988) have attempted to overcome ionic binding artifacts by succinylation of deglycosylated extensin, and noted that the treated proteins still retain an elongate form in the absence of carbohydrate. However, succinylation of the numerous lysine amino acid side-chains contributes another 3000 MW of mass to the peptide and may itself alter the molecules' shape due to the neutralization of the positive charges and the association of the succinate molecules with the nonpolyproline helix protein domains.

The pleated sheet method of negative staining (S. A. Smith and Seegan 1984) has been used to good advantage in the display of HRGPs from the *Chlamydomonas* cell wall (Grief 1987). Theoretically, this technique could provide greater resolution than rotary shadowed images and may be useful in imaging higher plant extensins. Sheets of freshly cleaved mica are carbon-coated and cut into small strips (5 × 30 mm) prior to use. Two small vessels, constructed from the center third of disposable Eppendorf tubes (1.5 ml), are filled with the diluted protein sample and the uranyl formate negative staining solution (Leberman 1965). The mica strip is submerged into the protein sample at a 30° angle, allowing the carbon film to float on the solution surface and to absorb suspended molecules. The mica and attached carbon film with bound molecules are carefully withdrawn and transferred to the negative stain and the process is repeated. The film is then gently compressed against the sidewall of the vessel to form pleats that trap the negative stain, and the crumpled carbon film is lifted up on a copper EM grid for subsequent analysis in the electron microscope.

5 In Vitro Cross-Linking of Extensin Monomers

In vitro cross-linking experiments, initiated by Everdeen et al. (1988), promise to answer some of the more recalcitrant questions relating to the assembly of the extensin network in the plant cell wall. In particular, this approach allows both the characterization of potential extensin-specific peroxidases and recovery of in vitro cross-linked products. Cells in early stationary phase are used to recover the wall-associated peroxidases with high enzyme activity (Everdeen et al. 1988). All steps in the isolation are performed at 4° C. The tissue culture cells are collected on a coarsely sintered funnel, washed in dH_2O, dried by aspiration and extracted twice (5 min each) with 250 ml of 50 mM $CaCl_2$. The peroxidase-rich eluate is pooled, degassed for 5 min to remove oxygen, and concentrated to a final volume of 10 ml using an Amicon ultrafiltration apparatus (TCF-10 / YM10 membrane). The concentrated eluate is then dialyzed overnight against 50 mM sodium citrate, pH 6.3. After centrifugation at 9000 rpm (Sorvall SS-34 rotor) for 15 min, the supernatant is filtered through a 0.22-μM Millipore filter and frozen in liquid nitrogen. Peroxidase activity is maintained only if the frozen enzyme is thawed rapidly. This concentrate constitutes the crude enzyme preparation used in all subsequent steps.

The details of the cross-linking assay are clearly described in Everdeen et al. (1988). Basically, the crude peroxidase preparation is combined with purified extensin substrate (tomato extensin P1a) in the presence of hydrogen peroxide, and the conversion of monomers to cross-linked oligomers monitored using a Superose-6 fplc column (Pharmacia 30 × 1 cm). The peroxidase shows a broad pH optimum between pH 5.5 and pH 6.5 and is dependent on hydrogen peroxide levels; at a monomer concentration of 3.3 mg/ml, the optimum peroxide concentration is 5–30 μM. When surveyed by rotary shadowing TEM, the resulting cross-linked oligomers appear as branched assemblies with open "pores" (Everdeen et al. 1988). This assay should allow the generation of more complex extensin assemblies than is currently recovered from the cell wall and perhaps lend insight into how these "higher order" structures are organized.

6 Quick-Freeze, Deep-Etch of Wall Assemblies

To understand how extensin participates in the cell wall, it would be ideal to study the molecules in their three-dimensional context. The quick-freeze, deep-etch method (Heuser 1983) offers a way to visualize macromolecular assemblies, and in particular, has proven useful in the study of the HRGP-rich cell wall of the alga *Chlamydomonas reinhardtii*. By thin-section EM analysis, the globular and fibrous protein components form a distinctly stratified cell wall (Roberts et al. 1972). Specimens prepared by quick-freeze, deep-etch show a remarkable "three-dimensional" topography, with each of the layers having a characteristic pattern (Goodenough and Heuser 1985). In fact, layers reconstituted in vitro as the result of self-assembly of particular glycoproteins regain their in vivo morphology. In this way, it has been possible to assign different protein components to their respective wall layer and begin to speculate on how this alga assembles its cell wall (Goodenough et al. 1986; Adair et al. 1987).

For deep-etching experiments, isolated algal cell walls are mixed with mica flakes and the resulting slurry is rapidly frozen (reviewed by Gilkey and Staehelin 1986), thus avoiding the need for chemical fixation. The frozen mixture is freeze-fractured in a Balzers freeze-etch apparatus at $-102°$ C and the surface ice sublimed at the same temperature for 3 min (Heuser 1983). The deep-etched surface is shadowed with platinum at an angle of 24° and stabilized with carbon at an angle of 65°. Conditions for cleaning the replicas vary with the type of tissue used and the freezing methodology employed (for instance, with or without mica flakes present). One possible method cleans the replicas in 100% bleach overnight to digest the protein and lipidic material. After gradually replacing the bleach with dH_2O, the samples are slowly brought up to 70% sulfuric acid and heated at 60° C (1 h) to remove cellulose remnants. The replicas are washed several times in dH_2O and picked up on EM grids. If mica flakes are used, they can be dissolved with full-strength hydrofluric acid overnight (Heuser 1983).

7 Immunolocalization of HRGPs

Immunolocalization studies suggest that there is spatial organization of components in the plant cell wall (Moore and Staehelin 1988; Stafstrom and Staehelin 1988). This technique uses antibodies together with colloidal gold-conjugated protein A to tag specific carbohydrate or glycoprotein antigens on thin-sectioned plant material (reviewed by Moore and Staehelin, this volume). Polyclonal antibodies against carrot extensin-1 (Stafstrom and Staehelin 1988) and extensin-2 (Swords and Staehelin, in prep.) have been used to probe a variety of plant tissues. In the case of extensin-1, the protein is localized in the cellulose-containing regions of the cell wall (Stafstrom and Staehelin 1988). Figure 3 illustrates extensin-1 immunolabeling in ethylene-treated green bean (*Phaseolus vulgaris* cv. Greensleeves) where the labeling invades the cell corner somewhat more than usual. However, a different pattern is seen in the distribution for extenin-2. Immunolabeling of carrot tissue culture cells, green bean stem ($\pm$ ethylene), and clover stem tissue with extensin-2 antibodies all show a concentration of antigen in middle lamella corner regions between cells (Fig. 4).

Cassab and Varner (1987) have introduced an interesting variation of the Western blot technique (Towbin et al. 1979) that uses the cut surface of plant tis-

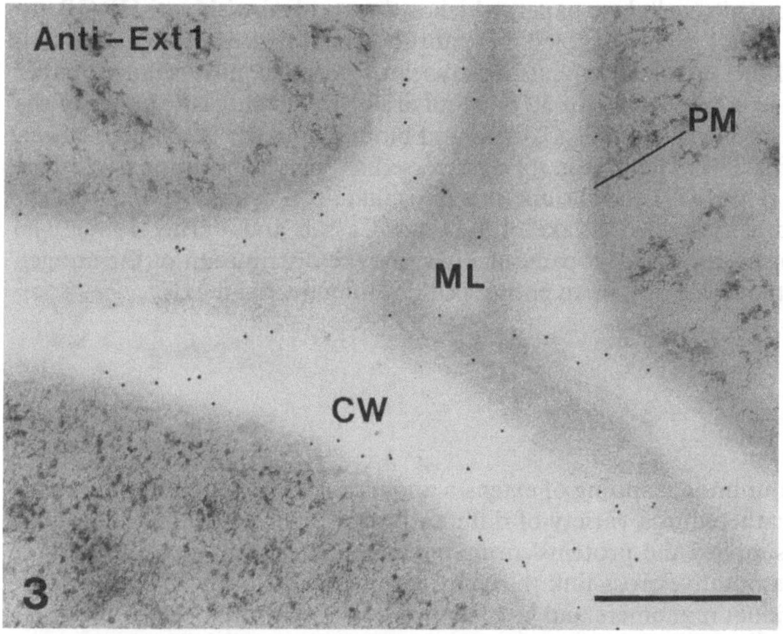

Fig. 3. Thin-section electron micrograph of parts of ethylene-treated green bean parenchyma cells immunolabeled with anti-extensin-1 antibodies. Most of the label is located over the cellulose-containing regions of the cell wall (*CW*) adjacent to the plasma membrane (*PM*); *ML* middle lamella. Bar line = 0.5 μm

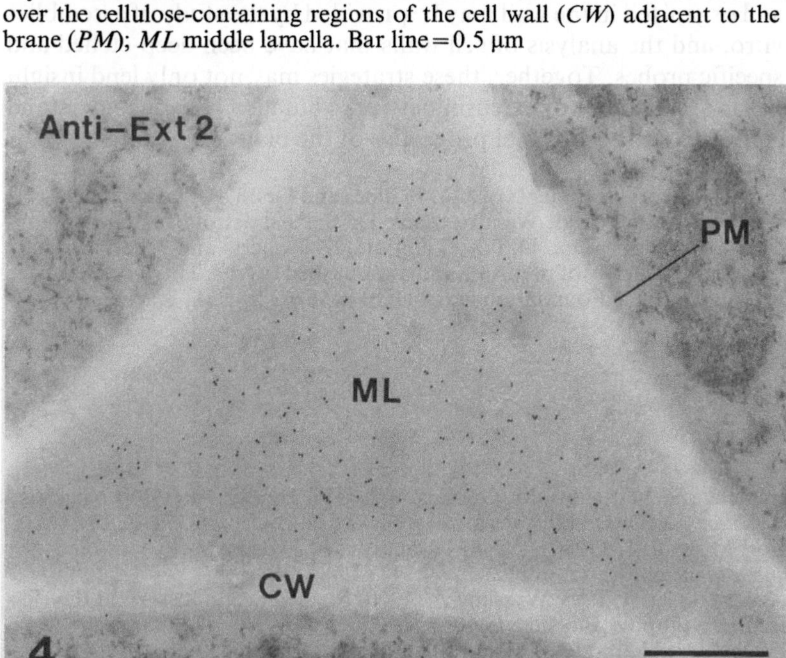

Fig. 4. Carrot suspension culture cells labeled with anti-extensin-2 antibodies show a concentration of label in the expanded middle lamella (*ML*) in cell corner regions; *CW* cell wall; *PM* plasma membrane. Bar line = 0.5 μm

sue printed onto nitrocellulose paper, which is then probed with anti-HRGP antibodies. This technique has enabled these authors to localize extensin in the coat, the hilum and the vascular elements of soybean seeds. The nitrocellulose paper is first soaked in 0.2 M $CaCl_2$ for 30 min and dried. The freshly cut surface of the soybean is washed briefly in dH_2O, dried and blotted onto the prepared nitrocellulose paper for 30 s. The blot is then processed as for a conventional Western blot (Towbin et al. 1979) and the alkaline phosphate-conjugated secondary antibody detected as described (Blake et al. 1984; Cassab and Varner 1987). The strength of this assay is that it can quickly show the distribution of the antigen in a complex tissue, such as in an entire seed or along the plant axis.

8 Conclusion

In summary, our understanding of extensin and its mechanism of assembly stems from results gathered in a variety of different ways. Biochemical and molecular studies have supplied the proteins' primary sequence, secondary conformation and revealed a putative cross-link that may act intermolecularly. Structural features of individual monomers and cross-linked complexes have been further analyzed using rotary shadowing techniques, and have given us a feeling for how a few molecules are cross-linked together. In the future, details of how extensin is assembled and organized in the wall may be provided by the study of assemblies catalyzed in vitro, and the analysis of cell walls that have been deep-etched and labeled with specific probes. Together, these strategies may not only lend insight into the molecular architecture of extensin networks but also help us understand how extensin influences the biological properties of the plant cell wall.

Acknowledgments. The helpful comments of J. A. Holmes and Drs. J. P. Stafstrom and P. J. Moore are gratefully acknowledged. We also thank Dr. J. P. Stafstrom for micrographs and Diane Lorenz for artwork. Drs. D. T. A. Lamport, K. Roberts and J. E. Varner are thanked for their generous supply of preprints and unpublished results. This work has been supported by National Science Foundation grant DCB 8615763 to L. A. Staehelin.

References

Adair SW, Steinmetz SA, Mattson DM, Goodenough UW, Heuser JE (1987) Nucleated assembly of *Chlamydomonas* and *Volvox* cell walls. J Cell Biol 105:2373–2382

Akiyama Y, Mori M, Kato K (1980) ^{13}C-NMR analysis of hydroxyproline arabinosides from *Nicotiana tabacum*. Agric Biol Chem 44:2487–2489

Albersheim P, Mühlethaler K, Frey-Wyssling A (1960) Stained pectin as seen in the electron microscope. J Biophys Biochem Cytol 8:501–506

Basile DV (1980) A possible mode of action for morphoregulatory hydroxyproline-proteins. Bull Torrey Bot Club 107:325–338

Blake MS, Johnston KH, Russell-Jones GJ, Gotschlich EC (1984) A rapid, sensitive method for detection of alkaline phosphatase-cojugated anti-antibody on Western blots. Anal Biochem 136:175–179

Bolwell GP (1984) Differential patterns of arabinosylation by membranes of suspension-cultured cells of *Phaseolus vulgaris* (French bean) after subculture or elicitation. Biochem J 148:427–435
Cassab GI, Varner JE (1987) Immunocytolocalization of extensin in developing soybean seed coats by immunogold-silver staining and by tissue printing on nitrocellulose paper. J Cell Biol 105:2581–2588
Cassab GI, Nieto-Sotelo J, Cooper JB, van Holst GJ, Varner JE (1985) A developmentally regulated hydroxyproline-rich glycoprotein from the cell walls of soybean seed coats. Plant Physiol 77:532–535
Chen J, Varner JE (1985) An extracellular matrix protein in plants: characterization of a genomic clone for carrot extensin. EMBO J 4:2145–2151
Chrispeels MJ (1969) Synthesis and secretion of hydroxyproline-containing molecules in carrots I. Kinetic analysis. Plant Physiol 44:1187–1193
Chrispeels MJ (1970) Synthesis and secretion of hydroxyproline-containing macromolecules in carrots II. in vivo conversion of peptidyl proline to peptidyl hydroxyproline. Plant Physiol 45:223–227
Cooper JB, Varner JE (1984) Cross-linking of soluble extensin in isolated cell walls. Plant Physiol 76:414–417
Cooper JB, Chen JA, van Holst GJ, Varner JE (1987) Hydroxyproline-rich glycoproteins of plant cell walls. TIBS 12:24–27
Drozdz M, Kucharz E, Szyja J (1976) A colorimetric micro-method for determination of hydroxyproline in blood serum. Z Med Labortech 17:163–171
Epstein L, Lamport DTA (1984) An intramolecular linkage involving isodityrosine in extensin. Phytochemistry 23:1241–1246
Esquerre-Tugaye MT, Lafitte C, Mazau D, Toppan A, Touze A (1979) Cell surfaces in plant-microorganism interactions II. Evidence for the accumulation of hydroxyproline-rich glycoproteins in the cell wall of diseased plants as a defense mechanism. Plant Physiol 64:320–326
Everdeen DS, Kiefer S, Willard JJ, Muldoon EP, Dey PM, Li X, Lamport DTA (1988) Enzymatic crosslinkage of monomeric extensin precursors in vitro. Plant Physiol 87:616–621
Fry SC (1982) Isodityrosine: a new crosslinking amino acid from plant cell wall glycoprotein. Biochem J 204:449–455
Gilkey JC, Staehelin LA (1986) Advances in ultrarapid freezing for the preservation of cellular ultrastructure. J Electron Microsc Tech 3:177–210
Goodenough UW, Heuser JE (1985) The *Chlamydomonas* cell wall and its constituent glycoproteins analyzed by the quick-freeze, deep-etch technique. J Cell Biol 101:1550–1568
Goodenough UW, Gebhart B, Mecham RP, Heuser JE (1986) Crystals of the *Chlamydomonas reinhardtii* cell wall: polymerization, depolymerization, and purification of glycoprotein monomers. J Cell Biol 103:405–417
Grief C (1987) Structural, chemical and immunological study of the cell wall of *Chlamydomonas reinhardii.* Thesis, University of East Anglia, Norwich
Hammerschmidt R, Lamport DTA, Muldoon EP (1984) Cell wall hydroxyproline enhancement and lignin deposition as an early event in the resistance of cucumber to *Cladosporium cucumerium.* Physiol Plant Pathol 24:43–47
Hayashi T, MacLachlan G (1984) Pea xyloglucan and cellulose. III. Metabolism during lateral expansion of pea epicotyl cells. Plant Physiol 76:739–742
Heckman JW Jr, Terhune BT, Lamport DTA (1988) Characterization of native and modified extensin monomers and oligomers by electron microscopy and gel filtration. Plant Physiol 86:848–856
Heuser JE (1983) Procedure for freeze-drying molecules adsorbed to mica flakes. J Mol Biol 169:155–195
Lamport DTA (1965) The protein component of primary cell walls. Adv Bot Res 2:151–218
Lamport DTA (1967) Hydroxyproline-O-glycosidic linkage of the plant glycoprotein extensin. Nature 216:1322–1324

Lamport DTA (1980) Structure and function of glycoproteins. In: Preiss F (ed) The biochemistry of plants. Academic Press, New York, pp 501–541
Lamport DTA (1986) Roles for peroxidases in cell wall genesis. In: Greppin H, Penel C, Gaspar T (eds) Molecular and physiological aspects of plant peroxidase. Universite de Geneve-Centre de Botanique, pp 199–208
Lamport DTA, Epstein L (1983) A new model for the primary cell wall: a concatenated extensin-cellulose network. In: Randall DD, Blevins DG, Larson RL, Rapp BJ (eds) Current topics in plant biochemistry and physiology. University of Missouri, Columbia, Missouri, pp 73–83
Lamport DTA, Katona L, Roetig S (1973) Galactoserine in extensin. Biochem J 133:125–131
Leach JE, Cantrell MA, Sequeira L (1982) Hydroxyproline-rich bacterial agglutinin from potato. Plant Physiol 70:1353–1358
Leberman R (1965) Use of uranyl formate as a negative stain. J Mol Biol 13:606
Mellon JE, Helgeson JP (1982) Interaction of a hydroxyproline-rich glycoprotein from tobacco callus with potential pathogens. Plant Physiol 70:401–405
Moore PJ, Staehelin LA (1988) Immunogold localization of the cell-wall-matrix polysaccharides rhamnogalacturonan I and xyloglucan during cell expansion and cytokinesis in *Trifolium pratense* L.: implication for secretory pathways. Planta 173:433–445
Mort AJ, Lamport DTA (1977) Anhydrous hydrogen fluoride deglycosylates glycoproteins. Anal Biochem 82:289–309
Mould PA, Holmes DF, Kadler KE, Chapman JA (1985) Mica sandwich technique for preparing macromolecules for rotary shadowing. J Ultrastruct Res 91:66–76
Roberts K, Gurney-Smith M, Hills GJ (1972) Structure, composition, and morphogenesis of the cell wall of *Chlamydomonas reinhardi* I. Ultrastructure and preliminary chemical analysis. J Ultrastruct Res 40:599–613
Roland JC, Vian B (1979) The wall of the growing plant cell wall: it's three dimensional organization. Int Rev Cytol 61:129–166
Sadava D, Chrispeels MJ (1973) Hydroxyproline-rich cell wall protein (extensin): role in the cessation of elongation in excised pea epicotyls. Dev Biol 30:49–55
Sadava D, Walker F, Chrispeels MJ (1973) Hydroxyproline-rich cell wall protein (extensin): biosynthesis and accumulation in growing pea epicotyls. Dev Biol 30:42–48
Selvendran RR (1985) Developments in the chemistry and biochemistry of pectic and hemicellulosic polymers. J Cell Sci Suppl 2:51–88
Smith JJ, Muldoon EP, Lamport DTA (1984) Isolation of extensin precursors by direct elution of intact tomato cell suspension cultures. Phytochemistry 23:1233–1239
Smith JJ, Muldoon EP, Willard JJ, Lamport DTA (1986) Tomato extensin precursors P1 and P2 are highly periodic structures. Phytochemistry 25:1021–1030
Smith SA, Seegan GW (1984) The pleated sheet: an unusual negative staining method for transmission electron microscopy of biological macromolecules. J Ultrastruct Res 89:111–122
Stafstrom JP (1986) Structure, function and cellular localization of extensin. Thesis, University of Colorado, Boulder
Stafstrom JP, Staehelin LA (1986a) The role of carbohydrate in maintaining extensin in an extended conformation. Plant Physiol 81:242–246
Stafstrom JP, Staehelin LA (1986b) Cross-linking patterns in salt-extractable extensin from carrot cell walls. Plant Physiol 81:234–241
Stafstrom JP, Staehelin LA (1987) A second extensin-like hydroxyproline-rich glycoprotein from carrot cell walls. Plant Physiol 84:820–825
Stafstrom JP, Staehelin LA (1988) Antibody localization of extensin in carrot cell walls. Planta 174:321–332
Stermer BA, Hammerschmidt R (1987) Association of heat shock induced resistance to disease with increased accumulation of insoluble extensin and ethylene synthesis. Physiol Mol Plant Pathol 31:453–461
Stuart DA, Varner JE (1980) Purification and characterization of a salt-extractable hydroxyproline-rich glycoprotein from aerated carrot discs. Plant Physiol 66:787–792

Toppan A, Roby D, Esquerre-Tugaye MT (1982) Cell surfaces in plant-microorganism interactions III. in vivo effect of ethylene on hydroxyproline-rich glycoprotein accumulation in the cell wall of diseased plants. Plant Physiol 70:82–86

Towbin H, Staehelin T, Gordon J (1979) Electrophoretic transfer of proteins from polyacrylamide gels to nitrocellulose sheets: procedure and some applications. Proc Natl Acad Sci USA 76:4350–4354

Traub W, Shmueli U (1963) Structure of poly-L-proline II. In: Ramachandran GN (ed) Aspects of protein structure. Academic Press, London, pp 81–92

Tyler JM, Branton D (1980) Rotary shadowing of extended molecules dried from glycerol. J Ultrastruct Res 71:95–102

Valent B, Albersheim P (1974) Structure of plant cell walls. V. On the binding of xyloglucan to cellulose fibers. Plant Physiol 54:105–109

van Holst GJ, Varner JE (1984) Reinforced polyproline II conformation in a hydroxyproline-rich cell wall glycoprotein from carrot root. Plant Physiol 74:247–251

van Holst GJ, Klis FM, Bouman F, Stegwee D (1980) Changing cell-wall compositions in hypocotyls of dark-grown bean seedlings. Planta 149:209–212

Vaughan D (1973) Effects of hydroxyproline on the growth and cell-wall protein metabolism of excised root segments of *Pisum sativum*. Planta 115:135–145

Wilson LG, Fry JC (1986) Extensin – a major cell wall glycoprotein. Plant Cell Environ 9:239–260

Methods for Analysis of Dietary Fibre

R. R. SELVENDRAN, A. V. F. V. VERNE, and R. M. FAULKS

1 Introduction

Dietary fibre (DF) was defined by Trowell et al. in 1976 as all the polysaccharides and lignin in the diet that are not digested by the endogenous secretions of the human digestive tract. This definition embraces the polysaccharides and lignin of plant cell walls (CW) and also the small amount of polysaccharide food additives that makes up less than 2% of fibre in our diet. These may be present as plant gums, algal polysaccharides, pectins, modified celluloses and modified starches. Plant CW are the major source of DF and thus for analytical purposes DF refers mainly to the non-starch polysaccharides (NSP) and lignin in the diet (Southgate 1976). The food additives are, however, commercially available and have similar structural features to CW polysaccharides: for these reasons they have been used as model compounds to study the action of DF.

Most of our DF intake comes from CW of vegetables, fruits, cereal products and other seeds. The CW composition depends on the stage of maturity at which the plant organ is harvested, and to some extent on the conditions under which it is stored. The principal DF components are complex polysaccharides, some of which are associated with polyphenolics (including lignin) and proteins. Non-carbohydrate components of CW are present in small amounts but some – lignin and phenolic esters in lignified wheat bran, cutin and waxes in leafy vegetables and suberin in roots and tubers – affect the properties and physiological effects of DF significantly.

The main components of cell walls from different tissue types of food plants and the structural features of some major cell-wall polymers are summarized in Tables 1 and 2 of Selvendran and O'Neill (1987). The parenchymatous tissues are the most important. They comprise the bulk of DF ingested from fruits, vegetables, seed cotyledons and cereal endosperm. Wheat bran contains the lignified, outer layers of the grain and the aleurone layer and makes a significant contribution to cereal-fibre intake. The relative amounts of the polymers in different tissue types are shown in Table 2 of Selvendran (1987). For an extended discussion on CW polymers of edible plants see Selvendran (1983, 1985), and for an account of the chemistry and properties of DF see Southgate (1976), Southgate and Englyst (1985), Selvendran (1984) and Selvendran et al. (1987). The above reviews give references to a number of major reviews and publications in the cell-wall area.

This chapter is not a comprehensive review but rather an assessment of some of the promising methods, drawing out their relative merits; the problems encountered and the ways these were overcome. Selvendran and Du Pont (1984) give an extended discussion of methods published up to 1983. Asp and Johansson (1984) have discussed some of the later methods with particular reference to gravimetric analysis (which can be coupled to carbohydrate analysis). The main concern of most procedures is to estimate total DF content and obtain an indication of the main DF polymers present from the constituent sugars. Now that the problems besetting empirical DF analysis have largely been solved, the emphasis should shift to the more difficult tasks of defining the constituent polymers, their physicochemical homogeneity, properties and effects on the digestive tract. Some of these aspects have been discussed in a recent review on DF (Selvendran et al. 1987). A detailed account of the methodology of isolation and analysis of CW (mostly from edible plant organs) is given in Selvendran et al. (1985) and Selvendran and O'Neil (1987). Selvendran and Stevens (1986) describe the applications of mass spectrometry for the examination of pectic polysaccharides. This chapter primarily aims to help food researchers and analysts as they select or devise a simplified method for DF analysis of a particular food product.

The problems associated with DF analysis are threefold: those arising during preparation of DF; those associated with determination of constituent carbohydrates and those encountered in lignin determination. The methods available for lignin estimation will be fully discussed in Sect. 2.3.

2 Problems Associated with DF Analysis

2.1 Fresh and Processed Foods

The major difficulty in DF analysis is removing starch quantitatively from fresh and processed foodstuffs. Plant storage organs are generally rich in both starch and proteins. The starch:protein:DF ratios in oats and potatoes are 7.1:1.6:1 and 15:0.7:1 respectively (Selvendran and Du Pont 1980a), showing that incomplete removal of starch can seriously affect DF estimates, giving inflated values. Complete removal of starch (and proteins) is hindered by incomplete disruption of tissues in food samples. Gelatinization of starch makes it more readily degradable by α-amylase. In our experience, ultraturrax treatment followed by ball-milling of the extractive-free residue in 90% aqueous alcohol breaks the cell structure down completely. This allows solvents and enzymes easy access to intracellular components. The merits of this procedure for preparing gram quantities of CW and DF from fresh and processed plant foods, which can then be used for detailed fractionation studies have been described before (Selvendran and Du Pont 1984; Selvendran and O'Neil 1987).

In processed foods there are additional problems with starch removal. Interactions of starch (and cell walls) with intracellular proteins, and retrogradation of starch render the starch less degradable by enzymes. Starch-lipid(-protein)-complexes are a particular problem with some cereal products. Heat treatment, such as extrusion cooking, appears to cause irreversible condensation between adjacent hydroxyl groups on some of the starch molecules and the complex formed is less readily degraded enzymically. Starch which is released into the aqueous medium but is only partially soluble in it undergoes retrogradation on cooling due to hydrogen bonding between regions of the glucan chains, and precipitation takes place. The precipitated starch is less readily degraded by α-amylase, and the problem is compounded in processed products which therefore have significant amounts of "resistant starch".

The bulk of the protein present is only removed in the gravimetric methods. DF prepared by other methods contains intracellular proteins which make estimation of lignin difficult (see later). Neutral detergent is a good solvent for intracellular proteins, nevertheless the residues after neutral detergent treatment (Neutral Detergent Fibre, NDF) find only limited application as fibre preparations for food samples. Neutral detergent fibre is, however, widely used in animal feed studies because grasses contain only small amounts of pectins and most of these are cross-linked and therefore insoluble (Van Soest 1967a, b; Robertson and Van Soest 1977, 1981). Hot neutral detergent solubilizes undefined amounts of pectic substances, β-glucans and arabinoxylans (of flours) and only partially solubilizes starch from starch-rich products (Selvendran and Du Pont 1984). The solubilized (and residual) starch impedes filtration and washing of the residue. Starch can be removed enzymatically from NDF to give "modified NDF" by most commercial starch-degrading enzymes, and this facilitates filtration. Cold neutral detergent solubilizes small amounts of cold-water-soluble pectic substances from fruit and vegetables and some β-glucans and arabinoxylans from cereals. The residue after treatment with cold neutral detergent is easily isolable by centrifugation or filtration, and the associated starch is quantitatively soluble in 90% (v/v) aqueous DMSO provided the tissue is completely disrupted (Ring and Selvendran 1978; Selvendran and O'Neill 1987). In addition to solubilizing starch, the DMSO also solubilizes the detergent which is associated with the cold-NDF residue. Fibre in faecal samples has bacteria associated with it. Slavin and Marlett (1983) have written that neutral detergent may be used to advantage for washing the bacterial proteins from the faecal fibre. Our experience with bacterially degraded vegetable and wheat bran samples showed that the faecal bacteria adhere very strongly to the degraded fibre and are not fully removed even by 1% sodium-dodecylsulphate in 90% (w/v) phenol at 70° (Stevens et al. 1988; Stevens and Selvendran 1988).

2.2 Determination of the Monomeric Composition

The major neutral sugars of DF preparations are, in order of decreasing abundance, glucose, xylose, arabinose, galactose, mannose and rhamnose. The uronic acids are galacturonic acid in parenchymatous tissues of vegetables and fruits, and glucuronic acid (and 4-O-Me-glucuronic acid) in lignified tissues, particularly those derived from cereals.

The monosaccharides of DF polymers are linked by glycosidic bonds of variable acid lability. No acid hydrolysis procedure will cleave every linkage and give a quantitative yield of each component. All hydrolyses cause some degradation. The following statements generally apply: furanosidic and deoxypyranosidic linkages cleave more readily than pyranosidic linkages; pentopyranosidic bonds cleave more readily than hexopyranosidic bonds; α-glycosidic linkages are less acid-resistent than their β counterparts and among pyranosidic linkages $1 \rightarrow 4$ is more labile than $1 \rightarrow 3$, which is more labile than $1 \rightarrow 2$. For further comments and relevant references see Aspinall (1973).

Uronic acids linked to neutral sugars yield acid-resistant aldobiouronic acids (uranosyl-sugar disaccharides). Examples are GalpA(1→2)Rhap from pectins and GlcpA(1→2)Xylp from acidic-xylans and -arabinoxylans. Some insoluble polysaccharides, such as cellulose and long-chain mannans (and acidic-xylans), require pretreatment with 72% H_2SO_4 before they can be completely hydrolyzed by 1 M acid. Pectins and pectic acids tend to precipitate in acid which could result in incomplete hydrolysis of the associated neutral sugars. Hydrolysis can be facilitated by prior treatment with polygalacturonase.

Polysaccharides are usually hydrolyzed by refluxing in dilute sulphuric acid, with or without preliminary dispersion in 72% H_2SO_4. 2 M TFA, 0.5 and M H_2SO_4 hydrolyze the bulk of the non-cellulosic carbohydrate polymers (NCP) and 5%–10% of the cellulose in cell walls. The cellulose can be quantitatively hydrolyzed after first dispersing the CW in 72% (w/w) H_2SO_4 for 2–3 h at 20° C (Saeman et al. 1963; Selvendran et al. 1979). Losses of xylose have been reported following 3 h dispersion, and 2 h has been preferred by Rasper et al. (1981). Concentrated H_2SO_4 sulphonates the cellulose, particularly at C_6, disrupting the hydrogen bonds between the microfibrils. The cellulose dissolves and is hydrolyzed to large oligosaccharides. The syrup is then diluted to molar acid concentration and heated at 100° C for 2–3 h to complete the hydrolysis. This "post-hydrolysis" releases mono-saccharides. Other polysaccharides are affected by the concentrated acid in a similar way.

Table 1 of Selvendran and DuPont (1984) indicates the relative effectiveness of 2 M TFA, 0.5 M H_2SO_4, M H_2SO_4 at 100° C for 2.5 h and Saeman hydrolysis conditions in hydrolyzing NCP and cellulose of potatoes and mature runner beans. In general, the first three conditions hydrolyze the bulk of the NCP, but M H_2SO_4 and (particularly) Saeman hydrolysis release significantly larger amounts of xylose from CW of heavily lignified tissues. Although Saeman hydrolysis conditions hydrolyze most of the cellulose, Theander and Westerlund (1986) have reported that 5 h post-hydrolysis is necessary to break it down completely. Some of the cellulose has been isolated as glucose-6-sulphate after 4 h post-hydrolysis (Hardell and Theander 1970). According to Neilson and Marlett (1983) HPLC has shown ~10% of cellulose to exist as cellobiose after 3 h post-hydrolysis. This incomplete hydrolysis after 3 h is a point to note. Theander and Westerlund (1986) reported that autoclaving at 125° C for 1 h in place of post-hydrolysis leaves only negligible amounts of cellobiose. Slavin and Marlett (1983) have observed significant losses (~25%–30%) in CW monosaccharides during 2 h M H_2SO_4 treatment at 100° C. This is at variance with the findings of Selvendran et al. (1979) and Englyst and Cummings (1984), who found little loss in sugars using the same conditions. Further investigation is probably required. The stability of the galacturonic acid-rhamnose linkage, GalpA(1→2)-Rhap, means that only 63% of rhamnose is released from pectins after 2 h hydrolysis (Selvendran et al. 1979). Xylose linked to glucuronic acid (GlcpA) in acidic xylans is likewise difficult to estimate quantitatively due to accumulation of GlcpA-(1→2)-Xylp in the hydrolysate (Whistler and Richards 1958). Quantitative estimation of uronic acids – primarily galacturonic acid, (galA) in pectins and glucuronic acid (glcA) in acidic-xylans and -arabinoxylans is difficult. For water-soluble pectic material this problem can be overcome by reducing the "carbodiimide activated" galacturonic acid residues to galactose using $NaBH_4$ or $NaBD_4$ (Taylor and Conrad 1972). Thus the glycosiduronic linkage is replaced by a more acid labile glycosidic linkage. Neutral sugars are then quantitatively released and estimated as their alditol acetates by glc. Native galactose can be distinguished (by GC-MS) from that derived by reduction if $NaBD_4$ is used. The efficiency of this reduction procedure may be low for some pectic polysaccharides (Aspinall and Jiang 1974; Aspinall and Fanous 1984).

In the DF context, the sugars released on acid hydrolysis have mainly been estimated by colorimetric methods, GC of alditol acetates or by HPLC. Our comments will concentrate on these methods.

Colorimetric Methods. A coloured derivative is formed in two stages. An initial reaction (with acid) produces a chromogen (usually a furfural derivative of the sugar). This is condensed with a chromophore (often a phenol or an aromatic nitrogen base) to give the coloured product. Colorimetric methods do not require prior release of the individual sugars. It should be noted that individual sugars give varying responses. See Selvendran and DuPont (1984) for further comments on the use of non-specific colorimetric methods. Uronic acid (UA) is usually estimated by the method of Blumenkrantz and Asboe-Hansen (1973). The DF (or CW) is dispersed in dilute acid before quantitative reaction with m-hydroxybiphenyl in H_2SO_4 containing sodium tetraborate to give a pink derivative with A^{max} 524 nm. There is negligible interference from pentoses or hexoses. Most colorimetric methods can be used to estimate individual sugars at concentrations as low as 5 μg/ml. The PAHBAH method (Lever 1972) is sensitive down to 1 μg/ml (Hudson et al. 1976). UA can also be estimated by a decarboxylation procedure, as used by Theander and Aman (1979), which relies on quantitative release of CO_2 on heating UA and their derivatives with hydroiodic acid.

GC of Alditol Acetates. The alditol acetate chromatogram shows a single peak for each sugar and the alditol acetates derived from plant CW are readily separable on commercially available liquid phases such as OV225 (Selvendran et al. 1979) and SP-2330 (Englyst et al. 1982). We use 3 m × 2.2 mm i.d. glass column containing 3% w/w OV225 at 200° C, whereas Englyst et al. (1982) use 2.1 m × 4 mm i.d. glass column packed with Supelcoport (100–200 mesh) coated with 3% SP2330 at 220° C. The detection limit using the packed column is approximately 0.1–0.5 μg, depending on the retention time of the alditol acetate. The above papers should be consulted for details of the procedures. Phthalic esters have been reported to interfere with glc determination of alditol acetates (Dudman and Whittle 1976). This can be avoided using high quality reagents or capillary columns with polar phases (Henry et al. 1983). Capillary GC is, however, more suited to separation of complex mixtures of partially methylated alditolacetates (PMAA) derived from methylation analysis. The detection limit of capillary GC may be ten times smaller than that of conventional GC.

HPLC. HPLC cannot, as yet, match the resolution of GC in separation of monosaccharides (Selvendran and DuPont 1984). No method has yet been developed to resolve all the neutral monosaccharides commonly found in DF preparations, although Slavin and Marlett (1983) and Neilson and Marlett (1983) have achieved good separation of most of the sugars and have detected cellobiose in NDF hydrolysates. We have obtained good resolution of a range of oligosaccharides from starch hydrolysates. We used a Waters carbohydrate column, 25 cm × 3.9 mm, eluting with Acetonitrile: water (82:18), at a rate of 1.8 ml/min. The system resolves rhamnose, arabinose, xylose and mannose but glucose and galactose co-elute. These could be separated by increasing the acetonitrile concentration of the mobile phase but this causes unacceptable peak broadening and long retention times. Monosaccharides have no specific absorption maximum and so refractive index detectors have to be used. These are only moderately sensitive (Pryde and Gilbert 1979) and the responses of individual sugars show up to tenfold variation. The detection limit for monosaccharide sugars is 50 µg/ml – minor components may not, therefore, be detected. Detection limits can be increased by derivatization; either pre-column (Lehrfield 1976) or post-column (Honda et al. 1981, 1983).

2.3 Lignin

Lignins are high-molecular-weight aromatic polymers formed by enzymatic dehydrogenation and subsequent polymerization of coniferyl, sinapyl and p-coumaryl alcohols (Freudenberg and Neish 1968; Sarkanen and Hergert 1971). They comprise, for the most part, variously linked phenylpropane residues, and the relative proportions of the residues vary in the lignins of different plants. Lignified tissues are not readily degraded by colonic bacteria and quantification of lignin is therefore of interest in the DF context. Most plant organs except for some cereals are, however, harvested and eaten before much lignification has taken place and lignin is regarded as a minor component of most foods. Nevertheless the polyphenolic content of parenchymatous tissues is ~5%–10% of the dry weight of the CW rising to 10%–20% in partially lignified tissues (Selvendran and DuPont 1984; Selvendran 1985). The polyphenolic material appears to cross-link the matrix polymers and so influence their solubility.

Quantitative estimation of lignin in DF preparations is difficult for a number of reasons: the detailed chemistry of lignin is uncertain; lignin is associated with other CW components and difficult to separate from them – it is, for example, covalently linked to hemicelluloses and interacts with proteins especially in partially lignified vegetables and certain cereal products; lignin is partially soluble in acid, the extent of solubility depending on the plant source; artifacts, such as Maillard reaction products produced during cooking and processing of foods, interfere with lignin estimation. Lignin is best estimated using extractive-free material, and we recommend that proteins should be removed using proteolytic enzymes or detergents. The following methods have been used to obtain estimates of lignin.

The Sulphuric Acid Method. Sulphuric acid is used to solubilize the components of a sample leaving behind "Klason lignin". The sample is first dispersed in 72% H_2SO_4 (in which cellulose and other polysaccharides are soluble) and then diluted and hydrolyzed at 100° C. The mass of the residue expressed as percentage of dry weight of the sample gives the Klason lignin content. The method works well for conifer woods, which contain predominantly guaiacyl lignins; less well for angiosperm woods which contain a fair proportion of partially acid-soluble syringyl lignins, and least well for lignins of herbaceous plants including those of monocotyledonous species. These contain significant amounts of acid-soluble lignins, sometimes as much as 50% is soluble (Lai and Sarkanen 1971). Theander and Westerlund (1986) have used the sulphuric acid method to determine the lignin content of some DF preparations.

Permanganate Oxidation Method. Potassium permanganate oxidizes lignin to CO_2 and dibasic acids. The associated hemicelluloses are also liable to oxidation and if not removed would produce erroneously high values for lignin. For this reason, lignin estimations are normally performed on acid detergent fibre (ADF). ADF is prepared by treating the sample with 0.5 M H_2SO_4 in 2% (w/v) cetyltrimethylammonium bromide at 100° C for 1 h. These conditions are not, however, adequate to remove all the hemicelluloses and pectic polysaccharides from some vegetable fibres (Selvendran et al. 1987): as much as 20% (in total) may remain. The residual hemicelluloses and pectins may be solubilized by permanganate and give elevated values for lignin (Selvendran and DuPont 1984). Cellulose has not been reported to interfere. According to the gravimetric method of Robertson and Van Soest (1981) the ADF is treated with $KMnO_4$ to remove lignin and the amount removed is determined by mass difference. The method works well for wheat bran ADF but mature runner bean ADF gives a high value because of the residual hemicelluloses and pectic substances (~20%) solubilized by the permanganate. The method is suitable for grasses since they contain only small amounts of pectin and most of the hemicelluloses are hydrolyzed. Lignin can also be estimated in ADF by back-titrating the excess $KMnO_4$ after reaction to find how much permanganate has been used.

Acetyl bromide Lignin. The lignin and associated polysaccharides in the sample are acetylated and made soluble by treating with acetyl bromide and acetic acid at 70° C. The excess reagent is destroyed and precipitated proteins are removed by filtration. The absorbance of the filtrate at 280 nm is proportional to the lignin content. The method was developed by Johnson et al. (1961) for wood and wood products. It was further developed and applied to forages by Morrison (1972a, b) who derived a conversion factor to give the lignin content of grasses. The main advantages of the method are that it is rapid and only requires a few milligrams of material. In our experience the method can only be used to compare the lignin contents of different organs from the same plant or samples from comparable species, since different species require different conversion factors (Selvendran, DuPont and Verne, unpublished results). Although protein does not appear to interfere with the estimation, we recommend prior removal of protein with proteolytic enzymes. The acetyl bromide method gives good agreement with Klason lignin for some products and can be scaled down for 5–10-mg samples: the Klason method requires a minimum of 100–500 mg. The acetyl bromide method may be useful in the DF context since the value obtained includes the bulk of the cross-linking ferulic acid and p-coumaric acid groups. It may, therefore, give a good estimate of cross-linking phenolic components. Selvendran and DuPont (1984) have used the sulphuric acid, permanganate and acetyl bromide methods to estimate lignin in: fine-milled bran, bran CWM, bran ADR, fine-milled oat hulls, CWM from parchment layers of runner bean pods and ADR of mature runner bean pods. The values obtained are presented and their significance discussed in the above review.

3 Isolation and Analysis of Milligram Quantities of DF: An Assessment of Different Methods

We shall first consider eight methods which give reasonably good estimates of DF in a range of plant foods, then describe four of them in greater detail and finally make some recommendations regarding choice of method for fresh and processed foods. The eight methods are: Southgate (1976, 1981), Theander and Aman (1979), Schweizer and Würsch (1979, 1981), Selvendran and DuPont (1980a, 1984), Englyst et al. (1982) and the modification by Englyst and Cummings (1984), Faulks and Timms (1985), Asp and Johansson (1981) and the modification by Asp et al. (1983), and Prosky et al. (AOAC method) (1984, 1985). The Asp and Prosky methods are gravimetric assays. Some of the above workers who formulated the early methods have progressively modified and improved their assays to cope with the requirements of different types of fresh and processed foodstuffs. Most of the methods except Faulks and Timms and Prosky et al. have already been discussed at some length by Selvendran and DuPont (1984). The reader is advised to consult this review which uses flow charts to compare the procedures. The modifications by Faulks and Timms (1985) and by Englyst and Cummings (1984) will be discussed later in this chapter.

Southgate was the pioneer of DF analysis. His method used glucoamylase to remove starch from alcohol extracted samples and in a later paper a preliminary gelatinization step was added. The starch-depleted residue was extracted with hot water. Non-cellulosic polysaccharides were determined colorimetrically after molar sulphuric acid hydrolysis, in both the aqueous extract and the insoluble residue. The cellulose content of the insoluble residue was determined after treatment with 72% H_2SO_4 and the mass of the final residue gave a measure of Klason lignin (Southgate et al. 1978). The earlier method worked well for starch-poor tissues but gave elevated values of DF for starch-rich foods, both fresh and processed (e.g. rye-flour, rye biscuit and potato powder), because starch was not completely removed. Complete removal of starch has proved to be very difficult, even following gelatinization, and the Southgate method is to be commended. The currently recommended method of Southgate has been used by Finglas and Faulks (1984) to determine the DF content of fresh and cooked potatoes, and the above reference should be consulted for details of the method. Many of the newer methods are modifications of Southgate's procedure in which (a) the starch is removed more effectively using alternative starch degrading enzymes and (b) the sugars are analyzed by GC.

The Theander and Aman method together with the Faulks and Timms modification will be discussed in detail later. Briefly, Theander and Aman removed the bulk of the starch using Termamyl, a heat-stable α-amylase at 80° C. Neilson and Marlett (1983) have found preliminary ultrasonication of the sample solution helpful. They report that residual starch (as large dextrins) represents only 2% –3% of the original starch content of most foods, but Faulks and Timms (1985) report much higher values. ~10% for residual starch from starch-rich foods. Theander and Aman removed the dextrins using amyloglucosidase. Faulks and Timms precipitated the Termamyl-treated residue with 80% (v/v) alcohol and dispersed it in DMSO which solubilizes starch and dextrins before adding buffer and the amyloglucosidase.

Schweizer and Würsch autoclaved the extractive-free material at 130° C for 20 min; removed the co-precipitated proteins with pepsin at pH 1.5, and then removed the starch by incubating with pancreatin (glucoamylase) for 18 h at 37° C. This method has been used to determine the DF content of a range of foods. The procedures for removing protein and starch appear to be very effective and it is likely that the enzymic degradation of proteins facilitates the subsequent starch hydrolysis.

Selvendran and DuPont ball-milled the extractive-free material to disrupt the tissue structure completely. Starch was removed after gelatinization by treatment with hog pancreatic α-amylase and pullulanase at 37° C for 17 h. Pancreatic α-amylase was chosen because it has no hemicellulase activity; pullulanase is needed to ensure "complete" hydrolysis of amylopectins. The small amount of residual starch in the final DF preparations of certain starch-rich products, such as potato and potato powder (Selvendran and DuPont 1980a), has been traced to precipitated dextrins. During starch hydrolysis dextrins of ~60 glucose residues tend to associated and come out of solution. This can be avoided by (a) using an excess of α-amylase; (b) ensuring that the temperature of the reaction mixture does not fall below 40° C and (c) more vigorous mixing as has been achieved by Englyst et al. (1982). Englyst continuously inverted the reaction tube during incubation. Our enzymic hydrolysis was carried out on a shaking water bath (at 37° C) which, being less vigorous, reduced but did not entirely prevent the formation of small amounts of "dextrin crystallites".

Englyst et al. (1982) have successfully removed most, if not all, of the starch from plant products including a number of starch-rich, fresh plant foods. They treated the gelatinized material with hog pancreatic α-amylase and pullulanase as we had done, but continuously inverted the reaction tube during incubation using a specially developed apparatus. The use of this mixing device and the precisely defined incubation temperature of 42° C suggest that careful work has gone into devising conditions under which starch could be more completely removed. The way in which the problem has been overcome by careful experimentation is commendable but some account in their methodology of the search for the solution would have been helpful. Chen and Anderson (1981) have used a similar procedure to obtain the DF content of a range of plant foods. Any resistant starch which was not degraded by α-amylase and pullulanase was, in the Englyst et al. (1982) procedure, measured as glucose after solubilization in alkali and hydrolysis with amyloglucosidase. In subsequent work (Englyst and Cummings 1984) hot DMSO was used to solubilize all the starch, including resistant starch, prior to enzymic treatment. Using this modification, continuous inversion of the reaction mixture was found to be unnecessary.

In the gravimetric methods, proteins and starch are enzymically digested in vitro and the mass of the undigested material (corrected for residual proteins and ash) gives the total DF. In nutritional studies the only enzymes used were pepsin and pancreatin. The principles of the "gravimetric method" were proposed by Williams and Olmsted (1935) and further developed by Hellendoorn et al. (1975), Asp and co-workers (1983) and several others. The first action AOAC method incorporates developments from several laboratories, as will become apparent later. Prosky (1986) lists the workers whose methods have been incorporated.

Asp and Johansson (1981) improved on the method of Hellendoorn et al. (1975) by including a gelatinization step before enzymic digestion by pepsin and pancreatin. Preliminary gelatinization allowed for more efficient degradation of starch. In a later paper Asp et al. (1983) used an initial treatment with Termamyl for 15 min at 100° to gelatinize and (partially) degrade starch in a single step. The suspension was then sequentially treated with pepsin and pancreatin in 0.1 M phosphate buffer at pH 1.5 and 6.8 respectively and filtered through a crucible containing Celite. The filtrate was fractionally precipitated with 80% (v/v) alcohol and both precipitates were weighed. It should be noted that the 80% alcohol causes co-precipitation of phosphate with the polymeric material. Both Schweizer et al. (1984) and Frølich et al. (1984) have studied this, and based on their combined experience a weaker phosphate buffer (0.05 M) was suggested for the second action AOAC method.

3.1 Observations on the Procedure of Theander and Co-workers and the Modification by Faulks and Timms

The original procedure of Theander and Aman (1979) has been slightly modified by Theander and Westerlund (1986). This method measures total DF as the sum of the water-soluble and water-insoluble DF fractions. The two fractions are analyzed for neutral sugars by GC of alditol acetates and for uronic acids by a rapid decarboxylation method.

The essential points to note are: (1) the dry plant material (or 80% ethanol-insoluble residue in the case of materials such as fruits and vegetables, which have a high water content) (1–3 g) is finely subdivided using a Cyclotec Mill (Tecator AB, Hoganas, Sweden) with a 0.5-mm screen; (2) the milled product is extracted with 80% ethanol and then refluxed with hexane to remove sugars, lipids and other extractives. In the earlier version chloroform was used instead of hexane. Extraction with hexane could be omitted for products containing less than 5% lipids. Light petroleum ether (bp 60° to 70°) is cheaper than hexane and can also be used to remove lipids; (3) starch is removed from the extractive-free residue by sequential treatments with Termamyl 120L (Novo AS) and amyloglucosidase (from Aspergillus Niger). In the original procedure the material was gelatinized at 85° C for 30 min before treating with Termamyl at 85° C for 45 min. In the revised method the gelatinized suspension is brought to 80°–90° C before adding Termamyl in order to minimize side-reactions, the reaction mixture is then heated in a boiling water bath (96° C) for 30 min. The suspension is then cooled to 60° C and treated with amyloglucosidase at 60° C for 16 h (overnight), although "complete" hydrolysis is achieved after 6 h. The hydrolyzed starch can be estimated as glucose by the glucose oxidase method. It is useful to note that both Termamyl and pancreatic α-amylase break starch down to maltose and dextrins only, and amyloglucosidase treatment is necessary to complete the hydrolysis to glucose; (4) the suspension is centrifuged and the supernatant precipatated with 4 vol, of absolute ethanol to give the water-soluble DF residue, which is also isolated by centrifugation. Both precipitates are washed with ethanol and acetone; (5) the neutral sugars, liberated by 12 M H_2SO_4 hydrolysis at 30° C for 1 h and 0.4 M H_2SO_4 hydrolysis at 125° C for a further 1 h, are determined by GC. CO_2 is released from samples containing 1–20 mg UA by refluxing with hydroiodic acid under nitrogen for 30 min, trapped in a cell containing dilute sodium hydroxide and measured potentiometrically; for details of the method see the original references.

In this procedure the bulk of the starch is hydrolyzed to low-molecular-weight dextrins by the Termamyl treatment. Termamyl is a heat-stable endo-α-amylase and under the conditions used does not release detectable amounts of sugars from barley β-D-glucans, barley straw arabinoxylans or cotton cellulose. The work of Englyst et al. (1982) suggests that Termamyl contains appreciable β-glucanase activity at 40° C. The hemicellulases are probably inactivated at 90° C, and therefore Theander and Westerlund (1986) recommend adding the Termamyl only after the suspension has reached 80°–90° C. The amyloglucosidase treatment at 60° C then completes the hydrolysis of the larger dextrins which would, otherwise, contaminate the final DF preparation(s). Incubating at 60° C during the amyloglucosidase step minimizes association of the dextrins which might cause precipitation and incomplete degradation. The starch content of food products can be estimated by glucose analysis of an aliquot of the supernatant. Theander and Westerlund have reported that the amyloglucosidase from Aspergillus Niger is free of hemicellulase activity. Boehringer amyloglucosidase E.C. 3.2.1.3. contains no detectable β-glucanase activity (Selvendran and DuPont 1980a). Since Termamyl and amyloglucosidase are used extensively in DF analysis, some additional checks on the absence of hemicellulase activity under the recommended conditions would be desirable. Newcomers to DF analysis are advised to check the enzyme-degraded materials for residual starch using iodine / potassium iodide before proceeding to analyze for sugars and uronic acids. It should be noted that in starch-rich processed foods a proportion of the retrograded starch will survive both enzyme treatments. Faulks and Timms have developed a more rapid procedure for DF analysis which removes both "soluble" and retrograded starch. Their method incorporates ideas from the methods of Theander and Aman (1979) and Englyst and Cummings (1984) and is shown in Fig. 1. The bulk of the starch is removed from the gelatinized material using Termamyl at 100° C for 15 min; further treatment did not increase starch removal. The cooled mixture is made up to 80% ethanol and the precipitate removed by centrifugation. The precipitate is then dispersed in DMSO at 100° C, suspended in buffer and incubated with amyloglucosidase at 37° C for 35 min: the DMSO solubilizes retrograded starch and both this and the residual dextrins are hydrolyzed. The extract is finally treated with 80% ethanol as before and the precipitate dried in the oven at 100° C. We have found (R. R. Selvendran and S. E. King, unpublished results) that freeze-drying the final precipitate renders it more easily dispersable in dilute acids, for

200 mg dry AIR

Add 10 ml 0.1 M Tris-maleate buffer, pH 6.7
Mix, 100° C, 10 min

Add 0.2 ml diluted (× 25) Termamyl 120 L
Continue heating and mixing, 15 min

Add 40 ml absolute EtOH
Cool, 0° C, 30 min

Centrifuge, discard supernatant

↓

Residue

Add 2 ml DMSO
Mix 100° C, 5 min

Add 8 ml 0.1 M acetate buffer pH 6 and
0.1 ml amyloglucosidase
Mix, 37° C, 35 min

Add 40 ml absolute EtOH
Cool, 0° C, 30 min

Centrifuge, discard supernatant

↓

Residue

Dry, 100° C

Add 2 ml 12 M H_2SO_4
Mix, 35° C, 1 h

Add 22 ml H_2O
Mix, 100° C, 2 h – remove 10 ml aliquot
after 1 h for UA analysis

↓

Measure total sugars colorimetrically
or individual neutral sugars by GC

Fig. 1. Procedure of Faulks and Timms

hydrolysis prior to sugar analysis, and the sample is more suitable for methylation analysis.

Faulks and Timms have found that with starch-rich products such as white flour starch hydrolysis is rapid but stops short of 100% (at ~90%), after 15 min. This may indicate that Termamyl is inhibited by its hydrolysis products since the 80% ethanol precipitate can be destarched further by a second 15-min Termamyl treatment. Faulks' and Timms' results support previous findings that amyloglucosidase treatment of the Termamyl-treated residue is necessary to effect "complete" removal of starch.

3.2 Observations on the Procedure of Englyst et al.

In the original procedure of Englyst (1981) starch was removed from the gelatinized food sample using amyloglucosidase for 16 h at 45°. In a subsequent paper Englyst et al. (1982) recommended the use of pancreatic α-amylase and pullulanase, for 16 h at 42° C, in place of amyloglucosidase. Pancreatic α-amylase and pullulanase were recommended by Selvendran and DuPont (1980 a, b) for effec-

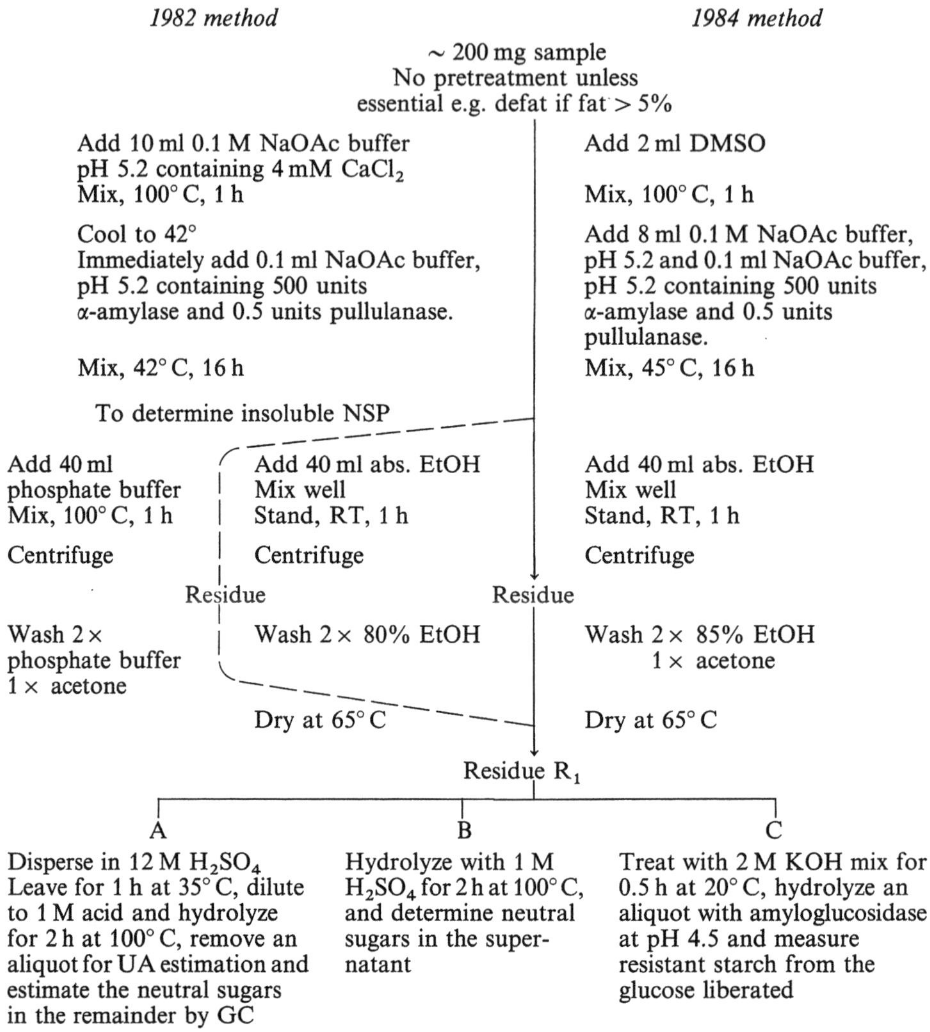

Fig. 2. Procedure of Englyst et al.

tive removal of starch from starch-rich foods. Englyst and Cummings (1984) have further modified their procedure by incorporating treatment with dimethyl sulphoxide (DMSO) for 1 h at 100° C before treatment with starch-degrading enzymes. Treatment with DMSO ensures solubilization of "resistant" starch. We shall first discuss the procedure of Englyst et al. (1982) for isolating and analyzing the non-starch polysaccharides (NSP) of plant foods. This method is shown schematically in Fig. 2, together with the amounts of material and reactants used.

The essential points to note are: (1) as far as possible the foods should be analyzed without any pretreatment. Foods with low water content may be ball-milled briefly (3 min), and those with high water content homogenized. Presumably, the authors meant milled, because ball-milling for 3 min will achieve no subdivision whatsoever; we have found that ball-milling the cell walls of wheat bran and apples in 90% ethanol for 15 h at 2° does not cause detectable degradation of cell-wall polymers (Selvendran and DuPont 1984). Treatment with acetone is recommended for removal of lipids, but in our view chloroform / methanol (1 : 1 v/v) or hexane / methanol is to be preferred for the quantitive removal of lipids. (2) The dry sample (100–200 mg) is taken up in acetate buffer (made 4 mM with $CaCl_2$ to stabilize and activate the α-amylase); gelatinized for 1 h at 100° C with continuous stirring; cooled to 42° C and treated with 0.1 ml enzyme solution for 16 h at 42° C (strength of enzyme solution = 5000 units of α-amylase and 5 units of pullulanase / ml of acetate buffer, pH 5.2). The need to use 500–800 units of α-amylase to "completely" hydrolyze starch under the recommended conditions has been confirmed (Englyst and Cummings 1984). The temperature of the incubation medium must not drop below 42° C. Ring et al. (1988) have shown that when gelatinized starch is stored at 20° C, retrogradation of amylose causes a significant proportion of it to become resistant to amylolysis with time. A large proportion of the retrograded (resistant) starch in processed foods is not enzymically degraded unless it is first solubilized by either alkali or (hot) DMSO. (3) The enzyme-treated suspension is treated with absolute alcohol to 80% (v/v) to precipitate the buffer-soluble and -insoluble NSP together. To isolate the insoluble NSP, the suspension is treated with phosphate buffer (40 ml, 0.2 M, pH 7) mixed for 1 h at 100° C, cooled, and centrifuged. (4) To obtain a value for resistant starch (starch not hydrolyzed by α-amylase and pullulanase) a portion of the residue (R_1) is dispersed in 2 M KOH, neutralized with 2 M HOAc and treated with amyloglucosidase (in acetate buffer containing 20 mM $CaCl_2$) for 1 h at 65° C and glucose determined in an aliquot of the supernatant; it may be an advantage to include 10 mM $NaBH_4$ in the 2 M KOH to eliminate "peeling" of the reducing ends of the starch molecules.

In a subsequent paper Englyst and Cummings (1984) have modified the earlier method by dispersing the food sample in DMSO (2 ml) for 1 h at 100° C, prior to treatment with buffer (8 ml, pH 5.2) containing the starch degrading enzymes for 16 h at 45° C; the enzymes are active in aq. DMSO.

The method has the advantage that (most of) the resistant starch is solubilized by hot DMSO along with the normal starch and can therefore be degraded enzymically under the conditions used, obviating the need to measure and correct for it separately. We are of the opinion that since resistant starch is not degraded by alimentary enzymes and will therefore survive into the large intestine to be degraded by colonic bacteria, it may be an advantage to obtain a measure of resistant starch in processed foods as in the earlier method.

There is still some debate as to whether resistant starch should be included under the broader definition of the term DF, and the reader is advised to consult the following papers for further information Berry (1986); Englyst and McFarlane (1986); Ring et al. (1988); Faulks et al. (1989).

3.3 Observations on the AOAC Method

The AOAC gravimetric method (Prosky et al. 1984) was used in an interlaboratory study to determine the DF content of a range of starch-rich (oats, potatoes, rice, rye bread and whole-wheat flour) and starch-poor (corn bran, wheat bran and iceberg lettuce) products. In general, where the DF content of the samples was greater than 7 g/100 g dry weight, there appears to be good correlation between the DF values obtained using the AOAC method and those of Theander and Aman (1979) and Englyst et al. (1982) (see Theander and Westerlund 1986; Cummings and Englyst 1987). But, for samples containing less than 5 g/100 g dry weight, the values obtained by the different methods varied; particularly where the samples were high in proteins (e.g. soya isolate) and starch (e.g. rice). It should be remembered, when making comparisons, that the Englyst procedure does not include values for lignin. One of us (RMF) was involved in the AOAC trial, and we shall make some comments based on our experience that may be useful to other workers.

The principle of the AOAC method as outlined by Prosky et al. (1984) is as follows: duplicate samples of dried food were defatted if they contained more than 5% fat; gelatinized in Termamyl to degrade the bulk of the starch and then digested with protease and amyloglucosidase to remove the bulk of the proteins and residual high-molecular-weight dextrins respectively. It is essential that the gelatinization should be performed at 95°–100° C. Using a water bath, the mixture only reaches 85° C and this seems to result in a slow filtration rate (B. W. Li, private comm.). The higher temperature may be attained using a steam bath. The soluble and insoluble DF were precipitated using 4 vol of 95% ethanol; filtered, washed sequentially with 75% ethanol, 95% ethanol and acetone; dried and weighed. One of the duplicates was analyzed for proteins and the other was incinerated at 525° C to determine the ash. The mass of the residue less the masses of protein and ash gave the total DF. Figure 3 gives an outline of the method. The contributions of Theander and Aman (1979), Asp and Johansson (1981), Asp et al. (1983), Schweizer and Würsch (1979), and other researchers are evident. It should be noted that the values for DF obtained by the above method would include "resistant starch" which could be a significant amount for some processed starch-rich foods. The "resistant starch" could be digested enzymically if the defatted samples are first dispersed in hot DMSO as in the Englyst and Cummings (1984) procedure.

In a subsequent collaborative study Prosky et al. recommended the use of a weaker phosphate buffer (0.05 M) to minimize the co-precipitation of phosphate. Acetate buffer does not co-precipitate and might be preferred, but the optimum pH for protease is ~7 and acetate is therefore unsuitable. Based on the work of Faulks and Timms (1985), a tris-maleate buffer (pH 6.7–7.5) may be recommended because there is no apparent co-precipitation with the DF when the mixture is treated with EtOH to 80% (v/v). Stevens and Selvendran (1980) have found, however, that on incubation of cabbage fibre preparations with Tris buffer, the triethylethanolamine bound strongly to the fibre and could not be removed even using phenol / acetic acid / water. The Tris presumably binds to the acidic, pectic polysaccharides. Tris bound to DF would give inflated values for nitrogen and hence incorrect estimates of residual protein.

Problems arose with certain products, depending on the types of DF polymers present, during filtration of the alcohol-treated suspension (~400 ml) through a Celite-bed. In order to retain the integrity of the Celite bed, it was found necessary

1984 original method	*1985 modifications*
Duplicate 1 g solid samples defatted with 40:60 petroleum ether	
Disperse in 50 ml 1 M phosphate buffer pH 6.0	50 ml 0.05 M phosphate buffer
Add 100 μl Termamyl 120 L	
Heat on shaking water bath 100°, 30 min	Preheat solution before adding Termamyl
Cool to room temperature	
Add 0.285 M NaOH to pH 7.5 (10 ml)	
Add 5 mg protease as a dry powder or a suspension in phosphate buffer at pH 7.5	
Incubate, 60°, 30 min	
At 60° add 0.329 M H_3PO_4 to pH 4.5±0.2 (10 ml)	
Add 0.3 ml amyloglucosidase	
Incubate, 60°, 30 min and treat with 4 volumes of 95% ethanol.	
Allow to precipitate at room temperature, 1 h	
Filter though pre-weighed sinter containing celite filter bed	
Residues 1 and 2	
Wash	
Dry, 105°, 18 h or under vacuum, 70°, 18 h	
Analyse residue 1 for ash	
Analyse residue 2 for protein	
DF = mean mass of dry residues 1 and 2 less mass of protein and mass of ash	

Fig. 3. AOAC procedure

to wet the Celite thoroughly with 75% EtOH and suck dry. Care must be taken during drying to ensure that the bed does not crack or peel away from the edges allowing the filtrate to pass straight through the glass sinter. The suspension derived from wheat bran filtered in 20–30 min, whereas that from oats (containing a significant amount of viscous β-glucans) took ~1.5–2 h. Filtration is made easier by allowing the suspension to settle for ~15 min before filtering. The longer the filtration time, the more phosphate is precipitated and the correction for ash thus needs to be increased with filtration time. We have found that when filtration becomes slow, scraping the surface of the Celite-bed gently to expose "new" Celite aids filtration. We prefer this to back-flushing the bed with nitrogen, which results in the loss of some DF polymers in the filtrate, especially whilst the bed is resettling. Co-precipitation of phosphate with DF also contributes to a slow filtration rate. The total amount of phosphate as NaH_2PO_4 in the revised assay medium is ~0.7 g. If only 5% of the phosphate co-precipitated with the fibre, that would amount to ~35 mg – a comparable mass with the DF component of a 1 g sample containing 3.5 g fibre / 100 g dry weight. The problems of co-precipitation and filtration are thus compounded in samples with a low DF content. In our hands, separate filtration of the insoluble DF and the buffer soluble DF (after precipitation with alcohol) minimized the filtration problems and gave lower correction factors for ash.

The analysts participating in the AOAC trial (Prosky et al. 1984) found most difficulty with starch removal and with the filtration step. Soya isolate is rich in protein and low in DF and proved particularly difficult to analyze. The problems were attributed to incomplete removal of proteins by protease. In foods, such as the soya isolate and rice, where DF represents only a small proportion of the total solids, small errors in nitrogen determination and the use of a standard protein conversion factor could result in inaccurate estimation of DF. The ash estimate may be inaccurate and if, as suggested above, the phosphate contribution alone is of the same order of mass as the total DF, errors in DF estimation could be large. In the second AOAC trial (Prosky et al. 1985) several changes were included, the most significant being use of the more dilute (0.05 M) phosphate buffer. The reproducibility was greatly improved over the previous trial, although DF values of soya isolate and rice were overestimated. This suggests that the method needs further improvement, particularly for products with low DF content.

Li and Andrews (1988) have devised a simplified enzymatic-gravimetric procedure. They do not remove protein but rely on the nitrogen determination to account for all the protein in the samples. The method uses a single enzyme – amyloglucosidase – instead of the three, reducing both cost and analysis time. A single 4 M acetate buffer, pH 4.8, avoids the three pH changes in the AOAC method.

Duplicate samples and two "blanks" were taken through the procedure. Samples containing more that 5% fat were defatted in hexane. Gelatinization was performed in aqueous suspension in an autoclave at 121° C for 1 h. The samples were cooled to 65°–70° C and starch was removed quantitatively by incubating with amyloglucosidase (No. A-9913, Sigma) in acetate buffer at 55° C for 2 h. The contents were transferred to beakers and treated with 95% ethanol to obtain 75% ethanol final concentration, and the precipitates allowed to settle for 1 h. The alcohol-insoluble precipitates were collected on fritted Celite crucibles as in the AOAC method, dried at 105° C overnight, cooled in a dessicator for 2 h and weighed. One crucible was ashed at 525° C and the other analyzed for total protein. The equations for calculating total DF are as follows:

$$\text{Blank} = \text{mg BR} - \left(\frac{\%\ \text{protein BR} + \%\ \text{ash BR}}{100}\right) \times \text{mg BR}$$

Total DF (%)

$$= \left\{\frac{\text{mg SR} - \left(\frac{(\%)\ \text{protein SR} + \%\ \text{ash SR}}{100}\right) \times \text{mg SR} - \text{av blank}}{\text{mg sample}}\right\} \times 100$$

where BR = blank residue and SR = sample residue.

The preparation of the crucibles should be noted carefully since this influences the speed of filtration and reproducibility of results. The Celite was made wet and redistributed using 78% EtOH to form the bed. It was sucked dry to form an even mat and then ashed at 525° C for 1 h at least 1 day before use, cooled to 100° C and then cooled to room temperature in a dessicator, releasing the pressure from time to time. The crucible was used straight from the dessicator. This method of preparing the crucibles could also be applied to the AOAC method (B. W. Li, private comm.).

3.4 Analysis of Sugars in DF Preparations

3.4.1 Hydrolysis Conditions as Described by Selvendran et al. (1979)

Two hydrolytic procedures are usually used in our lab to release sugars from DF preparations; (i) M H_2SO_4 hydrolysis for 2.5 h at 100° C and (ii) a modification of the Saeman et al. (1963) treatment with 72% (w/w) H_2SO_4 (12 M) for 3 h, followed by dilution to M acid and hydrolysis for 2.5 h at 100° C. The details and relative merits of these procedures have already been described and discussed. We have also treated DF and CW preparations with 72% H_2SO_4 at 20° C for 2 h prior to post-hydrolysis, following the paper by Rasper et al. (1981). Compared with our 3 h treatment, the 2 h pre-hydrolysis gave an increase (~10%) in the yield of xylose from DF preparations of cereal, but not vegetable products. The latter are usually low in lignified fibres, which have significant amounts of acidic xylans. Our 3 h treatment increased however, the yield of cellulose by ~8%–10% showing that the 2 h treatment with 72% H_2SO_4 may not be sufficient to solubilize cellulose completely. Prolonged exposure to 72% H_2SO_4, in order to release glucose quantitatively from cellulose, needs to be balanced against the losses it causes in other sugars, particularly the pentoses.

Using DF and CW preparations from samples containing significant amounts of lignified tissues (runner bean pods, cabbage leaves and wheat bran) and seed coats (pea hulls and soy bran), we found consistently higher values for xylose released under Saeman-hydrolysis conditions than under direct M H_2SO_4 hydrolysis: there was a small decrease in the level of arabinose. Samples containing negligible amounts of lignified tissues, such as DF from potato parenchyma or white wheat flour, showed consistently (slightly) higher values for arabinose and xylose released by M H_2SO_4 than by Saeman-hydrolysis. The following figures for xylose released under Saeman hydrolysis conditions as a percentage of xylose released by M H_2SO_4 illustrate the difference between lignified and unlignified tissues: pea hulls (no phloroglucinol-HCl detectable lignin) 178%; soya bran 181%; runner bean 115%; potato 105%; beeswing bran 101%; wheat bran 97%; wheat (coarse milled) 94%; wheat flour 98% (R. R. Selvendran and M. S. DuPont, unpublished results). The results suggest that two factors need to be borne in mind when selecting hydrolysis conditions for the quantitative release of sugars from DF preparations: (1) Saeman-hydrolysis solubilizes all the DF polymers but causes a small loss in the pentoses that are readily solubilized from CW of non-lignified tissues. The increased yield of xylose from lignified (vegetable) tissues compensates for the losses from non-lignified material. (2) M H_2SO_4 hydrolysis is probably preferable for the quantitative hydrolysis of non-cellulosic polysaccharides (NCP) from DF preparations containing little or no lignified fibre.

3.4.2 Hydrolysis Conditions as Described by Englyst et al. (1982)

Englyst et al. (1982) carried out the modified Saeman-hydrolysis as follows: the DF preparation (prepared from 100–200 mg of food sample) was dispersed in 1 ml of 12 M (72% w/w) H_2SO_4 and left at 35° C for 1 h to solubilize all the polysaccharides including cellulose. The slurry was diluted with 11 ml distilled water and hydrolyzed in a boiling water bath for 2 h with continuous mixing. One of us (RMF) has used the Saeman hydrolysis conditions just described (Method A) and hydrolysis with M H_2SO_4 for 2 h at 100° C (Method B) to test the release of sugars from NCP of a range of cereal based food samples. The sugars released were analyzed as alditol acetates. In every case it appears that (1) the amount of xylose released by method B was significantly greater (15%–20%), and sometimes up to 30%–40% greater than that released by method A; (2) the amount of arabinose released was 10%–15% greater using method B; (3) the galactose level was highly comparable using both methods of hydrolysis. (1) and (2) may indicate that the arabinoxylans from the DF of endospermous tissues are readily solubilized and partially degraded by the conditions of method A, whereas those from DF of lignified tissues are released more slowly and suffer smaller losses by degradation. The value A as a percentage of value B is given in Table 1 for xylose, arabinose and galactose of some of the breakfast cereals examined.

We have subsequently verified these observations using DF preparations from wheat bran and wheat flour (R. R. Selvendran and S. E. King, unpublished results). Treatment with 12 M H_2SO_4 at 35° C causes significant losses of pentoses, notably xylose, in the cereal products examined. This point should be considered when the procedure of Englyst et al. (1982) is used to determine NSP of DF preparations. We would recommend treatment with 12 M H_2SO_4 for 2 to 3 h at 20° C followed by dilution to molar acid strength and hydrolysis at 100° C for 2.5 h. The sulphuric acid may be removed by treatment with 0.25 M $Ba(OH)_2$ with continuous stirring (Selvendran et al. 1979) or with fast reacting solid $BaCO_3$ (Englyst et al. 1982).

3.4.3 Reduction and Acetylation of Sugar Mixtures

The procedures for reduction and acetylation of sugars, used routinely in our laboratory, are given in Selvendran et al. (1979). Englyst et al. (1982) have slightly modified the procedure and do not remove the sodium ions with Dowex 50 × 8 resin before co-distilling boric acid with methanol. More recently, Blakeney et al. (1983) have described a simple, rapid procedure for preparing alditol acetates from CW polysaccharides. Treatment with 72% H_2SO_4 at 20° C lasts only 45 min. The sugars liberated on acid hydrolysis are reduced to alditols by sodium borohydride in DMSO and acetylated in the presence of a 1-methyl-imidazole catalyst. Removal of borate is unnecessary and acetylation is completed in 10 min at room temperature. We have compared the Blakeney et al. method with our own, using a selection of CW preparations. We found: (1) Only 63% of cellulose was hydrolyzed to glucose after a 45-min treatment of CW from lignified tissues of runner bean pods with 72% H_2SO_4. The 3 h treatment gave 12% less xylose than the 45-min treatment, but five times as much arabinose. The bulk of the additional arabinose released by the longer treatment probably originated not in

Table 1. Sugars recovered from DF preparations by Method A expressed as a % of Method B[a]

Cereal products	Xylose	Arabinose	Galactose
All bran (Kelloggs, UK)	72	83	90
"Farm House" bran flakes	67	78	100
Branflakes (Kellogs, UK)	78	91	95
Shredded wheat	61	74	97
"Wheetabix" wheat biscuit	81	90	100
Round puffed wheat	82	89	94
Special K (Kelloggs, UK)	64	73	82
Oat-based cereal	73	85	100

[a] The conditions of hydrolysis are given under 3.4.2.

arabinoxylans, but in the small but significant amounts of pectins in runner bean pods that are heavily impregnated with lignin. The difference in arabinose values between the two treatments would not be significantly different for CW from parenchymatous tissues or even lignified tissues of cereal products (N. M. Rigby and R. R. Selvendran, unpublished results). Blakeney et al. used the virtually lignin-free CW of suspension-cultured Italian rye-grass endosperm for their studies. It is (even) possible that the release of glucose from the rye-grass cellulose was more or less quantitative. (2) The main advantages of the Blakeney et al. procedure are a) larger numbers of samples can be processed at one time since there are no filtration or precipitations steps and b) the method is twice as fast as others. (3) A disadvantage of the Blakeney et al. method is that the early running peaks on GC, such as those of rhamnose and fucose derivatives, become contaminated with some unknown compounds which have not been observed with the traditional method. Furthermore, a small amount of a compound (presumably DMSO) is extracted into the dichloromethane causing a tailing solvent peak, particularly with a packed column. The problem could be circumvented by using a capillary column or a good data-processing system coupled to the GC. We used a wall-coated open tubular OV225 capillary column, 15 m × 0.5 mm, 1.0 μm film thickness. This gives much better resolution and faster elution of the alditol acetates than the packed column (OV225, 4 m × 3 mm). The packed column is, however, more robust, and the top of the column can be repacked in the laboratory if it is contaminated after a series of injections. Following the procedures of Bittner et al. (1980) and Blakeney et al. (1983), Englyst and Cummings (1984) have used N-methylimidazole (NMetIm) to catalyze the acetylation of alditols. In this modified procedure they reduced the sugar mixture with $NaBH_4$ in aqueous ammonia at 40° C for 1 h, neutralized the ammonia with acetic acid (which also converts borate to boric acid) and acetylated using excess acetic anhydride in the presence of NMetIm at 20° C for 10 min. The excess acetic anhydride was then decomposed with water and 7.5 M KOH and the mixture cooled, partitioned with dichloromethane, vortexed, and centrifuged to separate the two phases. The lower, organic phase was drawn off and analyzed for alditol acetates. Partitioning the alditol acetates obviates the need to remove the borate (from $NaBH_4$) as boric acid.

3.4.4 Determination of Uronic Acid (UA) Content of CW and DF

In the DF context, total UA of preparations is best estimated by colorimetric methods or the accurate decarboxylation method described by Theander and Aman (1979). We have not used decarboxylation ourselves and are not in a position to comment on its merits. But Scott (1979) has made a comparison of UA values in natural products obtained by both colorimetry and decarboxylation.

In our earlier studies we determined the UA content of CW and DF preparations using an adaptation of the modified carbazole method of Bitter and Muir (1962), which is only applicable to UA and soluble UA-containing polymers. The pectic polymers were solubilized from CW (10 mg) by dispersing in 72% H_2SO_4 (0.82 g) at 20° C for 3 h, diluted to 6 ml (as in Saeman hydrolysis), filtered through Whatman GF/C glass fibre filter paper, and an aliquot of the filtrate taken for analysis. A small correction was made for the contribution from neutral sugars. In subsequent work we improved on this method in two ways: (1) the material dispersed in M acid, as above was heated at 100° C for 1–3 h before an aliquot was taken and analyzed. We recommend 1 h post-hydrolysis before filtration and UA analysis for CW and DF preparations from fruits and vegetables. After 0, 2, and 3 h post-hydrolysis of apple CW the UA content was, respectively, 89%, 95%, and 81% of the value after 1 h post-hydrolysis (Selvendran and DuPont, unpublished results). For DF from cereal products, which have relatively low levels of UA, post-hydrolysis may not be essential but 1 h is nevertheless recommended. (2) The m-phenylphenol (3-phenylphenol) method for the determination of UA (Blumenkrantz and Asboe-Hansen 1973) was later used instead of the carbazole method because it gives negligible interference from neutral sugars and no correction needs to be made for them. Scott (1979) has tested a range of phenolic reagents for hexuronic acid determinations. He has found both 3,5-dimethylphenol and 3-phenylphenol to be sensitive, but prefers the former. Colorimetric assay of UA runs into difficulty where DF preparations are incompletely soluble, but in general the values for UA are of the right order of magnitude.

Englyst et al. (1982) and Englyst and Cummings (1984) used a modification of the Scott (1979) procedure to estimate UA in DF samples. It is interesting to note that our procedure and the Scott (1979) method were developed independently and both involve dissolution of the acidic polysaccharide in strong acid before colorimetric assay. In the Scott procedure the sample (25–50 mg) is dispersed in 2–4 ml 72% H_2SO_4 at 50° C for 10 min, diluted with water to a volume estimated to contain 20–80 μg/ml UA, and an aliquot taken for analysis. In the modified procedure of Englyst and Cummings (1984) an aliquot of the sample dispersed in sulphuric acid is taken for UA analysis after 2 h post-hydrolysis at 100° C (see Fig. 2).

4 Concluding Remarks

Two main methods are recommended for analyzing DF to obtain information about the constituent polymers (1) Theander and co-workers' procedure has proved satisfactory for a number of food samples and its modification by Faulks and Timms has proved useful; but we now have reservations about the use of the Englyst et al. (1982) hydrolysis conditions, as explained above. The DF value obtained by Theander's method includes a contribution from resistant starch of processed foods. The resistant starch can be measured separately by dispersing the final residue in alkali and measuring the glucose released on treatment with glucanase-free amyloglucosidase, as described by Englyst et al. (1982). The Faulks and Timms modification gives a value for NSP which does not include resistant starch (for processed foods), because the resistant starch is solubilized using hot DMSO. Some useful hints have been given in Sect. 13.3.4 regarding the optimum conditions for hydrolysis of DF preparations and analysis of constituent sugars. (2) The procedures of Englyst et al. (1982) and Englyst and Cummings (1984) take longer to carry out but are more likely to give enzyme-degraded DF preparations in which all the NSP are retained because they have carefully chosen enzymes which have no hemicellulase activity. We do have some doubts, though, about the validity of the procedure used to hydrolyze the DF preparations, particularly those of cereal origin which are rich in xylose and arabinose. Recently, attention was drawn to the fact that when the enzymic digests of starch-rich foods are treated with ethanol to 75%–80% final concentration, in order to precipitate the DF-polymers small but significant amounts of glucose and other low-molecular-weight oligosaccharides are co-precipitated with the DF polymers. The co-precipitated glucose and oligosaccharides would give inflated values for the apparent glucan content of the DF preparations; the contaminants could, however, be removed by dialysis (J. A. Marlett, private comm.).

The AOAC method needs to be improved but the DF values may be more meaningful than Cummings and Englyst (1987) have given them credit. A modification of the first part of the AOAC method could be used to prepare gram quantitites of DF, virtually free from starch and the bulk of the intracellular proteins. Resistant starch from processed foods will, however, only be removed if treatment with hot DMSO is included. We have found that digestion with pronase removed intracellular proteins effectively (Stevens and Selvendran 1980). Alternatively, sequential extraction of the wet-ball-milled product with cold 1% detergent, phenol / acetic acid / water (2/1/1, w/v/v) and 90% aqueous DMSO gives DF from wheat bran, vegetables and unripe fruits cheaply, cleanly and in large quantities (Selvendran and DuPont 1984; Selvendran and O'Neill 1987). The (partially) purified, insoluble (and soluble) DF preparations could be sequentially extracted with aqueous inorganic solvents and the fractions further resolved by ion-exchange chromatography and examined by methylation analysis (Selvendran and O'Neil 1987; Selvendran and Stevens 1986).

The choice of sequential extraction procedures (see Fig. 10 of Selvendran and DuPont 1984, or Figs. 3 and 4 of Selvendran and O'Neil 1987) depends on the conditions used for starch (and protein) removal. If the material has already been gelatinized in buffer to remove starch, then some of the pectic polysaccharides would already have been degraded by transeliminative degradation and milder extraction conditions for depectination (Redgwell and Selvendran 1986; Selvendran and O'Neil 1987) need not be used. The recommended methods for preparing and analyzing relatively pure CW from vegetables (and unripe fruits) are described by Selvendran and O'Neil (1987). The review also outlines methods for analyzing CW of cereal products: the methods were largely developed by Stone and co-workers (Mares and Stone 1973; Bacic and Stone 1981).

In this chapter we have attempted to bring together some of the relevant information on polysaccharide analysis and those methods for DF analysis which are, in our opinion, reasonably reliable. We have refrained from making firm recommendations regarding the choice of a particular method because we believe there is still more work to be done, possibly on a collaborative basis.

References

Asp N-G, Johansson C-G (1981) Techniques for measuring dietary fiber: principal aims of methods and a comparison of results obtained by different techniques. In: James WPT, Theander O (eds) The analysis of dietary fiber in food. Marcel Dekker, New York, pp 173–189

Asp N-G, Johansson C-G (1984) Dietary fibre analysis. Rev Clin Nutr 54:735–752

Asp N-G, Johansson C-G, Hallmer H, Siljeström M (1983) Rapid enzymatic assay of insoluble and soluble dietary fiber. J Agric Food Chem 31:476–482

Aspinall GO (1973) Degradation of polysaccharides. In: Bentley KW, Kirby GW (eds) Techniques of chemistry, vol 4: Elucidation of organic structures by physical and chemical methods. Wiley-Interscience, New York, pp 379–450

Aspinall GO, Fanous HK (1984) Structural investigations on the nonstarchy polysaccharides of apples. Carbohydr Polymers 4:193–214

Aspinall GO, Jiang K-S (1974) Rapeseed hull pectin: Carbohydr Res 38:247–255

Bacic A, Stone BA (1981) Chemistry and organisation of aleurone cell wall components from wheat and barley. Aust J Plant Physiol 8:475–495

Berry CS (1986) Resistant starch: formation and measurement of starch that survives exhaustive digestion with amylolytic enzymes during the determination of dietary fibre. J Cereal Sci 4:301–314

Bitter T, Muir HM (1962) A modified uronic acid carbazole reaction. Anal Biochem 4:330–334

Bittner AS, Harris LE, Campbell WF (1980) Rapid N-methyl imidazole-catalysed acetylation of plant cell wall sugars. J Agric Food Chem 28:1242–1245

Blakeney AB, Harris PJ, Henry RJ, Stone BA (1983) A simple and rapid preparation of alditol acetates for monosaccharide analysis. Carbohydr Res 113:291–299

Blumenkrantz N, Asboe-Hansen G (1973) New method for quantitative determination of uronic acids. Anal Biochem 54:484–489

Chen WL, Anderson JW (1981) Soluble and insoluble plant fiber in selected cereals and vegetables. Am J Clin Nutr 34:1077–1082

Cummings JH, Englyst HN (1987) Development of methods for the measurement of dietary fibre in food. In: Mortin ID (ed) Cereals in a European context. Ellis Horwood, Chichester, UK, pp 188–220

Dudman WF, Whittle CP (1976) Interference by phthalic esters in the gas-chromatographic analysis of sugars. Carbohydr Res 46:267–272

Englyst HN (1981) Determination of carbohydrate and its composition in plant materials. In: James WPT, Theander O (eds) The analysis of dietary fiber in food. Marcel Dekker, New York, pp 71–93

Englyst HN, Cummings JH (1984) Simplified method for the measurement of total non-starch polysaccharides by gas-liquid chromatography of constituent sugars as alditol acetates. Analyst 109:937–942

Englyst H, Macfarlane GT (1986) Breakdown of resistant and readily digestible starch by human gut bacteria. J Sci Food Agric 37:699–706

Englyst HN, Wiggins HS, Cummings JH (1982) Determination of the non-starch polysaccharides in plant foods by gas-liquid chromatography of constituent sugars as alditol acetates. Analyst 107:307–318
Faulks RM, Timms SB (1985) A rapid method for determining the carbohydrate component of dietary fibre. Food Chem 17:273–287
Faulks RM, Southon S, Livesey G (1989) Utilisation of α-amylase resistant maize and pea starch in the rat. Br J Nutr (in press)
Finglas PM, Faulks RM (1984) Nutritional composition of UK retail potatoes, both raw and cooked. J Sci Food Agric 35:1347–1356
Freudenberg K, Neish AC (1968) In: Constitution and biosynthesis of lignin. Springer, Berlin Heidelberg New York
Frølich W, Schweizer TF, Asp N-G (1984) Minerals and phytate in the analysis of dietary fiber from cereals II. Cereal Chem 61:35–359
Hardell HL, Theander O (1970) Quantitative determination of carbohydrates in cellulosic materials. Losses as sulphates. Sven Papperstidn 73:291–293
Hellendoorn EW, Noordhuff MG, Slagman J (1975) Enzymatic determination of the indigestible residue (dietary fibre) content of human food. J Sci Food Agric 26:1461–1468
Henry RJ, Blakeney AB, Harris PJ, Stone BA (1983) Detection of neutral and amino sugars from glycoproteins and polysaccharides as their alditol acetates. J Chromatogr 256:419–427
Honda S, Takahashi M, Kakehik K, Ganno S (1981) Rapid automated analysis of monosaccharides by high-performance anion-exchange chromatography of borate complexes with fluorimetric detection using 2-cyanoacetamide. Anal Biochem 113:130–138
Honda S, Takahashi M, Shimada S, Kakehik K, Ganno S (1983) Automated analysis of alditols by anion-exchange chromatography with photometric and fluorometric postcolumn derivatisation. Anal Biochem 128:429–437
Hudson GJ, John PMV, Bailey BS, Southgate DAT (1976) The automated determination of carbohydrate. Development of a method for available carbohydrates and its application to foodstuffs. J Sci Food Agric 27:681–687
Johnson DB, Moore WE, Zank LC (1961) The spectrophotometric determination of lignin in small wood samples. Tappi 44:793–796
Lai YZ, Sarkanen KV (1971) Isolation and structural studies. In: Sarkanen KV, Ludwig CH (eds) Lignins: occurrence, formation, structure and reactions. Wiley-Interscience, New York, pp 165–240
Lehrfield J (1976) Separation of some perbenzoylated carbohydrates by high-performance liquid chromatography. J Chromatogr 120:141–147
Lever M (1972) A new method for colorimetric determination of carbohydrates. Anal Biochem 47:273–279
Li BW, Andrews KW (1988) Simplified method for the determination of total dietary fibre in foods. J Ass Off Anal Chem 71:1063–1064
Mares DJ, Stone BA (1973) Studies on wheat endosperm I composition and ultrastructure of the cell walls. Aust J Biol Sci 26:793–812
Morrison IM (1972a) A semi-micro method for the determination of lignin and its use in predicting the digestibility of forage crops. J Sci Food Agric 23:455–463
Morrison IM (1972b) Improvements in the acetyl bromide technique to determine lignin and digestibility and its application to legumes. J Sci Food Agric 23:1463–1469
Neilson MJ, Marlett JA (1983) A comparison between detergent and nondetergent analyses of dietary fiber in human foodstuffs, using high-performance liquid chromatography to measure neutral sugar composition. J Agric Food Chem 31:1342–1347
Prosky L (1986) Analysis of total dietary fiber: The collaborative study. In: Vahouny GV, Kritchevsky D (eds) Dietary fiber, basic and clinical aspects. Plenum, New York, pp 1–16
Prosky L, Harland BF (1985) Dietary fibre methodology. In: Trowell HC, Burkitt DF, Heaton KW (eds) Dietary fibre, fibre-depleted foods and diseases. Academic Press, London, pp 57–76

Prosky L, Asp N-G, Furda I, Devries JW, Schweizer TF, Harland BF (1984) Determination of total dietary fiber in foods, food products and total diets: interlaboratory study. J Assoc Off Anal Chem 67:1044–1052
Prosky L, Asp N-G, Furda I, Devries JW, Schweizer TF, Harland BF (1985) Determination of total dietary fiber in foods and food products: collaborative study. J Assoc Off Anal Chem 68:677–679
Pryde A, Gilbert MT (1979) Applications of high performance chromatography. Chapman & Hall, London
Rasper VF, Brillouet JM, Bertrand D, Mercier C (1981) Analysis of dietary fiber in faeces of rats fed with fiber-supplemented diets. J Food Sci 46:559–563
Redgwell RJ, Selvendran RR (1986) Structural features of cell wall polysaccharides of onion *Allium cepa*. Carbohydr Res 157:183–199
Reistad R, Fröhlich W (1984) Content and composition of dietary fibre in some fresh and cooked Norwegian vegetables. Food Chem 13:209–224
Ring SG, Selvendran RR (1978) Purification and methylation analysis of cell wall material from *Solanum tuberosum*. Phytochemistry 17:745–752
Ring SG, Gee JM, Whittam M, Orford P, Johnson IT (1988) Resistant starch: its chemical form in foodstuffs and effect in digestibility *in vitro*. Food Chem 28:97–109
Robertson JB, Van Soest PJ (1977) Dietary fiber estimation in concentrate feedstuffs. J Anim Sci 45:Suppl 1, 254
Robertson JB, Van Soest PJ (1981) The detergent system of analysis and its application to human foods. In: James WPT, Theander O (eds) The analysis of dietary fiber in food. Marcel Dekker, New York, pp 123–158
Saeman JF, Moore WE, Millet MA (1963) Sugar units present. Hydrolysis and quantitative paper chromatography. In: Whistler RL (ed) Methods in carbohydrate chemistry, vol 3, Chap 12. Academic Press, New York, pp 54–69
Sarkanen KV, Hergert HL (1971) Classification and distribution of lignins. In: Sarkanen KV, Ludwig CH (eds) Lignins: occurrence, formation, structure and reactions. Wiley Interscience, New York, pp 43–94
Schweizer TF, Würsch P (1979) Analysis of dietary fibre. J Sci Food Agric 30:613–619
Schweizer TF, Würsch P (1981) Analysis of dietary fiber. In: James WPT, Theander O (eds) The analysis of dietary fiber in food. Marcel Dekker, New York, pp 203–216
Schweizer TF, Frølich W, Del Vedovo S, Besson R (1984) Minerals and phytate in the analysis of dietary fiber from cereals I. Cereal Chem 61:116–119
Scott RW (1979) Colorimetric determination of hexuronic acids in plant materials. Anal Chem 51:936–941
Selvendran RR (1983) The chemistry of plant cell walls. In: Birch GG, Parker KJ (eds) Dietary fibre. Applied Science, London, pp 95–147
Selvendran RR (1984) The plant cell wall as a source of dietary fiber. Chemistry and structure. Am J Clin Nutr 39:320–337
Selvendran RR (1985) Developments in the chemistry and biochemistry of pectic and hemicellulosic polymers. J Cell Sci Suppl 2:51–88
Selvendran RR (1987) Chemistry of plant cell walls and dietary fibre. Scand J Gastroenterol Suppl 22:33–41
Selvendran RR, DuPont MS (1980 a) Simplified methods for the preparation and analysis of dietary fibre. J Sci Food Agric 31:1173–1182
Selvendran RR, DuPont MS (1980 b) An alternative method for the isolation and analysis of cell wall material from cereals. Cereal Chem 57:278–283
Selvendran RR, DuPont MS (1984) Problems associated with the analysis of dietary fibre and some recent developments. Dev Food Anal Tch 3:1–68
Selvendran RR, O'Neill MA (1987) Isolation and analysis of cell walls from plant material. In: Glick D (ed) Methods of biochemical analysis, vol 32. Wiley (Tube and Science), New York, pp 25–153
Selvendran RR, Stevens BJH (1986) Applications of mass spectrometry for the examination of pectic polysaccharides. Mod Methods Plant Anal New Ser (Linskens HF, Jackson JF, eds) 3:23–46

Selvendran RR, March JF, Ring SG (1979) Determination of aldoses and uronic acid content of vegetable fibre. Anal Biochem 96:282–292
Selvendran RR, Stevens BJH, O'Neill MA (1985) Developments in the isolation and analysis of cell walls from edible plants. In: Brett CT, Hillman JR (eds) Biochemistry of plant cell walls. SEB Seminar 28, Cambridge Univ Press, London, pp 39–78
Selvendran RR, Stevens BJH, DuPont MS (1987) Dietary fiber: chemistry, analysis and properties. In: Chichester CO, Mrak EM, Schweigert BS (eds) Advances in Food Research, vol 31. Academic Press, New York, pp 117–209
Slavin JL, Marlett JA (1983) Evaluation of high performance liquid chromatography for measurement of the neutral saccharides in neutral detergent fiber. J Agric Food Chem 31:467–471
Southgate DAT (1976) The chemistry of dietary fibre. In: Spiller GA, Amen RJ (eds) Fiber in human nutrition. Plenum, New York, pp 31–72
Southgate DAT (1981) Use of the Southgate method for unavailable carbohydrates in the measurement of dietary fiber. In: James WPT, Theander O (eds) The analysis of dietary fiber in food. Marcel Dekker, New York, pp 1–19
Southgate DAT, Englyst H (1985) Dietary fibre: chemistry, physical properties and analysis. In: Trowell H, Burkitt D, Heaton K (eds) Dietary fibre, fibre-depleted foods and disease. Academic Press, London, pp 31–55
Southgate DAT, Hudson GJ, Englyst H (1978) The analysis of dietary fibre, the choices for the analyst. J Sci Food Agric 29:979–998
Stevens BJH, Selvendran RR (1980) The isolation and analysis of cell wall material from the alcohol-insoluble residue of cabbage (*Brassica oleracea* var *capitata*) J Sci Food Agric 31:1257–1267
Stevens BJH, Selvendran RR (1988) Changes in composition and structure of wheat bran resulting from the action of human faecal bacteria in vitro. Carbohydr Res (in press)
Stevens BJH, Selvendran RR, Bayliss CE, Turner R (1988) The degradation of cell wall material of apple and wheat bran by human faecal bacteria in vitro. J Sci Food Agric 44:151–166
Taylor RL, Conrad HE (1972) Stoichiometric depolymerisation of polyuronides and glycosaminoglycuronans to monosaccharides following reduction of their carbodiimide-activated carboxyl groups. Biochemistry 11:1383–1388
Theander O, Aman P (1979) Studies on dietary fibre I. Analysis of water-soluble and water-insoluble dietary fibre. Swed J Agric Res 9:97–106
Theander O, Westerlund E (1986) Studies on dietary fiber 3. Improved procedures for analysis of dietary fiber. J Agric Food Chem 34:330–336
Trowell H, Southgate DAT, Wolever TMS, Leeds AR, Gassull MA, Jenkins DJA (1976) Dietary fibre redefined. Lancet 1:967
Van Soest PJ (1967 a) Development of a comprehensive system of feed analyses and its application to forages. J Anim Sci 26:119–128
Van Soest PJ (1967 b) Use of detergents in the analysis of fibrous feeds IV. Determination of plant cell wall constituents. JAOAC 50:50–55
Whistler RL, Richards GN (1958) Uronic acid fragments from slash pine (*Pinus elliottii*) and their behaviour in alkaline solution. J Am Chem Soc 80:4888–4891
Williams RD, Olmsted WH (1935) A biochemical method for determining indigestible residue (crude fibre) in faeces: lignin, cellulose and non-water soluble hemicelluloses. J Biol Chem 108:653–666

Methods Used in the Investigation of Insoluble Dietary Fiber

T. Yoshida and K. Kuwano

1 Introduction

Recent progress in research in the field of dietary fiber, mainly of plant origin, has led to a remarkable enrichment of our knowledge with respect to the physiological and nutritional functions of the fibrous components of food. It is well recognized that diets which are rich in dietary fiber are protective against a range of diseases, particularly those prevalent in Western countries.

Thus, analytical procedures and sample preparations for total content and the individual components of dietary fiber in foods and feeds are necessary to clarify the role of dietary fiber in animals, including man. In addition, investigations on their chemical and physical properties are basically important for studies on the physiological and nutritional function of dietary fiber. Therefore, the general methods for these studies are described.

Plant cell walls are a major component of dietary fiber, and they contain noncellulosic polysaccharides (hemicellulose), cellulose, lignin, pectin, gums, etc. (Southgate 1982). These indigestible components of plant foods are roughly divided into soluble and insoluble types, and hemicellulose, celluloses and lignin are classified as insoluble components (Hellendoorn 1978).

Recommended definitions for dietary fiber in the future will be toward chemical ones (Bingham 1987). If fiber is defined in terms of its chemical components, the principal one of which is nonstarch polysaccharide, lignin might not be included as a dietary fiber (Cummings 1987). However, lignin is traditionally one of the fibers in animal nutrition and is presumed to be indigestible by human enzymes. In this chapter, therefore, the term "insoluble dietary fiber" (IDF) is used principally to cover all of these insoluble components, and sometimes other types of IDF, e.g., chitin, are also included.

2 Sample Pretreatment for Determination and Preparation

Sample pretreatment is an important step in fiber analysis and preparation. If possible, samples should be prepared generally without any pretreatment.

Heat-drying of samples often produces browning substances or heat-denaturated substances, and these have a definite effect on fiber analysis. Van Soest (1965) reported that heat-drying of forage at temperatures above 50° C showed analytically significant increase in the yield of dietary fiber by the production of artifact lignin. Owing to the Maillard reaction products, vegetables dried at

Table 1. Effect of heating on IDF[a] values of mushrooms (% of dry matter)

Treatment	Sakurashimeji (*Hygrophorus russula*)	Mukitake (*Panellus serotinus*)	Urabenihoteishimeji (*Rhodophyllus crassipes*)
Control (Freeze-dried)	40.9	41.8	31.0
Heating of control (105° C, 6 h with air circulation)	43.7	43.4	33.0

[a] Insoluble dietary fiber.

Table 2. Effect of sterilization on IDF[a] values of a practical diet[b] for germfree rats and mice (% of dry basis)

Treatment	Hemicellulose (A)	Cellulose (B)	Lignin (C)	IDF (A+B+C)
Control (Pelletting with heat)	5.19	4.52	1.49	11.20
Autoclaving of control (121° C, 30 min)	5.34	4.88	1.55	11.77
Irradiation of control (5 Mrad)	5.28	4.51	1.47	11.26

[a] Insoluble dietary fiber; determined by modification of the method of Southgate (Mori and Aragane 1981).
[b] Diet L-485 (Practical diet based on corn-soybean meal, Kellog and Wostmann 1969).

100° C had two and a half to three times higher values of dietary fiber than those of the freeze-dried sample (Hoshi and Takehisa 1981). Tables 1 and 2 show that heating increased the dietary fiber values of mushrooms (Kurasawa et al. 1982) and of a mixed diet (Ishii et al. 1988).

Consequently, freeze-drying before fiber analysis is preferred to hot-air drying to facilitate treatment and prevent the alteration of high moisture samples, such as vegetables and fruits. Murakami and Takeyama (1978) observed 18%–96% accumulation of lignin in cut vegetables 1 day after cutting, which is a necessary procedure for freeze-drying the plant samples. Therefore, fresh samples should be treated rapidly to minimize the time between cutting and freeze-drying.

For the purpose of the simultaneous removal of the lipid, the color, and the substrates for polyphenoloxidase, acetone-dried powder obtained after repeated extraction and dehydration of the homogenized samples with three- to fourfold acetone is available for dietary fiber analysis not accompanied by other chemical components (see e.g., McConell et al. 1974). Southgate (1981) recommended extraction and dehydration of the samples with hot 85% (v/v) methanol followed by hot acetone to remove free sugars, lipid, most pigments, etc., but methanol is poisonous and penetrates the skin. Instead of methanol, therefore, 80% ethanol (v/v) is preferred (Theander and Westerlund 1986).

When there are problems in taking a representative sample, samples with a low moisture content may be powdered carefully for 2–3 min. Samples with a higher moisture content are homogenized and dried after freezing or by acetone, and then powdered (Englyst and Cummings 1984). Thereafter, these powdered materials are pretreated with hydrolytic enzymes to decompose the starch. Low-starch samples, such as hays (McQueen and Nicholson 1979), vegetables and fruits (Mori 1982) do not require amylase treatment. Selvendran et al. (1981) emphasized the importance of wet- and not dry-ball milling to decompose the plant cell substances accessible to solvents and enzymes.

Our procedure for the pretreatment of samples for IDF determination and preparation is as follows. All stages are performed at temperatures not exceeding 35° C.

1) Cereals, legumes, nuts and seeds.
   The samples are ground and sifted to pass a 1-mm screen using a Cyclotec mill (Tecator AB, Hoganass, Sweden), and defatted with ether if necessary.
2) Vegetables, fruits, fungi and algae.
   a) When components other than dietary fiber are analysed:
   The samples are freeze-dried, ground and sifted as described in 1), and defatted with ether if necessary.
   b) When only dietery fiber analysis or preparation of the IDF fraction to investigate its properties are required:
   The samples are homogenized with fourfold acetone, stirred for 30 min and filtered. Twofold acetone with respect to the fresh samples is added to the residue, stirred for 30 min and filtered. This procedure is repeated twice, and the residues are washed with ether equal in volume to the fresh samples, and air-dried on filter paper. Thereafter, the residues are ground and sifted the same as described in 1).
3) Common human diets.
   The same volumes of water are added to the samples and homogenized. The homogenized samples are freeze-dried, coarsely prepowdered, defatted, air-dried, reground, and sifted.
   All samples are stored in air-tight containers.

3 Analytical Methods for Determination

The crude fiber method (AOAC 1984) gave much lower values than the modern methods of fiber analysis, as shown in Table 3. These results were obtained from

Table 3. Comparison of fiber contents obtained by different analytical methods (% of dry matter)

Diet	Crude fiber	NDF[a]	ADF[b]	Hemicellulose	Cellulose	Lignin
L-485	3.88	15.46	6.78	8.68	4.14	2.64
L-488F	2.84	13.48	5.48	8.00	5.09	0.39

[a] Neutral detergent fiber.
[b] Acid detergent fiber.

the two kinds of autoclaved diet for laboratory animals (Yoshida and Ohkubo 1985); one was a practical diet based on corn-soybean meal (diet L-485, Kellogg and Wostmann 1969) and the other was a purified diet based on casein-corn starch (diet L-488F, Yoshida et al. 1982). The demand for accurate and rapid determinations of dietary fiber in foods has prompted investigations of alternatives to the crude fiber method. Since detailed methods of dietary fiber analysis are described by Dr. Selvendran, brief outlines of the three representative methods and applications of IDF analysis are shown in this section.

1) Detergent method of Van Soest
   This gravimetric method measures an insoluble residue after chemical solubilization of nonfiber constituents (Van Soest 1963 a, b; Van Soest and Wine 1967). Acid detergent fiber (ADF) gives an estimate of cellulose plus lignin. These two components may be estimated separately in the ADF fraction by removal of lignin with permanganate oxidation or of cellulose by strong sulfuric acid. Hemicellulose is estimated as the difference between neutral detergent fiber (NDF) and ADF.
   The ADF method is suitable for the determination and preparation of samples which contain mainly cellulose and lignin as IDF, and the NDF method is appropriate for those of total dietary fiber in samples which contain little water-soluble dietary fiber. Although there is a limitation to the applicable materials, the main advantages of the detergent method are simplicity and rapidity. ADF and NDF determinations can be completed in an 8-h day except for samples that contain significant amounts of starch (Baker et al. 1979).
   For determination of animal IDF such as chitin, which is a major component of the crab and shrimp exoskelton, the ADF method was acceptable (Stelmock et al. 1985). The NDF method modified by Kuwano et al. (1985) was also applied to the determination of antarctic krill chitin in the diet and faeces of rats (Kuwano et al. 1987).
2) Polysaccharide fractionation method of Southgate
   The Southgate method (1969) is based on fractionation into soluble and insoluble noncellulosic polysaccharides, cellulose and lignin, followed by quantitative analysis with colorimetry of the dietary fiber polysaccharides. Lignin is determined gravimetrically as Klason lignin (residue insoluble in 72% sulfuric acid). Before applying this method to dietary fiber determination, complete removal of starch, which is degraded by successive treatment with amylases and protease (Mori and Aragane 1981), is necessary.
   Complicated techniques and prolonged analyses, taking 6 days to complete the determination of one sample, are the inevitable disadvantages of this method. The Southgate method is suitable for fractionation and determination of the various fractions of dietary fiber with an appreciable accuracy, but cannot be applied for IDF preparation to investigate various properties of IDF. This method was refined by gas-liquid chromatography (Englyst et al. 1982; Englyst and Cummings 1984; Englyst and Hudson 1987).
3) Enzymatic gravimetric method
   Simple gravimetric methods using indigestible residue, after successive digestion with a thermostable alpha amylase, Termamyl 120 L (Novo Industrias,

Bagsvaerd, Denmark), protease and amyloglucosidase to remove the protein and starch, have been proposed (Asp et al. 1983; Prosky et al. 1984, 1985). If necessary, 95% ethanol is added to precipitate the soluble dietary fiber.

Careful treatment to avoid erroneously low values of IDF is required. Johansson et al. (1982) observed that recovery of hemicellulose was low with the enzymatic method using pepsin as a protease, evidently due to acid hydrolysis during the pepsin incubation step. Moreover, Furda (1981) showed that part of the hemicellulose was removed during boiling for 20 min with 0.005 M HCl. To dissolve protein in the samples, therefore, treatment by neutral protease is recommended and the pH should be higher than ca. 4.5 for digestion by other enzymes. Such mild conditions will also be useful for the collection of each fraction of IDF used for examination of their physicochemical properties.

4 Sample Preparation for the Investigation

As regards IDF sample preparation for IDF investigation, it is necessary to prepare ca. 1 g for the analysis of sugar components, 10 g or more for the examination of the physicochemical properties and 100 g or more for laboratory animal feeding studies.

4.1 Detergent Method

Because of the simple procedures involved, ADF and NDF methods (Van Soest 1963a, b) based on detergent treatment are recommended to prepare IDF samples. Moreover, since the same conditions as in the IDF assay are used for the procedure to produce the IDF samples, these methods can obtain IDF materials of the same composition as samples for the IDF assay.

Briefly, the procedure for preparation of IDF is as follows:

1) Preparation of the sample.
   See Sect. 2.
2) Detergent treatment.
   For details on the reagents for the ADF and NDF methods, see Chapt. by Dr. Selvendran.
   a) ADF sample.
   A pretreated sample of 0.5–1 g (0.5 g for low sp. gr. samples such as acetone powder of vegetables, and 1 g for high sp. gr. samples such as cereals) is boiled gently under reflux with a 500-ml flask in 100 ml of acid detergent solution for 1 h, filtered through a No. 2 glass filter, and washed with hot distilled water until no acidity is found in the filtrate.
   b) NDF sample
   For low-starch material, 0.5–1 g of pretreated sample [the same as described in a)] is added to 100 ml of the neutral detergent solution, 2 ml of decalin (decahydronapthalene) and 0.5 g of sodium sulfite. The solution is

refluxed for 60 min, timed from the onset of boiling. Thereafter the solution is filtered through a No. 2 glass filter and the residue is rinsed with 300 ml of hot distilled water.

In the case of high-starch material, Sakamaki et al. (1988) proposed the following procedures: 1 g of sample is weighed and placed in a 500-ml flask, to which 50 ml of 0.1 M phosphate buffer solution (pH 7.0) is added and dispersed. Next, 100 µl of Termamyl 120 L [see Sect. 3 (3)] solution is added to the flask, warmed under reflux, heated gently to boiling in 10–15 min and boiled for a total of 30 min over a weak flame to digest the starch. To the same flask, 50 ml of neutral detergent solution, twice the concentration shown by Van Soest (1963a), is added. Thereafter, according to Van Soest, 2 ml of decalin and 0.5 g of sodium sulfite are added to the solution. Following this treatment, the solution is boiled, filtered and rinsed by the same procedure as in the case of low-starch material. If appreciable starch remains, i.e., the iodine-starch reaction is positive, repeated Termamyl treatment is required.

3) Dehydration and powdering.
The ADF or NDF sample is rinsed thoroughly, collected, washed three times with acetone in a fourfold volume of the wet residue. After acetone treatment, the residue is washed with ether and air-dried on a filter paper. To avoid formation of a hard mass of IDF which affects the physical properties of IDF, it is necessary to dry the IDF sample completely. The air-dried sample is filtered through a 1-mm screen and stored. A Cyclotec mill (see Sect. 2) is useful for grinding and sifting the IDF sample.

4.2 Enzymatic Method

This method is based on enzymatic treatment of the defatted samples similar to that proposed by Prosky et al. (1984, 1985) for dietary fiber determination. As shown in Sect. 3 (3), care should be taken not to keep the samples under the condition of strong acidic digestion for a long time.

Our present procedure used for the preparation of IDF is as follows:

1) Pretreatment of the samples.
See Sect. 2.
2) Enzymatic (Termamyl, protease and amyloglucosidase) treatment.
The successive enzyme treatment of Prosky et al. (1984, 1985) is used for the elimination of protein and starch from the sample. For details of the enzymes and reagents, refer to Chapt. by Dr. Selvendran.

Termamyl Treatment. In a 200-ml beaker, 50 ml of phosphate buffer solution (pH 6.0) and 0.1 ml of Termamyl 120 L solution are added to 0.5–1 g of the pretreated sample [see Sect. 4.1 (2)]. Incubation in the boiling water bath is continued for a total of 30 min.

Protease Treatment. The solution is cooled to room temperature and adjusted to pH 7.5 ± 0.1 by adding 10 ml of 0.17 M NaOH. Then 0.1 ml of protease solution is added to the digest. The solution is incubated for 30 min at 60° C with continuous agitation.

Amyloglucosidase Treatment. After cooling of the solution, 10 ml of 0.21 M H_3PO_5 is added to adjust the pH to 4.5 ± 0.2. To this solution, 0.3 ml of amyloglucosidase solution is added, and then incubated for 30 min at 60° C with continuous agitation.

3) EDTA Treatment
   When removal of ash remaining in the IDF obtained by the method of Prosky et al. (1984, 1985) is necessary, EDTA treatment is useful. After amyloglucosidase digestion, EDTA (disodium salt) is added to the solution to make 0.1 M, and shaken for 30 min. Consequently, 35%–50% of the ash is removed from the residues.
   The beaker is covered with aluminum foil during the enzymatic and EDTA treatments.
4) Filtration and washing.
   After EDTA treatment, the residues are filtered through a No. 2 glass filter and washed with 20 ml of hot water. The washing is repeated four times.
5) Dehydration and powdering.
   See Sect. 4.1. (3).

4.3 Large-Scale Preparation

For animal feeding experiments, it is generally necessary to prepare a large amount of IDF. IDF obtained by the same methods as the IDF assay is preferable, but the amounts are usually unsatisfactory. Therefore, other procedures are necessary to treat the samples on a large scale, and the procedure used in our experiments is described. It is important to check the purity of the components of the IDF products obtained.

Samples, such as vegetables, are placed in a home disposal unit under running water, and the exhausted water is filtered through a cloth bag. Instead of a disposal unit, a mixer can also be used. After repeated washing with water (sometimes hot water), the residues are treated with ethanol, air-dried, ground and sifted through a screen (Takeda and Kiriyama 1979; Takeda et al. 1986).

Preparation of Antarctic Krill Chitin. To prepare chitin, the animal IDF, from antarctic krill, Kuwano et al. (1987) used a juice extractor and an ordinary washing machine.

1) Separation of the shell.
   Frozen block of raw antarctic krill is thawed at room temperature overnight and placed in a juice extractor fitted with a 0.5-mm screen to separate the shell from the meat under running water.
2) Washing of the shell.
   The crude shell is placed in a nylon bag (0.07-mm screen), washed under running water for 5 min and centrifuged for 2 min in a washing machine. This procedure is repeated five times.

3) Elimination of protein and ash.
   From this crude shell, pure chitin is prepared according to a modification of Hackman's method, using the treatment with 1 M HCl and 1 M NaOH (Muzzarelli 1977).
4) Decoloration, dehydration and powdering.
   The crude chitin is packed into a glass column (ca. 10 cm i.d. × 100 cm) and extracted by acetone. This decolored and dehydrated chitin is also obtainable from a batch procedure, but a large amount of acetone is required for the latter. After air-drying, the chitin is filtered through a 0.3–1.7-mm screen.
   Recovery of 0.5 kg of chitin is possible from 100 kg of raw antarctic krill.

5 Chemical and Physical Properties

Many physiological and nutritional effects have been attributed to fiber in the diet. However, a few reports have dealt with the actions of fiber components, such as metal binding due to the ion exchanging nature of fiber or possible adsorption of metabolites based on physical properties of fiber.

In studies where dried IDF is not needed, e.g., settling volume in water, water-holding capacity and measurements of adsorptive properties, it may be possible to apply directly the wet IDF obtained by omitting the dehydration with acetone shown in Sect. 4.1 (3) and Sect. 4.2 (5).

It is recommended that statistical processing of the measured values of the various chemical and physical properties in this section is carried out.

5.1 Analysis of Sugar Components

To obtain the informations on detailed physicochemical properties of IDF, it is useful to analyze the sugar components and quantities of IDF. A common method for this purpose is gas-liquid chromatography (GCL, e.g., Englyst and Hudson 1987). Recently, high performance liquid chromatography (HPLC) has occasionally been used for the same purpose.

5.1.1 Gas-Liquid Chromatography (GLC)

Since detailed methods of determination of the component sugars in dietary fiber using GLC are described by Dr. Selvendran, only simplifed procedures of GLC for the fractionation of component sugars in IDF are given here.

1) Preparation of IDF.
   See Sects. 2 and 4.
2) Extraction and acid hydrolysis.
   In general, hydrolysis with 5% or 0.5 M sulfuric acid for hemicellulose and extraction with 72% sulfuric acid followed by hydrolysis with dilution to 1 M sulfuric acid for cellulose are used.

3) Neutralization, deionization and drying.
   An aliquot is neutralized with solid barium carbonate. After centrifugation to eliminate barium sulfonate, excess barium ions in the supernatant are removed with a strong cation-exchange resin (H^+ type). The solution is evaporated and dried.
4) Derivatization and GLC.
   Derivatization of sugars to their corresponding alditol acetate compounds has made it possible to analyze quantitatively carbohydrates by using GLC (e.g., Collins and Yokoyama 1979).

5.1.2 High-Performance Liquid Chromatography(HPLC)

Many experiments have been conducted to evaluate HPLC as a method to measure sugar compositions of polysaccharides, and Hanai (1986) summarized HPLC analyses of sugars used in the food industry and biochemical research.

Pretreatments for HPLC analysis are the same as in Sect. 5.1.1 (1–3), but drying is unnecessary, and the sugar concentration detectable by the analysis should be obtained for the solution (10 μg or over for a refractive index detector and 500 ng or over for a fluorescence detector using a post-column label). Compared to GLC, much lower sensitivities are observed with HPLC: 1/1000 for the refractive index detection and 1/10–1/200 for the fluorescence detection.

Recently, highly sensitive and stable detection has been developed using an electrochemical detector (ECD), and sugar analysis by sensitive and selective HPLC has been performed (Watanabe 1984, 1985). This new method, using the redox reaction of bis(1,10-phenanthroline) copper as a marker (Fig. 1), has a sensitivity of 1 ng of glucose, which is the same as that of the flame-ionization detector method using a GLC capillary column, and an pretreatment of the sample is simpler than that for GLC.

$$CBP^{2+} \longrightarrow CBP^{+} \longrightarrow CBP^{2+}$$

reducing sugars, pH 10.9, 98.5° C, post-column (first step); electrochemical oxidation, +40 mV vs. Ag/AgCl (second step)

Fig. 1. Redox reaction of copper complex coupled with the reducing ability of sugar (Watanabe 1984); *CBP* bis (1,10-phenanthroline) copper (II)

5.2 Observations of the Surface

Scanning electron microscopy (SEM) is usually used for surface observations, e.g., the shape of the particles and microstructure of the surface, of IDF.

5.2.1 SEM Sample Preparation and Observation

Sample preparations are the most important procedures for SEM studies, and the main points are as follows.

1) Sample preparation.
   When samples for SEM observations are prepared by the methods described in Sect. 4, careful dehydration at the last stage is needed. Since a very high surface tension, 20–100 MPa, is observed during water evaporation from the surface of IDF (Goldstein et al. 1981), residual water in IDF should be avoided. If the normal shape of IDF for SEM studies is not maintained due to the weak framework of the sample, such as vegetable IDF, drying by the critical point method (Goldstein et al. 1981) after dehydration with acetone may be attempted instead of air-drying.
2) Coating and SEM observations.
   For details of the procedures of coating technique for SEM, refer to Goldstein et al. (1981). A large number of electron micrographs should be obtained and observed.

5.2.2 Shape of the Particle

Such observations are made under comparatively low magnification to obtain a scale of ×250–1000. For instance, Ayano et al. (1980) observed SEM (×500) of NDF and ADF from defatted rice bran, in which the NDF preparation was composed mainly of blocks of cytoplasm-free cell walls with a fairly integral structure and the ADF preparation was composed of fragments of ruptured cell walls. Moreover, they showed that the NDF preparation had a significant effect to prevent the elevation of serum cholesterol, while the ADF preparation had no effect.

5.2.3 Microstructure of the Surface

SEM observations of chitin, an animal IDF, by Kuwano et al. (1986) were conducted at a magnification of ×5000–10000. SEM photographs of chitins of antarctic krill, spiny lobster and crab are shown in Fig. 2. Crab chitin had a markedly different surface from the other two chitins which showed similar pictures. Other

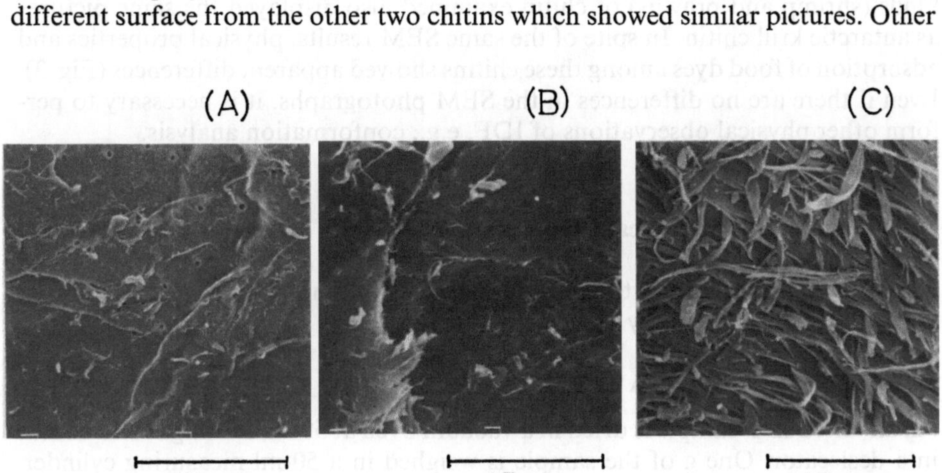

Fig. 2 A–C. Scanning electron micrographs of the surface microstructure of various chitin samples; **A** Antarctic krill; **B** Spiny lobster; **C** Crab

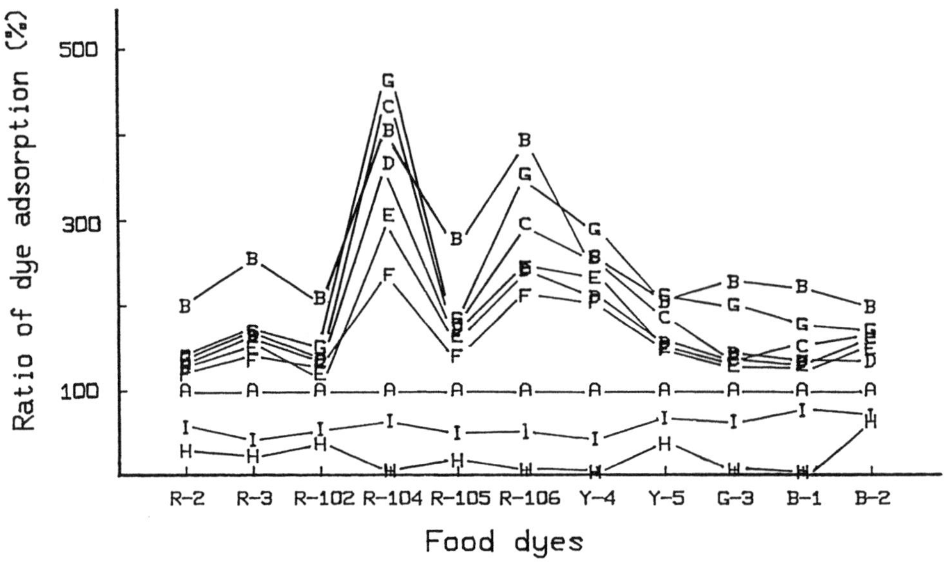

Fig. 3. Patterns of ratio of food dye adsorption of various chitins. (Ratio of dye adsorption of each chitin was calculated as % of dye adsorption in antarctic krill chitin.) *A* Antarctic krill; *B* Sakura shrimp; *C* Tenaga-ebi (*Macrobrachium nipponense*); *D* Kuruma prawn (small); *E* Kuruma prawn (medium); *F* Kuruma prawn (large); *G* Fleshy prawn; *H* Spiny lobster; *I* Crab. Food dyes in Japanese Standards of Food Additives by Food Sanitation Law: *R-2* amaranth; *R-3* erythrosine; *R-102* new coccine; *R-104* phloxine; *R-105* rose bengal; *R-106* acid red; *Y-4* tartrazine; *Y-5* sunset yellow FCF; *G-3* fast green FCF; *B-1* briliant blue FCF; *B-2* indigo carmine

kinds (shrimp and prawns) of chitin examined also displayed the same pictures as antarctic krill chitin. In spite of the same SEM results, physical properties and adsorption of food dyes among these chitins showed apparent differences (Fig. 3). Even if there are no differences in the SEM photographs, it is necessary to perform other physical observations of IDF, e.g., conformation analysis.

5.3 Measurement of Physical Properties

Among the physical properties of IDF, apparent volume, settling volume in water and water-holding capacity are usually measured.

5.3.1 Apparent Volume (AV)

An air-dried IDF sample is dried in a vacuum oven at 50° C overnight and cooled in a desiccator. One g of the sample is weighed in a 50-ml measuring cylinder. After the cylinder is shaken vigorously until the decrease of the sample volume is stopped, the volume is read as AV (ml/g) (Kuwano et al. 1986). The apparent sp.gr. (g/ml) is calculated from the AV.

Different AV values appear according to the particle size of the samples. Therefore, AV values should be compared with the IDF samples of the same particle size.

5.3.2 Settling Volume in Water (SV)

One half to 1 g of IDF sample is dried and cooled as in AV measurement, transferred to a beaker, and mixed with deionized water (ca. 100 ml for cereal IDF and ca. 200 ml for vegetable IDF). After air in the IDF are eliminated using an aspirator, the beaker is allowed to stand for 24 h of hydration. The water with IDF is transferred to a cylinder of appropriate volume and allowed to stand for 24 h. The volume of IDF in the water is read as SV (ml/g of dry IDF). Instead of the SV, the term "swelling" is used in some cases (Rasper 1979).

With respect to the SV and water-holding capacity (see Sect. 5.3.3), Takeda and Kiriyama (1982), who observed the particle size of IDF preparations by SEM, showed SV values to be higher in NDF than in ADF. Although a high correlation is observed between SV and water-holding capacity of IDF, it is possible that these values are related to independent physiological effects (Takeda and Kiriyama 1982). Thus, both values should be measured.

5.3.3 Water-Holding Capacity (WHC)

Because different values of WHC are obtained by different methods of measurement, comparison of WHC values should be made using the same measurement method.

a) Measurement by centrifugation method.
   This method is called "water retention" by Rasper (1979). In a 25-ml centrifuge tube, 0.2–0.5 g of IDF is weighed, hydrated for 24 h with deionized water at 20° C, and centrifuged at 14000 g for 1 h. The WHC (g water / g of dry IDF) is measured as the amount of water retained by weighing the hydrated sample (McConnell et al. 1974). Robertson and Eastwood (1981) recommended centrifugation at 6000 g for 15 min.
b) Measurement under natural acceleration of gravity.
   One g of air-dried sample is placed on a filter paper fitted in a polyethylene container (6 cm in diameter and 3 cm in depth) with a 0.7-mm screen wire-bottom of stainless steel, and the container with the sample is placed in a large dish containing sufficient amount of deionized water to immerse the sample for 12 h. Instead of this procedure, Kuwano et al. (1986) proposed a measuring method by which the hydrated IDF after SV measurement is transferred to a funnel with a filter paper made by glass fiber (Toyo Filter Paper GA 100) attached. The hydrated IDF with the container or glass filter is removed and allowed to stand for 10 min to drain excess water, and then weighed after blotting the water adhering to the wall. By substracting the weight of the container fitted with moistened filter paper or the moistened glass fiber filter paper and the weight of the sample applied (1 g of dry IDF) from the final weight of the container or glass filter with the wet sample, the WHC is obtained (Takeda and Kiriyama 1979).

Insted of the WHC, the term "water adsorption" which is measured by a Millipore filtering membrane (AA 0.80 μm) is used by Rasper (1979).

c) Measurement by the osmolality method.
Dried IDF (0.2–0.5 g) is presoaked and transferred to a dialysis bag of 19-mm diameter Visking tubing (Gallenkamp, volume 5 ml). For 24–48 h the bag is placed in a 250-ml conical flask containing 100 ml of a solution of known osmotic potential using polyethyleneglycol 6000. Afterward, the contents are weighed and dried (105° C) overnight, and the WHC is calculated. This measurement is also appliable to the WHC measurement of soluble dietary fiber (Robertson and Eastwood 1981).

5.4 Determination of Adsorptive Properties

Adsorptive properties provide much interesting information with relation to the in vivo nutritional effects and other physicochemical properties of IDF. Among the adsorptive properties of IDF, two studies by our group are shown in here.

5.4.1 Ion Exchange Capacity

In a beaker, 0.2–0.5 g of dried IDF is weighed and immersed in deionized water as shown in Sect. 5.3.2. After filtering through a No. 2 glass filter the IDF is returned to the beaker and stirred with 0.5 M HCl in a five to tenfold volume of the wet IDF, followed by filtration through the same glass filter. Washing with deionized water is continued until a neutral filtrate is obtained, and the residue is transferred to a beaker with a small amount of deionized water. Titration with 0.1 M NaOH is conducted with continuous stirring after adding 50 ml of 1 M NaCl, and the titration curve of pH change of the solution is plotted using a pH meter. Cation exchange capacity (meq/g of dry IDF) is obtained from the titration curve (McConnell et al. 1974; Rasper 1979). Because of treatment with HCl

Table 4. Cation exchange capacity of various dietary fibers (meq/g of dry dietary fiber)

Dietary fiber sample	NDF[a]	ADF[b]	IDF[c]
Whole wheat	0.25	0.086	0.21
Whole rye	0.17	0.11	0.095
Whole rice	0.34	0.27	0.34
Soybean	0.32	0.46	0.47
Potato	0.095	0.26	0.77
Sweet potato	0.098	0.15	0.48
Carrot	0.12	0.66	0.82
Cabbage	0.22	0.65	0.91
Lettuce	0.38	0.75	0.94

[a] Neutral detergent fiber.
[b] Acid detergent fiber; prepared by detergent method (Sect.4.1).
[c] Insoluble dietary fiber; prepared by enzymatic method (Sect. 4.2).

in this method, the elimination of acid soluble fraction of IDF is probable (Takeda and Kiriyama 1982). Results observed from several dietary fiber are presented in Table 4 (Sakamaki et al. 1988).

For IDF which has anion exchange action, such as chitin and chitosan, the IDF is treated with 0.5 M NaOH instead of 0.5 M HCl, washed with deionized water, dispersed into 1 M NaCl and titrated with 0.1 M HCl.

5.4.2 Food Dye Adsorption

Dried IDF (50–100 mg) is swelled into deionized water (see Sect. 5.3.2) and dispersed in 10–20 ml of deionized water. After adding 10 ml of a mixed solution containing 11 kinds of coal-tar dyes (1 μM of each / ml), permitted use as food coloring in Japan, the IDF in the solution is allowed to stand overnight and the dyes are adsorbed. The IDF is separated from the solution by filtration, and washed thoroughly with deionized water. All filtrates containing the dyes not adsorbed on the IDF are collected, prepared in a constant volume. Each dye is determined using the appropriate measuring technique, e.g., HPLC (Kuwano and Mitamura 1986), and the weight of each dye adsorbed on the IDF (μM or mg/g of dried IDF) is calculated. Figure 3 (p. 270) shows the adsorption pattern of coal-tar dyes on various chitins in which the ratios of dye adsorption of various chitins are compared to that of antarctic krill chitin.

5.5 Conformation Analysis

Even if the IDF to be investigated consists of polysaccharides which have the same composition and combinations of monosaccharides, it is possible that each IDF has a different conformation and thus the physicochemical characteristics and nutritional properties are different from each other. Therefore, future subjects of study for IDF include clarification of the cross-linking of hemicellulose, cellulose and lignin molecules in IDF, and the interrelationship between these components and the physicochemical properties of IDF.

Generally used techniques for conformation analysis are Fournier transform-infrared spectrometry, X-ray diffraction, and ^{13}C NMR of solid state high resolution by the cross polarization-Magic angle spinning technique. Although there have been few experiments on the conformation of IDF, some related studies have been carried out for cellulose (Harada and Goto 1982), plant IDF, and chitin and chitosan (Muzzarelli 1977), animal IDF, because these are easily obtained as pure materials.

6 Nutritional Properties of Dietary Fiber

Dietary fiber continues to be of great interest to nutritionists since it has become accepted that a number of diseases prevalent in affluent societies may be caused by a lack of these substances in modern diets (Burkitt and Trowell 1975; Eastwood 1984). Since it is nearly impossible to show all methods for nutritional studies of dietary fiber, only several observations considered to be related to intake of dietary fiber are presented.

1) Increased faecal volumes and shortened intestinal transit time by IDF ingestion may be related to the prevention of diseases in the large bowel, including cancer and constipation (Burkitt and Trowell 1975).
2) Prevention of excessive energy intake by providing bulk, and modified blood glucose tolerance and hormonal response associated with high-fiber food may make dietary fiber a useful tool in obese or diabetic subjects (Ali et al. 1982; Anderson 1982).
3) Decrease of toxicity of dietary toxicants (Takeda and Kiriyama 1979), prevention of cholecystitis and gallstones (Burkitt and Trowell 1975), and alteration of the gut microflora (Mizutani et al. 1982) have also been observed with the intake of dietary fiber.
4) The hypocholesterolemic effect of soluble dietary fiber due to inhibition of mixed micelle formation in the intestine and the altered enterohepatic circulation of bile acids is recognized as a beneficial factor in coronary disease and arteriosclerosis (Tsuji et al. 1982).
5) In contrast to the positive effects of increased fiber in the diet, there have been reports of disadvantages with respect to the bioavailability of nutrients including vitamins (Kelsay 1982). A number of reports indicate that dietary fiber in the diet may result in decreased or negative mineral balances (Oku et al. 1982; Kuwano et al. 1987).

7 Summary

The crude-fiber method, involving extraction with dilute acid and alkali, results in large losses of dietary fiber components and gives lower fiber values than those obtained by modern methods. The choice of modern analytical methods depends on the need for detailed information on dietary fiber composition, and modifications of some of these methods are also utilized effectively for the preparation of IDF samples. In any case, samples should be treated carefully as mentioned in Sect. 2.

The principal chemical and physical properties of IDF are those of sugar components, shape of particle, density, hydration capacity, ion exchange capacity and conformation. Their roles in controlling the physiological and nutritional characteristics of IDF are intricate and investigations in this area should be completed.

References

Ali R, Staub H, Leveille GA, Boyle PC (1982) Dietary fiber and obesity. A review. In: Vahouny GV, Kritchevsky D (eds) Dietary fiber in health and disease. Plenum, New York, pp 139–149

Anderson JW (1982) Dietary fiber and diabetes. In: Vahouny GV, Kritchevsky D (eds) Dietary fiber in health and disease. Plenum, New York, pp 151–167

Asp N-G, Johansson C-G, Hallmer H, Siljestrom M (1983) Rapid enzymatic assay of insoluble and soluble dietary fiber. J Agric Food Chem 31:476–482

Association of Official Analytical Chemists (1984) Offical methods of analysis. AOAC, Arlington, Virginia, pp 160–162

Ayano Y, Ohta F, Watanabe Y, Mita K (1980) "Dietary fiber" fractions in defatted rice bran and their hypocholesterolemic effect in cholesterol-fed rats (in Japanese). J Jpn Soc Nutr Food Sci 33:283–291

Baker D, Norris KH, Li BW (1979) Food fiber analysis: advances in methodology. In: Inglett GE, Falkehag SI (eds) Dietary fibers: chemistry and nutrition. Academic Press, New York, pp 67–78

Bingham S (1987) Definitions and intakes of dietary fiber. Am J Clin Nutr 45:1226–1231

Burkitt DP, Trowell HC (eds) (1975) Refined carbohydrate foods and disease. Some implications of dietary fiber. Academic Press, New York

Collings GF, Yokoyama MT (1979) Analysis of fiber components in feeds and forages using gas-liquid chromatography. J Agric Food Chem 27:373–377

Cummings JH (1987) Dietary fiber. Am J Clin Nutr 45:1040–1044

Eastwood MA (1984) Dietary fiber. In: Olsen RE, Broquist HP, Chichester CO, Darby WJ, Kolbye AC Jr, Stalvey RM (eds) Present knowledge in nutrition. Nutrition Foundation, Washington, DC, pp 156–176

Englyst HN, Cummings JH (1984) Simplified method for the measurement of total non-starch polysaccharides by gas-liquid chromatography of constituent sugars as alditol acetates. Analyst 109:937–942

Englyst HN, Hudson GL (1987) Colorimetric method for routine measurement of dietary fibre as non-starch polysaccharides. A comparison with gas-liquid chromatography. Food Chem 24:63–76

Englyst HN, Wiggins HS, Cummings JH (1982) Determination of the non-starch polysaccharides in plant foods by gas-liquid chromatography of constituent sugar as alditol acetates. Analyst 107:307–318

Furda I (1981) Simultaneous analysis of soluble and insoluble dietary fiber. In: James WPT, Theander O (eds) The analysis of dietary fiber in food. Marcel Dekker, New York, pp 163–172

Goldstein JI, Newbury DE, Echlin P, Joy DC, Fiori C, Lifshin E (1981) Scanning electron microscopy and X-ray microanalysis. A text for biologists, materials scientists, and geologists. Plenum, New York, pp 461–494, pp 522–531

Hanai T (1986) Liquid chromatography of carbohydrates. In: Giddings JC, Grushka E, Cazes J, Brown PR (eds) Advances in chromatography, vol 25. Marcel Dekker, New York, pp 280–307

Harada H, Goto T (1982) The structure of cellulose microfibril in *Valonia*. In: Brown RM Jr (ed) Cellulose and other natural polymer systems. Biogenesis, structure, and degradation. Plenum, New York, pp 383–401

Hellendoorn EW (1978) Fermentation as the principal cause of the physiological activity of indigestible food residue. In: Spiller GA (ed) Topics in dietary fiber research. Plenum, New York, pp 127–168

Hoshi S, Takehisa F (1981) Examination of the dietary fiber procedure (in Japanese). J Jpn Soc Nutr Food Sci 34:583–588

Ishii K, Ohkubo M, Yamamoto Y, Yoshida T (1988) Effect of diet sterilization on digestibility of dietary fiber in germfree and conventional mice (in Japanese). J Tachikawa College Tokyo 21:85–89

Johansson C-G, Åman P, Asp M-G, Theander O (1982) Enzymatic and neutral detergent fibre methods for dietary fibre analysis. Swed J Agric Res 12:157–161
Kellogg TE, Wostmann BS (1969) Stock diet for colony production of germfree rats and mice. Lab Anim Care 19:812–814
Kelsay JL (1982) Effects of fiber on mineral and vitamin bioavailability. In: Vahouny GV, Kritchevsky D (eds) Dietary fiber in health and disease. Plenum, New York, pp 91–103
Kurasawa S, Sugawara T, Hayashi J (1982) Proximate and dietary fiber analysis of mushrooms (in Japanese). J Jpn Soc Food Sci Technol 29:400–406
Kuwano K, Mitamura T (1986) Ion-pair high performance liquid chromatographic separation and determination of food coaltar dyes (in Japanese). J Food Hyg Soc Jpn 27:278–282
Kuwano K, Tanimaru E, Sakamaki C, Tsukui A, Mitamura T (1985) Pretreatment of neutral detergent fiber (NDF) determination by enzyme and NDF content of vegetables (in Japanese). J Tokyo Kaseigakuin College 25:43–46
Kuwano K, Sakamaki C, Mitamura T, Yoshida T (1986) Physical and chemical properties of antarctic krill chitin (in Japanese). J Agric Chem Soc Jpn 60:913–920
Kuwano K, Sakamaki C, Mitamura T, Yoshida T (1987) Effect of the Antarctic krill chitin on the availability of dietary protein and minerals in growing rats (in Japanese). J Jpn Soc Nutr Food Sci 40:213–219
McConnell AA, Eastwood MA, Mitchell WD (1974) Physical characteristics of vegetable foodstuffs that could influence bowel function. J Sci Food Agric 25:1457–1464
McQueen RE, Nicholson JWG (1979) Modification of the neutral-detergent fiber procedure for cereals and vegetables by using α-amylase. J Assoc Off Agric Chem 62:676–680
Mizutani T, Benno Y, Mitsuoka T (1982) Effect of dietary fiber on tumorigenesis and longevity: with special reference to the fecal microflora. Nutr Rep Int 26:289–296
Mori B (1982) Contents of dietary fiber in some Japanese foods and the amount ingested through Japanese meals. Nutr Rep Int 26:159–166
Mori B, Aragane K (1981) On the determination of dietary fiber and the ingested amount through meals under administration on catering service (in Japanese). J Jpn Soc Nutr Food Sci 34:97–104
Murakami H, Takeyama S (1978) Accumulation of lignin in cut edible parts of vegetables (in Japanese). J Jpn Soc Nutr Food Sci 31:91–94
Muzzarelli RAA (1977) Chitin. Pergamon, Oxford, pp 45–51, pp 89–94
Oku T, Konishi F, Hosoya N (1982) Mechanism of inhibitory effect of unavailable carbohydrate on intestinal calcium absorption. J Nutr 112:410–415
Prosky L, Asp N-G, Furda I, DeVries JW, Schweizer TF, Harland BF (1984) Determination of total dietary fiber in foods, food products, and total diets: interlaboratory study. J Assoc Off Anal Chem 67:1044–1052
Prosky L, Asp N-G, Furda I, DeVries JW, Schweizer TF, Harland BF (1985) Determination of total dietary fiber in foods and food products: collaborative study. J Assoc Off Anal Chem 68:677–679
Rasper VF (1979) Chemical and physical properties of dietary cereal fiber. Food Tech 33(1):40–44
Robertson JA, Eastwood MA (1981) A method to measure the water-holding properties of dietary fibre using suction pressure. Br J Nutr 46:247–255
Sakamaki C, Kuwano K, Mitamura T, Yoshida T (1988) Effect of preparation method on the physical and chemical properties of dietery fiber (in Japanese). J Tokyo Kaseigakuin College 28:67–71
Selvendran RR, Ring SG, DuPont MS (1981) Determination of the dietary fiber content of the EEC samples and a discussion of the various methods of analysis. In: James WPT, Theander O (eds) The analysis of dietary fiber in food. Marcel Dekker, New York, pp 95–121
Southgate DAT (1969) Determination of carbohydrates in foods. II. Unavailable carbohydrates. J Sci Food Agric 20:331–335

Southgate DAT (1981) Use of the Southgate method for unavailable carbohydrates in the measurement of dietary fiber. In: James WPT, Theander O (eds) The analysis of dietary fiber in food. Marcel Dekker, New York, pp 1–19
Southgate DAT (1982) Definition and terminology of dietary fiber. In: Vahouny GV, Kritchevsky D (eds) Dietary fiber in health and disease. Plenum, New York, pp 1–7
Stelmock RL, Husby FM, Brundage AL (1985) Application of Van Soest acid detergent fiber method for analysis of shellfish chitin. J Dairy Sci 68:1502–1506
Takeda H, Kiriyama S (1979) Correlation between the physical properties of dietary fibers and their protective activity against amaranth toxity in rats. J Nutr 109:388–396
Takeda T, Kiriyama S (1982) Physico-chemical properties of dietary fiber (in Japanese). In: Innami S, Kiriyama S (eds) Dietary fiber. Daiichi Shuppan, Tokyo, pp 55–79
Takeda T, Nakajima A, Kiriyama S (1986) Effect of dietary fiber on utilization of heat-damaged casein in rats (in Japanese). J Agric Chem Soc Jpn 60:927–933
Theander O, Westerlund EA (1986) Studies on dietary fiber. 3. Improved procedures for analysis of dietary fiber. J Agric Food Chem 34:330–336
Tsuji K, Nakagawa Y, Iwao H, Okamatsu H, Yatake T (1982) Influence of a microbial polysaccharide produced by *Bacillus polymyxa* No. 271 on cholesterol metabolism in rats. Nutr Rep Int 26:231–238
Van Soest PJ (1963a) Use of detergents in the analysis of fibrous feeds. I. Preparation of fiber residues of low nitrogen content. J Assoc Off Anal Chem 46:825–829
Van Soest PJ (1963b) Use of detergents in the analysis of fibrous feeds. II. A rapid method for the determination of fiber and lignin. J Assoc Off Anal Chem 46:829–835
Van Soest PJ (1965) Use of detergents in analysis of fibrous feeds. III. Study of effects of heating and drying on yield of fibre and lignin in forages. J Assoc Off Agric Chem 48:785–790
Van Soest PJ, Wine RH (1967) Use of detergents in the analysis of fibrous feeds. IV. Determination of plant cell-wall constituents. J Assoc Off Anal Chem 50:50–55
Watanabe N (1984) Amperometric detection of muramic acid in high-preformance liquid chromatography with a post-column reaction. J Chromatogr 316:495–500
Watanabe N (1985) Amperometric detection of reduced carbohydrates in high-performance liquid chromatography using an amino-bounded column and acetonitrile-water as the eluent. J Chromatogr 330:333–338
Yoshida T, Ohkubo M (1985) Effect of age and diet on the digestibility of gross components and dietary fiber in rats. Agric Biol Chem 49:235–238
Yoshida T, Shinoda S, Matsumoto T, Watarai S (1982) Feed digestibility and mineral balance of the diet of young mice kept in mouse cages inside or outside an isolator, using varied concentrations of sodium phytate. J Nutr Sci Vitaminol 28:401–410

Measurement of Lint Production in Cotton and Factors Affecting Yield

R. Wells

1 Introduction

Cotton is an indeterminate plant with no morphological limit to flower production or yield (Hearns and Constable 1984). Unfortunately, cotton is grown commercially in an imperfect environment which is suboptimal with respect to temperature, moisture, light, pests, etc. during some portion of a fruiting cycle. Such limitations reduce yield potential and reduce profitability for the producer. To assess the effects of these yield-limiting factors (or yield-producing factors for the optimists) biological scientists must analyze plant growth and yield in controlled experiments. The following chapter will describe analytical methods utilized to determine lint production and components of cotton yield. In addition, procedures used to estimate cotton growth and fruiting patterns as affected by genotype and/or environment will be addressed.

2 Description of Growth Patterns

Successful analysis of cotton yield and yield components requires a fundamental knowledge of how the cotton plant grows. Cotton was originally a perennial in a subtropical environment. Now it is grown as an annual in a temperate zone. Its indeterminate growth habit means vegetative and reproductive development occurs simultaneously (Mauney 1984). The following section will discuss vegetative and reproductive growth as seperate entities. However, it's important to remember that these processes are directly linked and are not mutually exclusive.

2.1 Vegetative Development

Germination of the cotton seed occurs above 15° C with a linear increase in germination rate with increased temperature. Maximum hypocotyl elongation occurs at approximately 34° C, as described by Arndt (1945). Both early and late growth of the plant is directly related to temperature. Hesketh et al. (1972) reported that the number of days from planting to cotyledons with length × width values of 10 cm^2 dropped from 12 to 5 as the temperature increased from 22° to 30° C. Similarly, the interval between successive leaves (length × width = 10 cm^2) dropped from 6.5 to 2.5 as the temperature increased from 18° to 30° C.

The production of leaf area is an important determinant of early cotton growth induced by both temperature (Potter and Jones 1977) and genotype (Muramoto et al. 1965; Wells and Meredith 1986). Increased early leaf area would allow heightened light interception before the plants become large enough to compete for space with their neighbors. Cotton plants in relatively unstressed environments can produce leaf area indices (LAI) of more than 6 m^2 leaves m^{-2} ground (Hearn and Constable 1984; Ashley et al. 1965). Canopy leaves are found on both vegetative and reproductive branches.

The vegetative branches (monopodia) and reproductive branches (sympodia) are located on a main axis. Lower nodes on the main stem develop branches that are monopodia, or those which develop continuously from an apical bud until some limiting factor causes an interruption in growth. The main axis and the vegetative branches have a spiral phyllotaxy, with each successive leaf being rotated 135° from the position of the preceding leaf. Rotation associated with the leaf arrangement can be either clockwise or counter-clockwise. At the base of the main stem leaves, two, or occasionally three, buds can be found. The first of these buds is called the first axillary, and is in the axil of the prophyll (an inconspicuous, stipule-like leaf). Since the prophyll has no internode, this bud resides nearly in the axis of the main stem leaf and has been called an extra axillary bud. Virtually all branching occurs from the first axillary bud except in cases where either the first axillary bud is aborted, or vigorous growth is possible, as in the case of widely spaced plants (Tharp 1965; Mauney 1984).

The sympodia arise from the main axis at higher nodes on the plant than the monopodia and are more horizontal relative to the main axis. As the sympodium starts to grow, a floral bud is the first structure to be produced. The floral bud (or square) is carried away from the main stem as the branch grows, and a leaf emerges beside it. The growing point is terminated by each leaf and flower produced. Growth of the sympodium is continued by development of the first axillary at the base of the last leaf produced. This pattern of node production continues and gives the sympodium a "zig-zag" appearance (Mauney 1984).

The maximal vegetative development is usually reached between 90 and 100 days after planting (Constable and Gleeson 1977; Wells and Meredith 1984a). This pattern can be lengthened if reproductive development is interrupted by some stress (e.g., drought, insects) keeping the plant in a vegetative mode for a longer period.

2.2 Reproductive Development

From planting to the first square about 35 days are required at a mean daily temperature of 26° C. First square to first flower takes approximately another 27 days. From that point it would require about 3 days for each flower vertically up the main axis, and about 8 days horizontally out the sympodium. Temperature can greatly alter this pattern by increasing the interval between events with decreasing temperatures, and the reciprocal with increasing temperatures (Hesketh et al. 1972). Stresses caused by pests, moisture deficits, light reductions, etc. can induce shedding of fruiting forms, thereby creating variations in reproductive patterns (Guinn 1982). Alterations in fruiting patterns can cause large changes in growth and makes cotton one of the most dynamic systems of any agronomic crop.

The number of fruiting sites on a plant is related to the number of main stem nodes (Hearn 1969a). Flowering occurs over a 35–40-day period for most modern cultivars (Wells and Meredith 1984b). During this period, the percentage of flowers which are retained on the plant declines in linear fashion (Verhalen et al. 1975), indicating the reduction over time of both the individual leaf (Wells 1988) and whole plant (Wells et al. 1986) assimilatory capacities. Eventually the production of fruiting sites ceases as the plant stops growing; an event known as "cut-out".

Seed and fiber dry-weight accumulation is completed by approximately 42 days after anthesis (Schubert et al. 1973; Leffler 1976). Desiccation and opening of the boll requires about 2 or 3 days more. Based on the flowering period, percentage of retained bolls, and the environmental conditions during fruit maturation, the boll opening period can require several months to complete. The first fruiting cycle can be followed by another in some environments with long growing seasons, such as California and Arizona.

2.3 Earliness in Maturity Patterns

The trend toward early maturity is described as "earliness" in cotton literature. Earliness can be induced by environmental effects (e.g., temperature, moisture), production practices (e.g., planting date, defoliation) and breeding programs. Earliness as a genotypic trait is quite valuable because it allows avoidance of late-season insect pressure (Adkisson et al. 1982) and other yield-limiting environmental factors (Hearn and Constable 1984).

Early-maturing genotypes are often referred to as determinate although all cotton genotypes are morphologically indeterminate. Determinate types "cut-out" sharply and do not promptly begin another cycle of growth (Hearn and Constable 1984). Indeterminate types, on the other hand, flower over a longer period of time and exhibit a slower termination of growth at the end of the season. Breeding programs in the United States have relied on lint yield as the main selection criterion and have resulted in genotypes with shortened, more compact flowering and boll setting periods (Wells and Meredith 1984b). The relationship between yield and earliness is possibly related to a more rapid transition to reproductive growth, and a better coordination of fruit development with maximum vegetative assimilatory activity (Hearn 1969b). Differences in plant maturity are consistent with the observed changes in reproductive/vegetative ratios related to the shift toward earlier-flowering patterns (Hearn and Constable 1984; Wells and Meredith 1984b).

Generally, the components of earliness are (1) the time to first fruiting branch, (2) the rate of square production, (3) the percentage of retained fruit and (4) the length of the boll maturation period (Hearn and Constable 1984). Many of the traits associated with earliness have been examined by Hesketh et al. 1975.

3 Determination of Yield and Yield Components

Determining the total yield of seed cotton (unginned cotton, containing both seeds and lint) is relatively simple, and requires the harvest of a length of row by either hand or machine. If all bolls in the row had matured prior to harvest, the weight of the seed cotton is the yield and only conversions to some suitable unit of expression (per ha or m^2) are required. Much more information can be gained, however, if the harvest is handled in a number of other ways.

3.1 Hand Harvesting

For small plot research (row lengths from a few to many meters in length), hand harvesting is the most convenient method. The seed cotton is removed from the open boll while leaving the burs (dried carpel walls) behind. Since this method does not destroy the plant, multiple harvests of the same plot are possible. Multiple harvests taken from the same plot are called stratified harvests, and allow the researcher to discern information about the pattern of boll development over the course of the season (Meredith and Bridge 1973). Stratified harvests are especially attractive because the cotton plant is highly responsive to environmental factors, and alterations to the fruiting pattern readily occur (see Sect. 2.2). Using stratified harvests the investigator can identify maturity differences, seasonal changes in yield components and fiber properties (Wells and Meredith 1984c) and changes in fruit related to competition among developing bolls (Guinn and Mauney 1984). If row lengths are not long, and/or if stratified harvests occur close enough in time, the whole row can be harvested, ginned and used to compute yield components and fiber properties (Meredith and Bridge 1973). If the row length is large, however, and stratified harvests are few, a sub-sample must be taken because there will be too much seed cotton for the researcher to handle conveniently.

3.1.1 Sub-Sampling for Yield

Because research gins are comparatively small in relation to commercial units, relatively small seed cotton samples are required to avoid ginning periods of extended length. Large plots can be sub-sampled to meet this sample size requirement. Once ginned, the calculated yield components can be applied to seed cotton weights from the total row length. Usually, the sub-sample is between 25 and 100 bolls, but the actual number is dependent on the requirements of the research. The boll number taken is recorded, allowing calculation of certain parameters which will be discussed below. After the sub-sample is harvested, the remaining seed cotton should be harvested and the weight recorded. The number of bolls of this portion is not recorded, greatly reducing the mental concentration required during the harvest.

3.1.2 Selective Versus Nonselective Sampling

Taking a sub-sample of bolls can be accomplished in two different ways, one being better from the research standpoint than the other. The first method requires harvesting the sample randomly from the row by removing bolls from any position within the row. This sampling method represents a selective sampling because the sampler has the freedom to select the type of boll and position on the plant. Another method is to begin sampling at a point in the row and continue taking all bolls from that point onward until the required number of bolls is harvested. The latter method is a nonselective sampling because the sampler has no freedom to choose what bolls will or will not be taken (Meredith et al. 1975), and the boll appearence and position can not affect what bolls are sampled. Selectivity in sampling usually results in an overestimation of the weight per boll and other yield and fiber components (Culp and Harrel 1972; Meredith et al. 1975). Samplers tend to unintentionally select the largest and most eye-appealing bolls, thus biasing the sample. Meredith et al. (1975) state, "a standard sampling method that minimizes the sampler's freedom to choose bolls and approximately samples the entire crop is certainly preferred over selective sampling for critical genetic, cultivar, or agronomic experiments on yield, yield components and fiber properties."

3.2 Determination of Yield Components

Yield components can be determined from ginned seed cotton if the number of bolls in the sample are known (Meredith and Bridge 1973; Meredith et al. 1975). Components of yield can be determined as follows:

1) Lint Percentage (lint %) = lint weight ÷ (lint weight + seed weight) × 100
   OR = lint index ÷ (lint index + seed index)
2) Lint Yield = (lint % ÷ 100) × total seed cotton
3) Weight per Boll = sample seed cotton ÷ no. bolls in sample
4) Bolls m^{-2} = seed cotton m^{-2} ÷ weight per boll
5) Seed Index = weight of 100 seeds
6) Lint Index = lint % × [seed index ÷ (100-lint %)]
7) Seeds $Boll^{-1}$ = (100-lint %) × (boll weight ÷ seed index)

The yield components and their relationship to yield is complex, involving interactions with genotype and environment. Each component can give information about seasonal trends, especially when combined with stratified harvests. Meredith and Bridge (1973) reported a decline in weight per boll and seed index at each successive harvest during reproductive growth, with an especially noticeable drop in these parameters as the plants cut-out. A single harvest, or even a few, would not detect these trends. Stratified harvests can be useful in the detection of many seasonal variations in yield components.

3.3 Machine Harvesting

Many experiments are harvested by mechanical harvester. Labor requirements are reduced, but other problems are possible concerning experimental design and information gained.

There are two basic types of mechanical harvester, the stripper type and the spindle type. The stripper type is a nonselective harvester that strips the entire plant of opened and unopened bolls, thus allowing its use only once per plot. The entire crop must be mature at the time of harvesting (Anonymous 1965). Stripper harvesting would only be suitable for experiments which require information concerning total yield. Sub-samples can be taken and ginned. Modifications of stripper harvesters for plot work are possible (Brashears and Ray 1966).

The spindle type harvester, on the other hand, is selective because only locks of mature, opened bolls are harvested, leaving the burs and most other extraneous matter behind (Anonymous 1965). Spindle harvesters can therefore make multiple passes over plots with relatively little physical damage to the plants.

To maintain a high-quality lint sample, plots are usually chemically defoliated or desiccated before mechanical harvest. Both types of harvester introduce more trash than hand harvesting. The spindle type averages about 6% trash in the seed cotton and the stripper type averages about 25% trash (Pendleton and Moore 1968). Carefully harvested hand-picked samples, on the other hand, have less than 1% trash. Machine harvesting requires longer rows, usually 15 to 30 m in length, and wider end alleys (1 to 2 m), allowing clear demarcations between plots. Turnrows are required at the end of the field to allow clear turning for another pass through the field.

The high trash content of machine-harvested samples creates several problems when determining both yield components and fiber properties. One problem is that trash inflates the weight of seed cotton and, when multiplied by the lint percentage determined from a hand-harvested sub-sample, the amount of lint is overestimated. An alternative method is to gin the entire machine-harvested sample or an appropriately sized sub-sample, but the number of bolls is not known, and the weight per boll can not be calculated.

Most machine-harvested samples are large and require ginning in modified commercial gins with lint cleaners which alter several quality factors of the fiber (Williford et al. 1987). Hand-harvested samples, however, tend to have higher grades than those harvested by spindle type harvesters even when both samples are ginned in commercial gins (Williamson et al. 1966). In fiber tests from hand- and machine-harvested cotton at one 1969 Tennessee location, micronaire and length (2.5 SL) were 5.8% and 16.1% greater in hand-harvested samples than in machine-harvested samples (Graves 1971). This difference may also reflect the sampling technique as discussed in Sect. 3.1.2. Spindle type machine harvesting would simulate nonselective harvesting, because all opened bolls regardless of their condition would be included in the fiber sample.

3.4 Sampling Technique for Fiber Quality Measurements

The quality of lint declines as the exposure to the environment lengthens (Barker et al. 1979). Certain experiments concerning differences in fiber quality among treatments or genotypes should be planned with the effect of weathering in mind. Keeping weathering of fiber samples to a minimum is favorable for breeding and other studies where the treatment effects could be confounded with effects of weathering. If weathering is undesirable, stratified harvests during the boll maturation period would be appropriate. Weekly or bi-weekly harvests would identify discrete groups of bolls which are of similar age, thereby reducing the likelihood of weathering of the whole crop. Simpson and Hertel (1946) warned that weathering may mask genotypic differences, since variation due to location or date of harvest may be much larger than the effects of genotype. Other studies may require weathering as a treatment like that reported by Williford et al. (1987). In such cases, comparative harvests must be taken to test appropriately the hypothesis upon which the experiment is based.

4 Analysis of Cotton Plant Growth

The growth and yield of cotton is affected by many environmental and genetic factors. The best way to assess growth is through classical growth analysis and has been used to study many species of plants (Kvet et al. 1971; Watson 1952). This section will provide a cursory view of general growth analysis. More thorough discussions of the subject are available (Kvet et al. 1971; Radford 1967).

4.1 Collection of Primary Data

Data can be collected by either nondestructive or destructive means. Nondestructive measurements, such as plant height, main stem node number or fruiting form counts, can give valuable information, but no information about dry matter or leaf area changes over time can be ascertained. To measure these parameters of growth, destructive sampling of plants is required and is usually considered a requirement for "classical" growth analysis.

More than one plant should be harvested at any sampling. The actual number harvested will depend on the ability to physically handle and separate into parts for subsequent analyses. The two ways to harvest the plants are: (1) harvest individual plants which are randomly selected from the row or plot, and (2) harvest a section of row which would contain the required number of neighboring plants. Each method would require a uniform number of plants per length of row to allow the expression of the computed parameters on a land area basis. Such uniformity in plant density is most easily attained by planting an excess number of seed and thinning to a predetermined density after the plant stand is established (Wells and Meredith 1984a).

A number of plant harvests during the season is desirable. The most desirable experimental design is randomized blocks with all sampling dates and treatment combinations in completely randomized, bordered plots. If block size is limited by soil type, a split plot can be used with sample dates as whole plots and treatments as sub plots (Carter et al. 1983). Once the plants are harvested they must be separated into the parts required for analysis. These usually consist of leaves, stems (+ petioles), floral buds (squares), immature bolls and mature bolls (those cracked open more than 1 cm). After separation of the plant into the different parts, they are placed in labeled paper bags and dried at 70° C. Once dried, the weights are recorded. If required, a sub-sample of leaves should be taken to determine the specific leaf weight (SLW), or the dry weight of leaf per unit area. The leaf area in the total sample is equal to the total leaf dry weight divided by the

Table 1. Sample record sheet for primary plant data

Date: ____________ Location: ________________________

Plant no./sample: ________________________________

		Plot no.				
		1	2	3	4	5 . . . . . .
Treatment						
Replication						
Plant height	(cm)					
Main stem nodes	(no.)					
Veg. branches	(no.)					
Leaf DW	(g)					
Stem DW	(g)					
Immature boll DW	(g)					
Immature bolls	(no.)					
Mature boll DW	(g)					
Mature bolls	(no.)					
Square DW	(g)					
Squares	(no.)					
Leaf Sub-sample						
DM	(g)					
Area	(cm^2)					
no.						

SLW. A sub-sample for leaf area determination allows a greater number of samples to be measured with the area meter. The measurement of leaf area is usually the most time-limiting factor in growth analysis. Values for each plot should be recorded during the separation and weighing on a prepared data sheet (Table 1).

4.2 Growth Analysis Formulae (Kvet et al. 1971; Radford 1967)

4.2.1 Growth

Crop Growth Rate (CGR): The amount of dry matter produced per unit ground area.

$$CGR = dw/dt$$
$$CGR = W_2\text{-}W_1 / p(t_2\text{-}t_1)$$

Where, W_1 = dry weight at the first harvest; W_2 = dry weight at the second harvest; t_1 = time of the first harvest; t_2 = time of the second harvest; p = ground area.

CGR is the slope of a line drawn between the dry weights determined at t_1 and t_2. Growth rates can be calculated for every measurable plant part (Constable and Gleeson 1977).

Relative Growth Rate (R): Blackman (1919) was the first to express plant growth as having features analogous to the rate of interest in monetary growth. The efficiency index governing capital investments is identical to R. At any instant in time.

$$R' = (1/W)\, dW/dt.$$

The mean relative growth rate, R, over time interval t_2 to t_1 is

$$R = 1/t_2 = t_1 \int_{t_1}^{t_2} dW/W = \ln W_2 - \ln W_1/(t_2 - t_1) \quad [\text{weight weight}^{-1}\ \text{time}^{-1}].$$

The relative growth rate can be determined for the whole plant (R_W), or any plant part, such as stems (R_S), leaves (R_L) or leaf area (R_A).

Net Assimilation Rate (E): The expression of dry matter growth on a leaf area basis. The leaf area (A) is an estimate of the assimilatory potential of a plant or plant community. The estimate of E is therefore an assessment of the assimilatory efficiency. At any instant in time,

$$E' = 1/A\, (dW/dt).$$

The mean net assimilation rate is

$$E = 1/(t_2 - t_1) \int_{W_1}^{W_2} dW/A.$$

Integration of this expression requires that either the relationship between W and A be known, or the relationships between W and t, and A and t be known. If the ratio of R_W and R_A are linearly related as in the study by Potter and Jones (1977), then,

$$dW/dA = (W_2 - W_1)/A_2 - A_1$$

and

$$E = 1/t_2 - t_1 \int_{t_1}^{t_2} (1/A)(dA/dt)(dW/dA)dt$$

$$E=(1/t_2-t_1)[(W_2-W_1)/A_2-A_1]\int_{A_1}^{A_2} dA/A$$

$$E=(W_2-W_1/A_2-A_1)(\ln A_2-\ln A_1/t_2-t_1) \quad [\text{weight area}^{-1}\ \text{time}^{-1}].$$

Williams (1946) warned that if either W or A are increasing exponentially, they must be increasing by the same exponent. If W and A are related in a quadratic relationship, then,

$$W=k+(W_2-W_1/A_2^2-A_1^2)\,A^2$$

and

$$dW/dA=2A(W_2-W_1)/[(A_2-A_1)(A_2+A_1)]$$

$$E=1/t_2-t_1(2A/A)[W_2-W_1/(A_2+A_1)(A_2-A_1)]\int_{t_1}^{t_2}(dA/dt)dt$$

$$E=2(W_2-W_1)/[(A_2+A_1)(t_2-t_1)] \quad [\text{weight area}^{-1}\ \text{time}^{-1}].$$

Within large plants, mutual shading of leaves occurs and decreases the importance of total leaf area because not all leaves are photosynthetically active (Wells and Meredith 1988). For this reason, the relative contribution of leaf area to the production of photosynthate declines. Net assimilation rates reflect this trend and decline as growth progresses (Watson 1952).

4.2.2 Leaf Area and Related Values

Leaf Area Index (LAI): A measure of the leaf area (m^2) per unit ground area (m^2).

$$LAI = A/G$$

where, A = area of leaves and G = ground area from which the leaf sample was taken.

Maximum LAI of cotton is usually between 2 and 6 m^2 leaves m^{-2} ground. The actual value is closely tied to environmental conditions. The amount of leaf area, especially at lower LAI's, is linearly related to the capacity to intercept solar radiation up to a LAI of about 4 (Wells et al. 1986).

Leaf Area Duration (LAD): This measure introduces a time factor, because it estimates how long plants maintain their leaf area. For plant stands or communities,

$$LAD = \int_{t_1}^{t_2} A\,dt.$$

It is relatively easy to integrate the area under the curve of A over time. The relationship between A and t is given by the relative leaf area growth rate (R_L).

$$LAD=1/R_L\int_{t_1}^{t_2}(dA/dt)dt=A_2-A_1/R_L$$

$$LAD=[A_2-A_1/\ln A_2-\ln A_1](t_2-t_1) \quad [\text{area time}^{-1}].$$

Leaf Area Partitioning Coefficient (LAP): The LAP coefficient represents the amount of leaf area gained per unit leaf dry weight gained over a given period of time, usually a day (Potter and Jones 1977).

$$LAP = (dA/dt)/(dW/dt).$$

From the relationships for R_W and R_A

$$R_w = (\ln W_2 - \ln W_1)/t_2 - t_1 \quad \text{or} \quad W_2 = W_1\, e^{R_{w}t}$$

and

$$R_A = (\ln A_2 - \ln A_1)/t_2 - t_1 \quad \text{or} \quad A_2 = A_1\, e^{R_{A}t}.$$

Since $y = b\, e^{kx}$ and $dy/dx = kb\, e^{kx}$,

$$LAP = [(R_A)(A_1)e^{R_{A}t})]/[(R_W)(W_1)e^{R_{w}t})] \quad [(\text{area time}^{-1})(\text{weight time}^{-1})^{-1}].$$

The magnitude of LAP indicates the proportion of dry matter which is partitioned as leaf area. Increased leaf area during seedling growth has been identified as a valuable asset during early growth in cotton (Muramoto et al. 1965; Ashley et al. 1965; Wells and Meredith 1986).

Leaf Area Ratio (F): At any instant in time F is the ratio of leaf area per unit plant dry weight

$$F' = A/W.$$

The relationship among relative growth rate (R_W), net assimilation rate (E) and leaf area ratio (F) at any instant in time is

$$R'^{W} = E' + F'$$

This association only is true for mean values if both A and W are increasing exponentially and have the same exponents. It is best to determine F for each date of harvest separately.

4.3 Nondestructive Methods for Assessing Reproductive Development

4.3.1 Fruiting Form Counts

Flower counts are the simplest means of evaluating reproductive development. Flower counts provide information about when flowering started, the intensity or rate of flowering and the duration of the flowering period. For example, flower counts indicated that modern cultivars (since 1950) have a more intense flowering period, while obsolete cultivars (before 1920) flower less intensely initially, but have a longer flowering duration (Wells and Meredith 1984b).

Usually, weekly counts of white blooms from an undisturbed row are sufficient to monitor seasonal flowering trends. It is important to remember that white flowers occur on the day of measurement and pink flowers are from at least 1 day before. Whether all flowers present are counted, or only white ones, depends on the researcher. The interpretation of the data, however, will rely on the count method used. The number can be reported on a ground area basis.

4.3.2 Flower Tagging

Cotton is a dynamic crop species because it readily adjusts to the environment in which it is growing. The ability to study a square or boll from emergence to abscission/maturation is a valuable tool. The physical labeling of flowers, without damage to the organ is the best means of accomplishing this task. Each flower emerging on a specific date is labeled with a different color or shaped tag, and gives the researcher a reference point for determining age of development. Tagging of flowers has been used in the study of dry matter accumulation by bolls of different ontogenetic origins (Leffler 1976), the effect of blooming date on boll retention by the plant (Verhalen et al. 1975) and the estimation of progress in lint production after boll setting is initiated (Wanjura and Newton 1981). Tagging immature bolls of a certain size has also been used to follow reproductive development. Cathey et al. (1982) used this technique by tagging bolls to follow boll dehiscence and opening in response to various applied desiccant chemicals.

4.3.3 Plant Diagrams

The production of fruiting sites is a predictable phenomenon (see Sect. 2.2). The ultimate yield is dependent on the number of sites which produce bolls. Plant diagraming or mapping is one means of determining which sites contribute to the production of lint (Munro and Fairbrother 1969). Plants are harvested as in Sect. 4.1, and removed to a site for analysis. Individual plants are examined for (a) total main stem nodes, (b) the position of monopodia (node position), (c) the position of the sympodia, (d) the fruiting forms present (square, boll, or missing) at each site, and (e) the fruiting forms present on secondary sympodia (sympodia on monopodium branches). A system of notation will be required to record the data for each plant analysis. Munro and Fairbrother (1969) suggest the following: number the main stem nodes (0, 1, 2, ...) including the cotyledonary node; successive primary sympodia are lettered A, B, C, ..., including the node number where each is found; successive monopodia are lettered Z, Y, X, ..., including the node where they reside. The type of fruiting form at each position on the sympodium would be indicated by some symbol. A hypothetical data sheet could look like the following:

Main stem node

12–H	5–A 1b, 2–3x, 4b	
11–G	4–Y	
10–F 1s	3–Z	s = square
9–E 1s, 2s	2–	b = boll
8–D 1b, 2s, 3s	1–	f = flower
7–C 1x, 2f, 3s	0–COTYLEDON	x = missing or shed
6–B 1b, 2x, 3b, 4s		

The secondary sympodia could be mapped in the same manner.

Data can be summarized and frequency patterns can be formulated for node position or plant zone. The ability to identify aborted fruit is a definite strength of this analytical technique. Kerby and Buxton (1981) found that there is an enhanced likelihood of arborting fruiting forms on a sympodium when bolls are already present by using a similar plant diagraming system.

5 Resource Allocation

5.1 Lint Yield Determination

Every field experiment planned is limited to some extent in size and scope. The cost of land, materials and labor limits the number of treatments, genotypes, replications and locations which can be effectively handled. The dilemma facing scientists who work among large environmental effects is how to design research which answers the questions at hand, yet is manageable in size. It is a quest for precision in a world of limited resources. The differences among genotypes evaluated in only one environment are confounded with variation related to genotype × environment interactions (Miller et al. 1962). Valid genotypic recommendations are only reliable if sufficient environments are included in a breeding area to reduce such variation. Adequate numbers of environments can be included in one of the following ways: (1) a number of locations in one year, (2) a number of years in one location and (3) a combination of years and locations. Another less often used technique in cotton is planting on different dates at the same location (Wells and Meredith 1984a, 1986). If the dates are far enough apart in time (10–20 days), the patterns of plant growth are quite different.

The expected variance of a genotype mean (V_x) from a replicated number of tests at different locations and years was expressed by Miller et al. (1962) as:

$$V_{\bar{x}} = \sigma_e^2/rpy + \sigma_{gpy}^2/py + \sigma_{gp}^2/p + \sigma_{gy}^2/y$$

where,

σ_e^2 = plot variance
σ_{gpy}^2 = genotype × year × location variance
σ_{gp}^2 = genotype × location variance
σ_{gy}^2 = genotype × year variance
r = no. of replications
p = no. of locations
y = no. of years.

The variance of genotype lint yield means can be predicted for combination of replicates, locations and years, if the variance components are estimatable for a breeding area.

For a hypothetical breeding area, the following variance components are available:

$\sigma_e^2 = 3.97$
$\sigma_{gpy}^2 = 1.34$
$\sigma_{gp}^2 = 0.07$
$\sigma_{gy}^2 = 0.01$

The following case scenarios are possible:

Case 1: replicates = 4, years = 1, locations = 1;

$$V_{\bar{x}} = 3.97/(4)(1)(1) + 1.34/(1)(1) + 0.07/(1) + 0.01/(1) = 2.41$$

Case 2: replications = 4, years = 1, locations = 2;

$$V_{\bar{x}} = 3.97/(4)(1)(2) + 1.34/(2)(1) + 0.07/(2) + 0.01/(1) = 1.18$$

Case 3: replications = 4, years = 2, locations = 2;

$$V_{\bar{x}} = 3.97/(4)(2)(2) + 1.34/(2)(2) + 0.07/(2) + 0.01/(2) = 0.62$$

For cases 2 and 3, the variance of the genotype mean is reduced by the addition of years and/or locations. Addition of replications would also reduce $V_{\bar{x}}$ but the effect is greater if the number of years or locations is increased. Four additional replicates in case 1 would reduce the $V_{\bar{x}}$ to 1.92, while Case 2 demonstrates that an additional location with 4 replications would lower the $V_{\bar{x}}$ to 1.18. The effect of increasing the number of environments would be similar where treatments other than genotype were examined.

5.2 Growth Analysis Studies

Number of environments greatly affects the ability to detect true differences in growth analysis studies. Carter et al. (1983) examined five studies in soybeans and calculated a mean coefficient of variation of 25.6% for dry-weight measurements. On a single sampling date and assuming no treatment × environment interactions, 3, 16, and 65 replications would be required to detect 50%, 20%, and 10% treatment differences, respectively. Further, moderate treatment × environment interactions (5%–10% of error variance) were evident indicating the need for multiple environments to reliably test treatment effects. The authors present the following formulae to calculate usable combinations of environments, replicates and sampling dates for growth analysis studies:

$$E = [(2t_{\alpha/2} + t_{\beta})^2(\sigma_e^2/RS + \sigma_{TE}^2)]/d^2$$

Where,

$t_{\alpha/2}$ = t-value associated with the significance level of the t-test (usually 0.05);
t_{β} = t-value associated with the probability of accepting a false hypothesis based on the t-test;
σ_e^2 = error variance;
R = number of replications;
S = number of sampling dates;
σ_{TE}^2 = error associated with treatment × environment interaction;
d = true difference between treatments;

These relationships are graphically presented in Fig. 1 for a σ_{TE}^2 which is 10% of σ_e^2. Identification of a 15% treatment difference would require s/d of 1.5, and would neccessitate 4 environments if $R \times S = 18$. On the other hand, if $R \times S = 6$,

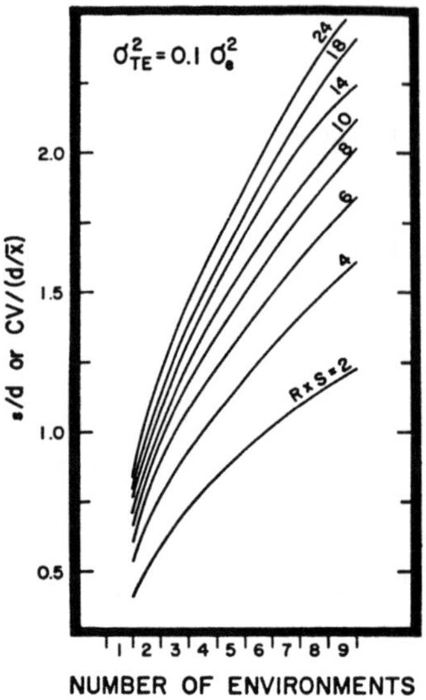

Fig. 1. Relationship among numbers of environments, replications, sampling dates, standard deviation, and absolute treatment difference necessary to show specified differences. $\sigma^2_{TE} = 0.1\ \sigma^2_e$; $\alpha = 0.05$; *R* replication no.; *S* sampling date no.; *s* standard deviation. (After Carter et al. 1983)

the number of environments would jump to 7. In either case, the allocation of resources for growth analysis studies is large, especially if the differences among the treatments are not large.

References

Adkisson PL, Niles GA, Walker JK, Bird LS, Scott HB (1982) Controlling cotton's insect pests: a new system. Science 216:19–22

Anonymous (1965) Mechanized harvesting of cotton. Southern Cooperative Series, Bull No 100

Arndt CH (1945) Temperature-growth relationships of the roots and hypocotyls of cotton seedlings. Plant Physiol 20:200–220

Ashley DA, Doss BD, Bennett OL (1965) Relation of cotton leaf area index to plant growth and fruiting. Agron J 57:61–64

Barker GL, McClendon RW, Colwick RF, Jones JW (1979) Relationship between cotton lint color and weather exposure. Trans ASAE 22:470–474

Blackman VH (1919) The compound interest law and plant growth. Ann Bot 33:353–360

Brashears AD, Ray LL (1966) Cotton stripper for harvesting research plots. USDA-ARS 42-122

Carter TE Jr, Burton JW, Cappy JJ, Isreal DW, Boerma HR (1983) Coefficients of variation, error variances, and resource allocation in soybean growth analysis experiments. Agron J 75:691–696

Cathey GW, Luckett KE, Rayburn ST Jr (1982) Accelerated cotton boll dehiscence with growth regulator and desiccant chemicals. Field Crops Res 5:249–256

Constable GA, Gleeson AC (1977) Growth and distribution of dry matter in cotton (*Gossypium hirsutum* L.). Aust J Agric Res 28:249–256

Culp TW, Harrel DC (1972) Variation in fiber and yarn properties of identical checks in yield tests and nursery plots of upland cotton, *Gossypium hirsutum* L. Crop Sci 12:355–358

Graves CR (1971) Performance of field crop varieties. Tenn Agric Exp Stn Bull 474:67–83
Guinn G (1982) Causes of square and boll shedding in cotton. USDA Bull no 1672
Guinn G, Mauney JR (1984) Fruiting of cotton II. Effects of plant moisture status and active boll load on boll retention. Agron J 76:94–98
Hearn AB (1969a) Growth and performance of cotton in a desert environment I. Morphological development of the crop. J Agric Sci 73:65–74
Hearn AB (1969b) Growth and performance of cotton in a desert environment II. Dry matter production. J Agric Sci 73:75–86
Hearn AB, Constable GA (1984) Cotton. In: Goldsworthy PR, Fisher NM (eds) The physiology of tropical field crops. John Wiley, New York, p 495
Hesketh JD, Baker DN, Duncan WG (1972) Simulation of growth and yield in cotton II. Environmental control of morphogenesis. Crop Sci 12:436–439
Hesketh JD, Lane HC, Alberte RS, Fox S (1975) Earliness factors in cotton: new comparisons among genotypes. Cotton Grow Rev 52:126–135
Kerby TA, Buxton DR (1981) Competition between adjacent fruiting forms in cotton. Agron J 73:867–871
Kvet J, Ondok JP, Necas J, Jarvis PG (1971) Methods of growth analysis. In: Sestak Z, Catsky J, Jarvis PG (eds) Plant photosynthetic production: manual of methods. W Junk, The Hague, p 343
Leffler HR (1976) Development of cotton fruit I. Accumulation and distribution of dry matter. Agron J 68:855–857
Mauney JR (1984) Anatomy and morphology of cultivated cottons. In: Kohel RJ, Lewis CF (eds) Cotton. Am Soc Agron, Madison, WI, p 59
Meredith WR Jr, Bridge RR (1973) Yield, yield component and fiber property variation of cotton (*Gossypium hirsutum* L.) within and among environments. Crop Sci 13:307–312
Meredith WR Jr, Ramey HH Jr, Bridge RR (1975) Determination of yield, yield components, and fiber properties as influenced by selective and nonselective samplings. Crop Sci 15:432–435
Miller PA, Robinson HF, Pope OA (1962) Cotton variety testing: additional information on variety × environment interactions. Crop Sci 2:349–352
Munro JM, Fairbrother HG (1969) Composite plant diagrams. Cotton Grow Rev 46:261–282
Muramoto H, Hesketh J, El-Sharkawy M (1965) Relationships among rate of leaf area development, photosynthetic rate, and rate of dry matter production among American cultivated cottons and other species. Crop Sci 5:163–166
Pendleton AM, Moore VP (1968) Ginning for quality preservation. In: Elliot FC, Hoover M, Porter WK Jr (1968) Advances in production and utilization of quality cotton. Iowa State Univ Press, Ames, IA, p 467
Potter JR, Jones JW (1977) Leaf area partitioning as an important factor in growth. Plant Physiol 59:10–14
Radford PJ (1967) Growth analysis formulae – their use and abuse. Crop Sci 7:171–175
Schubert AM, Benedict CR, Berlin JD, Kohel RJ (1973) Cotton fiber development – kinetics of cell elongation and secondary wall thickening. Crop Sci 13:704–709
Simpson DM, Hertel KL (1946) Environmental modification of fiber properties as a source of error in cotton experiments. J Agric Res 73:97–111
Tharp WH (1965) The cotton plant: how it grows and why its growth varies. USDA ARS, Agr Handbook no 178
Verhalan LM, Marnaghani R, Morrison WC, McNew RW (1975) Effect of blooming date on boll retention and fiber properties in cotton. Crop Sci 15:47–52
Wanjura DF, Newton OH (1981) Predicting cotton crop boll development. Agron J 73:476–481
Watson DJ (1952) The physiological basis of variation in yield. Adv Agron 4:101–145
Wells R (1988) Response of leaf ontogeny and photosynthetic activity to reproductive growth in cotton. Plant Physiol 87:274–279

Wells R, Meredith WR Jr (1984 a) Comparative growth of obsolete and modern cotton cultivars I. Vegetative dry matter partitioning. Crop Sci 24:858–862
Wells R, Meredith WR Jr (1984 b) Comparative growth of obsolete and modern cotton cultivars II. Reproductive dry matter partitioning. Crop Sci 24:863–867
Wells R, Meredith WR Jr (1984 c) Comparative growth of obsolete and modern cotton cultivars III. Relationship of yield to observed growth characteristics. Crop Sci 24:868–872
Wells R, Meredith WR Jr (1986) Heterosis in upland cotton I. Growth and leaf area partitioning. Crop Sci 26:1119–1123
Wells R, Meredith WR Jr (1988) Heterosis in upland cotton II. The relationship of leaf area to plant photosynthesis. Crop Sci 28:522–525
Wells R, Meredith WR Jr, Williford JR (1986) Canopy photosynthesis and its relationship to plant productivity in near-isogenic cotton lines differing in leaf morphology. Plant Physiol 82:635–640
Williams RF (1946) The physiology of plant growth with special reference to the concept of net assimilation rate. Ann Bot NS 10:41–72
Williamson EB, Shaw CS, Looney ZM, Shanklin EH (1966) Comparative effects of mechanical picking spindles and handpicking on cotton quality and spinning performance in Mississippi, 1960–1963. USDA Market Res Report no 730
Williford JR, Meredith WR Jr, Anthony WS (1987) Factors influencing yield and quality of Mississippi Delta cotton. Trans ASAE 30:311–316

Analysis of Carbohydrates Conferring Hardness on Seeds

J. S. G. Reid

1 Introduction

In relation to seeds the term "hardness" has two distinct connotations. It may refer to the physical resistance of a seed to mechanical deformation, or it may be used in the more specialized context of a seed which, in contrast to other seeds in a population, will not imbibe when exposed to water. Hardness of the latter type is akin to dormancy in that it prevents or delays germination. It is common amongst legumes, has survival value in a wild population and is undesirable in crop seeds. In this chapter "hardness" is used in its simplest sense – as an expression of a seed's resistance to mechanical deformation.

In seeds, as in other plant parts, the mechanical properties of tissues are controlled by the cell walls. Hard seeds possess massively thickened cell walls in a tissue which envelops the embryonic axis – usually in the endosperm or the storage cotyledons. This chapter is concerned with methodological approaches to the study of the polysaccharides which comprise the bulk of these walls.

Although these polysaccharides are responsible for the hardness of non-imbibed seeds, they have other biological functions which may be equally or more important in the reproductive strategy of the plant. Some are hydrophilic, leading to extensive swelling of the seed on imbibition. In the case of fenugreek (*Trigonella foenum-graecum* L., an endospermic legume) the galactomannan in the endosperm cell walls is the molecular basis of the seed's ability to imbibe copiously and to deploy the imbibed water to protect the germinating embryo from desiccation, should drought follow imbibition (Reid and Bewley 1979). By contrast, the mannan in the date *Phoenix dactylifera* L. endosperm (Keusch 1968) prevents extensive softening of the endosperm tissue on imbibition. Furthermore, it appears that massive carbohydrate deposits in endosperm or cotyledonary cell walls are invariably mobilized after germination, constituting a major carbohydrate reserve for the germinated seedlings. For this reason they are normally termed "cell-wall storage carbohydrates" (for reviews see Meier and Reid 1982; Reid 1985a).

Considered from the standpoint of seed hardness, the principal cell-wall storage carbohydrates of seeds are the mannan group, the xyloglucans and the galactans. The mannans comprise three structurally related types of polysaccharide, all of which share the structural feature of a linear $(1\rightarrow4)$-β-linked D-mannan backbone. The backbone may be unsubstituted or very lightly ($\sim$4%) substituted (the "pure" mannans) or relatively heavily (30% to 100%) substituted (the galactomannans) by α-D-galactopyranosyl residues linked $(1\rightarrow6)$ to mannose. The backbone may also contain $(1\rightarrow4)$ β-linked D-glucopyranosyl residues (the glucomannans). The linear cellulose-like pure and glucomannans resist hydration and remain hard on seed imbibition, whilst the branched galactomannans are water-imbibition molecules. All three types of mannan are typical of seed endosperms. The xyloglucans are structurally analogous to the galactomannans, have similar physical properties and may have similar overall biological roles. Structurally they consist of a (1→4)-β-linked D-glucan (cellulosic) backbone which carries α-D-xylopyranosyl and β-D-galactopyranosyl (1→2) α-D-xylopyranosyl substituents attached to C-6 of glucose. Xyloglucans are typically found in storage cotyledons. Galactans are of more restricted occurrence than the other groups. They are the principal polysaccharide constituents of the thickened cotyledonary cell walls of *Lupinus* species.

This chapter describes methodological approaches to the subcellular localization, quantitative and compositional analysis, and biosynthesis of cell-wall storage polysaccharides. The determination of their molecular structure has been carried out by standard techniques for the study of cell-wall carbohydrates (Selvendran 1989; elsewhere in this volume).

The cell-wall storage carbohydrates as a whole have been reviewed (Meier and Reid 1982; Reid 1985a) as have individual structural types: mannans and glucomannans (Bewley and Reid 1985); galactomannans (Dea and Morrison 1975; Reid 1985b).

2 Cytochemical Localization

In studies of the mobilization of cell-wall storage polysaccharides following germination, or their biosynthesis during seed development, it is necessary to observe and analyze changing distributions of polysaccharide within seed tissues. This section describes suitable methodologies at the light and electron microscopic levels.

An important technical problem encountered when dealing with dehydrated tissue blocks is the extreme hardness (!) of the cell walls, leading to difficulty both in obtaining adequate penetration of the walls by resins and in cutting. Dehydration may be avoided by sectioning unfixed frozen tissue using a cryostat. However, losses of hydrophilic cell-wall polysaccharides can then be a problem if the sections are placed in aqueous fixatives. Conventional aldehyde fixation procedures do not cross-link polysaccharides! These problems are highlighted in the experimental protocols which follow.

2.1 Light Microscopy

All known cell-wall storage polysaccharides have 1,2 glycol groupings which are periodate-reactive and may therefore be stained using periodic acid-Schiff (PAS) techniques [see Eqs. (1) and (2)]. In addition, the xyloglucans undergo a specific staining reaction with relatively concentrated iodine – potassium iodine reagents (Kooiman 1960), the molecular basis of which is not understood.

Cryostat-sectioning of galactomannan-containing tissue is described in conjunction with PAS-staining, followed by an iodine-staining procedure for xyloglucans.

2.1.1 Cryostat Sectioning and Periodic Acid-Schiff (PAS) Staining

Any fresh imbibed seed may be sectioned in the frozen state using a cyrostat, i.e. a microtome contained in a thermostated deep-freeze cabinet. The cryo-sections from seeds which contain water-insoluble polysaccharides (mannans, glucomannans) may be fixed conventionally, and then treated with *aqueous* periodic acid in the first step of the staining procedure described here. The sections from seeds with water-soluble polysaccharides, such as galactomannans and xyloglucans, should not be exposed to aqueous solutions until the periodate-reaction is complete. Microtome sections from fixed resin-embedded tissue blocks may be PAS-stained using aqueous periodic acid without loss of polysaccharide; the polymerized resin apparently prevents loss of polysaccharide from the sections.

Cryostat Sectioning of Fenugreek Seeds and PAS-staining of Galactomannan-containing Cell Walls (Reid 1971; Reid and Meier 1973). The galactomannans present in the cell-walls of legume seed endosperms are water-soluble. Care must therefore be taken to prevent their loss from fresh cryo-sections into aqueous fixatives or reagents. For this reason the frozen sections are placed into 40% ethanol and the periodate reaction is carried out using 40% ethanol as solvent. The ethanol prevents solubility losses of galactomannan from the section whilst allowing the reaction to take place. The polyaldehydic oxo-polysaccharide is water-insoluble, and can subsequently be stained with Schiff's reagent in aqueous solution [Eqs. (1) and (2)].

$$R{-}\underset{OH}{\overset{H}{\underset{|}{\overset{|}{C}}}}{-}\underset{H}{\overset{OH}{\underset{|}{\overset{|}{C}}}}{-}R' \quad + IO_4^- \rightarrow R{-}CHO + OHC{-}R' \quad + IO_3^- + H_2O \qquad (1)$$

1, 2 glycol groups in polysaccharide (soluble)	polyaldehydic oxo-polysaccharide (insoluble)

aldehyde	+ Schiff's reagent	→	aldehyde	+	*p*-rosaniline	(2)
	(*p*-rosaniline, decolorised with SO_2, soluble)		· bisulphite compound		(red; insoluble)	

Whole fenugreek seeds (imbibed, germinated or developing; diameter 4 mm or less) are placed on the metal sample holder of a cryostat. We normally use a Bright's (Bright Instrument Co, Huntingdon, UK) cryostat incorporating a Cambridge Rocking Microtome, and Ames OCT Compound (Ames Co Elkhart, Indiana, USA) as a medium to surround and orientate the sample prior to freezing. The base of the sample holder is then immersed in liquid nitrogen until the medium and sample are frozen. Frozen sections (15 to 30 μm thickness) are taken at a temperature of about −15° C, and placed or allowed to fall onto cold microscope slides. Section stretching and adhesion are achieved by gently warming the reverse of the slide, underneath the section, with a finger. The slides are then placed into a dish containing cold, vacuum degassed 40% (v/v) ethanol. The dish is then brought to room temperature.

Staining (at room temperature) is carried out by placing the slides sequentially in periodic acid (0.5% w/v) in 40% ethanol (5 to 15 min), 30%, 20%, 10% ethanol, water (5 min each), Schiffs reagent (BDH chemicals, Poole, UK), (5 to 20 min), SO_2-saturated water, prepared by adding 10 ml 10% w/v potassium metabisulphite and 10 ml 1 M HCl to 200 ml water (3 × 5 min) and finally in water 2 × 5 min. The exact times of treatment with periodic acid and with Schiff's reagent must be determined empirically.

Surface-water is drained from the specimens and they are enclosed in glycerine jelly for observation. Glycerine jelly is a mixture of glycerine (7 parts by weight), gelatine (1 part) and water (6 parts) containing 1% phenol as preservative. At room temperature it is a gel but it can be liquefied by warming to about 45° C. It may be obtained commercially (e.g. from R. A. Lamb, North Acton, London, UK). In practice, cover slips carrying a drop of liquefied glycerine jelly are warmed on an electric hot plate, and the slides carrying the specimens are inverted over them. Galactomannan is stained an intense purple-red. Other periodate-reactive polysaccharides may also be detected by this procedure.

2.1.2 Specific Iodine-Staining of Xyloglucans

The procedure described here may be used either on methacrylate-embedded sections or on cryostat sections. Cryostat sections should be treated directly with the stain. The reagent (Mitchell's reagent; Vogel and Schleiden 1839; Kooiman 1960) is prepared by dissolving iodine (0.3 g) and potassium iodide (1.6 g) in water (100 ml). The solution is placed directly on the object and covered with a cover glass. Coloration of xyloglucans is almost instantaneous, giving a clear blue colour in the cell walls. Starch granules in the cytoplasm are also stained by the reagent, giving an opaque brown-violet colour. The staining reaction is reversible, but semi-permanent preservation of specimens can be achieved by sealing the edges of the cover glass with wax or silicone grease.

2.2 Electron Microscopy

The electron microscope is necessary to examine the details of intramural polysaccharide digestion following germination and to ascertain the subcellular sites of polysaccharide biosynthesis during seed development. We find that the massively thickened cell walls of seeds can be embedded successfully using either traditional hydrophobic resins or modern hydrophilic media, such as L. R. White (London Resin Co, London, UK). Sufficient time must, however, be allowed for the media to penetrate the walls. Dehydrated cell walls are extremely hard and a diamond knife is essential to obtain reasonable sections.

Successful electron staining of polysaccharides can be obtained using Thiéry's (1967) excellent extension of periodate methodology to the electron microscope level, the periodic acid – thiocarbohydrazide – silver proteinate (or PATAg) technique. This deposits very fine silver grains on periodate-reactive structures. Techniques for the specific localization of individual polysaccharides based on immunocytochemistry and enzyme-gold labelling (Bendayan 1984) techniques are being developed at present (e.g. Reis et al. 1987). The following protocols illustrate PATAg staining of cell-wall storage polysaccharides (Goldberg and Roland 1970; Meier and Reid 1977) and enzyme-gold localization of xyloglucans (Reis et al. 1987).

2.2.1 Periodic Acid, Thiocarbohydrazide, Silver Proteinate Staining

The protocol is essentially that of Thiéry (1967). Sections are treated with periodic acid, and the aldehyde functions formed in the cell wall are reacted with thiocarbohydrazide [Eq. (3)] or thiosemicarbazide [Eq. (4)].

$$\mathrm{RCHO} + \mathrm{NH_2 \cdot NH \cdot \overset{\overset{\displaystyle S}{\|}}{C} \cdot NH \cdot NH_2} \rightarrow \mathrm{RC{=}NH \cdot NH \cdot \overset{\overset{\displaystyle S}{\|}}{C} \cdot NH \cdot NH_2} + \mathrm{H_2O} \quad (3)$$

thiocarbohydrazide

$$\mathrm{RCHO} + \mathrm{NH_2 \cdot NH \cdot \overset{\overset{\displaystyle S}{\|}}{C} \cdot NH_2} \rightarrow \mathrm{RC{=}NH \cdot NH \cdot \overset{\overset{\displaystyle S}{\|}}{C} \cdot NH_2} + \mathrm{H_2O} \quad (4)$$

thiosemicarbazide

A silver salt is then reduced by the immobilized thio-compound, to give a deposit of metallic silver on the specimen. In practice the non-stoichimetric "silver proteinate" gives excellent resolution. Conventional fixation and embedding techniques may be used. Controls must be carried out in which the periodic acid oxidation step is omitted.

Sections on gold grids (copper dissolves in the reagents!) are floated on 1% aqueous periodic acid for 20 min, washed thoroughly with distilled water, and transferred to thiocarbohydrazide (2% in 20% acetic acid) for up to 72 h. Some polysaccharides react rapidly (30 to 40 min) others very slowly. Grids are then washed thoroughly with 10% acetic acid followed by water, before transfer to 1% silver proteinate (Roques SA, Paris, France) in double-distilled water for 30 min in the dark.

2.2.2 Localization of Xyloglucan by Enzyme-Gold Cytochemistry

Enzymes linked to colloidal gold particles have been used successfully for the cytochemical localization of macromolecules (Bendayan 1984). The linkage of the enzymes to the gold is achieved by using the enzyme protein at a pH equal to or lower than its isoelectric point to protect a negatively charged gold colloid from coagulation. Each gold particle is effectively surrounded by a thin layer of protein, which apparently retains its substrate-binding properties, thus allowing the substrate to be localized in ultrathin sections. It is possible in practice to prepare monodisperse gold sols with particle diameters ranging from 5 nm to about 30 nm. For each sol and protein the optimum conditions (pH, protein concentration) for stabilization must be found experimentally, as must the optimum concentration and pH of the complex, for treatment of the sections. Carbohydrates and glycoproteins may also be localized using lectin-gold complexes prepared in the same way (Roth 1983). A number of good examples of the specific staining of plant cell-wall polysaccharides have been reported (e.g. Vian et al. 1983; Ruel and Joseleau 1984).

The following protocol describes the cytochemical localization of the cell-wall storage xyloglucan in tamarind-seed cotyledons (*Tamarindus indica* L.) by making use of the specificity of a β-D-galactosidase-gold complex towards the terminal non-reducing β-D-galactopyranosyl residues in the molecule (Reis et al. 1987). The enzyme chosen by the authors was the β-D-galactosidase from *E. coli*. This enzyme does not hydrolyze the β-D-galactosidic linkage of tamarind-seed xyloglucan (Edwards et al. 1988), but is clearly able to bind to the molecule.

Localization of Xyloglucan in Tamarind (Tamarindus indica L) Seed Cotyledons Using a β-D-Galactosidase-gold Complex (Reis et al. 1987). A colloidal gold sol (particle diameter 15 nm) is first prepared according to Frens (1973). A solution (0.01% w/v; 100 ml) of tetrachloroauric acid (Merck, Darmstadt, Germany) is heated to boiling and 4 ml of a solution (1% w/v) of trisodium citrate is added to it rapidly. Reduction of the gold salt to a colloidal gold sol occurs during about 5 min of further boiling, giving a deep reddish-orange colour. If necessary, the particle size of the gold sol may be checked using the electron microscope. The pH of the gold sol may be checked on a small sample (~10 ml) using a glass electrode, after adding 2 drops of a solution (1%) of polyethyleneglycol 20000 to stabilize the sol and prevent its coagulation onto the electrode.

An aliquot (10 ml) of the gold sol (adjusted to pH 4.7) is stabilized using the β-D-galactosidase from *E. coli* (Boehringer, Mannheim, Germany) by adding a volume of the enzyme solution (dialyzed against distilled water and adjusted to pH 4.7) corresponding to 0.5 mg protein. The mixture is then centrifuged at 8000 g, to give a colourless supernatant and a dark-red pellet (precipitated gold) covered by a red, mobile "pool" of sedimented but non-pelleted material (the stabilized colloidal gold particles). This mobile "pool" is carefully collected and diluted with phosphate-buffered saline pH 7.2 containing 0.02% polyethylene glycol 20000. This is the β-D-galactosidase complex for use in labelling experiments. The final dilution required to give even, specific labelling must be determined empirically.

Glutaraldehyde-fixed epon-araldite embedded sections (mounted on gold or nickel grids) are incubated on the gold complex for 30 min and washed thoroughly. This method apparently gives specific labelling of xyloglucan. Specificity in enzyme–gold technology must be checked by carrying out control experiments – e.g. by incubating at low temperatures, at unsuitable pH values for enzyme binding, in the presence of competing substrates, and with sections which do not contain the substrate to be localized.

3 Quantitative and Compositional Analysis

This section outlines different methodological approaches to the quantitative analysis of cell-wall storage polysaccharides in seeds. Depending on the system being studied and on the demands of the experiment, several different approaches are possible. It is known, for example, that galactomannan accounts for almost all the dry weight of some legume-seed endosperms (Reid 1971; Leung et al. 1981). An acceptable analysis of changes in galactomannan content in such seeds may therefore be obtained by manual isolation of the endosperms, extraction of these with aqueous ethanol to remove low-molecular-weight compounds, and dry-weight determination. Should compositional analysis be required, it may be carried out by hydrolyzing samples either of the extracted freeze-dried endosperm or of purified polysaccharide obtained from it. For other systems, for example cotyledonary storage tissue which usually contains important macromolecular reserves other than cell-wall polysaccharide, this approach is inappropriate. The cell-wall storage polysaccharides may, however, be extracted with alkali, purified free of low-molecular-weight material, protein and contaminating polysaccharides, and weighed; samples may be subjected to compositional analysis. If quantitative extraction of cell-wall storage polysaccharides from the tissue proves impossible, the whole tissue may be ground and extracted with aqueous methanol or ethanol to give an "alcohol-insoluble residue (AIR)". Weighed samples of this may be hydrolyzed directly with addition of an internal reference standard to give quantitative and compositional analysis of the cell-wall storage polysaccharide. There are occasions when specific aspects of the molecular structure of a cell-wall storage polysaccharide must be analyzed in relation to a developmental process. Such changes are best analyzed enzymatically.

The following protocols illustrate (i) the quantitative analysis of galactomannan by gravimetric analysis of isolated seed endosperms, and its compositional analysis following acid hydrolysis; (ii) the quantitative and compositional analysis of xyloglucan following quantitative extraction from the tissue and purification; (iii) the analysis of the cell-wall storage polysaccharides in the "alcohol-insoluble residues" of lupin-seed cotyledons and (iv) the enzymatic analysis of galactomannans.

3.1 Galactomannan by Gravimetric and Compositional Analysis

In fenugreek (*Trigonella foenum-graecum* L.) (Reid 1971) guar [*Cyamopsis tetragonoloba* (L.) Taub] (McClendon et al. 1976) and other leguminous seeds, galactomannan is the major constituent of the endosperm tissue. Most of the endosperm is a mass of non-living tissue "filled" with galactomannan of cell-wall origin (Meier and Reid 1977). This is surrounded by a thin, living aleurone layer. This type of endosperm anatomy is easily recognized by light microscopic examination of the imbibed seeds (Sect. 2.1 and Reid 1971). Other leguminous seeds, such as Carob (*Ceratonia siliqua* L.), have a less specialized endosperm anatomy (Seiler 1977) with living cells present throughout the endosperm. The errors involved in gravimetric galactomannan analysis of such seeds are greater.

3.1.1 Analysis of Galactomannan in Fenugreek (*Trigonella Foenum-graecum* L.)

Replicate (the number of replicates depending upon the requirements of the experiment) samples of six imbibed, germinated, (Reid 1971) or developing (Campbell and Reid 1982) seeds are dissected manually to remove the endosperms. The endosperm is easily recognized as the colourless translucent material between the yellow testa and the embryo. Dissection is easily accomplished using a binocular microscope or a large hand-lens; there is no adhesion of the endosperm either to the embryo or the testa. The replicate endosperm samples are dropped into glass centrifuge tubes (10 ml capacity) containing 70% v/v methanol (5 ml) at 70° C, and allowed to extract for 15 min with occasional agitation. The tubes are centrifuged (1500 g; 10 min) and the supernatants are discarded, unless they are required for analysis of low-molecular-weight carbohydrates (Reid 1971; Spyropoulos and Reid 1988). The pellets are re-extracted twice in the same way and the final pellets are suspended in water (0.2 ml) and freeze-dried. Secondary drying is carried out in a vacuum desiccator over P_2O_5. Samples are weighed and galactomannan is expressed in mg per seed on the assumption that it contributes all the dry weight of the extracted endosperm.

Compositional analysis of the galactomannan is carried out by total acid hydrolysis and quantitative analysis of the monosaccharides released. The following procedure is the one followed routinely in the author's laboratory. It incorporates the 72%/4% H_2SO_4 hydrolysis technique of Saeman et al. (1945) and a variant of the alditol acetate procedure for gas-liquid chromatographic analysis of the neutral monosaccharides released. It should be noted that acidic sugars are not analyzed in this procedure. They are absent from most cell-wall storage polysaccharides, but if they are present they must be analyzed by a separate procedure (see Sect. 3.3.1).

Duplicate samples (about 5 mg) of the polysaccharide are weighed accurately into 10-ml glass centrifuge tubes. To each tube is added 72% w/w sulphuric acid (7.2 g concentrated H_2SO_4 added to 2.61 g water), (0.1 ml), and a glass rod for mixing. The tubes are warmed at 30° C for 45 min with intermittent use of the glass rod to ensure wetting of the solid and eventually its complete dissolution. At the end of that time water (1.2 ml) is added to each tube followed by 0.5 ml of a solution of myoinositol (10.00 mg ml^{-1}) to serve as an internal standard for gas-liquid chromatographic analysis. The mixture is stirred and the glass rod removed. Tubes are capped with marbles and autoclaved at 120° C for 60 min. After ice-cooling of the tubes, the contents are neutralized by the gradual addition of solid barium carbonate, with thorough mixing. Neutrality is tested using Congo red paper. Water (3.0 ml) is added to each tube, and they are mixed and centrifuged (1500 g; 15 min). If the supernatants are still cloudy, they are transferred to polycarbonate centrifuge tubes and spun at 15000 g (15 min). Supernatants are transferred to small pear-shaped ground-glass-necked flasks. To each is added 1 M ammonia (0.5 ml) followed by 0.5 ml of a freshly prepared solution (20 mg ml^{-1}) of sodium borohydride in 1 M ammonia.

The flasks are allowed to stand for 1 h and excess borohydride is decomposed by adding glacial acetic acid until hydrogen evolution ceases (only a few drops). Each flask is taken to dryness on a rotary evaporator, and the glassy residue is taken up in methanol (at least of HPLC grade; stored and handled in glass apparatus) (0.5 ml). The methanol is removed on the evaporator and the residue is taken up in a further portion (0.5 ml) of methanol. Three further additions and evaporations of methanol are carried out. To the residue (usually white and powdery) is added (in a fume hood) acetic anhydride (0.5 ml) taking care to wash all solids to the bottom of the flask. The flasks are placed in an oil bath (behind a safety screen) at 130° C for 3 h. Stoppers are inserted only after the flasks have reached temperature equilibrium with the bath and removed before the flasks are cooled. The flasks are cooled and their contents analyzed directly by GLC. These mixtures may be stored indefinitely in capped, polyethylene tubes. GLC analysis may be carried out using any routine packed-column instrument with flame-ionization detector(s) and (preferably) glass column(s). In our laboratory, analyses are carried out using columns (2 m × 4 mm) packed with ECNSS-M (3% on Gas-Chrom Q) at 180° C. Samples (5 μl) are injected on-column.

The rationale behind the above procedure is explained in Table 1 and a typical separation is illustrated in Fig. 1.

Table 1. Rationalization of acid hydrolysis and alditol-acetate formation

Step	Purpose
72% H_2SO_4 digestion	Solubilization and partial hydrolysis
4% H_2SO_4 digestion	Total acid hydrolysis
$BaCO_3$	Neutralization using a sparingly soluble salt to give an insoluble product ($BaSO_4$)
$NaBH_4$	Reduction of monosaccharide (multiple ring and anomeric forms) to alditol (single compound)
Acetic acid	Decomposition of $NaBH_4$ to boric acid
Methanol addition and evaporations	Removes boric acid as the volatile methyl ester leaving alditol acetates and sodium acetate
Acetic anhydride	Formation of alditol acetates (sodium acetate as catalyst)

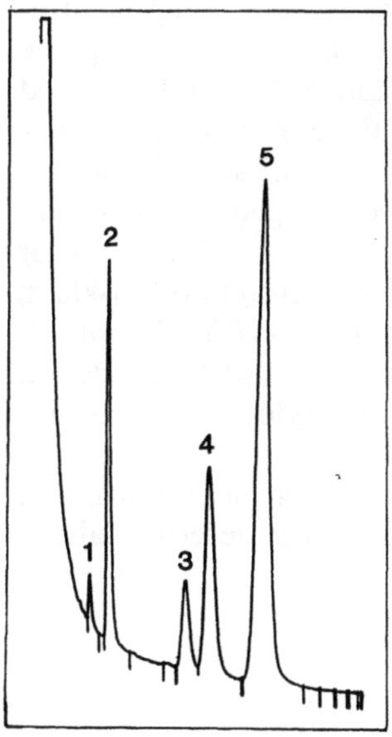

Fig. 1. A typical GLC separation of neutral sugars released on acid hydrolysis of a xyloglucan. Numbers indicate alditol acetates of: *1* arabinose; *2* xylose; *3* galactose; *4* glucose; *5* myoinositol (internal standard)

3.2 Xyloglucan by Extraction and Purification

Storage xyloglucans are present mainly in cotyledonary tissues. When extracted they are water-soluble, but it is difficult to attain quantitative extraction of them from the tissue using water. Quantitative extraction of xyloglucans (and most other cell-wall storage polysaccharides) may be achieved using aqueous alkali. To avoid alkaline oxidation of the polysaccharide (Aspinall et al. 1961), sodium borohydride is incorporated into the extraction solution. Other polysaccharides including starch are also solubilized in this way, but xyloglucans may be selectively purified by precipitating them from solution as their copper (II) complexes. Xylans and soluble mannan-type polysaccharides also form insoluble copper complexes, but we have not found these to be significant contaminants in our xyloglucan preparations.

3.2.1 Analysis of Xyloglucan in Nasturtium (*Tropaeolum majus* L.)

Each replicate sample of ten cotyledon pairs from imbibed or germinated seeds (Edwards et al. 1985) with testae attached (fresh weight when imbibed but ungerminated ~2.7 g) is ground thoroughly in a mortar with sand (5 g) and water (10 ml). The contents of the mortar are transferred to a 100-ml glass centrifuge tube containing sodium borohydride (50 mg). Quantitative transfer is assured by rinsing the mortar with 2 M sodium hydroxide (90 ml). The contents of the tube are stirred and alkaline extraction is allowed to proceed at 100° C for 1 h with occasional stirring. After centrifugation the supernatant is poured into glacial acetic acid: ethanol (1:10 v/v) (200 ml). This achieves neutralization of the sodium hydroxide, precipitation of the crude xyloglucan and decomposition of the borohydride (caution – hydrogen is evolved, plus hydrogen sulphide from endogenous sulphur-compounds). The precipitated xyloglucan is allowed to coagulate and is collected by centrifugation (1000 g). Prior to purification it is suspended in water (80 ml) and brought into solution by adding 10 ml of the alkaline component of Fehling's solution (Fehling's B) and heating at 100° C for 30 min. On cooling, the copper-containing component of Fehling's solution (Fehling's A; 10 ml) is added and the mixture is stirred vigorously. Any precipitate of the blue xyloglucan copper complex is collected by centrifugation (1400 g; 10 min) and dissolved in an appropriate volume (20 to 150 ml) of a mixture (1:4 by vol) of glacial acetic acid and water. The acid-decomplexed xyloglucan is recovered by adding 2 vol of ethanol and collecting the (colourless) precipitatte by centrifugation (1400 g; 10 min). The precipitated xyloglucan is resuspended in 65% (v/v) ethanol, allowed to stand overnight, collected by centrifugation, dissolved or suspended in water, freeze-dried, and weighed. Samples of the purified xyloglucan are subjected to compositional analysis (Sect. 3.1.1).

Experiments involving multiple extractions of the seed tissue have shown that 94% of the xyloglucan is extracted in a single alkali-treatment as described above (Edwards et al. 1985).

3.3 Lupin-Seed Cotyledonary Polysaccharides by Hydrolysis of Alcohol-Insoluble Residues

The thickened cotyledonary cell walls in *Lupinus* species contain a highly insoluble polysaccharide material which is very rich in galactose, and also contains arabinose, xylose and galacturonic acid. To assess the role of the material following germination in *Lupinus angustifolius* L., the amount and composition of it was analyzed at various stages of germination and seedling development (Crawshaw and Reid 1984). Because of its insolubility and the presence of large protein and lipid reserves the analysis had to be carried out by hydrolysis of alcohol-insoluble residues, with the application of a correction for starch, which is formed transitorily following germination (Crawshaw and Reid 1984). Uronic acids were determined separately.

3.3.1 Analysis of *Lupinus angustifolius* L. Cotyledons (Crawshaw and Reid 1984)

Replicate samples of ten cotyledons are removed from imbibed seeds, or seedlings and homogenized with methanol (50 ml AristaR quality, BDH chemicals, Poole, UK). Following centrifugation (3000 g; 15 min) the pellet is extracted with boiling 70% (v/v) methanol (20 ml) for 10 min. The extract is centrifuged (3000 g; 15 min), the supernatant decanted and the pellet re-extracted with two further portions (20 ml) of boiling 70% methanol. The pellet is then extracted with 30 ml of chloroform: methanol (2:1 v/v) for 15 min, freeze-dried and weighed. This material contains all the polysaccharide present in the original cotyledons. Quadruplicate samples of it (about 5 mg) are weighed exactly, hydrolyzed with acid and subjected to GLC analysis exactly as described in Sect. 3.1.1. The GLC trace is used to obtain compositional analysis, and also (using the internal standard of myoinositol) to calculate the weight of neutral sugar residues in each sample hydrolyzed. This figure includes glucose from starch, but not uronic acids.

To assess the contribution of starch-glucose to the results obtained as above (and to assay starch in the cotyledons!) a second set of quadruplicate samples is treated with α-amylase to remove the starch and then hydrolyzed with acid. Each accurately weighed sample (in a 10-ml centrifuge tube) is boiled with 1.5 ml of calcium chloride solution (3×10^{-3} M, pH 7) to gelatinize any starch present. α-Amylase from hog pancreas (Sigma, Poole, UK) (0.1 mg ml^{-1}) is added and the mixture incubated for 15 min at 30° C. Methanol is added to give a final methanol concentration of 70% and the mixture is boiled for 15 min and centrifuged (3000 g; 15 min). The supernatant, containing mono- and oligosaccharides released from starch, is discarded and the 70% methanol extraction is repeated twice. The pellet is re-hydrated, freeze-dried, and subjected to hydrolysis and GLC analysis as above.

Uronic acids are determined separately on the original alcohol-insoluble residues. Replicate samples (100 mg) of the material are suspended in water (25 ml) using a sonicator (MSE Crawley, UK; 15 s pulses for 15 min), and the fine suspension is diluted to 50 ml. Uronic acid levels are determined spectrophotometrically on aliquots (0.2 ml) using the method of Blumenkranz and Asboe-Hansen

(1973). It should be noted that when the tetraborate-sulphuric acid reagent is added, the suspended material dissolves. Because of the errors involved in handling suspensions, quadruplicate samples should always be analyzed by this method.

3.4 Enzymatic Analysis

Studies of the hydrolysis in vivo of galactomannans and xyloglucans, and screening studies of microbial enzyme sources, have led to the identification and purification of enzymes which are capable of being used for the enzymatic analysis of cell-wall storage polysaccharides. Of relevance to galactomannan analysis are *endo*-(1→4)-β-D-mannanases (EC 3.2.1.78) of differing specificities (McCleary 1979), α-D-galactosidases (EC 3.2.1.22) capable of removing the terminal non-reducing α-D-galactopyranosyl residues from galactomannans (McCleary et al. 1981; McCleary 1983), β-D-mannosidases (EC 3.2.1.25) (McCleary 1983) or exo-β-D-mannanases. The *endo*-β-D-mannanases have been used successfully to obtain information on the distribution of galactose substituents along the mannan backbones of different galactomannans (McCleary et al. 1985). These and other structural studies have been described in detail by McCleary and Matheson (1986). The use of enzymes to analyze ^{14}C-labelled galactomannans synthesized in vitro is described in Sect. 4.

In the case of the xyloglucans an endo-(1→4)-β-D-glucanase, a β-D-galactosidase (EC 3.2.1.23) and an α-D-xylosidase have been shown to participate in the hydrolysis of nasturtium seed xyloglucan in vivo (Edwards et al. 1985). The endo-β-D-glucanase has been purified to homogeneity and has been shown to be a novel enzyme, possessing apparently complete specificity towards xyloglucans (Edwards et al. 1986). The β-D-galactosidase has also been purified and has been shown to be capable of removing the terminal non-reducing β-D-galactopyranosyl residues from xyloglucans (Edwards et al. 1988). Both show promise for xyloglucan analysis.

4 Biosynthesis

Most of the analytical techniques described in Sects. 2 and 3 were developed to study cell-wall storage polysaccharide mobilization following germination, and the accumulation of the polysaccharides in developing seeds. The biosynthesis of these molecules in vitro has received attention only recently and only in the context of the galactomannans (Campbell and Reid 1982; Reid et al. 1987; Edwards et al., 1989). Galactomannan analysis carried out on developing fenugreek and guar seeds has established the patterns of galactomannan deposition in vivo (Edwards et al., 1989). Particulate enzyme preparations obtained from endosperms actively depositing galactomannan exhibit high activities of a GDP-mannose-linked D-mannosyltransferase and a UDP-galactose-linked D-galactosytransferase (Reid et al. 1987). Their cooperative interactions in vitro to synthesize galactomannans of the legume-seed type have been demonstrated (Reid et al. 1987; Edwards et al., 1989). The methods used to investigate galactomannan biosynthesis are reported here for two reasons. They provide an example of plant cell-wall polysaccharide biosynthesis in vitro in which the glycosyltransferase enzymes are present at high activity, and they illustrate the potential use of enzymes to characterize radiolabelled products formed in vitro.

4.1 Galactomannan Biosynthesis in Vitro (Reid et al. 1987; Edwards et al., 1989)

Particulate enzyme preparations have been isolated from developing fenugreek and guar-seed endosperms during the period of galactomannan deposition in vivo. In the presence of GDP-mannose and a divalent metal cation (Mg^{2+}, Mn^{2+} or Ca^{2+}) these preparations catalyze for formation of (1→4)-β-D-mannan. In the presence of UDP-galactose, GDP-mannose and Mn^{2+}, galactomannans are formed. They have been shown by enzymatic analysis to have the legume-seed type structure – a linear (1→4)-β-linked D-mannan backbone substituted by single-unit α-D-galactopyranosyl side-chains attached 1→6 to mannose. The transfer of D-galactosyl residues from UDP-galactose to galactomannan has been shown to be absolutely dependent upon the simultaneous transfer of D-mannosyl residues from GDP-mannose. D-Mannan sequences preformed in situ using the mannosyltransferase in the absence of UDP-galactose could not subsequently become galactose-substituted. The degree of galactose substitution of the galactomannan formed in vitro can be manipulated by varying the rate of D-mannosyltransfer via GDP-mannose substrate concentration at constant (saturating) UDP-galactose levels. These studies have enabled a tentative model for galactomannan biosynthesis to be proposed (Edwards et al., 1989), and they are allowing the investigation of the molecular basis of the genetic control of the degree of galactose substitution of galactomannans in seeds. The full experimental basis of the above conclusions is outlined elsewhere (Edwards et al., 1989). The following protocols describe the isolation of particulate transferase from fenugreek and guar endosperms, the conditions for achieving mannan or galactomannan biosynthesis in vitro, and the enzymatic analysis of labelled products formed in vitro.

4.1.1 Particulate Enzyme Preparations from Developing Fenugreek and Guar-seed Endosperms

Young pods from fenugreek and guar plants (Edwards et al., 1989) with seeds at a stage of development corresponding to maximal galactomannan deposition are chosen (seed fresh weight 25 to 35 mg for fenugreek, 90 to 110 mg for guar). Endosperms are isolated from developing seeds manually and ground in a glass homogenizer with isolation buffer (50 mM Tris-HCl pH 7.5 containing 5 mM $MgCl_2$ and 5 mM DTT). Homogenates are centrifuged (40000 g, 60 min) and the pelleted material is resuspended in isolation buffer (50 μl per fenugreek endosperm or 150 μl per guar endosperm). All isolation procedures following seed dissection are carried out at or below 4° C. Fresh particulate enzymes must be prepared for each experiment.

4.1.2 Incubation of Particulate Enzymes with Sugar Nucleotide Substrates

Incubations are carried out in polypropylene screw-capped microcentrifuge tubes (1.5 ml capacity) (Sarstedt Ltd, Leicester, UK) at 30° C, normally for 1 h. The incubation mixture (total volume 100 μl) comprises particulate enzyme suspension (50 μl), DTT (2.5 mM), $MgCl_2$ (2.5 mM), $MnCl_2$ (5 mM, only when UDP galactose is a substrate), GDP-mannose (normally 80 μM) and/or UDP galactose (800 μM) and 25 μM Tris-HCl pH 7.5. When radiolabelled sugar nucleotides are used (NEN Research Products, Stevenage, UK) their specific activities are adjusted by dilution with the corresponding non-labelled nucleotides. Normally UDP-U-^{14}C-D-galactose is used at a specific radioactivity of 23 Bq $nmol^{-1}$, and GDP-U-^{14}C-D-mannose at 46 Bq $nmol^{-1}$. At the end of the incubation period glacial acetic acid (50 μl) is added to each tube (to decompose any Mn^{2+}-galactomannan complex) and the capped tubes are heated to 100° C for 2 min. To each tube a solution (100 μl; 2 mg ml^{-1}) of guar or fenugreek galactomannan is added (as co-precipitant) followed by methanol to give a final methanol concentration of 70% by volume. Tubes are capped, mixed and heated at 70° C for 5 min. Precipitates (of galactomannan plus any 70% methanol insoluble product of the enzymatic reaction) are recovered by centrifugation (15000 g; 5 min) and supernatants are discarded. Pellets are washed twice with 70% methanol (2 × 750 μl) at 70° C as above. The pellets remaining after the final incubation are freeze-dried prior to enzymatic or other analysis (Edwards et al. 1988) or dissolved in 12 M HCl (15 μl) by heating at 70° C for 10 min prior to liquid scintillation counting. Scintillation fluid (1 ml, Optiphase X–LKB Instruments, Ltd, South Croydon, UK) is added directly to the microcentrifuge tube, giving a clear stable solution. The microcentrifuge tubes fit exactly inside polyethylene microvials (LKB) for the scintillation counter.

When the sole sugar nucleotide substrate in the incubation is GDP-mannose, the product is 1→4 β-linked D-mannan. When the sole substrate is UDP-galactose only very low levels of incorporation are achieved, and the only labelled monosaccharide released on acid hydrolysis of the product is glucose. When UDP-galactose and GDP-mannose are present in the incubation mixture, the product is a galactomannan. By carrying out replicate incubations, some containing labelled GDP-mannose and unlabelled UDP-galactose and others with labelled UDP-galactose and unlabelled GDP-mannose, plus controls with labelled UDP-galactose alone, the mannose-galactose ratio in the galactomannan product may be calculated.

4.1.3 Enzymatic Analysis of Galactomannans

For the enzymatic analysis of galactomannans or radiolabelled products formed in vitro in the previous section two highly purified enzymes are used in our laboratory; an endo-(1→4)-β-D-mannanase from *Aspergillus niger* (McCleary and Matheson 1983) and an α-D-galactosidase from germinated guar seeds (McCleary 1983). Both were purified exactly as described by the original authors and were obtained as ammonium sulphate suspensions. The suspensions were dialyzed against ammonium acetate buffer (100 mM, pH 5.0) and frozen in 1 ml aliquots. When assayed (pH 5.0) using carob seed galactomannan as substrate (10 mg ml^{-1}) the endo-β-D-mannanase (0.14 mg protein ml^{-1} assayed according to Sedmak and Grossberg 1977) released reducing power (Halliwell and Riaz 1970) at an initial rate of 89 mmol mannose equivalents min^{-1} l^{-1} (mg protein)$^{-1}$. Under the same assay conditions the enzyme released 7.4×10^{-3} mmol D-galactose (determined as described by Edwards et al. 1985) min^{-1} l^{-1} (mg protein)$^{-1}$ indicating a very slight contamination by α-D-galactosidase (<0.01%). The α-D-galactosidase (protein content 0.09 mg ml^{-1}) released D-galactose at the rate of 17.3 mmol min^{-1} l^{-1} (mg protein)$^{-1}$. It had no endo-β-D-mannanase activity.

For analysis, galactomannans (0.2 mg) or labelled products formed in vitro containing unlabelled galactomannan (0.2 mg) as co-precipitant are freeze-dried in 1.5-ml microcentrifuge tubes. Aliquots (50 μl) of α-D-galactosidase and/or endo-β-D-mannanase are added to each tube and the total volume is adjusted to 100 μl. After incubation at 30° C with intermittent mixing for 4 h, the enzymes are denatured by heating at 100° C for 3 min. Any insoluble material is sedimented (15000; 5 min) and the supernatant is analyzed by thin-layer chromatography (TLC) on Kieselgel 60 (E. Merck, Darmstadt, W. Germany) using double development in n-propanol : ethanol : water 7 : 1 : 2 by volume. Plates are scanned for radioactivity if appropriate, sprayed with sulphuric acid (5% v/v in ethanol) and heated at 100° C to reveal separated compounds.

Hydrolysis of galactomannans with α-D-galactosidase alone releases only galactose. Hydrolysis with both enzymes together releases galactose, mannobiose, mannose and mannotriose, the ratio of galactose to mannose and mannosaccharides being commensurate with the mannose : galactose ratio in the original galactomannan. Used alone the *endo-β*-mannanase gives a pattern of hydrolysis which is indicative of the degree of galactose substitution of the galactomannan substrate. With carob galactomannan (32% galactose-substituted) as substrate, mannobiose, mannose and mannotriose are released plus a compound migrating slightly more slowly than mannotriose; this has been identified as a monogalactosyl-substituted mannobiose (McCleary and Matheson 1983). Higher-molecular-weight saccharides are also released, plus some chromatographically immobile material.

With guar galactomannan (65% galactose-substituted) as substrate the degree of hydrolysis is less, since more of the hydrolysate is chromatographically immobile. The main lower saccharides released are mannotriose, mannose and galactosylmannobiose. With the highly galactose-substituted (97%) galactomannan of fenugreek as substrate, there is a very low degree of hydrolysis. Only traces of mannobiose, mannose and galactosylmannobiose can be detected on chromatograms. These results are illustrated in Fig. 2.

Enzymatic hydrolysis of labelled products formed in vitro from GDP-mannose and/or UDP galactose have been analyzed enzymatically, using the enzymes singly and in combination. This has enabled the positive identification of products formed in the presence of both substrates as galactomannans, and the assessment of their degrees of galactose substitution. A typical radiochromatogram of a mannose-labelled product is shown in Fig. 3.

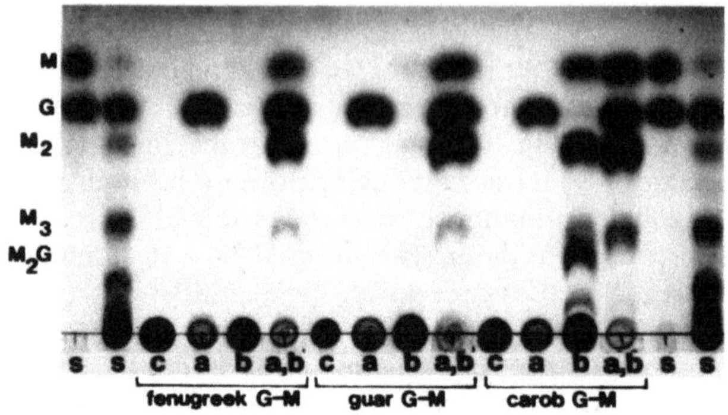

Fig. 2. Patterns of saccharides released from galactomannans by α-D-galactosidase (*a*) and *endo*-β-D-mannanase (*b*) used separately and together. *C* control with no enzyme; *G* galactose; *G–M* galactomannan; *M* mannose; M_2 mannobiose; M_3 mannotriose; M_2G galactosylmannobiose; *S* reference standards (from Edwards et al. 1989)

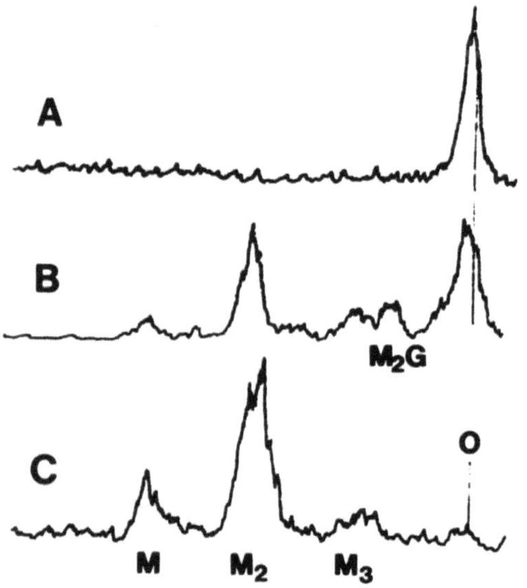

Fig. 3 A–C. Radio TLC traces of enzyme hydrolysates of a product formed in vitro from UDP-galactose (800 μM) and GDP-U-^{14}C-D mannose (80 μM) by the guar enzyme. The calculated degree of substitution of the hypothetical galactomannan product was 21%. **A** hydrolysis using α-D-galactosidase; **B** hydrolysis using *endo*-β-mannanase; **C** hydrolysis using both enzymes together; *O* chromatogram origin. The symbols, *M*, M_2, M_3, M_2G have the same significance as in Fig. 2. They are positioned to indicate the migration distances of reference standards (from Edwards et al. 1989)

5 Conclusion

Current trends in the analysis of cell-wall storage polysaccharides are towards the development of enzymatic techniques for quantitative and fine-structural analysis, and towards the use of specific biochemical probes for ultrastructural localization. These methodologies will almost certainly constitute a major part of any future revision of this chapter.

References

Aspinall GD, Greenwood CT, Sturgeon RJ (1961) The degradation of xylans by alkali. J Chem Soc 3667–3674

Bendayan M (1984) Enzyme-gold cytochemistry: a new affinity approach for the ultrastructural localmotion of macromolecules. J Electron Microsc Tech 1:349–375

Bewley JD, Reid JSG (1985) Mannans and glucomannans. In: Dey PM, Dixon RA (eds) Biochemistry of storage carbohydrates in green plants. Academic Press, London, pp 289–304

Blumenkranz N, Asboe-Hansen G (1973) New method for quantitative determination of uronic acids. Anal Biochem 54:484–489

Campbell JM, Reid JSG (1982) Galactomannan formation and guanosine 5′-diphosphate-mannose: galactomannan mannosyltransferase in developing seeds of fenugreek (*Trigonella foenum-graecum* L., Leguminosae). Planta 155:105–111

Crawshaw LA, Reid JSG (1984) Changes in cell-wall polysaccharides in relation to seedling development and the mobilisation of reserves in the cotyledons of *Lupinus angustifolius* cv. Unicrop. Planta 160:449–454

Dea ICM, Morrison A (1975) Chemistry and interactions of seed galactomannans. Adv Carbohydr Chem Biochem 31:241–312

Edwards M, Dea ICM, Bulpin PV, Reid JSG (1985) Xyloglucan (amyloid) mobilisation in the cotyledons of *Tropaeolum majus* L. seeds following germination. Planta 163:133–140

Edwards M, Dea ICM, Bulpin PV, Reid JSG (1986) Purification and properties of a novel xyloglucan-specific *endo*-(1→4)-β-D-glucanase from germinated nasturtium seeds (*Tropaeolum majus* L.). J Biol Chem 261:9489–9494

Edwards M, Bowman YJL, Dea ICM, Reid JSG (1988) A β-D-galactosidase from nasturtium (*Tropaeolum majus* L.) cotyledons – purification, properties and demonstration that xyloglucan is the natural substrate. J Biol Chem 263:4333–4337

Edwards M, Bulpin PV, Dea ICM, Reid JSG (1989) Biosynthesis of legume-seed galactomannans in vitro. Cooperative interactions of a GDP-mannose linked (1→4)-β-D-mannosyltransferase and a UDP-galactose linked α-D-galactosyltransferase in particulate enzyme preparations from developing endosperms of fenagreek (*Trigonella foenam-graecum* L.) and quar [Cyamopsis tetragonoloba (L.) Taub]. Planta in press

Frens G (1973) Controlled nucleation for the regulation of the particle size in monodisperse gold solutions. Nature Phys Sci 241:20–22

Goldberg R, Roland J-C (1970) Structure des polysaccharides de réserve accumulés dans les parois des graines d'*Asparagus officinalis* L. et d'*Endymion mutans* Dumort. C R Acad Sci D 270:702–705

Halliwell G, Riaz M (1970) The formation of short fibres from native cellulose by components of *Trichoderma koningii* cellulase. Biochem J 116:35–42

Keusch L (1968) Die Mobilisierung des Reservemannans im keimenden Dattelsamen. Planta 78:321–350

Kooiman P (1960) On the occurrence of amyloids in plants. Acta Bot Neerl 9:208–219

Leung DWM, Bewley JD, Reid JSG (1981) Mobilisation of the major stored reserves in the embryo of fenugreek (*Trigonella foenum-graecum* L. Leguminosae) and correlated enzyme activities. Planta 153:95–100

McCleary BV (1979) Modes of action of β-mannanase enzymes of diverse origin on legume seed galactomannans. Phytochemistry 18:757–763

McCleary BV (1983) Enzymic interactions in the hydrolysis of galactomannan in germinating guar: the role of exo-β-mannanase. Phytochemistry 22:649–658

McCleary BV, Matheson NK (1983) Action patterns and substrate-binding requirements of β-D-mannanase with mannosaccharides and mannan-type polysaccharides. Carbohydr Res 119:191–219

McCleary BV, Matheson NK (1986) Enzymic analysis of polysaccharide structure. Adv Carbohydr Chem Biochem 44:147–276

McCleary BV, Amado R, Waibel R, Neukom H (1981) Effect of galactose content on the solution and interaction properties of guar and carob galactomannans. Carbohydr Res 92:269–285

McCleary BV, Nurthen E, Taravel FR, Joseleau J-P (1983) Characterisation of the oligosaccharides produced on hydrolysis of galactomannan with β-D-mannanase. Carbohydr Res 118:91–109

McCleary BV, Clark AH, Dea ICM, Rees DA (1985) The fine structure of carob and guar galactomannans. Carbohydr Res 139:237–260

McClendon JH, Nolan WG, Wenzler HF (1976) The role of the endosperm in the germination of legumes: galactomannan, nitrogen and phosphorus changes in the germination of guar (*Cyamopsis tetragonoloba*; Leguminosae). Am J Bot 63:790–797

Meier H, Reid JSG (1977) Morphological aspects of the galactomannan formation in the endosperm of *Trigonella foenum graecum* L. (Leguminosae). Planta 133:243–248

Meier H, Reid JSG (1982) Reserve polysaccharides other than starch in higher plants. In: Loewus FA, Tanner W (eds) Plant carbohydrates I. Springer, Berlin Heidelberg New York, pp 418–471 (Encyclopedia of plant physiology, New Series, vol 13 A)

Reid JSG (1971) Reserve carbohydrate metabolism in germinating seeds of *Trigonella foenum-graecum* L. (Leguminosae). Planta 100:131–142

Reid JSG (1985a) Cell wall storage carbohydrates in seeds – biochemistry of the seed "gums" and "hemicelluloses". Adv Bot Res 11:125–155

Reid JSG (1985b) Galactomannans. In: Dey PM, Dixon RA (eds) Biochemistry of storage carbohydrates in green plants. Academic Press, London, pp 265–288

Reid JSG, Bewley JD (1979) A dual rôle for the endosperm and its galactomannan reserves in the germinative physiology of fenugreek (*Trigonella foenum-graecum* L.) an endospermic leguminous seed. Planta 147:145–150

Reid JSG, Meier H (1973) Formation of the endosperm galactomannan in leguminous seeds: preliminary communications. Caryologia Suppl 25:219–222

Reid JSG, Edwards ME, Dea ICM (1987) Biosynthesis of galactomannan in the endosperms of developing fenugreek (*Trigonella foenum-graecum* L.) and guar [*Cyamopsis tetragonoloba* (L.) Taub.] seeds. Food Hydrocolloids 1:381–385

Reis D, Vian B, Darzens D, Roland J-C (1987) Sequential patterns of intramural digestion of galactoxyloglucan in tamarind seedlings. Planta 170:60–73

Roth J (1983) The colloidal gold marker system for light and electron microscopic cytochemistry. In: Bullock GR, Petrusz P (eds) Techniques in immunocytochemistry, vol 2. Academic Press, New York, pp 217–284

Ruel K, Joseleau J-P (1984) Use of enzyme-gold complexes for the ultrastructural localisation of hemicelluloses in the plant cell wall. Histochemistry 81:573–580

Saeman JF, Buhl JL, Harris EE (1945) Quantitative saccharification of wood and cellulose. Ind Eng Chem Anal Ed 17:35–37

Sedmak JJ, Grossberg SE (1977) A rapid, sensitive and versatile assay for protein using Coomassie Brilliant Blue G250. Anal Biochem 79:544–552

Seiler A (1977) Galactomannanabbau in keimenden Johannisbrotsamen (*Ceratonia siliqua* L.). Planta 134:209–221

Spyropoulos CG, Reid JSG (1988) Water stress and galactomannan breakdown in germinated fenugreek seeds. Stress affects the production and the activities in vivo of galactomannan-hydrolysing enzymes. Planta 174:473–478

Thiéry J-P (1967) Mise en évidence des polysaccharides sur coupes fines en microscopie électronique. J Microsc 6:987–1018

Vian B, Brillouet J-M, Satiat-Jeunemaitre B (1983) Ultrastructural visualization of xylans in cell walls of hardwood by means of xylanase-gold complex. Biol Cell 49:179–182

Vogel T, Schleiden MJ (1839) Über das Amyloid, eine neue Pflanzensubstanz. Poggendorfs Ann Phys Chem 327–330

Methods Used in Monitoring and Controlling the Quality of Bread with Particular Reference to the Mechanical Dough Development Process

T. A. Mitchell

1 Introduction

The assessment of bread quality by bread consumers is highly subjective and is influenced by ethnic, cultural and personal preferences. The quality factors assessed and the standards applied may also vary with the type of bread being consumed. However in spite of these variables the ultimate measure of the quality of commercial breads must be the level of consumer satisfaction achieved by individual manufacturers. Where customers have a choice of brands and varieties, sales figures and changes in sales figures are indicators of consumer preferences and can be correlated with specific features of the finished product. The twofold challenge to bread bakers and cereal technologists is to translate consumer preferences as measured this way into objective measurable aspects of bread quality and to relate these measurements to processing factors which may be controlled so that bread with the most desirable consumer characteristics is produced. This chapter is concerned with monitoring and controlling the quality of bread made from wheaten flour by mechanical dough development (MDD) processes such as the Chorleywood bread process and variants of it. The author's personal experience has been in the development and use of the MDD process in New Zealand.

2 MDD Bread Processes

2.1 Origins of MDD Processes

MDD processes are so-called because manufacture starts with a brief period (2 to 5 min) of intensive mixing of the dough. The energy requirement is generally between 8 and 12 watt-hours per kilogram (wh/kg) of dough. This is about four to six times the amount of energy needed for dough-mixing in traditional straight dough processes and results in a rapid rise in dough temperature. The origin of these processes dates back to 1926 when Swanson and Working (1926) showed that a brief period of intensive mixing could achieve the same optimization of dough properties as the traditional long period of fermentation that was needed following less intensive mixing. A commercial process utilizing this finding was patented by J. C. Baker (1950) of the Baker Process Company. His process was called the Baker Do-Maker process and had the additional feature that dough production was continuous. The commercial production units marketed in North America by the company had a high production capacity so they were most suited to large plant bakeries. The bread produced was notable for the very fine uniform structure of the crumb. This particular feature rendered the bread unpalatable to many consumers especially those outside North America.

In spite of this, the unique feature of the Do-Maker process, namely the elimination of the long fermentation period in bread-making, was of such theoretical and practical importance that scientific investigations were started in several laboratories with the twin aims of understanding the new process and if possible modifying it so that it could be used in smaller bakeries and for the production of bread of a more open texture. Mitchell and Crawford (1960) showed that bread similar in texture to that of traditional bulk fermentation (BF) bread could be produced in the laboratory by a batch-wise MDD process. However this could only be achieved by the use of potassium iodate as a dough oxidant. This substance is not a legally permitted food additive in New Zealand. Subsequently, Mitchell (1963) designed a laboratory dough developer which produced bread identical in texture to commercial Do-Maker bread but without the use of potassium iodate. Development of commercial batch MDD breadmaking processes proceeded simultaneously in Australia and Britain. The Australian Brimec process did not proceed past initial commercial trials but the British Chorleywood process as described by Chamberlain Collins and Elton (1960) was widely adopted in Britain and elsewhere. This is a batch process producing bread similar in crumb structure to British BF bread. The dough oxidant used by Chamberlain et al. was 10 ppm of potassium iodate or, alternatively, 75 ppm of ascorbic acid (both on flour weight).

2.2 Operating Conditions for Batch MDD Processes

Chamberlain et al. found that the production of good quality bread by their process required a work expenditure in dough-mixing of 40 J/g of dough (11 wh/kg). Further it was necessary that this work be expended in a time not exceeding 5 min. Immediately following mixing the dough was divided and moulded and then given a 10-min period of "intermediate" proof. From that point the final moulding, final proofing and baking stages were similar to those used in BF processes.

Their dough formula differed in some respects from the formulae used in BF bread. First, they used either a low level of fast-acting dough oxidant (10 ppm potassium iodate) or a high level of slow-acting dough oxidant (75 ppm ascorbic acid), second they found fat or modified fat to be an essential ingredient, third they added 3.5% extra water (on flour weight) and fourth they increased yeast levels by 50%–100%. Mitchell and Crawford (unpublished work) had found in their laboratory process that there were significant differences between flours in the amount of work that was required in dough mixing to achieve best bread quality. New Zealand flours had generally required about half the level of work input that was needed in the Chorleywood process. The range for commercial flours was 4 to 8 wh/kg. Work inputs above these levels had given sticky over-mixed doughs and breads with reduced volume and coarse open texture.

Mitchell and Crawford also found (unpublished work) that New Zealand flours required high levels of dough oxidants. Each flour had a minimum requirement but there was a high degree of tolerance to levels in excess of that minimum. A combination of 50 ppm potassium bromáte and 100 ppm ascorbic acid (on flour weight) was found to be adequate for all normal commercial flours.

Kilborn and Tipples (1972) used two laboratory mixers of differing design to investigate the effects of rate of working and total work input in MDD dough mixing. Rate of working or mixing intensity was varied by varying the mixer speed and work input by varying the duration of mixing. They found that there was both a minimum rate of working ("critical speed") and a minimum amount of work necessary for the production of bread of high loaf volume. They also found that doughs gave bread of maximum volume when they were mixed to slightly past peak consistency as indicated by the mixing curve from their recording dough mixers. To achieve peak dough development as indicated by the mixing curve peak, different flours required different levels of work. Flours of high work input requirement tended to be more tolerant to mixing beyond the peak than those of low work input requirement. Kilborn and Tipples concluded that MDD dough mixers must be capable of a mixing intensity sufficiently high to be above the critical speed for all flours likely to be encountered and that they should have a reliable means of determining dough development, such as an energy- or torque-indicating device to ensure that optimum development (peak consistency) has been achieved.

The Farinograph is a recording dough mixer which has been used for many years to provide information on the mixing characteristics of flours. Its use for determining mixing requirements for MDD processes was studied by Tanaka and Tipples (1969) because they had observed that a flour-water farinogram does not necessarily reflect mixing requirement in their baking tests. Their investigation showed that some flours, notably those of low protein content, show farinogram peak development times which are markedly lower than would be expected from the optimum work input level for the GRL Chorleywood method. Mitchell has obtained analogous results with New Zealand flours (unpublished work). A series of 20 commercial flours was used. Work input for full development was calculated for each from the farinogram peak development time. For ten of the flours this calculated work input corresponded to the work input requirement for optimum bread quality. For the remaining ten flours optimum work input, as indicated by test-baking, ranged from one-half to twice that calculated from farinogram peak development time. One possible but unproven explanation for these anomalous results is that the standard Farinograph mixing speed (62 rpm) is below the critical speed for MDD dough development.

2.3 Commercial Practice in MDD Bakeries

Commercial Do-Maker plants first operated in the United States of America about 1955 and the Chorleywood Process was introduced in Britain about 1962. It has since spread to Australia and North America. In New Zealand batch MDD processes have been in commercial operation since 1965 and now account for about three-quarters of the bread produced in this country. Thus worldwide there is now a considerable amount of commercial expertise in the use of MDD processes.

2.3.1 Dough Consistency, Water Absorption and Work Input

New Zealand plant bakers have found that dough consistency must be carefully controlled. Doughs which are slightly slack because they contain too much water or sticky because they are overmixed tend to stick in bakery machinery particularly the make-up plant. Doughs which are slightly tight or undermixed give bread of lower quality than that possible and tight doughs also reduce bread yields. Flours which have good dough strength do not give sticky doughs when they are overmixed but overmixing does generate extra heat which must be removed by cooling the dough. The theoretical rise in dough temperature from the energy imparted by mixing is 1.33° C per watt hour per kilogram of dough. Thus the accurate control of the level of water addition and of work input are both regarded by bakers as important.

Baker's flour specifications usually specify a flour water absorption capacity of 60% of flour weight and there are sometimes price adjustments based on variations from this specification. The desired work input requirement is usually specified as about 8 wh/kg of dough. Bakers have found that flours with a lower optimum work input requirement than this may be able to produce high-quality bread but often lack tolerance to mixing. Flours of higher optimum work input requirement impose an additional cooling load at mixing. Work input requirement is primarily cultivar-dependent and New Zealand does not yet have many wheat cultivars which have both high MDD baking quality and medium work input requirements, so commercial flours frequently have an optimum work input requirement in excess of the specification. However, the specification is retained as a long-term guide to flour-millers and wheat breeders.

2.3.2 Measuring and Checking Work Input Levels

The majority of MDD dough mixers used in New zealand are of the single-speed Tweedy design originally used in the Chorleywood process. Work input measurement during mixing is by means of a kilowatt-hour meter built into an electrical circuit which includes provision for adjustable compensation of the no-load current of the motor and provision for setting a counter to stop the mixer at any desired level of work input. Adjustment of the no-load current is a simple procedure and correct adjustment is easy to observe. However, the calibration of the kilowatt-hour meter itself tends to drift with time, so that actual work input is not that which has been set or is indicated on the counter. Errors of as much as 50% have been found in commercial mixers so bakers have adopted the practice of using their dough-mixers as calorimeters to check the calibration of their work-meters.

The check is generally made in the middle of a production run. It is normal for the first few doughs in a run to make the dough ingredients hotter than for the rest of the run. This is done to compensate for the loss of heat from the doughs into the cold bowl. However, after three or four doughs have been mixed the system reaches a steady state in which heat flows from the bowl into the ingredients during the early stages of mixing and from the dough into the bowl in the later stages of mixing.

For the purposes of using the mixer as a calorimeter it is assumed that during the time of a mixing cycle the amount of heat lost from the exterior of the enclosed and polished stainless steel bowl is negligible in relation to the amount of heat generated in mixing the dough.

The usual procedure is to run the mixer just long enough to incorporate a batch of dough ingredients so that the temperature of the undeveloped dough can be measured. The mixer is then reset to the desired work input for full dough development and run until it again stops, when the dough temperature is again measured. The difference between the two measured dough temperatures is divided by 1.33 to determine the actual amount of work expended and this can then be compared with the setting used on the work input counter. When there are significant differences between these the kilowatt-hour meter is serviced or replaced.

An alternative to measuring the temperature of the undeveloped dough is to measure the temperatures of the flour and water before they enter the mixing chamber. The initial dough temperature can then be calculated from these temperatures and the specific heats of flour and water. Results are the same by either method.

2.3.3 Dough Temperature

The final dough temperature recommended to MDD bakeries by the New Zealand Wheat Research Institute (WRI) is 32° ± 0.5° C though some bakers prefer dough temperatures a few degrees lower. When dough temperatures are higher than that recommended, the crumb texture of the bread tends to be "woolly" rather than fine and delicate. Temperatures significantly lower than 32° C tend to give doughs which are under-oxidized (under-matured) and they also tend to increase final proof time or yeast level. They have the economic advantage however that doughs can hold more water.

Because of the effects of dough temperature on dough consistency and final proof time it is important that the targeted dough temperature, whatever it may be, be maintained within close limits. This temperature may be calculated from the temperatures (and specific heats) of the two major dough ingredients, flour and water. The calculation assumes specific heats of 0.5 and 1.0 calories per gram for flour and water, respectively, and proportions of 100 parts flour to 60 parts water in the dough. Under these conditions the initial dough temperature when the flour and water are placed in the mixer approximates to the mean of their individual temperatures. The final dough temperature is then higher by an amount equal to the temperature rise from the heat generated in the mixing. An example will illustrate the use of the calculation:

Target final dough temperature	32° C
Targeted work input 9 Wh/kg × 1.33	12°
Difference	20° C

This temperature, 20° C, is the targeted temperature for the unmixed dough

Actual flour temperature	24° C
Targeted unmixed dough temperature	20° C
Difference	4°
Targeted unmixed dough temperature	20° C
Less difference for flour	4°
Required water temperature	16° C

3 Laboratory Evaluation and Testing for the MDD Process

3.1 Bread Properties

Wheaten flour has the rather unusual property that it can be used to make foods such as breads and cakes which are dry or slightly moist foams as consumed. In the absence of strongly flavoured additives foods made from white wheaten flour have a bland or neutral flavour. It is this absence of strong flavours and the mouth-feel arising from the unique foam texture which makes freshly baked bread so universally acceptable as a food.

Variations in the foam structure with their attendant variations in palatability, and hence consumer acceptability, result from variations in bread processing conditions and in the properties of the dough ingredients, especially the flour. Direct quantitative measurements of palatability are not possible so cereal technologists use measurements and assessments associated with the foam structure as guides to consumer acceptability. Loaf volume, crumb texture (or grain), ovenspring, loaf maturity and crumb softness are examples of such measurements and assessments.

Changes in crumb firmness and palatability of bread (staling) occur within hours or days of baking. These changes are associated with changes in taste and mouth-feel caused by retrogradation of the starch and loss of volatile flavour components. The rate of staling is affected by physical storage conditions such as temperature and humidity. D'Appolonia and Morad (1981) in a review of staling implicated most of the major components of flour in the staling process. Some of the other dough ingredients notably water, fats, surfactants and sugars can modify the staling process or mask some of the effects. Measurements of crumb firmness by means of various instruments are most commonly used for the assessment of degree of staling.

Even though bread has a bland flavour, there are subtle variations in flavour which affect consumer acceptance. These variations may arise from variations in processing, variations in formulae and ingredient composition and contamination by moulds, insects or extraneous substances. Chemical determination of these flavour variations is impractical so they are assessed as aroma, taint, and off-flavour.

Both the visual appearance of the bread crust and its physical structure influence consumer acceptability of bread, so crust is assessed in terms of colour, glazing, surface texture, crust thickness, hardness and toughness.

The overall external appearance of the loaf also influences consumers, so this is assessed in terms of symmetry, boldness and ovenspring.

Cereal laboratories that are involved in assessing commercial bread in order to gauge consumer acceptability, commonly allot scores or points for a series of chosen properties. For example, WRI allots points for the following features (maximum points in brackets): crumb texture (15), crumb softness (15), crumb colour (15), crust texture (5), crust colour (5), ovenspring (5), maturity (15), shape (15) and wrapping (10). The sum of the maximum points is 100. The allotted points are usually summed to give a total score which then forms the basis for comparison of different loaves.

Clearly, loaf rankings depend to some extent on the maximum points and hence weightings allotted to each attribute when the scoring system was designed. The WRI system was designed in consultation with commercial bakers and therefore reflects their judgment of the comparative importance of the above attributes in influencing consumer acceptability of bread.

While total loaf scores are useful for the overall comparison of loaves, scores for individual features should also be compared because they can emphasize the presence of specific and unacceptable defects in particular loaves.

Comparisons of total scores may also be misleading where there are significant differences in bread type. Public perception of bread quality depends to some extent on the circumstances under which the product is offered for sale. For instance aroma, loaf symmetry and crust appearance are important in hot bread displayed unwrapped, whereas loaf softness, the presumption of fine crumb texture, and good keeping quality based on past experience, are important in sliced and wrapped bread.

Few of the properties of commercial bread listed above can be measured by instrumental methods so they must be assessed by staff who have been selected and trained for this work. With adequate training and experience staff can develop a high level of skill and consistency in making the assessments.

3.2 Ingredient Formulae for MDD Bread

A typical formula for commercial MDD bread is shown in Table 1.

Shortening is usually a commercial mixture which typically will provide (on flour weight) 0.25% sodium stearoyl 2-lactylate (SSL) and 0.2% diacetyl tartaric acid esters of distilled monoglycerides (TEM). The mixture may also contain skim-milk powder, enzyme-active soya flour and fungal enzymes. Distilled monoglycerides and animal or vegetable fats may also be included in the dough formula. Other ingredients which may be included separately in the dough formula are dried vital wheat gluten, sugar, malt flour, soya preparations, milk powders and fungistatic agents, such as sodium diacetate and acid calcium phosphate. Propionates are not permitted additives in New Zealand.

As stated in Sect. 2.3.1, the level of water chosen is the maximum which can be used without causing adhesion of doughs to the bakery machinery.

The level of yeast is selected so that doughs are fully proofed at the end of final proof. It depends on yeast activity, dough temperature, the level of yeast nutrients in the dough and the duration of final proof.

Clearly, flour is the major component of bread, making up as it does approximately 55% of total dough weight. It is also the ingredient which varies most in manufacturing performance. The principal source of this variation is in the wheat from which the flour was milled. Whereas all the other dough ingredients can be produced under rigidly controlled conditions wheat, of necessity, is grown under

Table 1. Typical formula for commercial MDD bread

Ingredient	Percentage[a] or ppm[b]	
Flour	100	
Water	55–63	
Yeast	2– 3	
Salt	1.5	
Shortening	1.2	
Fungal alpha-amylase		200
Ascorbic acid		100
Potassium bromate		50

[a] Ingredient weights are expressed as a percentage of the flour weight.
[b] Parts per million of flour weight.

varying cultural and physical conditions. Additionally, wheat cultivars vary in their suitability for bread manufacture but the choice of cultivars made by wheat-growers must consider other independent factors, such as yield and disease resistance. Monitoring the quality of wheats for baker's flour is a logical prerequisite to producing flour of high manufacturing quality.

3.3 Flour Properties

The chemical composition of flour, its physical properties and the physical properties of doughs made from it are commonly used in specifications which form the basis of sales transactions between bakers and flour millers. A typical specification, such as that of Schiller (1984), may contain 15 or more individual quantitative specifications with stated tolerances, and a further 5 or more individual qualitative specifications. Such specifications are useful for distinguishing flours into broad classifications, such as bread flours, cake flours or biscuit flours, for setting limits on properties which have the potential to disrupt manufacture or lower product quality, or in the case of moisture content for establishing the terms of trade. As Schiller points out:

> "They are designed to define the quality of the flour being used but not necessarily the products made from the flour."

He continues:

> "All too often, bakers will purchase flour by specifications and then evaluate the quality in terms of a finished baked product. Although this may be an idealistic approach, it is often misleading and disappointing."

The quantitative parameters included in Schiller's paper are moisture %, protein %, ash %, colour, enrichment, granulation, starch damage, Farinograph absorption, arrival time, peak time, departure time, stability time and mixing tolerance index, hot paste viscosity and extraneous matter. The same or similar parameters are common in many flour specifications.

Of the parameters listed one, flour protein content, has been shown by many workers to have a highly significant statistical correlation with loaf volume, which is generally regarded as an indicator of flour quality. However, the correlation is generally far from perfect, which means that variations in loaf volume are only partially linked with variations in protein content. Thus a series of flours will give a different ranking on protein content from that which they give on loaf volume.

Cereal chemists attribute much of the variability in the protein / loaf volume relationship to variations in the quality of the flour proteins. Major differences between wheat cultivars of the same protein content are commonly observed in wheat breeding programmes and have been linked to the composition of the protein. Other flour components, notably lipids and pentosans, have also been shown to cause significant variations in loaf volume and there are variable interactions between flour and the other dough ingredients.

It has been the author's experience that the manufacturing performance of baker's flours in MDD bakeries is not closely correlated with their performance in BF bakeries and individual wheat cultivars frequently show a consistently better performance in one process than the other.

Though there have been some significant advances in recent years in linking the composition of wheat protein in terms of protein sub-units with loaf volume and other indicators of bread quality there are still no chemical or physical methods of analysis which on their own can accurately and consistently predict the performance of bread flours in commercial bakeries. Though such tests are sometimes used for this purpose, their use requires the exercise of skilled judgments on test results for a number of factors, by a person experienced in the interpretation of the data and on flours with which he or she is familiar.

3.4 Performance Testing for MDD Processes

The flour specifications used by most New Zealand MDD bakeries require data on and set standards for three principal parameters. These are bread baking performance, flour water absorption, and work input requirement. Some also include a specification for work input tolerance. Additionally, specifications usually require data on and set limits for Falling Number (as an indicator of sprout damage), Kent-Jones flour colour grade and flour moisture content. A general statement requiring compliance with the Food Act 1981, including freedom from adulteration or contamination, is usually included also. A typical flour specification will include the requirements in Table 2.

3.4.1 WRI Mitchell Mixer

Measurements of baking scores, water absorption and work input requirement for the specification in Table 2 require use of the WRI Mitchell mixer. This mixer was designed by the author specifically for test-baking.

Descriptions of the basic mixing chamber have been published by Mitchell (1971 and 1984) and mixers have been manufactured in three sizes which mix flour weights of 50 g, 125 g and 1 kg respectively. Different mixing-cycle control circuits, different mixing speeds (65 and 150 rpm respectively) and different dough temperatures (26° and 32° C respectively) are used for BF and MDD test-

Table 2. Typical New Zealand specification for MDD bread flour

Test	Standard	Method
MDD bake score	23 ± 2 units	WRI MDD flour test
Water absorption	60%	WRI Mitchell mixer
	or 62%	Brabender farinograph
Work input requirement	8 to 12 wh/kg	WRI Mitchell mixer
Work input tolerance	Maximum variation 1 unit in bake score for ± 10% variation from optimum work input	
Flour colour grade	2.5 KJ units maximum	Kent-Jones CG method
Falling Number	250 s minimum	AACC 56-81B
Moisture Content	15% maximum	ISO712 (1979) E

baking procedures. The mixing chambers are temperature-controlled, either by circulating water from a thermostated reservoir through a jacket around the chamber or by means of individual electronic thermostats which sense the temperature of the chamber and control an electric heating cable wound around it.

The two novel features of the mixers are the automatic dough-ejection system and the automatic ingredient-dispensing system. It is these features which make possible the high rate of sample testing and ensure precision in both dough weights and ingredient additions.

Doughs are ejected from the mixing chamber at the end of the mixing cycle by means of a flat slotted plate which is housed above the two rotating mixing paddles while a dough is being mixed. At the end of the cycle the paddles (which are flat) are stopped by an electronic circuit in a particular position which enables them to pass through the two slots in the ejection plate as it descends the full length of the mixing chamber. As the ejection-plate moves down it strips all dough from the paddles and walls and moves it out of the bottom of the mixing chamber where it hangs suspended between the ejection plate and base-plate. The whole process is completed in about 1 s. During mixing the lower end of the mixing chamber is closed by the flat base-plate, which is ganged to the ejection-plate so that they move up and down in tandem. Motive power for this is provided by pneumatic cylinders. Once ejected, doughs are removed manually from their position between the two plates and the mixer is then ready to mix another dough. The whole process of emptying and restarting the mixer can be accomplished comfortably in about 1 min. The rest of the cycle requires 2 to 3 min.

The basic units of the ingredient-dispensing system are automatic pipettes each of which has a changeover valve to permit filling and emptying. Each mixer is fitted with four pipettes, three of which are of fixed volume and are used for measuring and dispensing the dough ingredients as solutions or, in the case of yeast, as a suspension. The fourth pipette is variable in volume and is adjusted by the operator to dispense water in quantities appropriate to the required water absorption. The changeover valves of all four pipettes are ganged together and operated by the electronic control circuit so that they fill and empty at the appropriate times in the mixing cycle.

The pipettes are filled from four reservoirs by four small submersible centrifugal pumps which run continuously. Overflow from the pipettes returns continuously to the reservoirs during the filling phase. The liquids being pumped and measured are water, yeast suspension, ascorbic acid solution and salt/sugar/potassium bromate solution. Four pumps supply the four 125-g mixers used in the MDD flour testing system and a further four supply the six 50-g mixers used in the MDD wheat-testing system.

Preweighed flour samples and powdered shortening are introduced into the mixing chamber by means of a funnel through the top. They are poured manually into the funnel and their flow into the chamber is assisted by means of a vibrator on the smaller mixers and an air-slide on the kilogram model.

Two types of electronic cycle control circuit are currently in use on the MDD mixers. Their functions are to control the sequence of operations in each mixing cycle and to generate a voltage output proportional to instantaneous mixer torque so that both optimum water absorption and optimum work input requirement can be either measured or set as required. For both types of control circuit the voltage output may be applied to a flat-bed chart recorder to generate a curve of torque versus time, similar to a farinogram or mixogram, and from this curve optima for the two required parameters may be determined. (See Sect. 3.4.3.) However, this is unnecessary with the control circuit used for 125-g and 1-kg mixers. This circuit is microprocessor-based and can determine these two parameters electronically. Optimum work input requirement (per kilogram of dough), optimum water absorption, time to peak, total time, and total work (per kilogram of dough) are each determined and held in memory or displayed on the numerical registers as required.

The torque at peak dough consistency and the work expended in reaching peak dough consistency are both determined from a moving average torque value. A torque-sampling rate of one reading per second is used and the continuously moving average always includes 40 readings. Thus in determining peak values mixing must continue for at least 20 s beyond the peak.

The microprocessor circuit also zeros the no-load torque at the start of mixing and it includes calibration and fault-diagnosis functions.

The manual functions performed by the mixer operator are:

1) Setting the work input counter and the water absorption pipette.
2) Initiating each mixing cycle by pressing the cycle control button. This closes the mixer by raising the base-plate to cover the bottom of the mixing chamber and it starts the drive motor and flour funnel vibrator.
3) Pouring the pre-weighed flour sample and powdered fat into the flour funnel.
4) Initiating actual mixing of the dough, when all the flour is in the mixing chamber, by again pressing the cycle control button. This action stops the flour funnel vibrator and causes the control circuit to dispense accurately measured quantities of the liquid ingredients from the automatic pipettes into the mixing chamber. If also energizes the work-measuring circuit which will stop the mixer at the work input level set on the work counter.
5) Removing the dough from the mixer at the end of the cycle. The cycle control circuit adjusts the position of the mixing paddles and opens the mixer so that this can be done easily and completely.

The WRI MDD flour testing system uses four 125-g mixers controlled by one operator and producing 30 doughs per hour. The MDD wheat testing system uses six 50-g mixers controlled by one operator and producing 60 doughs per hour.

3.4.2 Calibration of Mitchell Mixers for Optimum Test-Baking, Water Absorption and Work Input

Calibration of the torque axis of a recording dough mixer is normally done by mechanical means. For example, for the 300-g Farinograph bowl a known static torque of 500 g force-metres is applied to the torque-measuring system by applying 500 g force to the upper torque arm of the dynamometer at a distance of 20 cm from the drive axis. Because of the multiplier effect of the lower torque arm this effectively represents a distance of 1 m from the drive axis. The balance pointer is then adjusted to read 500 BU.

When correctly calibrated, standard peak dough consistency for the Farinograph is that consistency which gives a farinogram with the peak centred on the 500 BU line on the chart. Farinograph water absorption for any particular flour is that percentage of water on flour weight which gives a dough with a peak consistency of 500 BU.

Mitchell mixers are intended for test-baking and for this purpose the optimum water absorption for test-baking is required. This is the percentage of water (on flour weight) which gives the optimum baking result. It is determined by baking a series of loaves with incremental increases of 1% water absorption starting below the optimum and finishing above it. The maximum point on the plot of percentage water absorption against loaf volume or loaf score then indicates the optimum water absorption.

The primary torque-sensing load-cells of the 125-g and 1-kg Mitchell mixers are of the cantilever type which utilize strain gauges bonded to a metal lever. During manufacture these load-cells are each tested and adjusted to have standardized linear load/voltage output characteristics. Thus, the load-cells are interchangeable. Provision is made in each dough mixer for the load-cell and the following circuitry to be checked by the application of a static load.

During the design stages of each size of mixer the average load-cell output voltage at peak dough consistency is determined for a series of flours mixed at optimum test-baking water absorption. The microprocessor control is then programmed so that the numerical water absorption registers indicate as optimal the water absorption for any dough with a peak consistency which generates this output voltage. For doughs which are tighter than optimum the registers are programmed to read out the required extra percentage of water to bring the dough to optimum consistency, and for doughs which are slacker they indicate the percentage of water in excess of the optimum. The chart recorder circuit is likewise adjusted so that it has an output of 0.50 V at peak dough consistency when a dough is mixed at its optimum water absorption. Two output terminals are provided. One gives an undamped recorder trace similar to a farinogram. The other is heavily damped to give a near-smooth curve.

The relationship between machine water absorption as indicated by any recording mixer and optimum water absorption for baking is not completely constant. Flours with unusual characteristics may give anomalous machine results, so WRI has adopted the practice of checking test-baking doughs by hand-feel to ensure that they are normal. Though subjective, this is an accurate and consistent procedure. Where consistent abnormalities are encountered, the system is recalibrated by test-baking.

The difference between Farinograph water absorption and Mitchell mixer absorption is approximately 2% of water but this varies because of the different methods of calibration and operation. The notable operating difference is the inclusion of all the dough ingredients in the Mitchell mixer determination.

The torque-measuring systems of 50-g Mitchell mixers uses a commercial electronic joulemeter/wattmeter which senses torque from the motor current. It is manufactured by Unilab Ltd, Clarendon Road, Blackburn, England. It has the capability for measuring water absorption but is not used for this purpose in the WRI mass wheat testing method for which 50-g mixers are used.

Work input is calculated in watt-hours per kilogram by both types of electronic control by integrating instantaneous drive torque over the period of the mixing cycle. Accurate calibration of the 125-g and 1-kg mixers for work input depends on accurate calibration of the primary strain-gauge load-cells. As stated above, this calibration is carried out by the application of a known static force to the torque arm.

The 50-g mixers are calibrated for work input measurements against the 125-g mixers. The optimum work input for a few flours is determined on the 125-g mixers and then the 50-g mixers are adjusted electrically to give the same optima as indicated by their mixing curves.

For the three sizes of mixer, 50 g, 125 g and 1 kg, nominal dough weights of 80 g, 200 g and 1600 g are assumed in the electronic calculation of work input per kilogram. In the unusual circumstance that a substantially different dough weight is used, the work input per kilogram is recalculated by the operator.

3.4.3 Determination of Optimum Water Absorption and Optimum Work Input

Two modes of operation are programmed into 125-g and 1-kg mixers. Mode 1 operation is for determining water absorption and optimum work input, and the mixer assumes this mode when it is switched on. Mode 2 operation of the mixers is used for mixing doughs for test-baking.

In Mode 1 operation the mixers are programmed to set a target work input of 12 Wh/kg, so they stop when this amount of work has been expended in mixing dough. The operator has the option of using the preset target or keying in an alternative target work input. The choice depends on the anticipated optimum work input. For this optimum to be determined accurately it is essential in Mode 1 operation that the mixer continue mixing for at least 20 s after peak dough consistency has been reached. Provided this condition is met, the numerical register will read out optimum work input at the end of mixing and this optimum value will enter memory as the target work input in the next (Mode 2) mixing cycle. It remains in memory till the mixer is returned to Mode 1 or a different target is keyed in.

The water absorption used in Mode 1 is a matter of operator choice and depends on the anticipated optimum water absorption. The choice is not critical, so a value of 60% is often used. This value is set on the water-dispensing automatic pipette. At the end of mixing the numerical register displays a correction figure which is applied to the pipette setting to calculate the optimum water absorption. The pipette is then reset to this value for the Mode 2 operation.

The determination of both parameters requires only one Mode 1 mix because, unlike time-to-peak, optimum work input requirement is quite independent of water absorption.

If a mixing curve is required for any reason, it is obtained by connecting a flatbed chart recorder to the recorder output terminals on the mixer. For the graphical calculation of optimum work input the ratio of the area under the curve up to peak consistency and the area under the curve to the end of mixing is multiplied by the total (i.e. target) work input. In practice, the linear distance on the x-axis from the origin to the point below the peak and from the origin to the end of mixing can be used in place of measured areas without a significant change in the ratio.

Water absorption is calculated from the peak height (on the y-axis) of the mixing curve. Optimum absorption is set on 50 on a chart with 100 divisions. For a chart width of 250 mm the water absorption value set on the water pipette is corrected by 1% for each 10 mm by which the curve height deviates from the 50 line.

The dough formula used in Mode 1 operation is the same as that used in Mode 2 for test-baking. The formula for the MDD 125-g flour test-baking system is given in the next section. The short-term standard error for these two determinations are ± 0.5 Wh/kg for optimum work input and $\pm 0.3\%$ for water absorption. These levels of variability are undetectable in baking. The standard errors over a period of years for a long-term standard flour are ± 0.9 Wh/kg and $\pm 0.7\%$ respectively.

3.4.4 MDD Test-Baking for Flour Quality Evaluation

The dough formula, solution composition and pipette volumes for the WRI flour testing system are shown in Tables 3 and 4.

Twenty-five to 30 commercial flour samples are tested daily. They are mixed concurrently, but otherwise individual doughs pass sequentially through the system.

Three sub-samples are weighed from each flour sample and to each a volume of powdered fat (Wrifat) equivalent to 1.5 g is added. One sub-sample is taken for the Mode 1 operation and the mixer used is thereby programmed for Mode 2. The two other sub-samples are then mixed in the same mixer at optimum work input and water absorption. After removal from the mixer dough temperatures are checked by electronic thermometer to ensure that they are 32° C and each dough is then put into a covered bowl. The bowls are placed in a small, purpose-built, travelling proofer thermostated at 32° C, from which they re-emerge after 10 min intermediate proof. The rate of dough-mixing on the four mixers is regulated by the operator so as to obtain one dough every 2 min. Thus, there are normally five doughs in transit in the intermediate proofer.

Each dough is lightly moulded in a Mono moulder (manufactured by Ayres Jones Ltd, Swansea, UK) put into a baking pan and placed on a stirrup of the purpose-built travelling proofer-oven unit. This unit uses a single endless conveyor chain to carry the doughs through the final proofer and oven. A single time-clock is used to step forward both the intermediate proofer and the proofer-oven unit one stirrup every 2 min. The final proof cabinet is operated at 40° C and 85% –90% relative humidity. When full, 22 doughs are in transit through it, so the final proof time is 44 min. The oven is operated at 220° C without forced convection, and has an adjustable vent set so that the composition of the oven atmosphere is maintained at 20% steam and 80% air. This composition maximizes the loaf volume discrimination between poor and good quality flours. When full, 15 doughs are in transit through the oven so the baking time is 30 min. Six pans of water both precede and follow the sequence of doughs through the proofer-oven unit.

Table 3. Dough formula for WRI flour testing system

Ingredient	Percentage on flour weight	Weight
Flour	100	125 g
Water	Variable[a]	
Yeast	Variable[b]	
Salt	2	2.5
Wrifat[c]	1.2	1.5 g
Cane sugar	0.75	0.94 g
Potassium bromate	0.005 (50 ppm)	12.5 mg
Ascorbic acid	0.01 (100 ppm)	6.25 mg

[a] The amount of water is as determined in Mode 1 operation of the mixers.
[b] The amount of yeast is as calculated from the measured yeast activity as indicated by a gassing rate test. The standard activity for 3% yeast requirement in the test-bake (3.75 g) is 860 ml of gas at 1 h. Yeast level is reduced by 0.01% (on flour weight) per ml of gas above this activity and increased by 0.01% per ml of gas below this acticity.
[c] Wrifat is a mixture of 20% SSL, 15% TEM, 20% enzyme active soya flour, anti-caking agent and flour. It is measured by volume and added to the weighed flour samples.

Table 4. Solution composition and pipette volumes for test-baking

Ingredients	Weight (g)	Solution volume (ml)	Pipette volume (ml)
Salt	200	4000	50
Sugar	75		
Potassium bromate	0.5		
Ascorbic acid	5	2000	5
Yeast	450 (for 3%)	1800	15
Water	–	–	0–35[a]

[a] This pipette volume covers a range of water absorption from 52% to 80%.

Duplicate samples of a short-term (6 weeks) reference flour are baked both at the start and end of each run. Additionally, duplicate samples of a long-term (2 years) reference flour are baked at monthly intervals. All reference flours are stored deep-frozen and are thawed before use. Operation of the equipment at the standard rate requires two people.

3.4.5 Loaf Evaluation

The purpose of this test-baking system is to assess the relative manufacturing potential of commercial flours in MDD processes. It is convenient to use some of the bread properties discussed in Sect. 3.1 for the assessment of test-baking loaves, but there is no implication in this that commercial bread will give the same or related scores for those properties. It is not technically feasible to make a statistical correlation between commercial bread properties and those of test-baked bread, but what is more important, it is fallacious to assume that commercial bakers want or could obtain results identical to those obtained in test-baking.

Three quality parameters, loaf volume, crumb texture (grain) and crumb colour are assessed in the WRI evaluation of test-baked bread. The scores for two of these, volume and texture, are added to obtain the MDD loaf score. The third, crumb colour, is indicative of a specific defect and may have a specific limit so it is not added into the loaf score but is the subject of separate comment in the flour report.

These three parameters were adopted in 1952 after a statistical analysis of scores for a range of parameters showed that the scores fell into three groups within which they were highly correlated. Usage of all the parameters was providing redundant information, so volume, texture and colour were selected from within the three groups. Further studies have shown a highly significant correlation between volume and texture, but these can be varied independently, so real information would be lost if either were removed from the scoring system.

After baking, loaves are stored overnight (16–20 h) on open trays at 22° C and ambient humidity. Loaf volume is determined in a rape-seed displacement meter calibrated in volume units. Each unit of volume is 18 ml and the scale zero is 540 ml. The nominal loaf weight is 180 g, so at a volume of 540 ml the specific volume of the loaf is 3 ml/g. The volume scale was constructed from the formula

$$\text{Loaf volume units} = \frac{\text{Loaf volume in ml} - 540}{18}$$

Thus the volume increments by one unit for each increase of 0.1 in specific loaf volume. The advantage of deriving a scale from specific loaf volume is that the volume score is independent of loaf weight, and hence flour weight.

Crumb texture is assessed visually on a scale of 0–14 (increasing as texture improves). The loaf is cut transversely at the centre to make this assessment. Preserved bread and bread photographs are used as standards but trained operators are essential in obtaining consistent results. Agreement between trained operators is good, and long-term drift is checked by means of the long-term flour standards.

Loaf volume and crumb texture scores are summed to give the MDD loaf score. Scores for commercial flours range from 15 to 30 units. The standard error of the testing procedures is ± 0.7 units, for single determinations. Experience in the use of the results since the test was introduced in 1974 has shown that flours scoring less than 20 are not suitable for bread manufacture unless blended with better flour or supplemented with gluten. Between 20 and 25 they are capable of producing satisfactory bread with further obvious improvements above 25. Though a statistical correlation has not been attempted, there is a clear and very consistent relationship between the test-baking scores and the incidence of customer complaints to bakeries. Production staff in bakeries can discern differences in their product when the baking score of the flour changes by as little as 2 units.

Crumb colour in terms of greyness (not yellowness) is assessed on a scale of 0–5, with the score decreasing as greyness increases. A colour of 3 or less is usually unacceptable for the production of white bread but may be used in meal breads where bran is added in the bread factory.

Preharvest sprouting of wheat raises flour alpha-amylase levels and this sometimes causes increased crust colour and crumb doughiness in bread. Where required, crumb doughiness is assessed from the feel of the crumb when it is compressed between the thumb and forefinger. This method is subjective and requires a trained operator but is more relevant and consistent than machine methods of assessing crumb doughiness or alternative methods of assessing alpha-amylase in flour.

3.4.6 MDD Test-Baking for Wheat Quality Evaluation

New Zealand millers use wheat quality indexing systems for relating the price of wheat to its quality in terms of a series of parameters which are of importance either in the milling process or in determining the manufacturing quality of the flour. Where appropriate, these specifications include an MDD baking test, and the price paid to the farmer is adjusted on the basis of the baking score. A large proportion of the wheat harvested is tested on the WRI MDD harvest-testing system. The wheat samples are provided by grain brokers prior to sale or to confirmation of sale of each line of wheat.

The samples are milled in duplicate on Brabender Quadramat Junior mills which have been modified by the addition of an exhaust system which automatically cleans the mill and removes all the offal after each sample is milled. The flour samples are weighed immediately following milling and are stored in racks of ten in sample tins. Wrifat is measured into the tins in a separate operation before the flour goes into them. The rolls and bearings on the mills are replaced annually to maintain the flour yield at above 60%.

The baking equipment is very similar to that used in testing flours, the major difference being the smaller size of some parts of it, because the flour weight is 50 g rather than 125 g.

Prior to each harvest the electronic joulemeter / wattmeters on the six mixers are recalibrated. The optimum work-input for a test flour is determined accurately on the 125-g mixers and then this flour is used to generate a series of mixing curves from each 50-g mixer. The input voltage of each joulemeter is progressively adjusted by means of a potentiometer until the readout on the work counter at the peak of the mixing curve is the same as the optimum work input determined on the 125-g mixers.

At the beginning of each harvest several hundred samples are baked on both testing systems. The work input and water absorption levels used in both systems are those determined as optimum for each sample on the 125-g system. These values are accumulated and averaged for each major cultivar and the cultivar averages are then used on the 50-g system for the rest of the harvest, subject to regular checks.

The baking scores of the early samples are also compared and if necessary the 50-g system is adjusted to give the same results as the 125-g system.

The loaf-scoring system used for the 50-g test-bake is the same as that used for the 125-g test-bake. Loaf "volume" is determined by a device which measures loaf heights and is calibrated in loaf volume units to read the same as the 125-g rape-seed displacement meter. Crumb texture is scored in the same way as for the 125-g system. Crumb colour has no relevance for wheat samples, so is not assessed.

Serious specific defects of wheat, notably grain-drier damage, bug damage and sprout damage, each have characteristic effects in baked bread so are diagnosed as they occur in the wheat samples. The degree of sprout damage can be assessed manually from crumb doughiness but flour-millers prefer the Falling Number test for rapid objective assessments of sprout damage.

Several series of experiments have shown that commercial flours baked on the 125-g flour system give bake-score results close to those for the same wheats milled on the Brabender mill and baked on the 50-g wheat-testing system. The differences range from zero to a loss of 2 units when wheats are milled commercially. The differences appear to be relatively constant for each flour-mill, suggesting that they arise from differences in commercial milling techniques.

The standard error of the harvest-testing system is ± 1 bake-score unit for the duplicate test. The range of baking scores encountered is from -5 to above 30. Samples which score less than 15 have been found to give highly variable and unpredictable results when gristed with high-quality wheats, so they are not suitable for milling for bakers flour. Samples which score between 15 and 20 are not satisfactory for bread flour unless gristed with high-quality wheats.

Samples which score above 20 are gristed for milling for MDD bread flour.

3.4.7 Wheat Cultivar, Protein Content and Falling Number as Quality Indicators for the MDD Process

Cultivar, protein content and Falling Number are included in many New Zealand wheat quality specifications.

The wheat cultivar (or variety) from which flour is milled is the primary determinant of flour quality for MDD processes just as it is for other bread processes. Statistical analysis of the annual wheat harvest test results since 1981 has quantified the major and consistent differences between cultivars. Preferred or designated cultivars are the basis of wheat purchasing specifications in many wheatgrowing countries.

Wilson (1983) investigated the use of wheat protein content as a measure of MDD baking quality. Twenty thousand samples from three crop years had been analyzed for protein by near-infra-red analysis and tested for MDD baking score on the harvest-testing system. Wilson found that there was a statistically significant correlation between protein content and baking score with a correlation coefficient of 0.37. The correlation was improved when the results were analyzed by cultivar, district and year, and in each case significant differences were found. Wilson concluded that the prediction of baking score from protein content was no more accurate than guessing it.

Mitchell and Casutt (1983) reported a similar investigation involving fewer samples but more precise determinations of protein content and MDD baking score. For protein the Kjeldahl method was used, and for baking score the 125-g method was used, in both cases with replication. In addition, the determinations of protein content were made on flours, not wheatmeals. Fiftysix samples of New Zealand wheat bulked on either protein content or baking score, and 52 samples from bulk shipments of imported Australian wheat, were used in this study. Mitchell and Casutt concluded that variations in the protein content of flours milled from New Zealand and Australian wheats do not parallel variations in the baking quality of those flours. They concluded that the protein content of either wheat or flour is not a measure or indicator of baking quality.

Cawley and Mitchell (1968) found that Hagberg penetrometer estimates of sprout damage made on wheatmeals were frequently different from those made on flours milled from the same wheats. For the 1961 New Zealand wheat harvest about half of the 8000 samples tested showed much higher results on flours than wheatmeals. Cawley and Mitchell showed that this anomaly was caused by the inhibition of alpha-amylase by bran phytic acid. They also showed that the internal distribution of alpha-amylase in wheat is variable and the level may be very high in the bran. When the bran is removed in milling, the test result may be lower than on the wheatmeal.

These same variables also affect the Falling Number test. Results on wheatmeals may be higher than those on flour because of inhibition of alpha-amylase by phytic acid, or lower than those on flour because of high levels of alpha-amylase in bran. The conclusion is that Falling Number results on wheatmeals are not well correlated with those on flours.

Mitchell has also found a variable relationship (unpublished work) between Falling Number results on flour and MDD bread baking performance. Where the Falling Number result is above 200 s, sprout damage results are not found in either commercial or test-baked bread. For some samples in the 60–200-s range there are obvious sprout damage effects in bread but for others there is no apparent effect. For bread-baking purposes the flour Falling Number can be used as a warning test but it is not an infallible indicator of sprout damage.

References

Baker JC (1950) U.S. Patent 2,953,460

Cawley RW, Mitchell TA (1968) Inhibition of wheat alpha-amylase by bran phytic acid. J Sci Fd Agric 19:106–108

Chamberlain N, Collins TH, Elton GAH (1962) The Chorleywood bread process. Bakers Dig 36:52–53

D'Appolonia BL, Morad MM (1981) Bread staling. Cereal Chem 58:186–190

Kilborn RH, Tipples KH (1972) Factors affecting mechanical dough development. 1. Effect of mixing intensity and work input. Cereal Chem 49:34–47

Mitchell TA (1963) A small-scale pressurized dough developer for the laboratory. J Sci Fd Agric 14:239–244

Mitchell TA, Casutt VJ (1983) Correlation between flour protein and MDD baking score on segregated mixtures of wheat. In: Proc Wheat Res Inst Cereal Sci Conf 10–12 October 1983, Christchurch

Mitchell TA, Crawford CA (1960) Note on breadmaking without bulk fermentation. The effect of mixing and stage of addition of potassium iodate. NZ J Sci 3:290–292

Schiller GW (1984) Bakery flour specifications. Cereal Foods World 29:647–651

Swanson CO, Working EB (1926) Mechanical modification of dough to make it possible to bake bread with only the fermentation in the pan. Cereal Chem 3:65–83

Tanaka K, Tipples KH (1969) Relation between farinograph mixing curve and mixing requirements. Cereal Sci Today 14:296–300, 310

Wilson AJ (1983) Correlation between flour protein and MDD baking score on harvest samples of wheat. In: Proc Wheat Res Inst Cereal Sci Conf 10–12 October 1983, Christchurch

Analytical Methods for Gelation of Soybean Proteins

T. NAKAMURA

1 Introduction

Soybeans are used for various traditional Japanese foods such as tofu, yuba, miso and syoyu. Recently, however, soybean protein products have also found application as ingredients for the manufacture of fabricated and processed foods due to their excellent functional properties and nutritional value. The gel-forming ability of soybean proteins induced by heating is one of their most significant functional properties with respect to their use in food preparation. The behaviour of soybean proteins during gel formation, their chemical and structural properties responsible for gelling ability and physical properties of the gels formed are important matters that should be elucidated. The clarification of these points will facilitate the control of the gel formation process and gel properties, thus enhancing the scope of soybean protein application to foods.

Extensive studies have indicated the basic factors affecting gelation of soybean proteins (reviewed by Wolf 1970; Hermansson 1978; Saio and Watanabe

Table 1. Components of the ultracentrifuged fraction in soluble soybean protein

Fraction	Components	References
2S	Alcohol dehydrogenase	Leblova and Perglerova (1976)
	β-Amylase	Morita et al. (1976); Mikami et al. (1986)
	Chalcon-flavone isomerase	Boland and Wong (1975)
	Cytochrome c	Fridman et al. (1968)
	Trypsin inhibitor	Steiner and Frattali (1969)
7S	Acid phosphatase	Fujimoto et al. (1977)
	Agglutininase	Lis et al. (1966); Shaanan et al. (1984)
	α-D-Galactosidase	Harpaz et al. (1977)
	α-D-Mannosidase	Saita et al. (1971)
	Formate dehydrogenase	Farinelli et al. (1983)
	Lactate dehydrogenase	Barthova et al. (1973)
	Lipoxygenase	Diel and Stan (1978)
	Malate dehydrogenase	Bronovitskaya and Kretovich (1970)
	β-Conglycinin (7S globulin)	Thanh and Shibasaki (1976); Sykes and Gayler (1981)
	γ-Conglycinin	Koshiyama and Fukushima (1976); Yamauchi et al. (1985)
	Basic 7S globulin	Yamauchi et al. (1984)
11S	Glycinin (11S globulin)	Reviewed by Nielsen (1985); Brook and Morr (1985)
15S	Urease	Polacco and Havir (1979)

Table 2. Molecular species and subunit structures of the two major components in soybean proteins

	Molecular species	Subunits	References for primary structure of subunits: DNA	References for primary structure of subunits: Protein
β-conglycinin (7S globulin)	B0 β3	α′	Schuler et al. (1982a, b); Doyle et al. (1986)	
	B1 β2α′	α	Schuler et al. (1983); Doyle et al. (1986)	Coated et al. (1985)
	B2 β2α	β	Doyle et al. (1986)	
	B3 βαα′			
	B4 βα2			
	B5 α2			
	B6 α3			
Glycinin (11S globulin)	4 (AxBy)6[a]	A1aB1b[b]	Negoro et al. (1985); Utsumi et al. (1987a)	
		A1bB2	Negoro et al. (1986)[c]	
		A2 B1a	Marco et al. (1984); Momma et al. (1985a)	Staswick et al. (1984a)
		A3 B4	Fukazawa et al. (1985); Scallon et al. (1985)	Hirano et al. (1984)
		A4A5 B3	Momma et al. (1985b); Scallon et al. (1985, 1987)	Hirano et al. (1985)

[a] Four molecular species detected by Utsumi et al. (1981). Ax, By, and (AxBy) show an acidic subunit, a basic subunit and a intermediary subunit, respectively.
[b] Subunit naming is based on the identification by Moreira et al. (1979) and Staswick et al. (1981, 1984b). Pairing of subunits is on the basis of cDNA sequences data.
[c] Report presented at the 9th annual convention of the Molecular Bilogical Society of Japan (1986).

1978; Kinsella 1979; Peng et al. 1984). However, little is known regarding the process of gel formation at the molecular level and the mechanisms involved. Neither is the correlation between gel properties and protein structure well understood. Gelation studies on each isolated soybean protein as well as multiple components should be carried out for the clarification of these points. Soybean protein consists of four resolvable ultracentrifugal components, 2S, 7S, 11S, and 15S. The sedimenting components are not homogeneous, as shown in Table 1. The two major soybean protein components, 7S globulin (beta-conglycinin) and 11S globulin (glycinin), are comprised of several subunits (Table 2).

This chapter presents a methodological description of heat-induced gelation. Analytical methods for the thermal gelation of soybean proteins are also described.

2 Analysis of the Gelation Process and its Mechanism

Since the gelation proceeds continuously with heating, it may be considered a single-stage process. However, thermal gelling may be regarded as occurring in various stages. This being the case, it should be possible to analyze the process of gelation and determine precisely the mechanism involved.

2.1 Conformational Changes in Soybean Proteins

Denaturation and/or conformational changes in protein molecules by heating is a prerequisite for thermal gel formation.

2.1.1 Differential Scanning Calorimetry

The denaturation temperature and enthalpy of a protein can be determined by differential scanning calorimetry (DSC). In this method, temperature increases linearly. The energy for protein denaturation is measured against a reference sample and observed as an endothermic peak whose area corresponds to enthalpy change. A denaturation thermogram is usually obtained under conditions such that the protein solution sample is pressure-sealed in a hermetic pan, and reference material is a sealed pan containing the same buffer as the protein solution. The references describe the procedure and the principle in greater detail (Krishnan and Brandts 1978; Hinz 1986).

Hermansson (1979) found DSC applicable to the determination of the degree of denaturation of commercial soy protein. The investigation of Bikbov et al. (1983) on the thermal denaturation of two soybean protein components, 7S and 11S globulins, indicated the denaturation temperature of 11S to be higher than that of 7S. Moreover, this temperature has been noted to rise with increase in NaCl concentration and to change with pH (Danilenko et al. 1985, 1986; Hermansson 1979; Koshiyama et al. 1980–1981). Utsumi et al. (1987a) found two molecular species in 11S globulin, one dissociable and the other undissociable at low ionic strength ($\mu = 0.01$), and their denaturation temperatures to differ.

The temperature to bring about denaturation is distinct for each soybean protein component and varies according to the environment (e.g., pH, salt). These factors should thus be taken into consideration when conducting the thermal gelation of soybean proteins.

2.1.2 Spectrometric Analysis

Optical rotatory dispersion, circular dichroism (CD), infrared spectroscopy and Raman spectroscopy may be used to carry out the conformational analysis of proteins. It should be noted in particular that the CD spectra are characteristic of the various secondary structures of polypeptides and proteins.

Difference absorption spectrophotometry provides information on environmental change in aromatic amino acid residues of protein. Aromatic side-chain motion serves as a convenient probe in protein dynamics.

Significant data can also be obtained from fluorescence spectroscopy in regard to environmental change in aromatic amino acids, as well as information on the surface hydrophobicity of a protein by using external chromophores, the so-called fluorescent probes, i.e., 1-anilinonaphthalene-8-sulfonate (ANS).

Nuclear magnetic resonance (NMR) indicates the internal motility of proteins. NMR provides data that frequently indicate static and dynamic aspects of protein conformations, so it is widely applicable in the study of gelation.

Many studies have been carried out on thermal conformational changes in each soybean protein component (Bikbov et al. 1981; Catsimpoolas et al. 1971; Koshiyama et al. 1980–1981; Hashizume and Watanabe 1979; Matsudomi et al. 1985; Takagi et al. 1979a, b; Van Kleef 1986; Yamagishi et al. 1983). The data obtained indicate thermal conformational change to be less than that caused by strong denaturants (i.e., urea and guanidine hydrochloride). Thermally denatured soybean proteins have ordered secondary structures.

2.1.3 Enzymatic Analysis

The extent to which a protein is denatured depends upon its digestibility by proteases. Native 11S globulin is partially degraded by trypsin into an intermediate (Kamata et al. 1980). Richardson and Catsimpoolas (1979) showed that the rate and extent of tryptic hydrolysis of 11S globulin change prior to and following thermal denaturation.

Determination of the amino acid or deduced amino acid sequences of the subunits of major soybean proteins has been already made (Table 2). It should be possible to estimate the cleavage point of the primary structure of a protein by amino acid analysis of the digested fragments. This method may prove useful for investigating changes in the surface structure of protein molecules before and after thermal denaturation.

2.1.4 Immunological Analysis

This analysis serves not only to detect structural changes in antigens resulting from various treatments (Catsimpoolas and Meyer 1968; Iwabuchi and Shibasaki 1981; Iwabuchi and Yamauchi 1984; Koshiyama et al. 1980–1981) but also to determine soybean proteins in meat products (Griffiths et al. 1984; Hitchcock et al. 1981). However, this method gives no precise indication as to the kind of structural change.

2.2 Association–Dissociation Behavior

Thermal association and dissociation of soybean proteins at low protein concentration have been extensively studied but under conditions not permitting gel formation. Mori et al. (1982a, 1986) and Nakamura et al. (1985a) studied heat-induced gelling of 11S globulin in its early stages at high protein concentration, at which gel formation occurs by continued heating. The thermal gelation of 11S globulin appears to occur in three steps: (1) aggregation of 11S globulin molecules into strands (soluble aggregates) and liberation of some acidic subunits, (2) interactions of strands to produce a gel network, (3) increase in gel hardness possibly due to stabilization of gel networks. Step 1 must occur at or above the denaturation temperature (temperature-dependent stage). Step 2 depends on protein concentration. At low protein concentration, other reactions, such as the dissociation–association of acidic subunits from soluble aggregates and insolubilization–coagulation of basic subunits, are predominant.

Gelling is influenced by change in protein concentration, pH and heating conditions or by the addition of salts and reducing agents. Moreover, other reactions may also take place. Therefore, it is necessary to pay much attention to experimental conditions.

The separation and characterization of the associates and/or dissociates are discussed in the following. The references present the condition in detail.

2.2.1 Ultracentrifugal Analysis

Ultracentrifugal analysis has been used to determine both the degree of protein purity and macromolecular size in biochemical field. The sedimentation velocity method has been found particularly useful for determining the size of any protein as well as for indicating whether one or more components are present in solution, provided they differ in molecular size and shape. However, individual components cannot be collected by this method.

Wolf and Tamura (1969) studied the thermal association–dissociation of 11S globulin, using this method. And Hashizume et al. (1975) examined the effects of temperature and ionic strength on the ultracentrifugal behavior of soybean proteins.

2.2.2 Sucrose Density Gradient Centrifugation

This mode of analysis has been used to separate and purify cell components, such as polyribosomes (Larkins 1985). Sucrose gradients are prepared using an automatic mixing device. The degree of separation can be controlled, within certain limits of course, through the selection of a proper density gradient and centrifugation conditions.

11S globulin solutions heated for various minutes were analyzed by this method (Nakamura et al. 1984a). As shown in Fig. 1, soluble aggregates with molecular weights of 1.8×10^6 (peak I), 4.0×10^6 (peak II), 8.0×10^6 (peak III) were formed by heating for 15, 30, and 60 s, respectively.

2.2.3 Gel Chromatography (Size-Exclusion Chromatography)

Gel chromatography is a simple and inexpensive means for separating biological macromolecules according to size. The choice of gel matrix as a column-packing material depends on the molecular weights of the components. This method cannot be used in the following three cases: (1) components are so large that they cannot pass through the pores for separation and elution in the intercarrier volume of the column; (2) components are too small to permit separation, and (3) components interact on some occasion with the gel matrix. To analyze the heat-induced soluble aggregates of soybean protein, a resin for high-molecular-weight fractionation ($>10^6$) must be chosen. Table 3 shows commercially available gel filtration media. It is not an exhaustive list, but it does indicate most of those applicable for this purpose.

By high-performance size-exclusion chromatography good reproducibility is possible and little time is required. However, no suitable column is available for

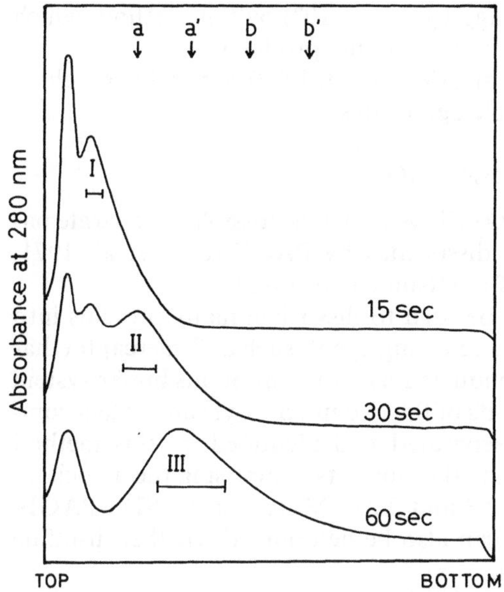

Fig. 1. Sucrose density gradient centrifugation of heated glycinin. Fifty-microliter aliquots of the glycinin solutions of 5% protein concentration in 35 mM potassium phosphate buffer (pH 7.6) containing 0.4 M NaCl was heated at 100° C in thin test tubes for the time periods as indicated. At the end of the heating period, the test tube was rapidly cooled to room temperature by immersing in tap water. The heat-treated glycinin solutions were centrifuged at 20° C in 12 ml of 15%–40% (w/v) linear sucrose gradient at 152,000 g for 60 min in a Hitachi RPS 40T rotor. After centrifugation, the gradient was examined for absorbance at 280 nm with an ISCO density gradient fractionator. The sedimentation positions of ribosome and TMV are indicated by *arrows*: *a* and *a'* monomer and dimer of the ribosome, respectively; *b* and *b'* monomeric and dimeric forms of TMV, respectively. The molecular weights of the monomeric forms of ribosome and TMV are 4.3×10^6 and 39.4×10^6, respectively. Peaks *I*, *II*, and *III* are discussed in the text. (From Nakamura et al. 1984a)

Table 3. Commercially available gel filtration media for high-molecular-weight proteins

Supplier	Trade name	Fractionation range for globular proteins (daltons)
Pharmacia (Sweden)	Sephacryl S-300	1×10^4–1.5×10^6
	Sephacryl S-400	2×10^4–8×10^6
	Sephacryl S-500	–
	Sephacryl S-1000	–
	Sepharose 6 prep grade	5×10^3–5×10^6
	Sepharose CL-4B	6×10^4–2×10^7
	Sepharose CL-2B	7×10^4–4×10^7
Toso (Japan)	TSKgel Toyopearl HW-65	5×10^4–5×10^6
	TSKgel Toyopearl HW-75	5×10^5–5×10^7
Bio Rad (USA)	Biogel A-5m	1×10^4–5×10^6
	Biogel A-15m	4×10^4–1.5×10^7
	Biogel A-50m	1×10^5–5×10^7
	Biogel A-150M	1×10^6–1.5×10^8

the analysis of high-molecular-weight aggregates ($>10^6$) by this method, which thus may be used only for the analysis of dissociating products.

Peng et al. (1982) and Yamagishi et al. (1981, 1983, 1987) conducted gel chromatography for the separation of soluble aggregates.

2.2.4 Polyacrylamide Gel Electrophoresis (PAGE)

In the absence of a detergent (Native-PAGE system), it is possible to separate native soybean proteins, associates and dissociates by PAGE (Saio et al. 1971; Hashizume et al. 1975; Mori et al. 1982a; Utsumi et al. 1985).

Sodium dodecyl sulfate (SDS) can be used to dissociate many proteins into constituent subunits in the presence of a reducing agent, such as 2-mercaptoethanol(2-ME). SDS-PAGE is generally conducted by a discontinuous buffer system (Laemmli 1970). The constituent subunits of the two major soybean protein components (7S and 11S globulins) are separated and identified by this method (Table 2). It is thus possible to determine the subunits comprising the associates and dissociates fractionated in Sect. 2.2.3 and 2.2.4. Moreover, by SDS-PAGE, in the absence of a reducing agent, it can also be determined whether disulfide bonds are present in these components.

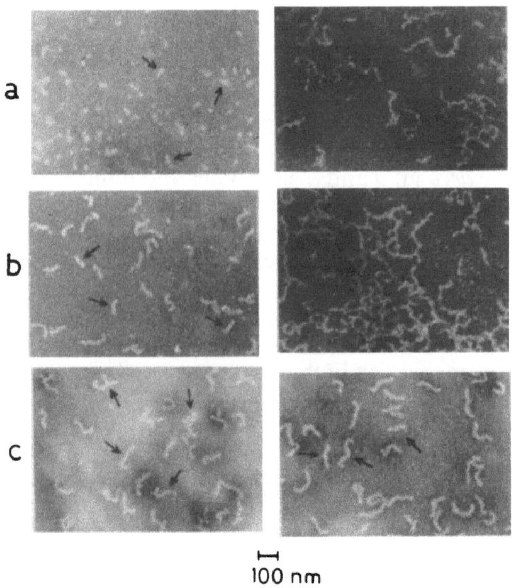

Fig. 2a–f. Transmission electron micrographs of negatively-stained materials. A solution of the protein was diluted with 35 mM potassium phosphate buffer (pH 7.6) containing 0.4 M NaCl to 0.01% protein concentration. A small drop was applied to a carbon-coated electron microscope grid and the grid was then floated on top of a 2% solution of potassium phosphotungstate (pH 7.0). Excess liquid was drained off with a filter paper and examined in a Hitachi H-700H electron microscope. **a** Peak *I* of Fig. 1; **b** Peak *II* of Fig. 1; **c** Peak *III* of Fig. 1; **d** and **e** unfractionated heated glycinin solutions for 2 min and 4 min, respectively; **f** soluble aggregate with a molecular weight of $4–8 \times 10^6$ obtained by sucrose gradient centrifugation of a heated glycinin solution for 10 min in the presence of N-ethylmaleimide. Typical strands are indicated by *arrows*. (From Nakamura et al. 1984a)

The size and subunit compositions of the constituent proteins of thermal protein gels have been found by two-dimensional electrophoresis on Native-PAGE to SDS-PAGE (Utsumi and Kinsella 1985). The two-dimensional electrophoresis serves quite well in identifying subunit constitution and types of bonds among the subunits.

2.2.5 Transmission Electron Microscopy

The size, shape, and direction of the soluble aggregates (as shown in Fig. 1) can be visualized by this method, using negative staining (Fig. 2). Nakamura et al. (1984a, 1986) noted that the soluble aggregates of 11S and 7S globulins formed at an early stage of heat-induced gelation and soluble macroaggregates resulted by further polymerization.

3 Gel Analysis

For analysis of solid-state gels, physiological rather than common biochemical methods are used. The state of water in a gel significantly affects its properties, but the present paper does not describe the analytical method.

3.1 Network Structure

Transmission electron microscopy has been used for the elucidation of gel network structures of soybean protein (Hermansson and Buchheim 1981; Hermansson 1985; Nakamura et al. 1984a, 1985b, 1986).

According to Hermansson (1985), the characterization of gels of soybean proteins by electron microscopy is a delicate procedure and artifacts may be easily produced, the most common arising from freezing of the gel. Ice and salt crystals may cause phase separation and disrupt the original network structure. During gel preparation, water must be removed or solidified without major change in the network structure. If this is done by dehydration, the structure must be chemically fixed to avoid possible change.

The preparation for characterizing the network structures of soybean protein gels by transmission electron microscopy is generally conducted as follows: (1) chemical fixation, (2) stepwise hydration by ethanol, (3) embedding in resin, (4) thin sectioning and (5) staining. Figure 3 shows the network structure of 11S globulin gels treated as above.

3.2 Gel Extraction

To elucidate the molecular forces or bonds in a gel network and identify the subunits present within it, the gel matrix is solubilized using various solvents (Aoki 1970; Utsumi and Kinsella 1985; Mori et al. 1986). As described by Utsumi and

a

b

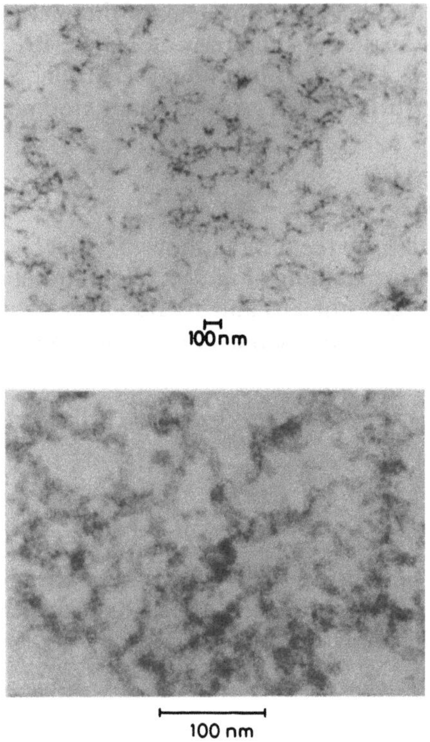

Fig. 3 a, b. Transmission electron micrographs of glycinin gel; **a** and **b** micrographs at a magnification of ×30000 and ×120000, respectively. (From Nakamura et al. 1984a)

Table 4. Instruments for rheological and textural measurements of soybean protein gels

Class	Instruments	References
Rheological measurements (Viscoelasticity)	Viscometer	Circle et al. (1964)
		Catsimpoolas and Meyer (1970)
		Shimada and Matsushita (1980)
		Umeya et al. (1980)
		Babajimopoulos et al. (1983)
		Beveridge et al. (1984)
		Varfolomeyeva et al. (1986)
	Instron UTM (Rheometer)	Kuwahata and Nakahama (1975)
		Furukawa et al. (1979)
		Van Kleef (1986)
	Penetrometer	Aoki (1965)
		Bikbov et al. (1981)
Textural measurements (Relative physical properties)	Texturometer	Saio et al. (1969)
		Yasumatsu et al. (1972)
		Furukawa et al. (1979)
		Shimada and Matsushita (1980)
		Utsumi et al. (1982)

Kinsella (1985), NaCl has stabilizing effects on hydrophobic interactions, while NaSCN has a destabilizing effect. Both neutralize charges in proteins. SDS destabilizes hydrophobic interactions, but not hydrogen bonding. Urea does both. Propylene glycol weakens hydrophobic interactions, but altering the dielectric constant it enhances hydrogen bonding and electrostatic interactions. 2-ME lessens the number of intermolecular disulfide bonds. By the analysis of solubilized materials and remaining gels it may be possible to determine the molecular forces or bonds which contribute most to the gel network, and the manner in which soybean protein forms the network structure.

3.3 Analytical Methods for Examining Rheological and Textural Properties of a Gel

Physical measurements of soybean protein gels are expressed either as rheological characteristic values, or relative experimental values. Table 4 shows typical instruments used for this purpose. The instruments are sometimes modified to suit individual preference.

4 Relationships Between Protein Structure and Gel Properties

The above sections gave an outline of analytical techniques of the gelling process and gel properties. To elucidate the basic structural factors of protein responsible for gel properties, it is necessary to investigate the formation process and properties of gels produced from soybean proteins with slightly different structures. The following sections describe the analytical methods for the gelation of such soybean proteins.

4.1 Native Proteins

4.1.1 Comparison of the Molecular Species of Various Subunits

Saio et al. (1969) studied the texture of tofu-gels made from various cultivars of soybean differing in their 11S to 7S protein content, and they found gel hardness to be in proportion to 11S.

As shown in Table 2, 7S globulin (beta-conglycinin) occurs as seven molecular species, each containing three kinds of subunits (Thanh and Shibasaki 1976; Geoffrey and Kenwyn 1981). 11S globulin (glycinin) is found in at least four molecular species (Utsumi et al. 1981). Moreover, the subunit compositions of 11S globulins from the seeds of various soybean cultivars vary according to the cultivar (Morit et al. 1981). Nakamura et al. (1984b) examined the gelation and gel properties of 11S globulins from five soybean cultivars differing in subunit compositions, and found that the constituent subunits do not contribute equally to the formation of the gel and its properties.

4.1.2 Comparison of Similar Storage Proteins from Various Beans and Cereals

Ma and Harwalkar (1987) investigated the association–dissociation of oat 11S globulin during heat treatment and the mechanism of thermal aggregation. A significant difference was noted in the heat coagulation of oat 11S globulin and soybean 11S globulin.

The primary structures of the subunits of 11S proteins isolated from pea (Lycett et al. 1984; Domoney et al. 1986), rape-seed (Simon et al. 1985), broad bean (Wobus et al. 1986), oat (Walburg and Larkins 1985) and soybean (Table 2) have been determined at the gene level. Differences in thermal association–dissociation between soybean 11S globulin and other 11S globulins can be perhaps elucidated on the basis of differences in amino acid sequences in 11S globulins.

4.2 Artificial Proteins

4.2.1 Reconstitution from Isolated Subunits

To determine the manner in which the subunits of 11S globulin contribute to the physical properties of a gel, Mori et al. (1982b) and Nakamura et al. (1985b) studied the formation of pseudo-11S globulins differing in subunit composition from native 11S globulin by reconstitution from combinations of native acidic and basic subunits or intermediary subunits. Gel hardness was also examined. Utsumi et al. (1983) found it possible to produce artificially reconstituted 11S globulins differing in subunits compositions from native 11S globulins, i.e., hybrid 11S globulins, using combinations of native acidic and basic subunits of soybean 11S globulins and broad bean 11S globulins. The gel properties of hybrid 11S globulins were examined and it became apparent that constituent subunits of 11S globulins vary in their contribution to gel properties.

4.2.2 Chemical Modification

The modification of protein properties by chemicals is frequently done in biochemical studies, in most cases to obtain information that may elucidate mechanisms of action of biologically active proteins, such as enzymes.

The general objectives of chemical modification of soybean proteins for food are the blocking of deteriorative reactions, and the improvement of physical and nutritional properties. Extensive studies on chemically modified soybean protein have been made by Kinsella and co-workers (Brinegar and Kinsella 1980; Damodaran and Kinsella 1982; Kim and Kinsella 1987), and Kito and co-workers (Haque et al. 1982; Hirotsuka et al. 1984). A comprehensive general textbook on the chemical modification of food proteins has been edited by Feeney and Whitaker (1977, 1982).

Little attention, however, has been directed to relationships between modified protein structure and gel properties. Virtually nothing is understood of the effects of chemical modification on the gelation mechanism.

4.2.3 Enzymatic Modification

Enzymes in vivo cause extensive post-translational modification of proteins, such as hydrolysis, hydroxylation, glycosilation, phosphorylation, acylation, methylation and the cross-linking of polypeptide chains. These modifications occur in an in vitro system. The reactions involved can be made to occur under mild conditions and the desired modification should be possible without damage to the protein structure. Several excellent reviews describe these reactions in detail (Feeney and Whitaker 1977, 1982).

Some reports are available on enzymatic modification of soybean proteins, as through the effects of hydrolysis by protease on functional properties (Puski 1975; Pour-EL and Swenson 1976), plastein reactions with papain (Yamashita et al. 1979) and immobilized alpha-chymotrypsin (Pallavicini et al. 1983), papain-catalyzed acylation (Sung et al. 1983) and cross-linking by transglutaminase (Ikura et al. 1980; Nio et al. 1985). In these studies attempts were made to improve the functional properties of soybean protein. It should be possible to elucidate the gelation mechanism by using enzymatic modification.

4.2.4 Protein Engineering

Methods for the mutagenesis of DNA have been developed to identify functionally important regions. By site-directed in vitro mutagenesis, it is possible to alter a defined site within a region of cloned DNA (Zoller and Smith 1983). This method has a wide application ranging from the definition of functional DNA sequences to the construction of new proteins. Kits for the site-directed mutagenesis of DNA as well as instruments for oligonucleotide synthesis are available.

DNA clones isolated by many investigators, which encode soybean storage proteins, are shown in Table 2. A gene encoding the alpha′-subunit of soybean 7S globulin has been expressed in seeds of transformed petunia plants (Beachy et al. 1985). Subunits of soybean 11S globulin have been synthesized from recombinant cDNAs using a combination of SP6 transcription and rabbit reticulocyte translation in vitro systems (Dickinson et al. 1987). However, the site-directed mutagenesis method is no longer used. In the near future, a modified soybean storage protein, some of whose amino acid residues are replaced with others, will be made. Its use should greatly facilitate clarification of the relationships between protein structure and gel properties.

Acknowledgments. The author is thankful to Drs T. Mori and S. Utsumi (The Research Institute for Food Science, Kyoto University) for permitting him to study and apply some of the analytical methods presented in this manuscript at their laboratory. Thanks are due to the Hohnen Oil Co., Ltd. for its cooperation in the preparation of this manuscript.

References

Aoki H (1965) Studies on the gelation of soybean protein. Part 1. On the methods determining the physical properties (chewy properties) of gels. J Jpn Soc Food Sci Tech 39:262–269

Aoki H (1970) Changes in soybean protein components during its gel formation process. J Agric Chem Soc Jpn 17:129–135

Babajimopoulos M, Damodaran S, Rizvi SSH, Kinsella JE (1983) Effects of various anions on the rheological and gelling behavior of soy proteins: thermodynamic observations. J Agric Food Chem 31:1270–1275

Barthova J, Hbras P, Leblova S (1973) Isolation and properties of plant lactate dehydrogenase. Collect Czech Chem Commun 38:2174–2180

Beachy RN, Chen ZL, Horsch RB, Rogers SG, Hoffmann NJ, Fraley RT (1985) Accumulation and assembly of soybean beta-conglycinin in seeds of transformed petunia plants. EMBO J 4:3047–3053
Beveridge T, Jones L, Tung M (1984) Progel and gel formation and reversibility of gelation of whey, soybean, and albumin protein gels. J Agric Food Chem 32:307–313
Bikbov TM, Grinberg VY, Schmandke H, Chaika TS, Vaintraub IA, Tolstoguzov VB (1981) A study on gelation of soybean globulin solutions 2, Viscoelastic properties and structure of thermotropic gels of soybean globulins. Coll Polymer Sci 259:536–547
Bikbov TM, Grinberg VY, Schmandke H, Chaika TS, Vaintraub IA, Tolstoguzov VB (1983) Studies on gelation of soybean globulin solutions. Coll Polymer Sci 261:346–358
Boland MJ, Wong E (1975) Purification and kinetic properties of chalcone-flavanone isomerase from soya bean. Eur J Biochem 50:383–389
Brinegar AC, Kinsella JE (1980) Reversible modification of lysine in soybean proteins, using citraconic anhydride: characterization of physical and chemical changes in soy protein isolate, the 7S globulin, and the 11S globulin. J Agric Food Chem 28:818–824
Bronovitskaya ZS, Kretovich VL (1976) Molecular weight of soy cotyledon malate dehydrogenase. Dokl Biochem 190:19–26
Brooks JR, Morr CV (1985) Current aspects of soy protein fractionation and nomenclature. J Am Oil Chem Soc 62:1347–1354
Catsimpoolas N, Meyer EW (1968) Immunochemical properties of the 11S component of soybean proteins. Arch Biochem Biophys 125:742–750
Catsimpoolas N, Meyer EW (1970) Gelation phenomena of soybean globulins. 1, Protein-protein interactions. Cereal Chem 47:559–570
Circle SJ, Meyer EW, Whitney RW (1964) Rheology of soy protein dispersions. Effect of heat and other factors on gelation. Cereal Chem 41:157–172
Coates JB, Medeiros JS, Thanh VH, Nielsen NC (1985) Characterization of the subunits of beta-conglycinin. Arch Biochem Biophys 243:184–194
Damodaran S, Kinsella JE (1982) Effect of conglycinin on the thermal aggregation of glycinin. J Agric Food Chem 30:812–817
Danilenko AN, Grozav EK, Bikbov TM, Grinberg VY, Tolstoguzov VB (1985) Studies on the stability of 11S globulin from soybeans by differential scanning microcalorimetry. Int J Biol Macromol 7:109–112
Danilenko AN, Bikbov TM, Burova TV, Grinberg VY, Tolstoguzov VB (1986) The effect of neutral salts on the conformational stability of 11S globulins from some seed using differential scanning microcalorimetry. Nahrung 30:257–262
Dickinson CD, Floener LA, Lilley GG, Nielsen NC (1987) Selfassembly of proglycinin and hybrid proglycinin synthesized in vitro from cDNA. Proc Natl Acad Sci USA 84:5525–5529
Diel E, Stan HJ (1978) Purification and characterization of two isoenzymes of lipoxygenase from soybeans. Planta 142:321–328
Domoney C, Barker D, Casey R (1986) The complete deduced amino acid sequences of legumin beta-polypeptides from different genetic loci in *Pisum*. Plant Mol Biol 7:467–474
Doyle JJ, Schuler MA, Godette WD, Zenger V, Beachy RN, Slightom JL (1986) The glycosylated seed storage proteins of *Glycine max* and *Phaseolus vulgaris*: structural homologies of genes and proteins. J Biol Chem 261:9228–9238
Farinelli MP, Fry DW, Richardson KE (1983) Isolation, purification, and partial characterization of formate dehydrogenase from soybean seed. Plant Physiol 73:858–859
Feeny RE, Whitaker JR (eds) (1977) Food protein: improvement through chemical and enzymatic modification. Adv Chem Ser 160. Am Chem Soc, Washington D.C.
Feeny RE, Whitaker JR (eds) (1982) Modification of proteins: food, nutritional, and pharmacological aspects. Adv Chem Ser 198. Am Chem Soc Washington D.C.
Fridman C, Lis H, Sharon N, Katchalski E (1969) Isolation and characterization of soybean cytochrome c. Arch Biochem Biophys 126:299–304
Fujimoto S, Nakagawa T, Ohara A (1977) Further studies on the properties of violet colored acid phosphatase from soybean. Chem Pharm Bull 25:3283–3288
Fukazawa C, Momma T, Hirano H, Harada K, Udaka K (1985) Glycinin A3B4 mRNA; cloning and sequencing of double-stranded cDNA complementary to a soybean storage protein. J Biol Chem 260:6234–6239

Furukawa T, Ohta S, Yamamoto A (1979) Texture-structure relationships in heat-induced soy protein gels. J Texture Stud 10:333–346
Geoffrey ES, Kenwyn RG (1981) Detection and characterization of a new beta-conglycinin from soybean seeds. Arch Biochem Biophys 210:525–530
Griffiths NM, Billington MJ, Crimer AA, Hitchock CHS (1984) An assessment of commercially available reagents for an enzyme-linked immunosorbent assay (ELISA) of soya protein in meat products. J Sci Food Agric 35:1255–1260
Haque Z, Matoba T, Kito M (1982) Incorporation of fatty acid into food protein: palmitoyl soybean glycinin. J Agric Food Chem 30:481–486
Harpaz N, Flowers HM, Sharon N (1977) alpha-D-Galactosidase from soybeans destroying blood group B antigens. Eur J Biochem 77:419–426
Hashizume K, Watanabe T (1979) Influence of heating temperature on conformational changes of soybean proteins. Agric Biol Chem 39:683–690
Hashizume K, Nakamura N, Watanabe K (1975) Influence of ionic strength on conformation changes of soybean proteins caused by heating, and relationship of its conformation changes to gel formation. Agric Biol Chem 39:1339–1347
Hermansson AM (1978) Physico-chemical aspects of soy proteins structure formation. J Texture Stud 9:33–58
Hermansson AM (1979) Methods of studying functional characteristic of vegetable proteins. J Am Oil Chem Soc 56:272–279
Hermansson AM (1985) Structure of soya glycinin and conglycinin gels. J Sci Food Agric 36:822–832
Hermansson AM (1986) Soy protein gelation. J Am Oil Chem Soc 63:658–666
Hermansson AM, Buchheim W (1981) Characterization of protein gels by scanning and transmission electron microscopy. J Coll Int Sci 81:519–530
Hinz H-J (1986) Thermodynamic parameters for protein-protein and protein-ligand interaction by differential scanning microcalorimetry. In: Hirs CH, Timasheff SN (eds) Methods Enzymol, vol 130. Academic Press, New York, pp 59–79
Hirano H, Fukazawa C, Harada K (1984) The complete amino acid sequence of the A3 subunit of the glycinin seed storage protein of the soybean (*Glycine max* (L.) Merrill). J Biol Chem 259:14371–14377
Hirano H, Fukazawa C, Harada K (1985) The primary structures of the A4 and A5 subunits are highly homologues to that of the A3 subunit in the glycinin seed storage protein of soybean. FEBS Lett 181:124–128
Hirotsuka M, Taniguchi H, Narita H, Kito M (1984) Functionality and digestibility of a highly phosphorylated soybean protein. Agric Biol Chem 48:93–100
Hitchcock CHS, Bailey FJ, Crimes AA, Dean DAG, Davis PJ (1981) Determination of soya proteins in food using an enzyme-linked immunosorbent assay procedure. J Sci Food Agric 32:157–165
Ikura K, Kometani T, Sasaki R, Chiba H (1980) Crosslinking of soybean 7S and 11S proteins by transglutaminase. Agric Biol Chem 44:2979–2984
Iwabuchi S, Shibasaki K (1981) Immunochemical studies of the effect of ionic strength on thermal denaturation of soybean 11S globulin. Agric Biol Chem 45:285–293
Iwabuchi S, Yamauchi K (1984) Effects of heat and ionic strength upon dissociation-association of soybean protein fraction. J Food Sci 49:1289–1294
Kamata Y, Kikuchi M, Shibasaki K (1980) Non-dissociative small fragments of glycinin-T. Agric Biol Chem 44:575–580
Kim SH, Kinsella EJ (1987) Surface active properties of proteins: effects of progressive succinylation on film properties and foam stability of glycinin. J Food Sci 52:1341–1343
Kinsella JE (1979) Functional properties of soy proteins. J Am Oil Chem Soc 56:242–258
Koshiyama I, Fukushima D (1976) Purification and some properties of gamma-conglycinin in soybean seeds. Phytochemistry 15:161–164
Koshiyama I, Hamano M, Fukushima D (1980–1981) A heat denaturation study of the 11S globulin in soybean seeds. Food Chem 6:309–322
Krishnan KS, Brandts JF (1978) Scanning calorimetry. In: Hirs CH, Timasheff SN (eds) Methods Enzymol, vol 49. Academic Press, New York, pp 3–14
Kuwahata M, Nakahama N (1975) Viscoelasticity of soybean gel. J Agric Chem Soc Jpn 49:129–134

Laemmli UK (1970) Cleavage of structural proteins during the assembly of the head of bacteriophage T4. Nature 227:680–682

Larkins BA (1985) Polysomes. In: Linskens HF, Jackson JF (eds) Modern methods of plant analysis, New Series, vol 1. Springer, Berlin Heidelberg New York, pp 331–352

Leblova S, Perglerova E (1976) Soybean alcohol dehydrogenase. Phytochemistry 15:813–815

Lis H, Sharon N, Katchalski E (1966) Soybean hemagglutinin; a plant glycoprotein. J Biol Chem 241:684–689

Lycett GW, Croy RRD, Shirsat AH, Boulter D (1984) The complete nucleotide sequence of a legumin gene from pea (*Pisum sativum* L.). Nucl Acids Res 12:4493–4506

Ma CY, Harwalkar VR (1987) Thermal coagulation of oat globulin. Cereal Chem 64:212–218

Marco YA, Thanh VH, Tumer NE, Scallon BJ, Nielsen NC (1984) Cloning and structural analysis of DNA encoding an A2B1a subunit of glycinin. J Biol Chem 259:13436–13441

Matsudomi N, Mori H, Kato A, Kobayashi K (1985) Emulsifying and foaming properties of heat-denatured soybean 11S globulins in relation to their surface hydrophobicity. Agric Biol Chem 49:915–919

Mikami B, Nomura K, Morita U (1986) N-terminal sequence of soybean beta-amylase. J Biochem 96:513–516

Momma T, Negoro T, Udaka K, Fukazawa C (1985a) A complete cDNA coding for the sequence of glycinin A2B1a subunit precursor. FEBS Lett 188:117–122

Momma T, Negoro T, Hirano H, Matsumoto A, Udaka K, Fukazawa C (1985b) Glycinin A5A4B3 mRNA: cDNA cloning and nucleotide sequencing of a splitting storage protein subunit of soybean. Eur J Biochem 149:491–496

Moreira M, Hermodson MA, Larkins BA, Nielsen NC (1979) Partial characterization of the acidic and basic polypeptides of glycinin. J Biol Chem 254:9921–9926

Mori T, Utsumi S, Inaba H, Kitamura K, Harada K (1981) Differences in subunit composition of glycinin among soybean cultivars. J Agric Food Chem 29:20–23

Mori T, Nakamura T, Utsumi S (1982a) Gelation mechanism of soybean 11S globulin: formation of soluble aggregates as transient intermediates. J Food Sci 47:26–30

Mori T, Nakamura T, Utsumi S (1982b) Formation of pseudoglycinins and their gel hardness. J Agric Food Chem 30:828–831

Mori T, Nakamura T, Utsumi S (1986) Behavior of intermolecular bond formation in the late stage of heat-induced gelation of glycinin. J Agric Food Chem 34:33–36

Morita Y, Yagi F, Aibara S, Yamashita H (1976) Chemical composition and properties of soybean beta-amylase. J Biochem 79:591–603

Nakamura T, Utsumi S, Mori T (1984a) Network structure formation in thermally induced gelation of soybean glycinin. J Agric Food Chem 32:349–352

Nakamura T, Utsumi S, Kitamura K, Harada K, Mori T (1984b) Cultivar differences in gelling characteristics of soybean glycinin. J Agric Food Chem 32:647–651

Nakamura T, Utsumi S, Mori T (1985a) Effects of temperature on the different stages in thermal gelling process of glycinin. J Agric Food Chem 33:1201–1203

Nakamura T, Utsumi S, Mori T (1985b) Formation of pseudo-glycinin from intermediary subunits and their gel properties and network structure. Agric Biol Chem 49:2733–2740

Nakamura T, Utsumi S, Mori T (1986) Mechanism of heat-induced gelation and gel properties of soybean 7S globulin. Agric Biol Chem 50:1287–1293

Negoro T, Momma T, Fukazawa C (1985) A cDNA encoding a glycinin A1a subunit of soybean. Nucl Acids Res 13:6719–6730

Nielsen NC (1985) The structure and complexity of the 11S polypeptides in soybeans. J Am Oil Chem Soc 62:1680–1686

Nio N, Motoki M, Takinami K (1985) Gelation of casein and soybean globulins by transglutaminase. Agric Biol Chem 49:2283–2286

Pallavicini C, Peruffo ADB, Finley JW (1983) Comparative study of soybean plasteins synthesized with soluble and immobilized alpha-chymotrypsin. J Agric Food Chem 31:846–848

Peng IC, Dayton WR, Quass DW, Allen CE (1982) Studies on the subunits involved in the interaction of soybean 11S protein and myosin. J Food Sci 47:1984–1990
Peng IC, Quass DW, Dayton WR, Allen CE (1984) The physicochemical and functional properties of soybean 11S globulin: a review. Cereal Chem 61:480–490
Polacco JC, Havir EA (1979) Comparisons of soybean urease isolated from seed and tissue culture. J Biol Chem 245:1707–1715
Poure-EL A, Swenson TS (1976) Gelation parameters of enzymatically modified soy protein isolates. Cereal Chem 53:438–456
Puski G (1975) Modification of functional properties of soy proteins by proteolytic enzyme treatment. Cereal Chem 52:655–664
Richardson DP, Catsimpoolas N (1979) The effect of thermal denaturation on the tryptic hydrolysis of glycinin. J Sci Food Agric 30:463–468
Saio K, Watanabe T (1978) Differences in functional properties of 7S and 11S soybean proteins. J Texture Stud 9:135–157
Saio K, Kajikawa M, Watanabe T (1969) Food processing characteristics of soybean 11S and 7S proteins. Part 1. Effect of differences of protein components among soybean varieties on formation of tofu-gel. Agric Biol Chem 33:1301–1308
Saio K, Kajikawa M, Watanabe T (1971) Food processing characteristics of soybean 11S and 7S proteins. Part 2. Effect of sulfhydryl groups on physical properties of tofu-gel. Agric Biol Chem 35:890–898
Saita M, Ikenaka T, Matsushima Y (1971) Isolation and characterization of alpha-D-mannosidase from soybean. J Biochem 70:827–833
Scallon BJ, Thanh VH, Floener LA, Nielsen NC (1985) Identification and characterization of DNA clones encoding group-II glycinin subunits. Theor Appl Genet 70:510–519
Scallon BJ, Dickinson CD, Nielsen NC (1987) Characterization of a null-allele for the Gy4 glycinin gene from soybean. Mol Gen Genet 208:107–113
Schuler MA, Schmitt ES, Beachy RN (1982a) Closely related families of genes code for the alpha and alpha-prime subunits of the soybean 7S storage protein complex. Nucl Acids Res 10:8225–8244
Schuler MA, Ladin BF, Pollaco JC, Freyer G, Beachy RN (1982b) Structural sequences are conserved in the genes coding for the alpha, alpha-prime and beta subunits of the soybeans 7S seed storage protein. Nucl Acids Res 10:8245–8261
Schuler MA, Doyle JJ, Beachy RN (1983) Nucleotide homologies between the glycosylated seed storage proteins of *Glycine max* and *Phaseolus vulgaris*. Plant Mol Biol 2:119–127
Shaanan B, Shoham M, Yonath A, Lis H, Sharon N (1984) Crystallization and preliminary x-ray diffraction studies of soybean agglutinin. J Mol Biol 174:723–725
Shimada K, Matsushita S (1980) Gel formation of soybean 7S and 11S protein. Agric Biol Chem 44:637–641
Simon AE, Tenbarge KM, Scofield SR, Finkelstein RR, Crouch ML (1985) Nucleotide sequence of a cDNA clone of *Brassica napus* 12S storage protein shows homology with legumin from *Pisum sativum*. Plant Mol Biol 5:191–201
Staswick PE, Hermodson MA, Nielsen NC (1981) Identification of the acidic and basic subunit complexes of glycinin. J Biol Chem 256:8752–8755
Staswick PE, Hermodson MA, Nielsen NC (1984a) The amino acid sequence of the A2B1a subunit of glycinin. J Biol Chem 259:13424–13430
Staswick PE, Hermodson MA, Nielsen NC (1984b) Identification of the cystines which link the acidic and basic components of the glycinin subunits. J Biol Chem 259:13431–13435
Steiner RF, Frattali V (1969) Purification and properties of soybean protein inhibitors of proteolytic enzymes. J Agric Food Chem 17:513–518
Sung HY, Chen HJ, Liu TY, Su JC (1983) Improvement of the functionality of soy protein by introduction of new thiol groups through a papain-catalyzed acylation. J Food Sci 48:708–711
Sykes GE, Gayler KR (1981) Detection and characterization of a new β-conglycinin from soybean seeds. Arch Biochem Biophys 210:525–530
Takagi S, Akashi M, Yasumatsu K (1979a) Determination of hydrophobic region in soybean globulin. J Jpn Soc Food Sci 26:133–138

Takagi S, Okamoto N, Akashi M, Yasumatsu K (1979b) Hydrophobic bonding and SS bonding in heat denaturation of 11S of soybean protein. J Jpn Soc Food Sci 26:139–144
Thanh VH, Shibasaki K (1976) Heterogeneity of beta-conglycinin. Biochim Biophys Acta 439:326–338
Umeya J, Yamauchi F, Shibasaki K (1980) Hardening and softening properties of soybean protein-water suspending systems. Agric Biol Chem 44:1321–1326
Utsumi S, Kinsella JE (1985) Structure-function relationships in food proteins: subunit interactions in heat-induced gelation of 7S, 11S, and soy isolate proteins. J Agric Food Chem 33:297–303
Utsumi S, Inaba H, Mori T (1981) Heterogeneity of soybean glycinin. Phytochemistry 20:585–589
Utsumi S, Nakamura T, Mori T (1982) A micro-method for measurement of gel properties of soybean 11S globulin. Agric Biol Chem 46:1923–1924
Utsumi S, Nakamura T, Mori T (1983) Role of constituent subunits in the formation and properties of heat-induced gels of 11S globulins from legume seeds. J Agric Food Chem 31:503–506
Utsumi S, Kohno M, Mori T, Kito M (1987a) An alternate cDNA encoding glycinin A1bBx subunit. J Agric Biol Chem 35:210–214
Utsumi S, Nakamura T, Harada K, Mori T (1987b) Occurrence of dissociable and undissociable soybean glycinin. Agric Biol Chem 51:2139–2144
Van Kleef FSM (1986) Thermally induced protein gelation: gelation and rheological characterization of highly concentrated ovalbumin and soybean protein gels. Biopolymers 25:31–59
Varfolomeyeva EP, Danilenko AN, Bikbov TM, Grinberg VY, Leontiev AL, Tolstoguzov VB (1986) The rheological properties of diluted solutions of 11S globulin isolated from soybeans by using selective thermal denaturation of 2S and 7S globulins. Nahrung 30:487–500
Walburg G, Larkins BA (1985) Isolation and characterization of cDNA encoding oat 12S globulin mRNA. Plant Mol Biol 6:161–169
Wobus U, Baumlein H, Bassuner R, Heim U, Jung R, Muntz K, Saalbach G, Weschke W (1986) Characteristics of two types of legumin genes in the field bean (*Vicia faba* L. var. minor) genome as revealed by cDNA analysis. FEBS Lett 201:74–80
Wolf WJ (1970) Soybean proteins: their functional, chemical and physical properties. J Agric Food Chem 18:969–976
Wolf WJ, Tamura T (1969) Heat denaturation of soybean 11S protein. Cereal Chem 46:331–344
Yamagishi T, Yamauchi F, Shibasaki K (1981) Electrophoretical and differential thermal analysis of soybean 11S globulin heated in the presence of N-ethylmaleimide. Agric Biol Chem 45:1661–1668
Yamagishi T, Tomisawa T, Yamauchi F (1983) Spectroscopic studies on the aggregates and the dissociates induced by heating soybean 11S globulin in the presence of N-ethylmaleimide. Agric Biol Chem 47:2475–2481
Yamagishi T, Takahashi N, Yamauchi F (1987) Covalent polymerization of acidic subunits on heat-induced gelation of soybean glycinin. Cereal Chem 64:207–212
Yamashita M, Arai S, Imaizumi Y, Amano Y, Fujimaki M (1979) A one step process for incorporation of L-methionine into soy proteins by treatment with papain. J Agric Food Chem 27:52–56
Yamauchi F, Sato K, Yamagishi T (1984) Isolation and partial characterization of a salt-extractable globulin from soybean seeds. Agric Biol Chem 48:645–650
Yamauchi F, Sato W, Kamata Y (1985) Subunit structure of gamma-conglycinin in soybean seeds. Phytochemistry 24:1503–1504
Yasumatsu K, Toda J, Wada T, Misaki M, Ishii K (1972) Studies on the functional properties of food-grade soybean products. Agric Biol Chem 36:537–543
Zoller MJ, Smith M (1983) Oligonucleotide-directed mutagenesis of DNA fragments cloned into M13 vectors. In: Wu R, Grossman L, Moldave K (eds) Methods Enzymol, vol 100. Academic Press, New York, pp 468–500

Techniques of Solar Crop Dryers

N. K. Bansal and M. S. Sodha

1 Introduction

Plant products have to pass through well-defined steps for appropriate storage. The amount of moisture content of the product is the most important factor in determining the extent of deterioration during storage. Drying of products from plants is an important process in preservation. It is during storage that most of the deterioration can take place. Traditionally, sun drying is carried out by spreading the product on the ground and exposing it to the sun during the day, and preferably covering it at night, and also for protection from rain, dust, etc. Open-air drying, although simple, has many inherent disadvantages. The product cannot be protected from dust and other elements when it needs to be dried over a large area. As a result of lack of control over the drying process in the traditional methods, the quality of the product suffers and part of it is lost as a result of fungal proliferation, infestation by parasites and rodents, and reduced germination.

There are recommended maximum drying temperatures and moisture content for different products, for storage over 1 year or more. The principal objective in a drying operation is to supply what is required in an optimal manner so as to obtain a product of the best quality with minimum overall expenditure of energy. Conventional drying techniques utilize large-scale fossil fuel-fired air dryers. In many cases, hot combustion gases are passed directly through the product to be dried, which often becomes contaminated by unburnt fuel, fumes and soot. Electrical heating of air for drying is preferred, but it is very expensive and not feasible in rural areas of developing countries.

Of the possible options among other energy sources, solar energy is a promising candidate to provide the required amount of energy for drying. This option is most suitable for developing countries because most of these countries lie within the belt of good solar radiation throughout the year (Sodha et al. 1987).

2 The Drying Process

2.1 Drying Parameters

The process of drying is controlled by the properties of the drying air, often known as the external parameters, and the properties of the product, i.e. the internal parameters.

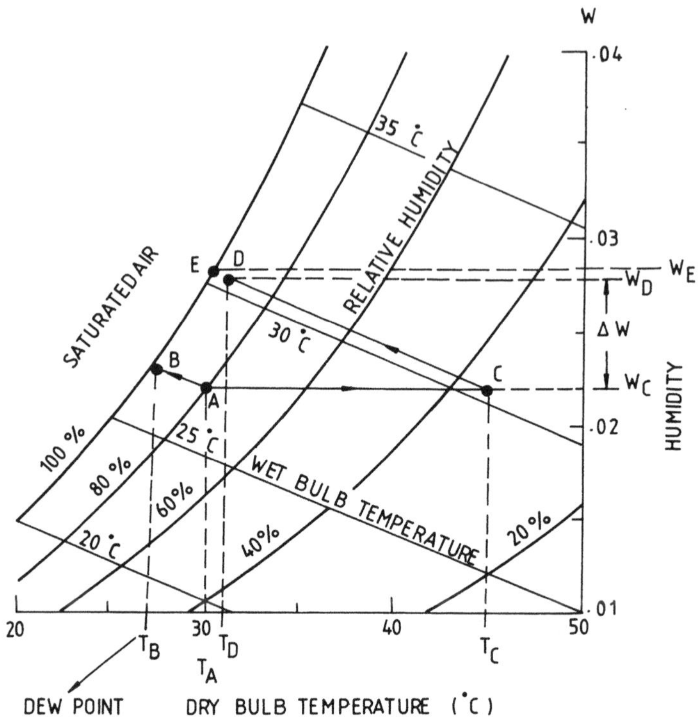

Fig. 1. Paths of temperature changes of air on the psychrometric chart

The external parameters are the dry- and wet-bulb temperatures, relative humidity, humidity ratio, temperature and enthalpy of the drying air. All these properties can be denoted on a chart called the psychrometric chart. In the process of drying, evaporation of water from the wet product is caused by heated or unheated air. The changes can be followed on the psychrometric chart by a line parallel to the wet-bulb lines, as shown in Fig. 1. If moist air is heated without the addition or removal of water, its humidity ratio remains constant, whereas the relative humidity decreases with an increase in the temperature. Line AC illustrates this phenomenon. This heated air, at dry-bulb temperature T_c, is now used in removing the moisture as well as in raising the product temperature. As a result of moisture evaporation from the product, the relative humidity of the drying air increases and the dry-bulb temperature decreases, which can be followed on a constant wet-bulb temperature line CD of constant enthalpy, the final temperature of the drying air at the exhaust being T_D. The total energy requirements for drying in the complete drying process can be calculated by the difference of the final and initial humidity ratios, i.e., $W_D - W_C = W$, as illustrated in the example in Sect. 4.

The internal parameters affecting the process of drying are moisture content, drying ratio and the latent heat. The moisture content is usually defined either on

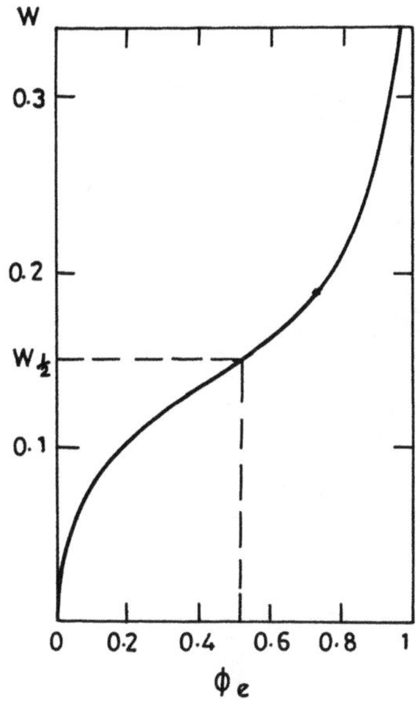

Fig. 2. Sorption isotherm of maize at 80° C

dry basis or wet basis, as follows:

$$M_{db} = \frac{w - d}{d} \times 100\% \tag{1}$$

$$M_{wb} = \frac{w - d}{w} \times 100\% \tag{2}$$

where w and d are the weights of the wet and dried samples, respectively.

Another parameter of importance in determining the parameters for drying is the equilibrium moisture content, M_e. The equilibrium moisture content is defined as the magnitude of the moisture content at the stage of equilibrium, i.e. when the rate of loss of moisture from the product to the surrounding environment equals the rate of moisture absorption by the product. The associated value of the relative humidity of the surrounding air is called the equilibrium relative humidity ϕ_e. The relationship between ϕ_e and M_e is denoted for a particular temperature on an equilibrium moisture curve known as sorption isotherm. The sorption isotherm varies with the product (and obviously also with the temperature); a typical example is given in Fig. 2 for maize. Empirically, the sorption isotherm satisfies the following relationship:

$$\phi_e = \exp(-CTM_e^n), \tag{3}$$

where ϕ_e is expressed as a fraction, temperature T in °K, M_e on a dry basis in %, and C and n are constants varying with the material.

2.2 Effect of Parameters

The rate of drying of a product depends on the air-flow rate and the temperature of the flowing air. Too rapid drying may result in the formation of hard surfaces

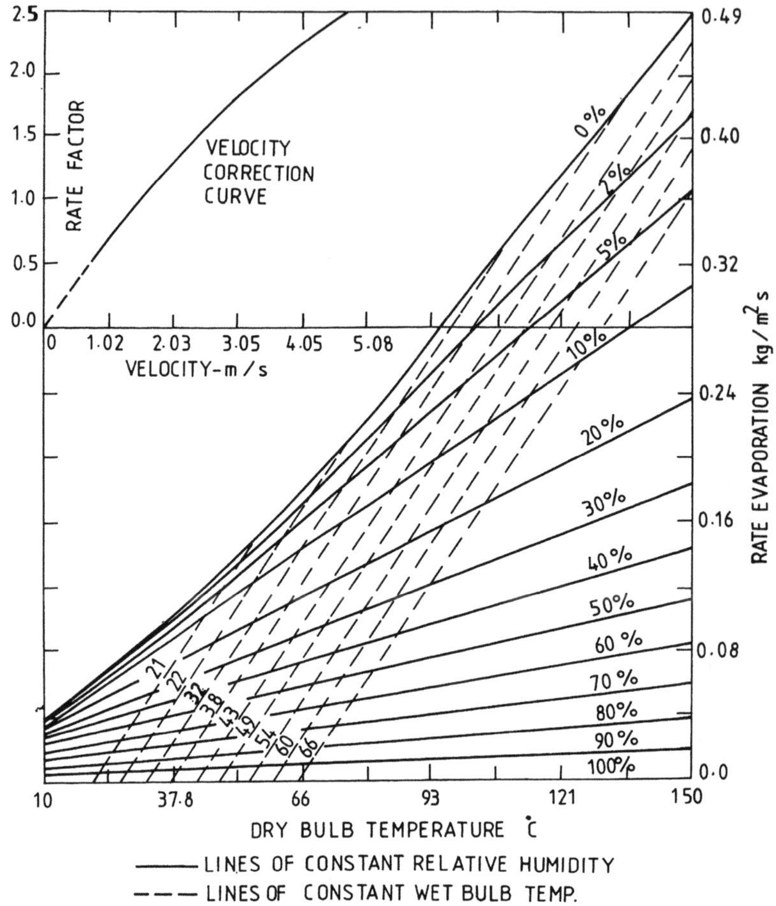

Fig. 3. Rate of evaporation chart. (After Ashrae 1976)

that will inhibit the migration of the moisture from the interior of the material surface. That is why the temperature of the drying air must be kept below some maximum value. Along with the temperature of the drying air, the flow velocity of the air also determines the drying time. The rate of moisture evaporation from the plant product is determined by a combination of the temperature, flow rate and humidity of drying air; their effect on the rate of evaporation is shown in Fig. 3.

3 Solar Drying Techniques

Depending on the mode of collection of solar energy and mode of heat transfer to the product, the dryers can be classified as direct and indirect with natural con-

Table 1. Classification of solar dryers

Heat collection mode	Heat transfer mode	Heating type	Drying space
		Ambient air	Racks, chambers
Direct	Natural convection	Preheated by solar	Chamber
	Forced circulation	Ambient	Bins
Indirect		Preheated by solar	Bins
	Natural	Preheated	Bins
	Forced	Preheated	Bins

vection and forced convection heat transfer modes. The formal classification of dryers along with their typical applications is given in Table 1. Detailed description of some of these dryers follows in this section.

3.1 Direct-Mode Solar Dryers

Direct-mode solar dryers essentially consist of a chamber covered with a transparent sheet. The evaporation of moisture takes place due to direct absorption of solar radiation by the plant product. The flow of air is either maintained by natural convection or by a blower (forced convection), which has to be driven electrically.

3.1.1 Natural Convection Dryers

Most natural convection solar dryers are made of either a cabinet or a chamber to keep the plant product. The cabinet / chamber is usually glazed and appropriate vents are provided to allow the flow of air by natural means.

Solar Cabinet Dryer. A solar cabinet dryer is usually made of a small wooden hot box with its length about three times its width (to minimize the shading effect of the side panels). The sides and the bottom of the cabinet are blackened and insulated, while the top is glazed by using one or two transparent sheets. The transparent glass is tilted to minimize the reflective losses (Fig. 4). The interior surfaces of the cabinet are painted black to absorb the solar radiation transmitted through the cover. This makes the temperature in the cabinet rise. The products to be dried are kept in simple trays having perforations / wire mesh at the bottom; the trays in turn are placed a few centimeters above the bottom in the assembly so that the product can receive the solar radiation directly. Holes are provided near the base and top of the rear vertical side (a few centimeters below the cover) to permit movement of air by natural convection. The ambient air enters the cabinet through holes at the bottom and escapes with the moisture vapors from the holes on the vertical side.

Some designs of cabinet dryers are given in Fig. 5. It is reported (Kapoor and Agarwal 1973; Ghosh 1973) that a cabinet dryer can at least halve the drying period in comparison to open-air drying. Improved product quality is of course and additional benefit.

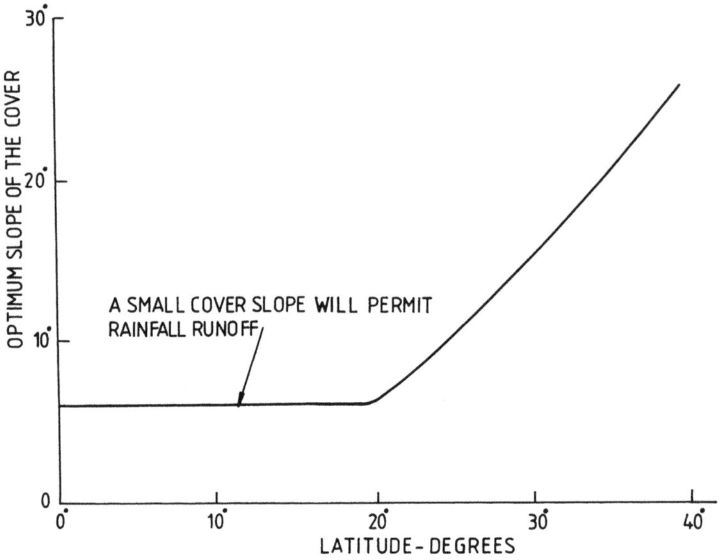

Fig. 4. Angle of slope of solar dryer covers as a function of the latitude of the location. (After Lawand 1975)

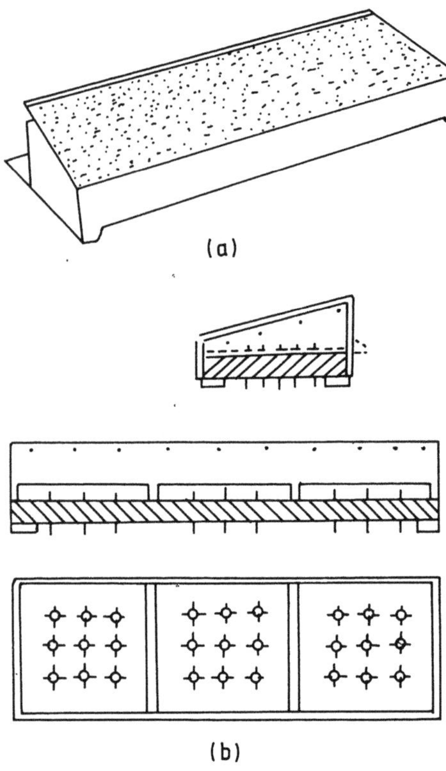

Fig. 5 a, b. Some variations of cabinet-type dryers; **a** Experimental solar dryer. The horizontal surface was a little more than 1 m^2. **b** Elevation and plan views of solar dryer. (After Lawand 1965, 1966)

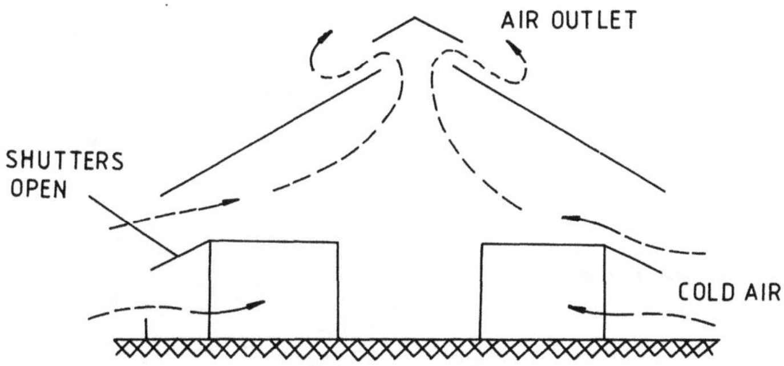

Fig. 6. Glass roof solar dryer. (After Lawand 1975)

Glass-roof Solar Dryer. This dryer is identical with a solar greenhouse. The drying unit consists of two parallel rows of drying platforms with slanted glass roofs, aligned lengthwise along a north-south axis. The drying platform is usually made of galvanized iron mesh laid over wooden beams. Air passes through the wire mesh, gets heated and picks up the moisture from the products spread over it. To facilitate the convection and ventilation, the inside surfaces of the dryer are painted black. Convection and ventilation are facilitated by the inside black surfaces and the openings that are provided on the eastern and western walls above and below the level of the platform containing the product. The schematics of a glass-roof solar dryer are shown in Fig. 6.

3.1.2 Forced Convection Dryers

The blower in the direct-mode solar dryer has to be operated by either electricity, mechanically or by a windmill, provided the wind speed is greater than 16 km/h. Direct-mode solar dryers include the downdraught dryer and the preheated circulation chamber dryer.

Downdraught Dryer. This dryer basically consists of a tray tilted at an angle of 20° from the horizontal. The tray is provided with a perforated metallic false base for keeping the product to be dried. This is further subdivided into four equal components by wooden partitions, as shown in Fig. 7. Both the ends are blocked and the air is driven by an exhaust fan (0.25 m^3/min) through the grain on the perforated tray. Temperatures in the range of 32°–50° C are achieved inside the drying space, and a batch of barley / wheat, weighing 23 kg, can be reduced from 21% moisture content (w.b.) to 15% moisture content (w.b.) successfully in 9 h.

Preheated Circulation Chamber Dryer. This dryer (Fig. 8) is basically used to dry fruits and vegetables. It consists of a slanted rectangular box covered with a glazed plate. The collection of solar energy and subsequent drying of the plant products is achieved in the same box. The air-heater unit is made of two absorber units, one of which is a corrugated iron plate blackened from the top and the other a perforated copper plate treated with selective coating to absorb maximum solar radiation and reduce heat losses. The product is put into lattice wire baskets fixed to the drying box section.

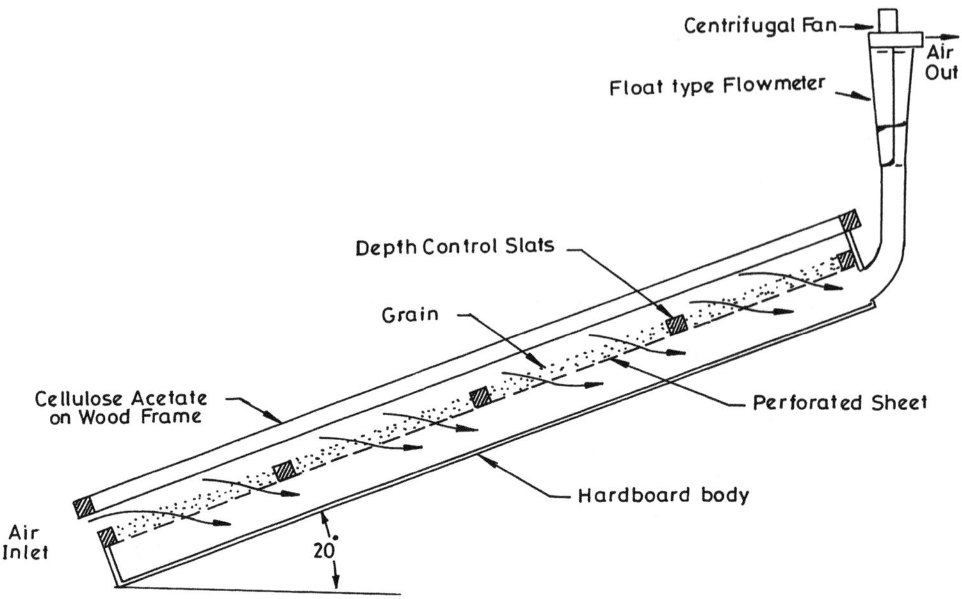

Fig. 7. The downdraught dryer. (After Lawand 1975)

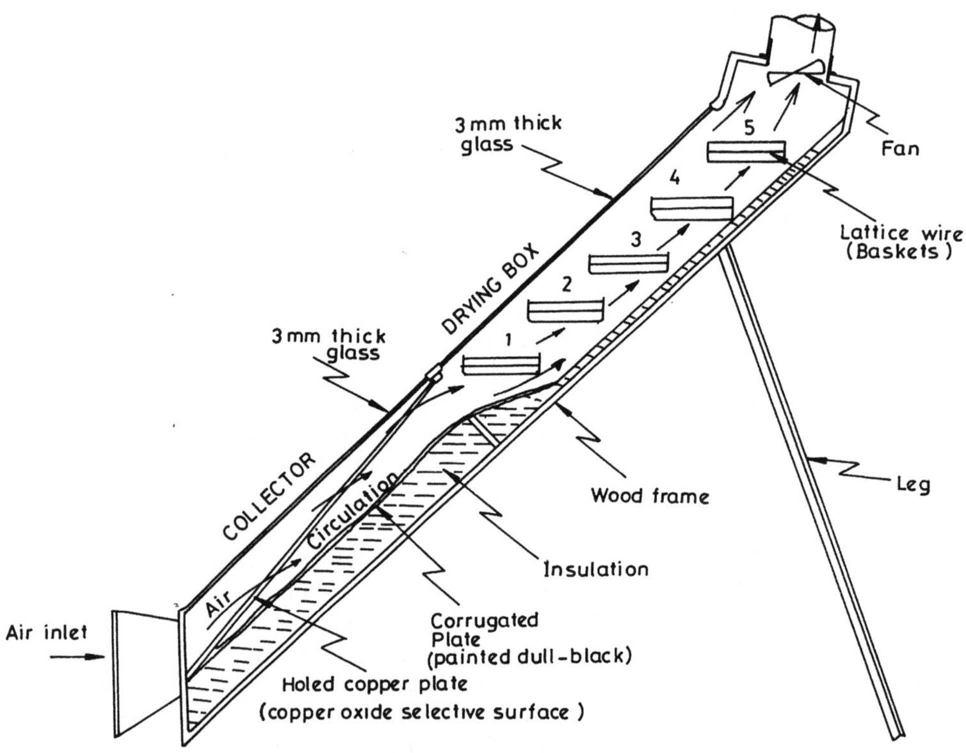

Fig. 8. Cross section of solar dryer. (After Lawand 1975)

3.2 Indirect-Mode Solar Dryers

In an indirect-mode solar dryer the plant product is not exposed to solar radiation directly. Air is first heated by using efficient and, if possible, low-cost solar air-heaters and is then passed into the chamber where the plant product is kept. An indirect-mode solar dryer, therefore, consists of the following components:

1) *Air-heater*: an air heater is used to raise the temperature of the drying air for drying purposes; a corrugated, galvanized iron plate type air-heater or inexpensive polymer collectors may be used for this purpose.
2) *Drying chamber*: the drying chamber is used as the platform for the products to be dried. It may be either tray type, bin type, tunnel type or belt type depending on the requirement and usage.

Some of the dryers require preheated air from an air-heater; simultaneously, the product absorbs solar radiation. These type of dryers are normally known as mixed-mode dryers.

3.2.1 Natural Convection Dryers

Depending upon the details of construction, the natural convection indirect – or mixed – mode dryer may be classified as a chimey-type dryer, a chamber-type dryer or a wind-ventilated dryer, as discussed in the following.

Chimney-Type Solar Dryer. A low-cost discontinuous-type solar dryer, based on natural convection, has been developed at the Asian Institute of Technology, Bangkok by Excell (1980) for drying paddy. The unit sketched in Fig. 9 consists of a solar air-heater, a cabinet for the rice bed and a chimney.

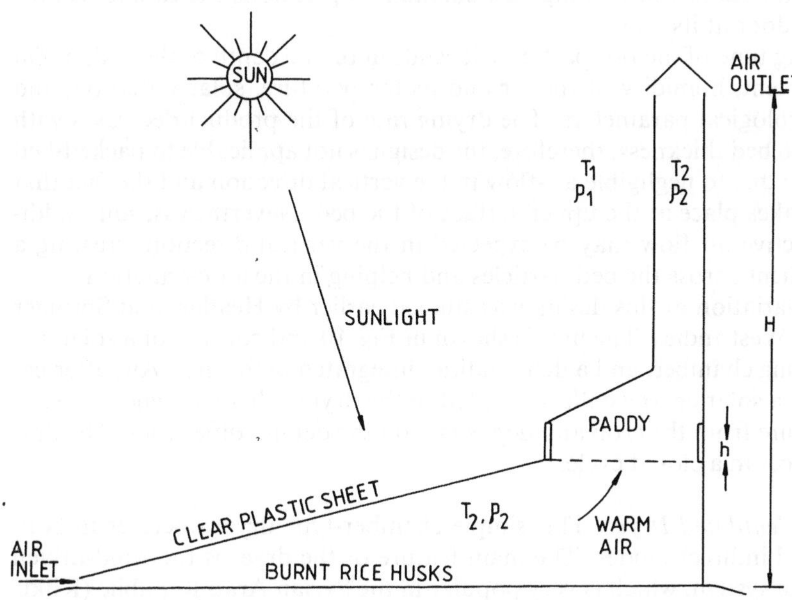

Fig. 9. Cross section of a natural convection solar dryer. (After Exell 1980)

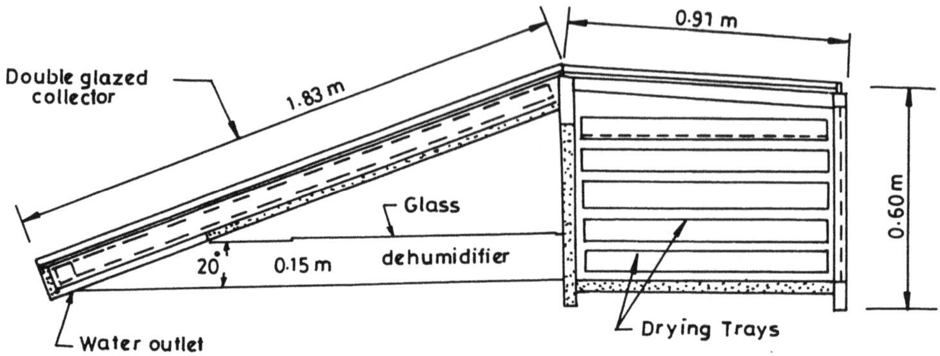

Fig. 10. Section view through solar dryer. (After Headley and Springer 1975)

A thick layer of burnt rice husk is laid on the ground and is glazed from the top with a transparent plastic sheet; this arrangement constitutes the solar air-heater. The solar radiation is transmitted through the plastic sheet and is absorbed by the blackened husk. Air gets heated through natural convection and rises. On its way it comes across the bed of rice particles. A chimney made of a bamboo frame covered with a dark plastic sheet to absorb the solar radition is provided at the top of the unit, giving rise to a tall column of warm air to increase the air-flow through the bed by natural convection. The chimney is partially covered with a cap from the top to protect the product from rain and other environmental degradation. To achieve an additional gain of solar energy, the drying cabinet may be provided with a transparent plastic cover on the front side. The dryer in this case operates in the mixed mode because the product also absorbs some radiation. For loading and unloading operations the cabinet is provided with a door at its rear.

The drying time of the rice particles depends upon the depth of the bed, initial moisture content, humidity of the air and its temperature, solar radiation, and other meteorological parameters. The drying rate of the product decreases with the increase in bed thickness; therefore, the design is not applicable to packed-bed configuration due to negligible air-flow in the vertical direction and the fact that convection takes place at the upper surface of the bed. Nevertheless, some additional convective air-flow may be expected in the vertical direction, creating a thermal gradient across the bed particles and helping in the air circulation.

A slight variation of this design was studied earlier by Headley and Springer (1975) in the West Indies. The unit is shown in Fig. 10 and consists of a solar air-heater, a drying chamber, and a dehumidifier integrated in the unit. Air, after being heated in a solar energy collector, is fed to the drying chamber where it picks up the moisture from the crop and deposits it on the dehumidifier roof. The drying takes places in a closed cycle.

Solar Wind-Ventilated Dryer. This simple chamber-type dryer operates in both the direct and indirect modes. The main feature of the dryer is the wind-driven air circulation system, which is very popular in the Syrian Arab Republic (Kilkis 1981) for drying a variety of agricultural products (ranging from fruits and vegetable to herbs and meat) on a small scale. This dryer differs from the shelf-type dryer in the sense that it does not require an electric fan for the air circulation and therefore has limited drying capacity (see Fig. 11).

The unit consists essentially of a solar air-heater integrated to the drying chamber at the base and a wind-driven fan at the top (Fig. 12). Air (preheated in

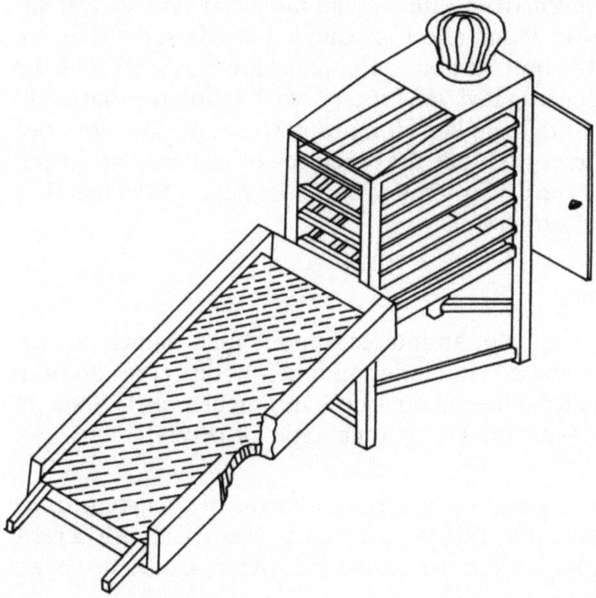

Fig. 11. Diagram of solar wind-ventilated dryer. (After Nahlawi 1966)

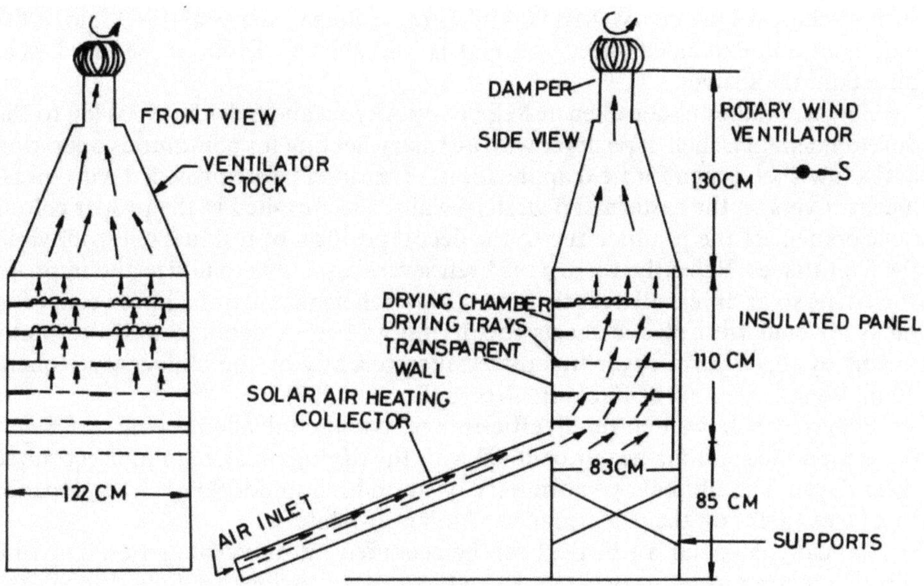

Fig. 12. Solar wind-ventilated dryer. (After Takla 1978)

the collector) is fed to the drying chamber where it picks up moisture from the products laid on the racks and gets exhausted through the openings located at the top of the rear wall of the chamber by natural convection. A black mesh screen midway between the translucent cover and the absorber provides effective air-heating due to the absorption of solar radiation by both mesh and hardboard. Additional drying is also achieved from direct solar radiation falling over the product through transparent side, front, and top panels. To reduce the heat losses, the north-facing wall and the bottom horizontal panel are blackend. The top of the dryer is provided with a corrugated vane rotary wind ventilator that sucks air from the ventilator stack. Okra, eggplant, tomato paste, and yams are dried successfully in this dryer. It is reported that the efficiency of the solar air-heater varies between 64%–84%, and the drying time of course is nearly 58% less than the time required by traditional sun drying.

3.2.2 Forced Convection Dryers

Apart from the essential components of an indirect-mode dryer, namely an air-heater and a drying chamber, a forced circulation indirect-mode dryer additionally consists of a blower to duct the heated air to the drying chamber. Some of this type of dryers are the shelf type, bin type, tunnel type, or belt type.

Shelf-type Solar Dryer. One of the earliest identified designs of shelf-type dryer was developed at the Brace Research Institute; it had a dryer box with several shelves holding the product. The shelves were provided with opaque side covers serving as an insulation.

Several variations of this configuration have been designed and tested. Figure 13 illustrates some of these designs. One version, developed at the Middle East Technical University (METU) in Turkey, during the year 1963 and 1966, had a slit and expanded sheet iron matrix and shelves. Fruits to be dried were placed on the shelves.

Further studies undertaken at METU by Akyurt and Selcuk (1973) led to the development of a shelf-type dryer with auxiliary heating for continuous operation and a low-cost design for field applications. The dryers were provided with translucent covers on the eastern and western walls. This resulted in the proper colour development of the product due to the decomposition of residual chlorophyll in the fruit tissues. Recently, there have been several attempts to utilize and improve shelf-type solar dryers all over the world. Reserchers in Australia have been using racks covered with plastic sheets for drying grapes. A detailed study to make raisins by this type of dryer was undertaken recently by the University Hohenheim, West Germany, at Kreta in Greece (Eissen 1983).

Experience shows that for an efficient and better controlled drying one has to use a separate solar air-heater coupled with the drying bin, i.e. an indirect-mode solar dryer. The ultimate performance of an indirect-mode solar dryer depends to a large extent on the performance of solar air-heaters.

Basically, all solar air heaters can be classified into two categories. The first type has a nonporous absorber in which the air-stream does not flow through the absorber plate. Air may flow above and/or below the absorber plate. The second

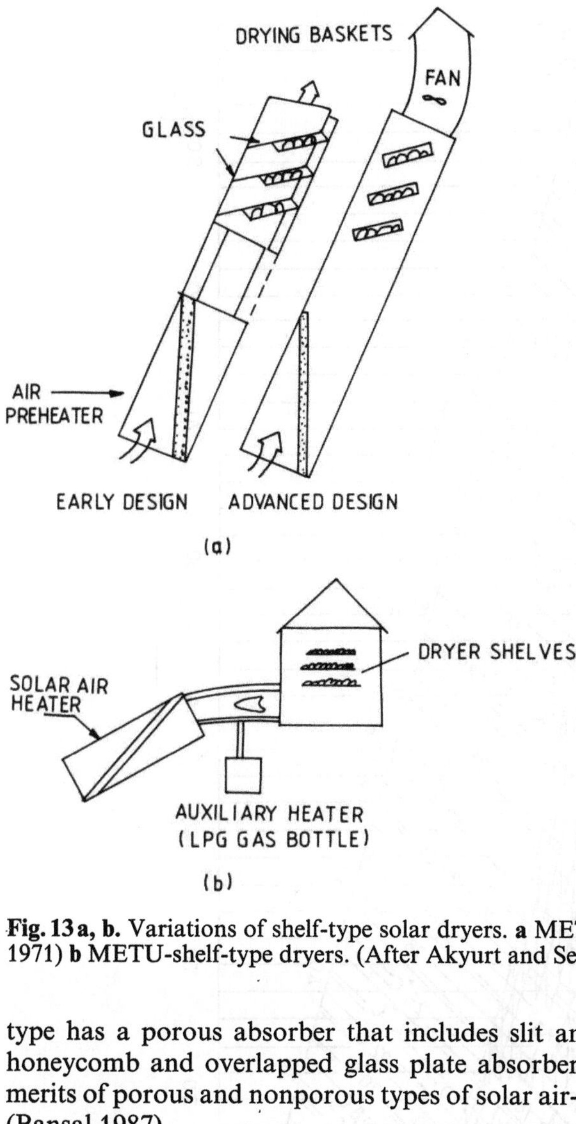

Fig. 13 a, b. Variations of shelf-type solar dryers. **a** METU shef-type dryers. (After Selcuk 1971) **b** METU-shelf-type dryers. (After Akyurt and Selcuk 1973)

type has a porous absorber that includes slit and expanded metal, transpired honeycomb and overlapped glass plate absorber. Various designs and relative merits of porous and nonporous types of solar air-heaters are discussed elsewhere (Bansal 1987).

4 Calculations for the Drying System

The physical conditions necessary to achieve the drying of food products is heat and ventilation. In arcid climates where the relative humidity is 35% or less and the ambient temperatures are above 35° C during the harvesting season, the ambient air can be utilized for drying. Figure 13, showing the enthalpy and humidity ratio on the psychrometric chart, permits to calculate the amount of vapor that can be taken by the ambient air. For example, ambient air at 35° C and 35% rel-

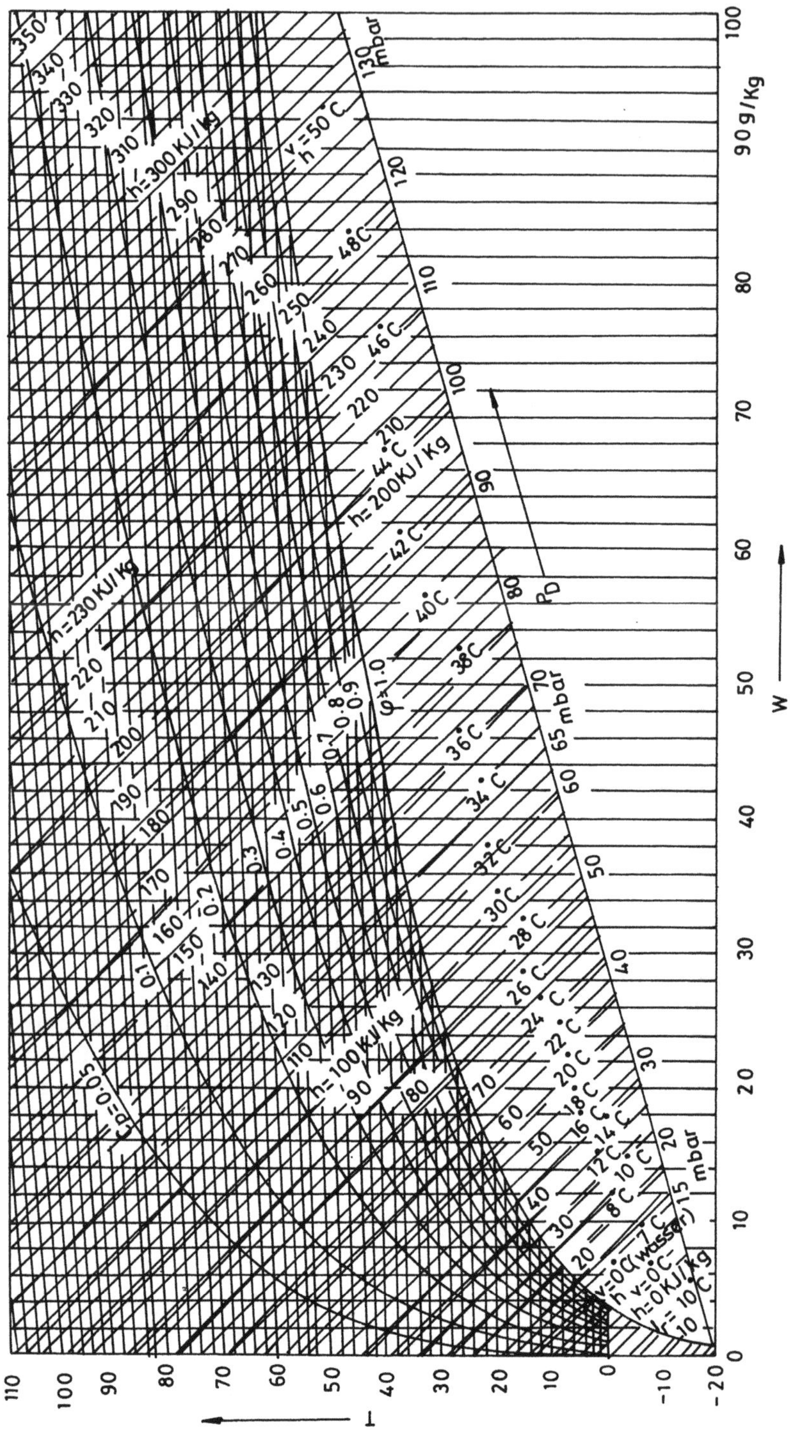

Fig. 14. h–W diagram

ative humidity can take a maximum of 5 g/kg of water vapor while coaling to 23° C (following the constant enthalpy line in Fig. 14).

To provide ventilation of air either through the product or over the product, one can utilize wind, solar radiation or electricity, as we have seen in the previous section.

4.1 Wind Ventilation

Wind ventilation should be considered in regions where the average wind velocity is of the order of 5 m/s or more. Both types of ventilations are possible. Narrow vertical bins are installed perpendicular to the main bin direction. The bottom should at least be 0.5 m above the ground. If the wind is to blow through the drying product, it should be able to overcome the resistance provided by the product. The pressure losses in the flow due to this resistance can be estimated by the following expression.

$$\Delta P_R = fh\frac{\varrho}{2}v^2, \tag{4}$$

where h is the thickness of the layer, ϱ and v the air density and velocity respectively, and f the specific air resistance. The value of f for a few tropical products are:

Material	f (1000/m)
Paddy	60
Maize	33
Cocoa Beans	60 wet
Peanuts	27 Dry
Peanuts in shells	2.5
Maize cobs	1.5

If the wind is to blow through the drying product at a wind speed of 5 m/s, ΔP should be below 15 N/m², or below 60 N/m² at a wind speed of 10 m/s. This type of ventilation therefore is applicable only for large-size materials, i.e. maize cobs, larger yam cubes, etc.

4.2 Natural Convection

Ventilation by natural convection makes use of the fact that hot air has a lower density than cold air and thus tends to rise.

The draught of air (Fig. 15) is established and it depends on the vertical difference in height between the air inlet and outlet H, the area of the inlet and the outlet and the corresponding densities ϱ_1 and ϱ_2 of air. The draught can be expressed as the pressure difference given by

$$\Delta P = gH(\varrho_1 - \varrho_2). \tag{5}$$

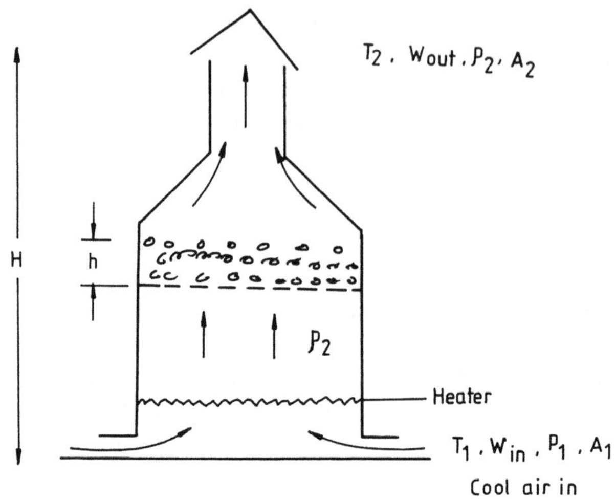

Fig. 15. Ventilation by Natural Convection

The densities are dependent on temperature as well as on the humidity ratio of the air as

$$\varrho = \varrho_o \left| \frac{1}{\left(1 + \frac{T}{273}\right)} \right| (1 + W), \tag{6}$$

where $g = 9.81$ m/s^2; H = height of hot air column; ϱ_1, ϱ_2 = air densities at temperatures T_1 and T_2 respectively (kg/m^3); $\varrho_0 = 1.293$ kg/m^3; T = temperature in °C and W = humidity ratio.

If the natural draught is able to overcome the resistance ΔP_R as expressed by Eq. (3) then the volumetric flow-rate of air can be expressed as

$$\dot{m} = a \frac{A_1 \cdot A_2}{(A_1 + A_2)^{\frac{1}{2}}} \left(\frac{\Delta P_R}{H} \right)^b \tag{7}$$

where $\dot{m}$ = volumetric flow-rate due to natural convection per unit cross-sectional area of the grain bed (m^3/s·m^2); A_1 = cross-sectional area at the inlet end; A_2 = cross-sectional area at the outlet end, and a, b are empirical constants. For rice, the values of the contants a and b have been determined to be 0.008 and 0.87 respectively.

4.3 Forced Ventilation

For cases where large quantities of material have to be dried, fan-forced ventilation is necessary. The fan has to be chosen according to the operational data for the dryer. The two specifications which are necessary for a fan are the blower capacity V given in m^3/h and the air pressure created by the blower. The electric power of the blower P_B is then given as $P = V\Delta P/\eta_b$, where η_b is efficiency of the blower / electric motor system.

4.4 Example: Calculation of Design Parameters

In this example, a procedure for the design of a drying system is presented for illustration purposes:

Procedure	*Example*
Initial data	
Type of drying material	Maize
Initial moisture content (w.b.)	$M_{in} = 35\%$
Final moisture content (w.b.)	$M_f = 15\%$
Maximum temperature allowed for the drying material	$T_{max} = 80°\,C$
Conditions of the ambient air	
Temperature	$T_{am} = 30°\,C$
Relative humidity	$= 0.5$
Rate of drying: this value commonly given as the amount of wet material dried in 1 h, dominantly influences the layout of the components of the dryer. The mass of water to be evaporated from the product is calculated from the formula	$m_w = \dfrac{W(M_{in} - M_f)}{(100 - M_f)}$
Mass of wet material (batch capacity)	$= 1000$ kg
Mass of water to be evaporated	$m_w = \dfrac{1000\,(35–15)}{(100–15)}$ $= 234$ kg
Drying time	$T_{dr} = 10$ h
Average drying rate	$\dot{m} = \dfrac{m_w}{t_{dr}}$ $= 23.4$ kg/h

Now we discuss the evaluation of various design parameters for achieving drying by using preheated air.

a) *Determination of operating temperatures*
Corresponding to the value of T_{am} and ϕ_{am}, find point 1 in the h-W diagramme

$T_a = 30°\,C$
$\phi_a = 0.5$
$W_1 = 13.5$ g/kg
0.0135
$h_1 = 65$ kJ/kg

Find point 2 in the h-W diagramme with $W_f = W_i$ and $T_2 = T_{max}$; read h_2

$W_2 = 13.5$ g/kg
$T_2 = 80°$ C
$h_2 = 115$ kJ/kg
from Fig. 3 for
$W_2 = 0.0135$,
$\phi_f = 0.5$

From the value of the final humidity ratio W_2, determine the value of the equilibrium (or final) relative humidity from the sorption isotherm of the product (Fig. 2) for maize.

Find point 3 in the h-W diagramme with ϕ_f and follow $h_f = h_2$ (the constant enthalpy line) to determine T_f and W_f

$\phi_f = 0.5$,
$h_f = 1$ J/kg
$T_f = 42°$ C
$W_f = 27$ g/kg
$= 0.027$

b) *Determination of blower capacity*
The maximum required air-flow rate determining the capacity of the blower can be determined by dividing the amount of water to be evaporated by the difference of the final and initial humidity ratios, i.e. capacity of the blower

$$m_i = m_w/(W_f - W_i)$$
$$= 23.4/(0.027–0.0135) \text{ kg/h}$$
$$= 1733 \text{ kg/h}$$

The density of air being 1.2 kg/m^3, the required volume capacity of the blower comes out to be 1400 m^3/h or 0.4 m^3/s.

c) *Determination of the required power of the heater*
The required power of the heater is determined by the formula

Q = difference in enthalpy of the drying air and the ambient air × rate of air flow

$$= (h_f - h_i) \cdot m_i/3600 = 25 \text{ kW}$$

d) *Calculation of required collector area/auxiliary fuel consumption*
The total energy requirement E_B for each batch is simply a product of the power of the heater and the drying time, thus

$$E_B = Q \times t_{dr}' \ (t_{dr} = 10 \text{ h})$$
$$= 25 \times 36000 = 900000 \text{ kJ}$$

Since the drying period is 10 h, the required energy can either be obtained in 1 day from solar energy collectors, from convectional fuel or from a mixture of both. Depending on the availability of solar radiation, the necessary solar energy collector area can be calculated from the formula $A_c = E_B/I\eta_c$; assuming an effective collector efficiency of about 30% and incident solar radiation of 11.0 MJ/m^2 day, the necessary collector are a comes out to be

$$A_c = 275 \text{ m}^2$$

The solar energy collector area is reduced to 164 m^2 for 50% collector efficiency and to only 136 m^2 for 60% efficiency.

e) *Determination of bin geometry*
Bin area. The desired air velocity through the maize is 0.1 m/s. The basic bin area can then be calculated as

$$\text{Basic area of the bin} = \frac{\text{Volume capacity of the blower V}}{\text{Air velocity}}$$

i.e. $A_B = 0.4/0.1 = 4\,m^2$

Layer thickness. The layer thickness is given by

$$\text{layer thickness d} = \frac{\text{Volume of a batch } V_B}{\text{area of bin } A_B}$$

The volume of the batch is easily determined by the initial mass of the product and the density of the bulk material, i.e.

$V_B = m_s/\varrho_B$

The bulk density for maize (ϱ_B) is 700 kg/m^3. Given $V_B = 1.4\,m^3$ and $d = 0.36$ m, the height of the bin should be about 0.20 m more, i.e. 0.56 m.

f) Determination of air resistance of layer
The air resistance of the layer is, as discussed earlier, given by

$P = fd\varrho_a\, v^2/2$ with $f = 33000\,m^{-1}$

$d = 0.36$ m, $\varrho_a = 1.2$ kg/m^3 and $v = 0.1$ m/s

$P = 71\ N/m^2$

The blower capacity determined earlier should therefore be increased to overcome this additional pressure; the actual power of the blower is calculated from the formula $P_B = m\Delta P/\eta_b$. For $\eta_b = 0.25$ and for the above calculations $P_B =$ 120 W.

5 Conclusions

The use of solar energy for the drying of various types of products has a large potential from a technical point of view. Numerous types of dryers have been designed and studied all over the world, yielding varying degrees of performance. The scope of solar drying has to be seen through a techno-economic analysis comparing the drying cost of various products with their market cost. For drying to be economically viable, it is essential that the cost for drying a product be much less than the cost of the product. For small quantities, in any case, solar drying is not only economically competitive but provides ideal solutions for drying various types of products, especially in regions of good solar radiation.

References

Akyurt M, Selcuk MK (1973) A solar dryer supplemented with auxiliary heating systems for continuous operation. Sol Energy 14:313–320
Ashrae (1976) Handbook industrial heating and drying. Am Soc Refrigerating and Airconditioning Engineers. New York

Bansal NK (1987) Solar crop drying. In: Garg HP (ed) Physics and engineering of solar energy. D. Reidel, Dordrecht, Holland
Eissen W (1983) Drying of grapes with solar energy. Dissertation, University of Hohenheim, Stuttgart, FRG
Exell RHB (1980) Basic design theory for simple rice dryer. Renew Energy Rev 1(2):1–14
Ghosh BN (1973) A new glass roof dryer for cocoa beans and other crops. Proc ISEC, Paris, France, V 30
Headley OSC, Springer BGF (1975) Natural convection solar dryers. Brace Res Inst, Tech Rep No. T-99, CS/EC/N-1
Kapoor SG, Agarwal HC (1973) Solar dryers for Indian conditions. Proc ISEC Paris, Paper No V 29
Kilkis B (1981) Solar energy-assisted crop and fruit drying systems: theory and applications, energy conservation and use of renewable energies in the bio-industries. Pergamon, Oxford, pp 307–333
Lawand TA (1965) A description of two solar agricultural dryers. Compl Bull 9; 51–56
Lawand TA (1966) A solar cabinet dryer. Sol Energy 10:158–164
Lawand TA (1975) A survey of solar agriculture dryers. Brace Res Inst Tech Rep No T-99
Lawand TA (1980) Agriculture and other low temperature applications of solar energy. In: Kreider JF, Kreith F (eds) Solar energy handbook. McGraw Hill, New York
Nahlawi M (1966) The drying of yams with solar energy. Brace Res Inst Tech Rep No T-270
Selak MK (1967) Final report of the solar fruit dryer. COMPLES Bulletin No 12, p 91
Sodha MS, Bansal NK, Kumar A, Bansal PK, Malik MAS (1987) Solar crop drying, vol I, II. CRC Press, Boca Raton, Florida, USA
Takla A (1978) Development of solar energy utilization in developing countries, technology for solar energy utilization. Dev Transf Technol Ser No 5, New York, pp 7–40

Subject Index

MIX
Papier aus verantwortungsvollen Quellen
Paper from responsible sources
FSC® C105338

If you have any concerns about our products,
you can contact us on
ProductSafety@springernature.com

In case Publisher is established outside the EU,
the EU authorized representative is:
Springer Nature Customer Service Center GmbH
Europaplatz 3, 69115 Heidelberg, Germany

Printed by Libri Plureos GmbH
in Hamburg, Germany